ALGORITHMS FOR LINEAR-QUADRATIC OPTIMIZATION

PURE AND APPLIED MATHEMATICS

A Program of Monographs, Textbooks, and Lecture Notes

MONOGRAPHS AND TEXTBOOKS IN PURE AND APPLIED MATHEMATICS

1. *K. Yano,* Integral Formulas in Riemannian Geometry (1970)
2. *S. Kobayashi,* Hyperbolic Manifolds and Holomorphic Mappings (1970)
3. *V. S. Vladimirov,* Equations of Mathematical Physics (A. Jeffrey, ed.; A. Littlewood, trans.) (1970)
4. *B. N. Pshenichnyi,* Necessary Conditions for an Extremum (L. Neustadt, translation ed.; K. Makowski, trans.) (1971)
5. *L. Narici et al.,* Functional Analysis and Valuation Theory (1971)
6. *S. S. Passman,* Infinite Group Rings (1971)
7. *L. Dornhoff,* Group Representation Theory. Part A: Ordinary Representation Theory. Part B: Modular Representation Theory (1971, 1972)
8. *W. Boothby and G. L. Weiss, eds.,* Symmetric Spaces (1972)
9. *Y. Matsushima,* Differentiable Manifolds (E. T. Kobayashi, trans.) (1972)
10. *L. E. Ward, Jr.,* Topology (1972)
11. *A. Babakhanian,* Cohomological Methods in Group Theory (1972)
12. *R. Gilmer,* Multiplicative Ideal Theory (1972)
13. *J. Yeh,* Stochastic Processes and the Wiener Integral (1973)
14. *J. Barros-Neto,* Introduction to the Theory of Distributions (1973)
15. *R. Larsen,* Functional Analysis (1973)
16. *K. Yano and S. Ishihara,* Tangent and Cotangent Bundles (1973)
17. *C. Procesi,* Rings with Polynomial Identities (1973)
18. *R. Hermann,* Geometry, Physics, and Systems (1973)
19. *N. R. Wallach,* Harmonic Analysis on Homogeneous Spaces (1973)
20. *J. Dieudonné,* Introduction to the Theory of Formal Groups (1973)
21. *I. Vaisman,* Cohomology and Differential Forms (1973)
22. *B.-Y. Chen,* Geometry of Submanifolds (1973)
23. *M. Marcus,* Finite Dimensional Multilinear Algebra (in two parts) (1973, 1975)
24. *R. Larsen,* Banach Algebras (1973)
25. *R. O. Kujala and A. L. Vitter, eds.,* Value Distribution Theory: Part A; Part B: Deficit and Bezout Estimates by Wilhelm Stoll (1973)
26. *K. B. Stolarsky,* Algebraic Numbers and Diophantine Approximation (1974)
27. *A. R. Magid,* The Separable Galois Theory of Commutative Rings (1974)
28. *B. R. McDonald,* Finite Rings with Identity (1974)
29. *J. Satake,* Linear Algebra (S. Koh et al., trans.) (1975)
30. *J. S. Golan,* Localization of Noncommutative Rings (1975)
31. *G. Klambauer,* Mathematical Analysis (1975)
32. *M. K. Agoston,* Algebraic Topology (1976)
33. *K. R. Goodearl,* Ring Theory (1976)
34. *L. E. Mansfield,* Linear Algebra with Geometric Applications (1976)
35. *N. J. Pullman,* Matrix Theory and Its Applications (1976)
36. *B. R. McDonald,* Geometric Algebra Over Local Rings (1976)
37. *C. W. Groetsch,* Generalized Inverses of Linear Operators (1977)
38. *J. E. Kuczkowski and J. L. Gersting,* Abstract Algebra (1977)
39. *C. O. Christenson and W. L. Voxman,* Aspects of Topology (1977)
40. *M. Nagata,* Field Theory (1977)
41. *R. L. Long,* Algebraic Number Theory (1977)
42. *W. F. Pfeffer,* Integrals and Measures (1977)
43. *R. L. Wheeden and A. Zygmund,* Measure and Integral (1977)
44. *J. H. Curtiss,* Introduction to Functions of a Complex Variable (1978)
45. *K. Hrbacek and T. Jech,* Introduction to Set Theory (1978)
46. *W. S. Massey,* Homology and Cohomology Theory (1978)
47. *M. Marcus,* Introduction to Modern Algebra (1978)
48. *E. C. Young,* Vector and Tensor Analysis (1978)
49. *S. B. Nadler, Jr.,* Hyperspaces of Sets (1978)
50. *S. K. Segal,* Topics in Group Kings (1978)
51. *A. C. M. van Rooij,* Non-Archimedean Functional Analysis (1978)
52. *L. Corwin and R. Szczarba,* Calculus in Vector Spaces (1979)
53. *C. Sadosky,* Interpolation of Operators and Singular Integrals (1979)

54. *J. Cronin,* Differential Equations (1980)
55. *C. W. Groetsch,* Elements of Applicable Functional Analysis (1980)
56. *I. Vaisman,* Foundations of Three-Dimensional Euclidean Geometry (1980)
57. *H. I. Freedan,* Deterministic Mathematical Models in Population Ecology (1980)
58. *S. B. Chae,* Lebesgue Integration (1980)
59. *C. S. Rees et al.,* Theory and Applications of Fourier Analysis (1981)
60. *L. Nachbin,* Introduction to Functional Analysis (R. M. Aron, trans.) (1981)
61. *G. Orzech and M. Orzech,* Plane Algebraic Curves (1981)
62. *R. Johnsonbaugh and W. E. Pfaffenberger,* Foundations of Mathematical Analysis (1981)
63. *W. L. Voxman and R. H. Goetschel,* Advanced Calculus (1981)
64. *L. J. Corwin and R. H. Szczarba,* Multivariable Calculus (1982)
65. *V. I. Istrăţescu,* Introduction to Linear Operator Theory (1981)
66. *R. D. Järvinen,* Finite and Infinite Dimensional Linear Spaces (1981)
67. *J. K. Beem and P. E. Ehrlich,* Global Lorentzian Geometry (1981)
68. *D. L. Armacost,* The Structure of Locally Compact Abelian Groups (1981)
69. *J. W. Brewer and M. K. Smith, eds.,* Emmy Noether: A Tribute (1981)
70. *K. H. Kim,* Boolean Matrix Theory and Applications (1982)
71. *T. W. Wieting,* The Mathematical Theory of Chromatic Plane Ornaments (1982)
72. *D. B.Gauld,* Differential Topology (1982)
73. *R. L. Faber,* Foundations of Euclidean and Non-Euclidean Geometry (1983)
74. *M. Carmeli,* Statistical Theory and Random Matrices (1983)
75. *J. H. Carruth et al.,* The Theory of Topological Semigroups (1983)
76. *R. L. Faber,* Differential Geometry and Relativity Theory (1983)
77. *S. Barnett,* Polynomials and Linear Control Systems (1983)
78. *G. Karpilovsky,* Commutative Group Algebras (1983)
79. *F. Van Oystaeyen and A. Verschoren,* Relative Invariants of Rings (1983)
80. *I. Vaisman,* A First Course in Differential Geometry (1984)
81. *G. W. Swan,* Applications of Optimal Control Theory in Biomedicine (1984)
82. *T. Petrie and J. D. Randall,* Transformation Groups on Manifolds (1984)
83. *K. Goebel and S. Reich,* Uniform Convexity, Hyperbolic Geometry, and Nonexpansive Mappings (1984)
84. *T. Albu and C. Năstăsescu,* Relative Finiteness in Module Theory (1984)
85. *K. Hrbacek and T. Jech,* Introduction to Set Theory: Second Edition (1984)
86. *F. Van Oystaeyen and A. Verschoren,* Relative Invariants of Rings (1984)
87. *B. R. McDonald,* Linear Algebra Over Commutative Rings (1984)
88. *M. Namba,* Geometry of Projective Algebraic Curves (1984)
89. *G. F. Webb,* Theory of Nonlinear Age-Dependent Population Dynamics (1985)
90. *M. R. Bremner et al.,* Tables of Dominant Weight Multiplicities for Representations of Simple Lie Algebras (1985)
91. *A. E. Fekete,* Real Linear Algebra (1985)
92. *S. B. Chae,* Holomorphy and Calculus in Normed Spaces (1985)
93. *A. J. Jerri,* Introduction to Integral Equations with Applications (1985)
94. *G. Karpilovsky,* Projective Representations of Finite Groups (1985)
95. *L. Narici and E. Beckenstein,* Topological Vector Spaces (1985)
96. *J. Weeks,* The Shape of Space (1985)
97. *P. R. Gribik and K. O. Kortanek,* Extremal Methods of Operations Research (1985)
98. *J.-A. Chao and W. A. Woyczynski, eds.,* Probability Theory and Harmonic Analysis (1986)
99. *G. D. Crown et al.,* Abstract Algebra (1986)
100. *J. H. Carruth et al.,* The Theory of Topological Semigroups, Volume 2 (1986)
101. *R. S. Doran and V. A. Belfi,* Characterizations of C*-Algebras (1986)
102. *M. W. Jeter,* Mathematical Programming (1986)
103. *M. Altman,* A Unified Theory of Nonlinear Operator and Evolution Equations with Applications (1986)
104. *A. Verschoren,* Relative Invariants of Sheaves (1987)
105. *R. A. Usmani,* Applied Linear Algebra (1987)
106. *P. Blass and J. Lang,* Zariski Surfaces and Differential Equations in Characteristic $p > 0$ (1987)
107. *J. A. Reneke et al.,* Structured Hereditary Systems (1987)
108. *H. Busemann and B. B. Phadke,* Spaces with Distinguished Geodesics (1987)
109. *R. Harte,* Invertibility and Singularity for Bounded Linear Operators (1988)

110. *G. S. Ladde et al.,* Oscillation Theory of Differential Equations with Deviating Arguments (1987)
111. *L. Dudkin et al.,* Iterative Aggregation Theory (1987)
112. *T. Okubo,* Differential Geometry (1987)
113. *D. L. Stancl and M. L. Stancl,* Real Analysis with Point-Set Topology (1987)
114. *T. C. Gard,* Introduction to Stochastic Differential Equations (1988)
115. *S. S. Abhyankar,* Enumerative Combinatorics of Young Tableaux (1988)
116. *H. Strade and R. Farnsteiner,* Modular Lie Algebras and Their Representations (1988)
117. *J. A. Huckaba,* Commutative Rings with Zero Divisors (1988)
118. *W. D. Wallis,* Combinatorial Designs (1988)
119. *W. Więsław,* Topological Fields (1988)
120. *G. Karpilovsky,* Field Theory (1988)
121. *S. Caenepeel and F. Van Oystaeyen,* Brauer Groups and the Cohomology of Graded Rings (1989)
122. *W. Kozlowski,* Modular Function Spaces (1988)
123. *E. Lowen-Colebunders,* Function Classes of Cauchy Continuous Maps (1989)
124. *M. Pavel,* Fundamentals of Pattern Recognition (1989)
125. *V. Lakshmikantham et al.,* Stability Analysis of Nonlinear Systems (1989)
126. *R. Sivaramakrishnan,* The Classical Theory of Arithmetic Functions (1989)
127. *N. A. Watson,* Parabolic Equations on an Infinite Strip (1989)
128. *K. J. Hastings,* Introduction to the Mathematics of Operations Research (1989)
129. *B. Fine,* Algebraic Theory of the Bianchi Groups (1989)
130. *D. N. Dikranjan et al.,* Topological Groups (1989)
131. *J. C. Morgan II,* Point Set Theory (1990)
132. *P. Biler and A. Witkowski,* Problems in Mathematical Analysis (1990)
133. *H. J. Sussmann,* Nonlinear Controllability and Optimal Control (1990)
134. *J.-P. Florens et al.,* Elements of Bayesian Statistics (1990)
135. *N. Shell,* Topological Fields and Near Valuations (1990)
136. *B. F. Doolin and C. F. Martin,* Introduction to Differential Geometry for Engineers (1990)
137. *S. S. Holland, Jr.,* Applied Analysis by the Hilbert Space Method (1990)
138. *J. Okniński,* Semigroup Algebras (1990)
139. *K. Zhu,* Operator Theory in Function Spaces (1990)
140. *G. B. Price,* An Introduction to Multicomplex Spaces and Functions (1991)
141. *R. B. Darst,* Introduction to Linear Programming (1991)
142. *P. L. Sachdev,* Nonlinear Ordinary Differential Equations and Their Applications (1991)
143. *T. Husain,* Orthogonal Schauder Bases (1991)
144. *J. Foran,* Fundamentals of Real Analysis (1991)
145. *W. C. Brown,* Matrices and Vector Spaces (1991)
146. *M. M. Rao and Z. D. Ren,* Theory of Orlicz Spaces (1991)
147. *J. S. Golan and T. Head,* Modules and the Structures of Rings (1991)
148. *C. Small,* Arithmetic of Finite Fields (1991)
149. *K. Yang,* Complex Algebraic Geometry (1991)
150. *D. G. Hoffman et al.,* Coding Theory (1991)
151. *M. O. González,* Classical Complex Analysis (1992)
152. *M. O. González,* Complex Analysis (1992)
153. *L. W. Baggett,* Functional Analysis (1992)
154. *M. Sniedovich,* Dynamic Programming (1992)
155. *R. P. Agarwal,* Difference Equations and Inequalities (1992)
156. *C. Brezinski,* Biorthogonality and Its Applications to Numerical Analysis (1992)
157. *C. Swartz,* An Introduction to Functional Analysis (1992)
158. *S. B. Nadler, Jr.,* Continuum Theory (1992)
159. *M. A. Al-Gwaiz,* Theory of Distributions (1992)
160. *E. Perry,* Geometry: Axiomatic Developments with Problem Solving (1992)
161. *E. Castillo and M. R. Ruiz-Cobo,* Functional Equations and Modelling in Science and Engineering (1992)
162. *A. J. Jerri,* Integral and Discrete Transforms with Applications and Error Analysis (1992)
163. *A. Charlier et al.,* Tensors and the Clifford Algebra (1992)
164. *P. Biler and T. Nadzieja,* Problems and Examples in Differential Equations (1992)
165. *E. Hansen,* Global Optimization Using Interval Analysis (1992)

166. *S. Guerre-Delabrière*, Classical Sequences in Banach Spaces (1992)
167. *Y. C. Wong*, Introductory Theory of Topological Vector Spaces (1992)
168. *S. H. Kulkarni and B. V. Limaye*, Real Function Algebras (1992)
169. *W. C. Brown*, Matrices Over Commutative Rings (1993)
170. *J. Loustau and M. Dillon*, Linear Geometry with Computer Graphics (1993)
171. *W. V. Petryshyn*, Approximation-Solvability of Nonlinear Functional and Differential Equations (1993)
172. *E. C. Young*, Vector and Tensor Analysis: Second Edition (1993)
173. *T. A. Bick*, Elementary Boundary Value Problems (1993)
174. *M. Pavel*, Fundamentals of Pattern Recognition: Second Edition (1993)
175. *S. A. Albeverio et al.*, Noncommutative Distributions (1993)
176. *W. Fulks*, Complex Variables (1993)
177. *M. M. Rao*, Conditional Measures and Applications (1993)
178. *A. Janicki and A. Weron*, Simulation and Chaotic Behavior of α-Stable Stochastic Processes (1994)
179. *P. Neittaanmäki and D. Tiba*, Optimal Control of Nonlinear Parabolic Systems (1994)
180. *J. Cronin*, Differential Equations: Introduction and Qualitative Theory, Second Edition (1994)
181. *S. Heikkilä and V. Lakshmikantham*, Monotone Iterative Techniques for Discontinuous Nonlinear Differential Equations (1994)
182. *X. Mao*, Exponential Stability of Stochastic Differential Equations (1994)
183. *B. S. Thomson*, Symmetric Properties of Real Functions (1994)
184. *J. E. Rubio*, Optimization and Nonstandard Analysis (1994)
185. *J. L. Bueso et al.*, Compatibility, Stability, and Sheaves (1995)
186. *A. N. Michel and K. Wang*, Qualitative Theory of Dynamical Systems (1995)
187. *M. R. Darnel*, Theory of Lattice-Ordered Groups (1995)
188. *Z. Naniewicz and P. D. Panagiotopoulos*, Mathematical Theory of Hemivariational Inequalities and Applications (1995)
189. *L. J. Corwin and R. H. Szczarba*, Calculus in Vector Spaces: Second Edition (1995)
190. *L. H. Erbe et al.*, Oscillation Theory for Functional Differential Equations (1995)
191. *S. Agaian et al.*, Binary Polynomial Transforms and Nonlinear Digital Filters (1995)
192. *M. I. Gil'*, Norm Estimations for Operation-Valued Functions and Applications (1995)
193. *P. A. Grillet*, Semigroups: An Introduction to the Structure Theory (1995)
194. *S. Kichenassamy*, Nonlinear Wave Equations (1996)
195. *V. F. Krotov*, Global Methods in Optimal Control Theory (1996)
196. *K. I. Beidar et al.*, Rings with Generalized Identities (1996)
197. *V. I. Arnautov et al.*, Introduction to the Theory of Topological Rings and Modules (1996)
198. *G. Sierksma*, Linear and Integer Programming (1996)
199. *R. Lasser*, Introduction to Fourier Series (1996)
200. *V. Sima*, Algorithms for Linear-Quadratic Optimization (1996)
201. *O. Redmond*, Number Theory (1996)
202. *J. K. Beem et al.* Global Lorentzian Geometry: Second Edition (1996)

Additional Volumes in Preparation

ALGORITHMS FOR LINEAR-QUADRATIC OPTIMIZATION

Vasile Sima
Research Institute for Informatics
Bucharest, Romania

Marcel Dekker, Inc. **New York • Basel • Hong Kong**

Lc
Mathematical optimisation

ISBN: 0-8247-9612-8

The publisher offers discounts on this book when ordered in bulk quantities. For more information, write to Special Sales/Professional Marketing at the address below.

This book is printed on acid-free paper.

Marcel Dekker, Inc.
270 Madison Avenue, New York, New York 10016

Current printing (last digit):
10 9 8 7 6 5 4 3 2 1

PRINTED IN THE UNITED STATES OF AMERICA

Preface

Linear-quadratic optimization is one of the most exciting topics in the control engineering literature of recent decades. The interest in linear optimal dynamical systems, and especially in those with quadratic cost functional, can be explained by both the richness of properties these systems possess, and the pragmatic value and physical significance of the available results. *Linear-quadratic optimization* is not only a well-established theory, but also a powerful design tool, useful for time-invariant, as well as variable dynamical systems.

Algorithms for Linear-Quadratic Optimization is intended to provide theoretical, algorithmic and computational support for solving the most frequently encountered linear-quadratic optimization problems. It is as self-contained as possible and includes both theoretical and practical topics. Theoretical treatment beneficially combines with an insight into algorithmic techniques, with relevant numerical issues, sophisticated software tools, implementation details that improve performance, practical recommendations, and solution of specific problems. The algorithms are presented in a simple and concise informal language, a refined MATLAB notation, which facilitates computer implementation. Algorithmic templates are used rather than descriptions of source codes, to offer customization flexibility to both casual users and high-performance specialists, and to ensure the generality and independence of a specific computer language. Recent advances in numerical linear algebra and associated mathematical software are taken into account.

For each particular numerical technique the following are discussed: mathematical description of the method and algorithm; computational issues; applicability and limitations. Comments on algorithmic and implementation issues facilitate better understanding of the details that decisively contribute to high performance. Many worked-out examples, based on published industrial models, are used for illustrative purposes.

Various algebraic matrix Riccati equations solvers are described. Newton's methods, Schur and generalized Schur vector methods, matrix sign function and structure-preserving methods are all dealt with. Advantages and disadvantages of each solution technique are noted. These high-level algorithms are based on the low-level numerical linear algebra building blocks, well known and generally available in computer libraries, such as BLAS (level I–III), EISPACK and LINPACK, as well as their successor, LAPACK. A comparative evaluation of the algorithms and implemented solvers discussed, for some test problems or application-oriented numerical examples, is appended.

The book is interdisciplinary, signalling an overview of recent advances in control and systems theory, numerical linear algebra, numerical optimization, scientific computations, and software engineering. Numerous references to current research literature and textbooks are made.

The book could serve as a reference to the algorithmic developments in the linear-quadratic optimization domain. Owing to its methodological guidance, it could also be used as a handbook for practitioners. The book is directed not only to researchers and specialists working in the areas of optimization, mathematical programming, and control theory, but also engineers, mathematicians, economists, students, and all those who use the systems approach in their work. A background in numerical linear algebra and control and systems theory is useful.

I would like to thank all those who have helped in writing this book. My special thanks go to Andreas Varga, Volker Mehrmann, and Peter Benner. I acknowledge the Commission of the European Communities, for granting a research fellowship in the Federal Republic of Germany, during the tenure of which I obtained some of the results discussed in the book. Last, but not least, I would like to thank the Marcel Dekker staff for their help and advice in the production of the book.

Vasile Sima

Contents

ALGORITHMS FOR LINEAR-QUADRATIC OPTIMIZATION

1

Linear-Quadratic Optimization Problems

The linear optimal dynamical systems, and especially those with quadratic cost functional, seem to be one of the most exciting subjects in control engineering literature of the last decades. There are thousands of papers and dozens of books devoted to this topic. (The categorized and indexed bibliography on the linear-quadratic-Gaussian problem in (Mendel and Gieseking, 1971) includes over 900 references, and many other references have been meantime added.) Some reference books are (Anderson and Moore, 1971, 1979; Åström, 1970; Athans and Falb, 1966; Bittanti, Laub, and Willems, 1991; Kalman, Falb, and Arbib, 1969; Kwakernaak and Sivan, 1972; Mehrmann, 1991; Sage and White, 1977). Many other books, like (Barnett, 1975; Brocket, 1970; Casti, 1977; Comte, Perdon, and Wyman, 1991; Kailath, 1980; Petkov, Christov, and Konstantinov, 1991; Rubio, 1971; Shinners, 1992), contain chapters or sections dealing with linear optimal systems.

Linear-quadratic optimization is not only a full theory, covering both time-invariant and variable dynamical systems, but also a powerful design tool. This approach offers, for instance, an elegant solution to the optimal systems stabilization and control, using a *state-feedback control law*. The optimal filtering problem can be solved as a dual of an optimal control problem. Some reference tracking and measurable disturbance rejection problems can be solved by simple theoretical and procedural extensions. The results of an optimal design are generally better suited in practice than those obtained by other techniques; for example, the pole assignment approach (Wonham, 1979) sometimes delivers gain matrices with too large magnitude, which may lead to large inputs in the controlled systems. Other approaches, as for instance the H_∞ theory (Francis, 1987), are largely dependent, from a procedural point of view, on the numerical techniques of the linear-quadratic optimization.

This chapter gives an overview of the main topics in linear-quadratic optimization. First, the basic theory is briefly presented and the standard linear-quadratic optimization problem for a constant continuous-time or discrete-time linear dynamical system with quadratic performance index is stated; its closed-form solution is given in terms of the unique symmetric positive semi-definite solution of an algebraic matrix Riccati equation. Then, several problems reducible to the standard problem are discussed; they include output regulation, proportional-plus-integral control, improving the tracking properties, and direct rejection of measurable disturbances. Topics in H_∞ and H_2 theories are discussed as well. Some basic numerical linear algebra algorithms of special interest in linear-quadratic optimization, and the associated software support are then summarized. They include, for instance, the computation and application of standard plane rotations, and solving systems of linear algebraic equations (including in the least squares sense). References are made to the BLAS and LAPACK public-domain software packages. Some relevant numerical issues (numerical stability, conditioning, etc.) are also commented. Various classes of algorithms for solving algebraic matrix Riccati equations and related problems are finally overviewed. The overview includes, beside the algorithms to be described in the next chapters, algorithms that determine the stationary solution of difference Riccati equations, quadratically convergent doubling algorithms, special algorithms for state estimation, and defect correction algorithms.

1.1 Standard Linear-Quadratic Optimization Problems

This section reviews the basic theory of linear-quadratic optimization, its main topics and results. This background is essential in understanding or developing powerfull algorithms for solving problems in this domain. Going through a full theory, including detailed proofs, has not been intended, since reference material is abundant and widely available. Both continuous-time and discrete-time systems are studied. Optimal estimation of discrete-time systems is discussed as well. The main structural properties of constant linear dynamical systems are presented. Finally, the standard linear-quadratic optimization problem for a constant continuous-time or discrete-time linear dynamical system, with positive (semi-)definite weighting matrices in the quadratic performance index, is stated.

1.1.1 Basic theory

Optimal control of continuous-time systems

Let the continuous-time linear dynamical system described by[1]

$$\dot{x}(t) = A(t)x(t) + B(t)u(t), \tag{1.1}$$

where $t \in [t_0, T]$ denotes the time variable, $x(t) \in \mathbb{R}^n$ is the state vector, $u(t) \in \mathbb{R}^m$ is the control (input) vector, and $A(t)$ and $B(t)$ are real matrices of suitable dimensions, and let the quadratic performance index (the cost functional) be

$$J(x_0, u(\cdot), t_0) = \int_{t_0}^{T} [x(t)^T Q(t)x(t) + u(t)^T R(t)u(t)]\, dt + x(T)^T F x(T), \tag{1.2}$$

where $x_0 := x(t_0)$ is given, and $Q(t) = Q(t)^T \geq 0$, $R(t) = R(t)^T > 0$, and $F = F^T \geq 0$ are positive (semi-)definite weighting matrices. Assume that the elements of $A(t)$, $B(t)$, $Q(t)$, and $R(t)$ matrices are continuous functions of t.[2]

The *continuous-time linear-quadratic control problem* consists in finding a control trajectory, $\{u^*(t), t \in [t_0, T]\}$, minimizing the J functional in (1.2) in terms of u, subject to the dynamical constraint (1.1).

This special optimal control problem has a closed-form solution that can be reached using the maximum principle (Pontryagin et al., 1962), combined with the Euler-Lagrange equations. The short derivation below uses the Hamilton-Jacobi equation (see, for example, Anderson and Moore (1971), for more details).

---

[1] It can be proven that (1.1), together with an *output* formula, defines a unique dynamical system as conceived by the general mathematical theory of dynamical systems (see, for instance, (Kalman, Falb, and Arbib, 1969)).

[2] The continuity assumption could be suppressed.

Consider the general nonlinear control problem for the system

$$\dot{x}(t) = f(x, u, t), \qquad x_0 := x(t_0) \text{ given}, \tag{1.3}$$

and the cost functional

$$J(x_0, u(\cdot), t_0) = \int_{t_0}^{T} \ell(x(\tau), u(\tau), \tau)\, d\tau + h(x(T)), \tag{1.4}$$

where f, ℓ and h are smooth functions in their arguments, and ℓ and h are non-negative. Generally, the optimal control (trajectory) u^* should belong to some particular set (such as the set of piece-wise continuous functions, the set of square integrable functions, etc.). Let $u[a, b]$ denote $u(t)$, for $t \in [a, b]$, and

$$J^*(x(t), t) := \min_{u[t,T]} J(x(t), u(\cdot), t). \tag{1.5}$$

Then, for each $t \in [t_0, T]$ and $t_1 \in [t, T]$, we obtain

$$\begin{aligned} J^*(x(t), t) &= \min_{u[t,t_1)} \left\{ \min_{u[t_1,T]} \left[\int_t^{t_1} \ell\, d\tau + \int_{t_1}^{T} \ell\, d\tau + h(x(T)) \right] \right\} \\ &= \min_{u[t,t_1)} \left[\int_t^{t_1} \ell(x(\tau), u(\tau), \tau)\, d\tau + J^*(x(t_1), t_1) \right]. \end{aligned} \tag{1.6}$$

Taking $t_1 = t + \Delta t$, for small Δt, and using the Taylor series expansion in (1.6), for $\Delta t \to 0$, the following functional form of the Hamilton-Jacobi equation is easily derived

$$\frac{\partial J^*}{\partial t} = -\min_{u(t)} \left[\ell(x(t), u(t), t) + \frac{\partial J^*}{\partial x}^T f(x(t), u(t), t) \right], \tag{1.7}$$

where the arguments $x(t)$ and t of the partial derivatives of J^* have been suppressed.

Let $\bar{u}(x(t), (\partial J^*/\partial x), t)$ be the control minimizing the right-hand side of (1.7). It follows that $\bar{u}(x(t), (\partial J^*/\partial x), t)$ minimizes $J(x(t), u(t), t)$ at t. Then, (1.7) implies the partial differential equation below, called the *Hamilton-Jacobi equation*,

$$\frac{\partial J^*}{\partial t} = -\ell(x(t), \bar{u}(x(t), \frac{\partial J^*}{\partial x}, t), t) + \frac{\partial J^*}{\partial x}^T f(x(t), \bar{u}(x(t), \frac{\partial J^*}{\partial x}, t), t), \tag{1.8}$$

with boundary condition

$$J^*(x(T), T) = h(x(T)). \tag{1.9}$$

The solution of (1.8) with (1.9) shall lead to $J^*(x(t),t)$, hence,

$$\widehat{u}(x(t),t) := \bar{u}(x(t), \frac{\partial J^*}{\partial x}(x(t),t),t), \tag{1.10}$$

and using the optimality principle (Bellman, 1957; Bellman and Dreyfus, 1962), the optimal control is

$$u^*(t) = \widehat{u}(x(t),t).$$

Note that $\widehat{u}(x(t),t)$ in (1.10) does not depend on t_0. Hence, $\widehat{u}(x(t),t)$ minimizes $J(x(\tau),u(\cdot),\tau)$ for any $\tau < T$. Also note that the optimal control at t is a *state-feedback control*, i.e., it depends on $x(t)$. This is of special importance for practical applications.

The results above are summarized and made explicit in the following theorem.

Theorem 1.1 *Let the dynamical system (1.3) and the cost functional*

$$J(x(t),u(\cdot),t) = \int_t^T \ell(x(\tau),u(\tau),\tau)\,d\tau + h(x(T)). \tag{1.11}$$

Assume that: (i) f, ℓ, and h, are continuously differentiable functions in their arguments; (ii) there is an absolute minimum, with respect to u, for the function $\ell(x,u,t) + \lambda^T f(x,u,t)$, of the form $\bar{u}(x(t),\lambda,t)$; (iii) $\bar{u}$ is continuously differentiable in its arguments, and (iv) $J^*(x(t),t)$ is a solution to the Hamilton-Jacobi equation (1.7) (or (1.8) and (1.9)). Then, $J^*(\cdot,\cdot)$ is the optimal performance index for (1.11), and (1.10) is the optimal control for the class of problems with performance index $J(x(\tau),u(\cdot),\tau)$.

Conversely, assume that: (i) f, ℓ, and h, are continuously differentiable functions in their arguments; (ii) there is an optimal control, and the corresponding minimal value of (1.11), $J^*(x(t),t)$, is twice differentiable in its arguments; (iii) the function $\ell(x,u,t)+(\partial J^*/\partial x)^T f(x,u,t)$ has an absolute minimum with respect to u, of the form $\widehat{u}(x(t),t)$; and (iv) $\widehat{u}(\cdot,\cdot)$ is differentiable in x and continuous in t. Then, $J^*(\cdot,\cdot)$ satisfies the Hamilton-Jacobi equation (1.7) (or (1.8) and (1.9)).

Now, back to the linear-quadratic optimization problem (1.1) and (1.2). It can be easily shown that, in this case, $J^*(x(t),t)$ is a quadratic form in x. Hence,

$$J^*(x(t),t) = x(t)^T X(t) x(t), \tag{1.12}$$

where $X(t) = X(t)^T$. (Otherwise, $X(t)$ may be replaced by $(X(t) + X(t)^T)/2$.) From the Hamilton-Jacobi equation (1.7), after completing the squares, it follows that

$$\begin{aligned} x(t)^T\dot{X}(t)x(t) &= -\min_{u(t)}\left[(u(t)+R(t)^{-1}B(t)^TX(t)x(t))^TR(t)(u(t)+\right.\\ &\quad R(t)^{-1}B(t)^TX(t)x(t))+\\ &\quad x(t)^T(A(t)^TX(t)+X(t)A(t)-\\ &\quad \left.X(t)B(t)R(t)^{-1}B(t)^TX(t)+Q(t))x(t)\right]. \end{aligned} \tag{1.13}$$

The optimal control is then

$$\bar{u}(t) = -R(t)^{-1}B(t)^TX(t)x(t), \tag{1.14}$$

and the corresponding Hamilton-Jacobi equation (which will hold for each $x(t)$) becomes the *differential Riccati equation*

$$\begin{aligned} -\dot{X}(t) &= A(t)^TX(t)+X(t)A(t)-\\ &\quad X(t)B(t)R(t)^{-1}B(t)^TX(t)+Q(t), \end{aligned} \tag{1.15}$$

$$X(T) = F. \tag{1.16}$$

The existence of optimal performance, $J^*(x(t),t)$, for $t \leq T$, is proved by contradiction. Assuming that $J^*(x(t),t)$ does not exist, the equation (1.15) with boundary condition (1.16) has not a solution $X(t)$, for all $t \leq T$. But the theory of differential equations, with the continuity assumption on the elements of $A(t)$, $B(t)$, $Q(t)$, and $R(t)$ matrices, guarantees the existence of a solution of (1.15), (1.16) in a neighbourhood of T. Hence, there is a finite time T' such that $X(t)$ exists for all $t \in (T',T]$, but $X(t)$ does not exist for $t \leq T'$. (Some elements of $X(t)$ tend to infinity when t tends to T'.) As, for $\varepsilon > 0$, $\varepsilon < T-T'$,

$$0 \leq J^*(x(T'+\varepsilon),T'+\varepsilon) = x(T'+\varepsilon)^TX(T'+\varepsilon)x(T'+\varepsilon),$$

it follows that $X(T'+\varepsilon) \geq 0$. This implies that there is an index i such that $X_{ii}(T'+\varepsilon) \to \infty$, when $\varepsilon \to 0$. Therefore, $J^*(e_i,T'+\varepsilon) \to \infty$, when $\varepsilon \to 0$. Now, letting $\Phi(t,\tau)$, the transition matrix of the dynamical system (1.1),[3] setting $u(t) \equiv 0$ and choosing the unit vector e_i as initial state at time $T'+\varepsilon$, it follows that $x(\tau) = \Phi(\tau,T'+\varepsilon)e_i$, and

$$\begin{aligned} J(e_i,0,T'+\varepsilon) &= \int_{T'+\varepsilon}^{T} e_i^T\Phi(\tau,T'+\varepsilon)^TQ(\tau)\Phi(\tau,T'+\varepsilon)e_i\,d\tau+\\ &\quad e_i^T\Phi(T,T'+\varepsilon)^TF\Phi(T,T'+\varepsilon)e_i. \end{aligned}$$

[3] The *transition matrix* of the system (1.1) is $\Phi(t,\tau) := \Xi(t)\Xi(\tau)^{-1}$, where $\Xi(t)$ is the unique matrix satisfying $\dot{\Xi}(t) = A(t)\Xi(t)$, with initial condition $\Xi(0) = I_n$ (Barnett, 1975). For the time-invariant case, with $A(t) \equiv A$, $\Phi(t,\tau) := \exp(A(t-\tau))$.

Hence,

$$\lim_{\varepsilon \to 0} J(e_i, 0, T' + \varepsilon) < \infty,$$

which contradicts the inequality $J^*(e_i, T' + \varepsilon) \leq J(e_i, 0, T' + \varepsilon)$.

Therefore, the optimal performance index for the optimal control problem with initial time t and initial state $x(t)$ is $x(t)^T X(t) x(t)$, where $X(t)$ is given by the solution of the differential Riccati equation (1.15) with boundary condition (1.16). The matrix $X(t)$ exists for all $t \leq T$. The optimal control for arbitrary initial time is given by the state-feedback law

$$u(t) = -R(t)^{-1} B(t)^T X(t) x(t), \tag{1.17}$$

for any t in the optimization time horizon.

The results can be extended over an infinite time horizon ($T = \infty$). An additional assumption is needed for reaching a finite value of J in (1.2).[4] Specifically, it is assumed that, for arbitrary $x(t)$ at t, there is a control $u(t)$, depending on $x(t)$ and t, such that $u[t, t_1)$, with finite $t_1 > t$, transfers the state $x(t)$ to zero. Such structural properties will be discussed later on in this subsection. Let $X(t, T)$ be the solution of the Riccati equation (1.15) with boundary condition $X(T, T) = 0$. Based on the above assumption, it can easily be proved that, for all $t \geq t_0$, there is the limit

$$\bar{X}(t) := \lim_{T \to \infty} X(t, T),$$

and $\bar{X}(t)$ is a solution of (1.15). Moreover, $x(t)^T \bar{X}(t) x(t)$ is the optimal performance index, when the initial time is t and the initial state is $x(t)$. The optimal control at time t (for arbitrary initial time) is uniquely defined by (1.17) with $X(t)$ replaced by $\bar{X}(t)$. The solution $\bar{X}(t)$ is called a *stationary solution* of the Riccati equation (1.15) with condition $\bar{X}(\infty) = 0$.

The results can be easily applied to *time-invariant* (or *constant*) case, when A, B, Q, and R, are constant matrices. Then, the limit $\bar{X}(t)$ above is constant (and also equals the limit of $X(t, T)$, when t tends to $-\infty$). Moreover, the Riccati equation (1.15) becomes an algebraic quadratic matrix equation, since the left-hand side is identically zero. This case will be discussed in § 1.1.2.

Optimal control of discrete-time systems

Discrete-time systems frequently occur in practice. Some systems, such as many economic systems, are inherently discrete; other discrete-time systems

---

[4] For the infinite time horizon problem, the matrix F is usually set to zero.

are obtained by discretizing continuous-time systems, in order to use real-time computer control algorithms. Let the discrete-time linear dynamical system described by

$$x(t+1) = A(t)x(t) + B(t)u(t), \qquad x_0 := x(t_0), \tag{1.18}$$

where $x(t) \in \mathbb{R}^n$ is the state vector, $u(t) \in \mathbb{R}^m$ is the control (input) vector, and $A(t)$ and $B(t)$ are real matrices of suitable dimensions, and let the quadratic performance index (the cost functional) be

$$J(x_0, u(\cdot), t_0) = \sum_{t=t_0}^{T-1} [x(t)^T Q(t)x(t) + u(t)^T R(t)u(t)] + x(T)^T F x(T), \tag{1.19}$$

where $Q(t) = Q(t)^T \geq 0$, $R(t) = R(t)^T > 0$ and $F = F^T \geq 0$ are weighting matrices.

The *discrete-time linear-quadratic control problem* consists in finding a control trajectory, $\{u^*(t),\ t = t_0{:}T-1\}$, minimizing the J functional in (1.19) in terms of u, subject to the dynamical constraint (1.18).

Standard mathematical programming (see, for example, (Gill, Murray, and Wright, 1981)) can be applied for the numerical solution of this optimal control problem. But it is preferable to use its closed-form solution, that can be obtained using, for instance, the discrete-time maximum principle or dynamic programming (Bellman, 1957; Bellman and Dreyfus, 1962; see (Kwakernaak and Sivan, 1972), for a proof for a slightly modified problem). The solution is given by the following theorem based on (Dorato and Levis, 1971).

Theorem 1.2 *The discrete-time optimal control problem defined above has a unique solution, given by the following state-feedback control*

$$u(t) = H(t)x(t), \qquad t = t_0{:}T-1, \tag{1.20}$$

where

$$H(t) = -(R(t) + B(t)^T X(t+1)B(t))^{-1} B(t)^T X(t+1)A(t), \tag{1.21}$$

and $X(t+1)$ *satisfies the* difference Riccati equation

$$\begin{aligned} X(t) &= A(t)^T X(t+1)A(t) - A(t)^T X(t+1)B(t)(R(t) + \\ &\quad B(t)^T X(t+1)B(t))^{-1} B(t)^T X(t+1)A(t) + Q(t) && (1.22) \\ &= A(t)^T (X(t+1)^{-1} + B(t)R(t)^{-1}B(t)^T)^{-1} A(t) + Q(t) && (1.23) \\ &= (A(t) + B(t)H(t))^T X(t+1)A(t) + Q(t) && (1.24) \\ &= (A(t) + B(t)H(t))^T X(t+1)(A(t) + B(t)H(t)) + \\ &\quad H(t)^T R(t)H(t) + Q(t), \qquad t = T-1{:}-1{:}t_0, && (1.25) \end{aligned}$$

with $X(T) = F$. *The minimal value of the* J *functional (1.19) is* $x_0^T X(0)x_0$.

Remark 1.1 Obviously, the form (1.23) exists only when $X(t+1)$ is non-singular. The cost functional (1.19) in (Dorato and Levis, 1971; see also (Caines and Mayne, 1970)) includes two *coupling terms*, $x(t)^T S(t)u(t) + u(t)^T S(t)^T x(t)$, and the corresponding relations (1.21) and (1.22) are more complicated. Generally, the optimal problem can be reformulated, so that the coupling terms should be dismissed (see Problem 1.1 in § 1.2.1). □

Theorem 1.2 also applies to the time-invariant case, but the control law remains variant, for finite final time. The difference Riccati equation is solved in backward time, and the control is applied in real (forward) time. The $H(t)$ matrices could be precomputed and stored. There is no need to store $X(t)$.

Remark 1.2 The non-negativity assumptions on the Q, R, and F matrices can be relaxed (Clements and Anderson, 1977). Necessary and sufficient conditions for the existence of a solution will then be:

$$\begin{aligned} R + B^T X(t+1)B &\geq 0; \\ \mathcal{N}(R + B^T X(t+1)B) &\subseteq \mathcal{N}(A^T X(t+1)B), \end{aligned}$$

for $t = t_0{:}T-1$, where $\mathcal{N}(M)$ is the null space of the matrix M, i.e. $\mathcal{N}(M) = \{x, x \in \mathbb{C}^n \colon Mx = 0\}$, when $M \in \mathbb{C}^{m \times n}$. The inverses in (1.21) and (1.22) are then replaced by Moore-Penrose generalized inverses (see, for example, (Nashed, 1976)). The solution of the Riccati equation can be reduced to the solution of similar problems of less dimensions, using the so-called *conjugate directions* of the Riccati equation. □

For the infinite time horizon case, a *stationary solution* of the Riccati equation is obtained, based on more restrictive assumptions. For instance, the following theorem can be proven (Kwakernaak and Sivan, 1972).

Theorem 1.3 *Assume that the* $A(t)$, $B(t)$, $Q(t)$, *and* $R(t)$ *matrices are bounded for* $t \geq t_0$ *and*

$$Q(t) \geq \alpha I_n, \quad R(t) \geq \beta I_m, \qquad \alpha, \beta > 0, \quad t \geq t_0.$$

If the system (1.18) is either controllable or exponentially stable[5]*, then the solution* $X(t)$ *of equation (1.22) with final condition* $X(T) = 0$ *converges to a matrix sequence* $\bar{X}(t) \geq 0$*, when* T *tends to* ∞*, and* $\bar{X}(t)$ *is a solution of equation (1.22).*

The feedback control law $u(t) = \bar{H}(t)x(t)$, where $\bar{H}(t)$ is obtained when replacing $X(t+1)$ in (1.21) by $\bar{X}(t+1)$, is the *stationary state-feedback control law*. The time-invariant case, with $T = \infty$, will be discussed in § 1.1.2.

[5] See the paragraph titled "Structural properties", for definitions (Definitions 1.1 and 1.3).

Optimal estimation for discrete-time systems

Consider the discrete-time linear dynamical system described by

$$\begin{aligned} x(t+1) &= Ax(t) + v(t), \qquad t \geq t_0, \qquad x_0 := x(t_0), \\ y(t) &= Cx(t) + e(t), \end{aligned} \tag{1.26}$$

where $x(t) \in \mathbb{R}^n$ is the state vector, $y(t) \in \mathbb{R}^l$ is the output vector, A and C are real matrices of suitable dimensions, and the *stochastic noise processes* $\{v(t)\}$ and $\{e(t)\}$ are independent, Gaussian vector sequences, with zero mean and known covariance matrices, that is

$$\begin{aligned} \mathcal{E}\{v(t)\} &= 0, \qquad \mathcal{E}\{v(t)v(t)^T\} = Q, \\ \mathcal{E}\{e(t)\} &= 0, \qquad \mathcal{E}\{e(t)e(t)^T\} = R, \qquad \mathcal{E}\{v(t)e(t)^T\} = 0, \end{aligned} \tag{1.27}$$

with $Q \geq 0$, $R > 0$, and x_0 is a Gaussian stochastic vector, independent of $v(t)$, and $e(t)$, for $t \geq t_0$, satisfying

$$\mathcal{E}\{x_0\} = M, \qquad \mathcal{E}\{x_0 x_0^T\} = X_0. \tag{1.28}$$

(See, for example, (Chung, 1974), for the theory of probability and stochastic processes.) The A, B, Q, and R matrices in (1.26) and (1.27) could depend on t, but this dependence was not indicated, for convenience.

The *discrete-time optimal estimation problem* consists in finding an estimate, $\tilde{x}(t+1)$, of the state $x(t+1)$, based on the available output information $Y(t) = \{y(t_0), y(t_0+1), \ldots, y(t)\}$ (or $Y(t+1)$), minimizing the quality criterion (the performance index)

$$\mathcal{E}\{g(a^T(x(t+1) - \tilde{x}(t+1)))\}, \tag{1.29}$$

where a is an n-dimensional constant vector, and $g(x)$ is an even and non-decreasing function in x.

The solution of this optimal estimation problem is given recursively by the theorem below (Åström, 1970).

Theorem 1.4 *The estimate of the state $x(t+1)$ of the system (1.26), based on $Y(t)$, minimizing the criterion (1.29), subject to conditions (1.27), and (1.28), is the* conditioned mean, $\tilde{x}(t+1) = \mathcal{E}\{x(t+1) \mid Y(t)\}$, *and satisfies the recursive equation:*

$$\tilde{x}(t+1) = A\tilde{x}(t) + K(t)(y(t) - C\tilde{x}(t)), \qquad \tilde{x}(t_0) = M, \tag{1.30}$$

where,

$$K(t) = AX(t)C^T(R + CX(t)C^T)^{-1}, \tag{1.31}$$

and $X(t) := \mathcal{E}\{(x(t) - \widetilde{x}(t))(x(t) - \widetilde{x}(t))^T\}$, *given by*

$$\begin{aligned} X(t+1) &= AX(t)A^T - AX(t)C^T(R + CX(t)C^T)^{-1}CX(t)A^T + Q \\ &= (A - K(t)C)X(t)A^T + Q \qquad (1.32) \\ &= (A - K(t)C)X(t)(A - K(t)C)^T + K(t)RK(t)^T + Q, \end{aligned}$$

with initial condition X_0.

The matrix $K(t)$ in Theorem 1.4 is usually called the *Kalman gain*. It is apparent that $X(t)$ does not depend on *observations* in $Y(t)$, so that $X(t)$ and $K(t)$ can be precomputed; $K(t)$ only should be stored.

It can be shown that $\widetilde{x}(t+1)$ is a *sufficient statistic* for the conditioned probability distribution of $x(t+1)$, given $Y(t)$. It can also be interpreted as a projection of the state $x(t+1)$ onto the linear subspace generated by the observations $Y(t)$ (see, for instance, (Luenberger, 1969), for such an approach.) Theorem 1.4 gives, in fact, an algorithm for computing the conditioned distribution, since a Gaussian stochastic process is completely defined by its mean vector and covariance matrix. Denoting $\widetilde{y}(t) = y(t) - \mathcal{E}\{y(t) \mid Y(t-1)\}$, the *innovation process*, it follows that the innovations $\widetilde{y}(t)$ and $\widetilde{y}(\tau)$ are independent for $t \neq \tau$, and

$$\mathcal{E}\{\widetilde{y}(t)\widetilde{y}(t)^T\} = R + CX(t)C^T.$$

Theorem 1.4 gives an *optimal predictor*. An *optimal filter* will result if using the following formulas (Åström, 1970):

$$\widehat{x}(t+1) = A\widehat{x}(t) + K(t+1)(y(t+1) - CA\widehat{x}(t)), \qquad \widehat{x}(t_0) = M, \quad (1.33)$$

where,

$$\begin{aligned} K(t) &= X(t)C^T(R + CX(t)C^T)^{-1}, \\ X(t) &= AS(t-1)A^T + Q, \qquad (1.34) \\ S(t) &= X(t) - K(t)CX(t), \qquad S(t_0) = X_0. \end{aligned}$$

Remark 1.3 Theorem 1.4 can be generalized for the case when the noise processes $\{v(t)\}$ and $\{e(t)\}$, $t \geq t_0$, have constant, but unknown means. These means are considered as additional state variables. Moreover, when the system (1.26) includes a deterministic control signal $Bu(t)$, this signal is also introduced into the estimator (in (1.30), or in (1.33)). □

Remark 1.4 Of special importance for practical applications of modern control theory is *system identification*, i.e., model determination using the available input-output data. Optimization techniques have been widely used; see, for instance, (Goodwin and Payne, 1977; Ljung, 1987; Ljung and Söderström, 1983; Söderström and Stoica, 1989). Recently, algebraic, subspace projection techniques have been proposed (Van Overschee and De Moor, 1994; Verhaegen, 1993a, 1993b, 1994; Verhaegen and Dewilde, 1992a, 1992b), that can directly find out the A, B, and C matrices from the $\{u(t)\}$ and $\{y(t)\}$ records. □

The above discussion on optimal control and optimal estimation of discrete-time systems reveals the close connection between the two problems. This is made explicit in the next result (Åström, 1970).

Theorem 1.5 (Duality of optimal control and estimation) *The optimal estimation problem for the system (1.26), for* $t = t_0{:}T-1$, *subject to the conditions (1.27), and (1.28), is equivalent to the optimal control problem for the system*

$$z(t) = A^T z(t+1) + C^T u(t+1), \tag{1.35}$$

with the performance index

$$\sum_{t=t_0}^{T-1} [z(t)^T Q z(t) + u(t)^T R u(t)] + z(t_0-1)^T X_0 z(t_0-1). \tag{1.36}$$

From a mathematical point of view, taking Theorem 1.5 into account, it is sufficient to have tools for solving the optimal control problem. However, the direct implementation of the optimal estimator recursive formulas can induce some numerical difficulties. It would be advisable to use special techniques for sequential estimation, like those described, for instance, in (Bierman, 1977). Specialized algorithms based on such techniques are given in Subsection 1.5.3.

According to the *separation theorem* (Åström, 1970), the optimal control and optimal estimation problems can be solved independently. The stochastic state-feedback control law is obtained by replacing $x(t)$ in (1.20) by its estimate, $\tilde{x}(t)$ (or $\hat{x}(t)$). The combined control-estimation problem is a *linear-quadratic Gaussian optimization problem*, or simply an *LQG problem*. So, in this case, the control configuration must include both a regulator (or controller), and an estimator (filter or predictor). This mixture is often called *compensator*.

Similar results have also been proven for continuous-time systems.

Structural properties

Main structural properties of linear dynamical systems are briefly discussed here, for reference purposes. Details can be found, for instance, in (Barnett,

1975; Casti, 1977; Kailath, 1980; Kwakernaak and Sivan, 1972; Rubio, 1971). Let the system described by

$$\delta(x(t)) = A(t)x(t), \qquad t \geq t_0, \qquad x_0 := x(t_0), \tag{1.37}$$

where $x(t) \in \mathbb{R}^n$, and $\delta(x(t))$ denotes either $\dot{x}(t)$ (the differential operator), or $x(t+1)$ (the difference operator). The zero state, called the *origin*, is an *equilibrium point* of the system (1.37), since $x(t_1) = 0$, $t_1 \geq t_0$, implies $x(t) = 0$, for any $t \geq t_1$. Non-zero equilibrium points are similarly defined, and can be translated to the origin.

Definition 1.1 The origin is a *stable* equilibrium point of the system (1.37) if, given t_0, for any $\varepsilon > 0$ there is a real number $\eta(\varepsilon, t_0) > 0$, so that $\|x_0\| < \eta(\varepsilon, t_0)$ implies $\|x(t)\| < \varepsilon$ for $t \geq t_0$.[6] If $\eta(\varepsilon, t_0)$ does not depend on t_0, then the origin is *uniformly stable.*

The origin is an *asymptotically stable* equilibrium point of (1.37) if it is stable and if for any $\varepsilon > 0$ and any solution $x(t; t_0, x_0)$ of the system with initial condition x_0, there is a number $N(\varepsilon, x_0, t_0)$, so that $\|x(t)\| < \varepsilon$ for $t > N(\varepsilon, x_0, t_0)$. (Equivalently, $\lim_{t\to\infty} \|x(t)\| = 0$.) If $N(\varepsilon, x_0, t_0)$ does not depend on t_0, the origin is *uniformly asymptotically stable.* (Number $N(\varepsilon, x_0, t_0)$ is taken as integer for discrete-time systems.)

The origin is an *exponentially stable* equilibrium point of (1.37) if there are two strictly positive constants, α and M, not depending on t_0, so that

$$\|x(t; t_0, x_0)\| \leq M e^{-\alpha(t-t_0)}, \qquad t \geq t_0. \tag{1.38}$$

The origin is an *unstable* equilibrium point of (1.37) if each solution with non-zero initial state has the following property: given a positive arbitrarily large number N, there is a number $\tau(N, x_0, t_0)$ so that $\|x(\tau)\| > N$. (Number τ is taken as integer for discrete-time systems.) □

Remark 1.5 For brevity, the system (1.37) is said to be stable, asymptotically stable, etc., if the origin has that property. □

The stability of a constant linear system is related to the location of its spectrum, $\{\lambda_i\} \subset \lambda(A)$. Specifically, a constant system is asymptotically stable if and only if, for $i = 1{:}\,n$,

$$\begin{aligned} \Re(\lambda_i) &< 0, && \text{for continuous-time systems;} \\ |\lambda_i| &< 1, && \text{for discrete-time systems.} \end{aligned}$$

---

[6] An arbitrary vector norm in $\mathbb{R}^n$ is used; the matrix norms are those for linear applications.

The stability domain in the complex plane is denoted by $\mathbb{C}^-$. The eigenvalues belonging to $\mathbb{C}^-$ are called *stable eigenvalues*. The remaining eigenvalues are *unstable*, and they belong to $\mathbb{C}^+ := \mathbb{C} \setminus \mathbb{C}^-$.

An important stability property is stated in the following theorem.

Theorem 1.6 *The system*

$$\delta(x(t)) = Ax(t), \qquad t \geq t_0, \qquad x_0 := x(t_0), \tag{1.39}$$

is asymptotically stable if and only if for any given symmetric positive definite matrix G, there is a positive definite matrix F, the unique solution of a Lyapunov equation:

$$A^T F + FA = -G, \qquad \textit{for continuous-time systems}; \tag{1.40}$$

$$A^T FA - F = -G, \qquad \textit{for discrete-time systems}. \tag{1.41}$$

The system (1.39) is exponentially stable if and only if it is asymptotically stable.

Definition 1.2 Consider the system (1.39) and let $\mathcal{N}_j$ be the null space of the matrix $(A - \lambda_j I_n)^{m_j}$, where λ_j is an eigenvalue of A with multiplicity m_j, $j = 1\colon k$, $m_1 + \cdots + m_k = n$. The *stable subspace* of the system is the *real* subspace given by the direct sum[7] of the null spaces $\mathcal{N}_j$ corresponding to stable eigenvalues. The *unstable subspace* is similarly defined, taken the unstable eigenvalues. □

Note that the stable subspace $\mathcal{S}$ of a system (1.39) is A-*invariant*, that is $A\mathcal{S} \subset \mathcal{S}$. It can be proven that if $x_0 \in \mathcal{N}_j$, then the corresponding solution of (1.39) satisfies $x(t) \in \mathcal{N}_j$, for $t \geq t_0$. Such a solution is called a *mode*.

Consider now the controlled system S described by

$$\delta(x(t)) = Ax(t) + Bu(t), \qquad t \in \mathbf{T}, \qquad x_0 := x(t_0), \tag{1.42}$$

where $\mathbf{T}$ is a subset of the real or integer numbers, depending on the type of the system (continuous-time or discrete-time, respectively).

Definition 1.3 A *phase* of the system S is an element of the Cartesian product $\mathbf{T} \times \mathbb{R}^n$.

The phase (t, x) of S is *reachable* or *accessible* (from the origin) if and only if there is a time t_1, $t_0 \leq t_1 < t$, and a control $u(\cdot)$ that transfers $(t_1, 0)$ in (t, x).

[7] A linear subspace, $\mathcal{N}_3$, is a *direct sum* of the linear subspaces $\mathcal{N}_1$ and $\mathcal{N}_2$ if any vector $x_3 \in \mathcal{N}_3$ can be uniquely written as $x_3 = x_1 + x_2$, where $x_1 \in \mathcal{N}_1$ and $x_2 \in \mathcal{N}_2$.

The phase (t, x) of S is *controllable* (to the origin) if and only if there is a time $t_2 > t$ and a control $u(\cdot)$ that transfers (t, x) in $(t_2, 0)$.

The system S or the matrix pair (A, B) is *completely reachable (controllable)* in $t \in \mathbf{T}$ if any phase (t, x), with fixed t, is reachable (controllable). If t is not specified, it is assumed that the property holds for all t.

The *accessibility (controllability) subspace* of S is the linear subspace of the reachable (controllable) states of S. □

Remark 1.6 Accessibility is equivalent to controllability for (constant) continuous-time systems or discrete-time systems with A nonsingular. □

Theorem 1.7 *The system (1.42) is completely reachable if and only if the* controllability matrix,

$$\Gamma = \begin{bmatrix} B & AB & \cdots & A^{n-1}B \end{bmatrix}, \tag{1.43}$$

has rank n.

Theorem 1.8 *The system (1.42) is completely controllable if and only if*

$$\mathcal{R}([\,B \;\; AB \;\; \cdots \;\; A^{r-1}B\,]) \supseteq \mathcal{R}(A^r), \qquad \forall r, \quad 0 < r \leq n-1, \tag{1.44}$$

where $\mathcal{R}(M)$ denotes the subspace generated by the columns of matrix M.

Theorem 1.9 *The accessibility subspace of the system (1.42) is generated by the columns of the matrix Γ in Theorem 1.7, and is included in the controllability subspace. The controllability subspace of (1.42) is A-invariant.*

There are many other characterizations of accessibility and controllability (see, for instance, (Anderson and Moore, 1979)).

The accessibility or controllability property is invariant under coordinate transformations in the state space. Let T be a linear nonsingular transformation, $T : x \to \widehat{x} = Tx$. Transformation T induces a *similarity transformation* of the A and B matrices, defined by

$$\widehat{A} = TAT^{-1}, \qquad \widehat{B} = TB. \tag{1.45}$$

Theorem 1.10 *Let the system (1.42). There is a nonsingular state transformation*

$$\widehat{x}(t) = T^{-1}x(t) \tag{1.46}$$

such that the transformed state vector satisfies the relation

$$\delta(\widehat{x}(t)) = \begin{bmatrix} \widehat{A}_{11} & \widehat{A}_{12} \\ 0 & \widehat{A}_{22} \end{bmatrix} \widehat{x}(t) + \begin{bmatrix} \widehat{B}_1 \\ 0 \end{bmatrix} u(t), \qquad \widehat{A}_{11} \in \mathbb{R}^{r \times r}, \tag{1.47}$$

where the pair $(\widehat{A}_{11}, \widehat{B}_1)$ is completely controllable, and $r = rank(\Gamma)$.

Definition 1.4 The system (1.42), or the pair (A, B), is *stabilizable* if the unstable subspace is included in the controllability subspace. □

Remark 1.7 It can be shown that (A, B) is stabilizable if $\text{rank}([\,\lambda I - A \;\; B\,]) = n$, for all $\lambda \in \mathbb{C}^+$. Clearly, any (constant) asymptotically stable system is stabilizable. Any completely controllable system is stabilizable.□

Theorem 1.11 *Let the system (1.42) and the transformed system (1.47) obtained from (1.42) using a nonsingular transformation (1.46). Then, the system (1.42) is stabilizable if and only if the matrix $\widehat{A}_{22}$ is asymptotically stable.*

Consider now the "observed" system S described by

$$\begin{aligned} \delta(x(t)) &= Ax(t), \qquad t \in \mathbf{T}, \qquad x_0 := x(t_0), \\ y(t) &= Cx(t), \end{aligned} \tag{1.48}$$

where $x(t) \in \mathbb{R}^n$, and $y(t) \in \mathbb{R}^l$. Let $x(s; t, x_1)$ be the state of S at time s for "initial" state x_1 at time t, and $y(s; t, x_1)$ be the corresponding output.

Definition 1.5 Two phases (t, x_1) and (t, x_2) of S belong to the same *reconstruction class* if and only if

$$y(s; t, x_1) = y(s; t, x_2), \qquad \forall s \leq t.$$

The phases (t, x_1) and (t, x_2) belong to the same *observation class* if and only if

$$y(s; t, x_1) = y(s; t, x_2), \qquad \forall s \geq t.$$

The phase (s, x) is *unreconstructible* if and only if it belongs to the reconstruction class of $(t, 0)$, that is

$$y(s; t, x) = 0, \qquad \forall s \leq t,$$

and (s, x) is *unobservable* if and only if it belongs to the observation class of $(t, 0)$, that is

$$y(s; t, x) = 0, \qquad \forall s \geq t.$$

The system S or the matrix pair (C, A) is *completely reconstructible (observable)* at $t \in \mathbf{T}$ if and only if it lacks any unreconstructible (unobservable) phase (t, x), $x \neq 0$. If t is not specified, it is assumed that the property holds for all t.

The *unreconstructibility (unobservability) subspace* of the system S is the linear subspace of the unreconstructible (unobservable) states of S. □

Remark 1.8 Reconstructibility is equivalent to observability for continuous-time systems or discrete-time systems with A nonsingular. □

Theorem 1.12 *The system (1.48) is completely reconstructible if and only if the* observability matrix,

$$\Theta = \left[\begin{array}{cccc} C^T & A^T C^T & \cdots & (A^T)^{n-1} C^T \end{array}\right]^T, \tag{1.49}$$

has rank n.

Theorem 1.13 *The unreconstructibility subspace of the system (1.48) is the null space of the observability matrix* Θ*. The unobservability subspace of (1.48) is A-invariant.*

Theorem 1.14 *Let the system (1.48). There is a nonsingular state transformation*

$$\widehat{x}(t) = Tx(t), \tag{1.50}$$

such that the transformed state and output vectors satisfy the relations

$$\begin{aligned} \delta(\widehat{x}(t)) &= \begin{bmatrix} \widehat{A}_{11} & 0 \\ \widehat{A}_{21} & \widehat{A}_{22} \end{bmatrix} \widehat{x}(t), \qquad \widehat{A}_{11} \in \mathbb{R}^{r \times r}, \\ \widehat{y}(t) &= \begin{bmatrix} \widehat{C}_1 & 0 \end{bmatrix} \widehat{x}(t), \end{aligned} \tag{1.51}$$

where the pair $(\widehat{C}_1, \widehat{A}_{11})$ *is completely reconstructible, and* $r = \text{rank}(\Theta)$.

Definition 1.6 The system (1.48), or the pair (C, A), is *detectable* if the unreconstructibility subspace is included in the stable subspace. □

Remark 1.9 The pair (C, A) is detectable if and only if (A^T, C^T) is stabilizable. Clearly, any asymptotically stable system is detectable. Any completely reconstructible (observable) system is detectable. □

Theorem 1.15 *Let the system (1.48) and the transformed system (1.51), obtained from (1.48) using a nonsingular transformation (1.50). Then, the system (1.48) is detectable if and only if the matrix* $\widehat{A}_{22}$ *is asymptotically stable.*

1.1.2 Standard control problems

Based on the previously presented results, we now state the standard control problem and its solution for continuous-time and discrete-time systems. Let the time-invariant continuous-time linear dynamical system described by

$$\dot{x}(t) = Ax(t) + Bu(t), \qquad x_0 := x(0), \tag{1.52}$$

where $x(t) \in \mathbb{R}^n$ is the state vector, $u(t) \in \mathbb{R}^m$ is the control (input) vector, and A and B are real matrices of suitable dimensions, and let the quadratic performance index (the cost functional) be

$$J(x_0, u) = \int_0^\infty [x(t)^T Q x(t) + u(t)^T R u(t)]\, dt, \tag{1.53}$$

where $Q = Q^T \geq 0$ and $R = R^T > 0$ are weighting matrices. (The nonnegativity assumptions on the matrices Q and R can be removed; in this case, the matrix inverses are replaced by generalized inverses.) In the formula (1.53), $x(t)$ is the solution of the vector differential equation (1.52), corresponding to a given control trajectory $\{u(t),\, t \geq 0\}$ and a given initialization x_0.

The *standard (continuous-time) linear-quadratic optimization problem* consists in finding a control trajectory $\{u(t), t \geq 0\}$, minimizing the J functional in (1.53) in terms of u, subject to the dynamical constraint (1.52).

Under certain conditions, as those to be specified later, there is a unique control trajectory,

$$u(t) = Hx(t), \qquad H := -R^{-1}B^T X, \tag{1.54}$$

that solves the standard linear-quadratic problem, where $X \in \mathbb{R}^{n \times n}$ is the unique symmetric positive semi-definite solution of the *algebraic matrix Riccati equation*

$$A^T X + XA - XBR^{-1}B^T X + Q = 0. \tag{1.55}$$

The control trajectory defined by (1.54) is a *state-feedback control law*, and, replacing it in (1.52), it can be shown to provide a stable *closed-loop system*, that is $\lambda(A + BH) \subset \mathbb{C}^-$. In this way, the optimal control law in (1.54) solves the stabilizing problem, and the solution $X \geq 0$ of the Riccati equation (1.55) is called a *stabilizing solution*. The matrix H, defined in (1.54), is the *gain matrix of the optimal regulator*, or *optimal stabilizing matrix*. By a suitable selection of the weighting matrices Q and R in (1.53), it is possible to have an appropriate behaviour of the closed-loop system, since both the dynamical properties and the magnitudes of the state and control variables are influenced.

All above may be similarly translated for discrete-time systems, described by

$$x(t+1) = Ax(t) + Bu(t), \qquad x_0 := x(0), \tag{1.56}$$

and the quadratic performance index

$$J(x_0, u) = \sum_{t=0}^{\infty} [x(t)^T Q x(t) + u(t)^T R u(t)]. \tag{1.57}$$

Under certain conditions there is a unique control law,

$$u(t) = Hx(t), \qquad H := -(R + B^T XB)^{-1} B^T XA, \tag{1.58}$$

minimizing the functional J in (1.57) in terms of u, where X is the unique symmetric positive semi-definite solution of the algebraic matrix Riccati equation

$$X = A^T XA - A^T XB(R + B^T XB)^{-1} B^T XA + Q. \tag{1.59}$$

The equations (1.55) and (1.59) will be referred, for convenience, as *continuous-time* and *discrete-time* Riccati equation, respectively. It is worth mentioning that these equations appear not only in the context presented above, but also in numerous procedures for analysis, synthesis and design of linear-quadratic Gaussian control and estimation systems, as well as in other branches of applied mathematics. The control law (1.54) or (1.58) is said to be a *proportional control*, by extending the terminology from the traditional single-input single-output control systems.

We shall now present several conditions for the existence and uniqueness of a positive semi-definite solution of the algebraic matrix Riccati equations (1.55) or (1.59). For convenience, in the discrete-time case we shall assume that the matrix A is nonsingular. (This is the case of any system obtained by discretizing a continuous-time system without time delay.) Let us define the following extended matrices (for continuous-time and discrete-time systems, respectively):

$$G = \begin{bmatrix} A & -F \\ -Q & -A^T \end{bmatrix}, \tag{1.60}$$

$$G = \begin{bmatrix} A + FA^{-T}Q & -FA^{-T} \\ -A^{-T}Q & A^{-T} \end{bmatrix}, \tag{1.61}$$

where $F := BR^{-1}B^T$. Necessary and sufficient conditions for the existence and uniqueness of positive semi-definite solutions of algebraic matrix Riccati equations are proved, for instance, in (Bucy, 1975; Caines and Mayne, 1970, 1971; Casti, 1977; Kučera, 1972; Lancaster, Ran and Rodman, 1986, 1987; Lancaster and Rodman, 1980; Martensson, 1971; Molinari 1973a, 1973b, 1977; Payne and Silverman, 1973; Willems, 1971, 1974; Wonham, 1968).

Theorem 1.16 *Equation (1.55), or equation (1.59), respectively, has a stabilizing solution $X \geq 0$ if and only if the following conditions are met:*

(i) the pair (A, B) is stabilizable;

(ii) the matrix G is a dichotomic matrix, *that is $\Re(\lambda_i) \neq 0$ for G in (1.60), or $|\lambda_i| \neq 1$ for G in (1.61), respectively, where $\lambda_i \in \lambda(G)$, $i = 1{:}\,2n$.*

Theorem 1.17 *The conditions (i) and (ii) above imply the existence and uniqueness of a solution $X \geq 0$ of equation (1.55), or equation (1.59), if and only if the following extra condition holds:*

(iii) the pair (D, A) is detectable, where the matrix D is a maximal rank factor of Q, that is $Q = D^T D$ and rank(D) = rank(Q).

Theorem 1.18 *Equation (1.55), or equation (1.59), respectively, has a unique stabilizing solution $X \geq 0$ if and only if (i) and (iii) hold. If (iii) is strengthened to complete reconstructibility, then the solution $X > 0$ is obtained.*

In the sequel, we shall assume that such necessary and sufficient conditions for the existence and uniqueness of a solution $X \geq 0$ of the equations (1.55) or (1.59) are met.

1.2 Problems Reducible to Standard Linear-Quadratic Optimization Problems

To illustrate the wide applicability of the algorithms to be investigated, we shall now present several problems reducible to the standard linear-quadratic optimization problem. A detailed control system design example is included.

1.2.1 Related problems and control schemes

Problem 1.1 (Adding coupling terms in cost functional) The optimal problem for the system (1.52) or (1.56), with the quadratic performance index including terms of the form $x(t)^T Su(t) + u(t)^T S^T x(t)$ under the integral in (1.53), or the sum in (1.57), respectively, may be reduced to the standard problem for the system $(\bar{A}, B)$ with the performance index $(\bar{Q}, R)$ (that is the index (1.53) or (1.57) with weighting matrices $\bar{Q}$ and R), where

$$\bar{A} := A - BR^{-1}S^T, \quad \bar{Q} := Q - SR^{-1}S^T, \quad \text{if } Q - SR^{-1}S^T \geq 0. \quad (1.62)$$

Indeed, denoting $\bar{u} := u + R^{-1}S^T x$, we have,

$$\dot{x} = Ax + Bu = \bar{A}x + BR^{-1}S^T x + Bu = \bar{A}x + B\bar{u},$$

and, after some simple manipulations,

$$x^T Qx + x^T Su + u^T S^T x + u^T Ru = x^T \bar{Q}x + \bar{u}^T R\bar{u}.$$

After computing $\bar{H}$ (using $\bar{A}$ and $\bar{Q}$), the optimal gain matrix is given by $H = \bar{H} - R^{-1}S^T$. By duality, Problem 1.1 corresponds, for the filtering case, to systems with correlated noise.

Problem 1.2 (Weighting control $u(t-1)$) The optimal problem for a discrete-time system (1.56) with cost functional (Kwakernaak and Sivan, 1972)

$$J(x_0, u) = \sum_{t=0}^{\infty} [x(t)^T Qx(t) + u(t-1)^T Ru(t-1)],$$

has the solution $\bar{X} = X - Q$ and $\bar{H} = H$, where X and H are matrices corresponding to the standard problem. This result follows simply by adding Q on both sides of the Riccati equation considered in the cited reference.

Problem 1.3 (Output regulation) Let us assume that the system output is defined by the l-dimensional vector

$$y(t) = Cx(t), \tag{1.63}$$

and that we want to minimize a functional obtained when replacing the term $x(t)^T Qx(t)$ in (1.53) or (1.57) by $y(t)^T \tilde{Q}y(t)$. The optimal problem obtained in this manner may be reduced to the standard problem by taking $Q := C^T \tilde{Q}C$.

Problem 1.4 (Proportional-plus-integral control) Generally, the solution of Problem 1.3 does not ensure a zero steady-state regulation error, when the desired outputs, called *references* or *setpoints* are not zero. This deficiency can be overcome by using some error integrators (Foulard, Gentil and Sandraz, 1977; Sandraz, 1973), defined as

$$v(t) = \int_0^t e(t)dt, \quad \text{or} \tag{1.64}$$

$$v(t+1) = v(t) + e(t), \tag{1.65}$$

for continuous-time systems and discrete-time systems, respectively, where $e(t) = r(t) - y(t)$, $r(t)$ being the references vector. The components of error "integrals" $v(t)$ are taken as additional state variables. The solution of Problem 1.4 is obtained by solving the standard problem for an extended system $(\bar{A}, \bar{B}, \bar{C})$, with the performance index $(\bar{Q}, R)$, defined by

$$\bar{A} = \begin{bmatrix} A & 0 \\ -C & P \end{bmatrix}, \quad \bar{B} = \begin{bmatrix} B \\ 0 \end{bmatrix},$$
$$\bar{C} = \begin{bmatrix} C & 0 \\ 0 & I \end{bmatrix}, \quad \bar{Q} = \begin{bmatrix} \widetilde{Q} & 0 \\ 0 & \widehat{Q} \end{bmatrix}, \tag{1.66}$$

where $P = 0$, in the continuous-time case, or $P = I_l$, in the discrete-time case, respectively, and P, $\widetilde{Q}$ and $\widehat{Q}$ have order l. In (1.66), $\widehat{Q}$ is the weighting matrix of the output of error integrators. Note that matrix $\bar{A}$ is unstable (even if A is stable), and the outputs of error integrators are also outputs of the extended system, so that the detectability or observability properties should be inherited, due to the special structure of the matrices. If the extended system is stabilizable, and the original system is detectable, one can solve Problem 1.3 for the extended system (with $\widetilde{Q} \leftarrow \mathrm{diag}(\widetilde{Q}, \widehat{Q})$), obtaining the *proportional-plus-integral control law*

$$\begin{aligned} u(t) &= \bar{H}\bar{x}(t) + H_a r(t) =: Hx(t) + H_i v(t) + H_a r(t), \\ \bar{x}(t) &:= (x(t)^T, v(t)^T)^T. \end{aligned} \tag{1.67}$$

The matrix $\bar{H}$ is computed with a relation (1.54) or (1.58), corresponding to the extended system, and the matrix H_a is determined by a simple matrix relation containing $\bar{X}$, the solution of the Riccati equation for the standard problem defined by the matrices in (1.66). The control law is suboptimal if $r(t)$ is variable in time.

The proportional-plus-integral (PI) control structure just described is depicted in Fig. 1.1. The control signal is determined by the functional blocks in the left-hand side of the block-diagram, based on the formulas above, and it is applied to the controlled process. The closed-loop, feedback configuration is visible. A *state estimator* was included, for generality; it determines estimates of the state ($\widehat{x}$) and filtered output ($\widehat{y}$), using the measured process variables.

Problem 1.5 (Improving reference tracking) The proportional-plus-integral control can provide good regulation properties, but generally it denies a good behaviour on changes of references, especially when the process dynamics is slow and/or includes time delays (also called *dead times*). For such dynamics, large error signals would be integrated for quite long time intervals after a reference change, producing large, undesired control values. The tracking properties can be improved, without disturbing the regulation properties, using an asymptotically stable reference model, to be constructed based on the essential dynamics of the process (see § 1.2.2 for

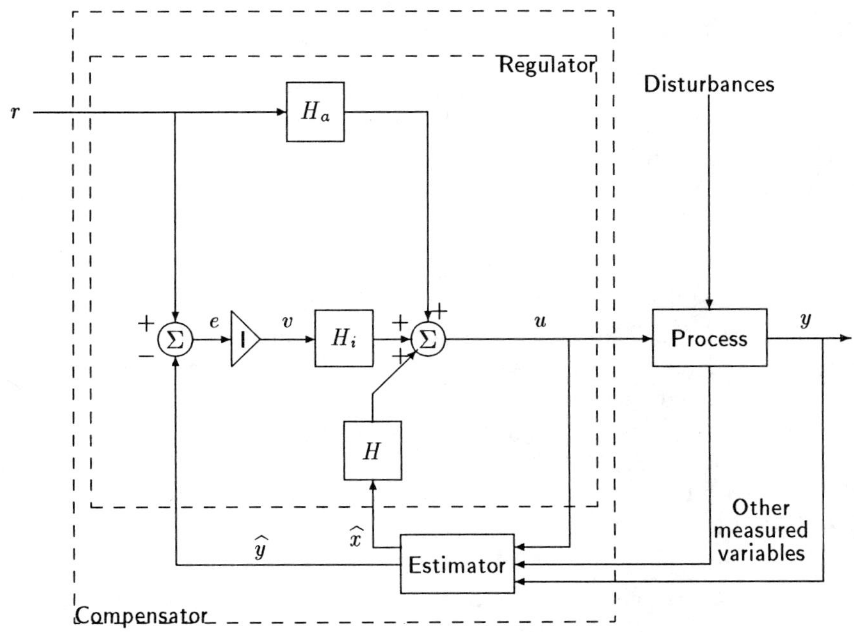

Fig. 1.1 Proportional-plus-integral (PI) control structure; see text for description.

an example). Let the reference model be defined as a system (A_r, B_r, C_r), with asymptotically stable matrix A_r, input r, state $x^{(r)}$, and output $y^{(r)}$. It is possible to extend the system $(\bar{A}, \bar{B}, \bar{C})$ in Problem 1.4, and consider the standard problem for the resulted system $(\bar{\bar{A}}, \bar{\bar{B}}, \bar{\bar{C}})$, where

$$\bar{\bar{A}} = \begin{bmatrix} \bar{A} & 0 \\ 0 & A_r \end{bmatrix}, \qquad \bar{\bar{B}} = \begin{bmatrix} \bar{B} \\ 0 \end{bmatrix}, \qquad \bar{\bar{C}} = \begin{bmatrix} \bar{C} & 0 \\ 0 & C_r \end{bmatrix}. \tag{1.68}$$

The definitions (1.68) and Theorems 1.11 and 1.15 in § 1.1.1 show that the inclusion of a reference model neither affects the detectability (observability), nor does it affect the stabilizability (the uncontrollable part remains stable, if the initial system has been stabilizable). It is not necessary to solve the standard problem for the extended system. Indeed, it can be proved that the optimal control law is

$$u(t) = \bar{\bar{H}}\ \bar{\bar{x}}\ (t) + H_a r(t) = \bar{H}\bar{x}(t) + H_r x^{(r)}(t) + H_{cr} r(t), \tag{1.69}$$

where $\bar{H}$ is the same matrix as in (1.67), H_{cr} is computed using a simple matrix relation, and H_r is determined as the solution of a Sylvester equation. For instance, in the discrete-time case, matrix H_r satisfies the equation

$$H_r = EH_r A_r + F, \tag{1.70}$$

where $E := (\bar{A} + \bar{B}\bar{H})^T$ and A_r have stable eigenvalues only, and F is a matrix that depends on $\bar{X}$, the solution of the Riccati equation for the standard problem defined by the matrices in (1.66). Note that the vector $v(t)$ involved in (1.69) is obtained by integrating the error between the model output and the process output, $e(t) = y^{(r)}(t) - y(t)$.

The proportional-plus-integral control structure with reference model is depicted in Fig. 1.2. The block-diagram is slightly more complicated than that in Fig. 1.1, since it includes the reference model and an extra term in the control law.

Problem 1.6 (Direct rejection of measurable disturbances) Let us consider that (A_p, B_p, C_p) is a stable system driven by the measurable disturbance vector p, with state $x^{(p)}$, and whose output $y^{(p)}(t)$ is added to the output of the system (A, B, C). Constant (in time) measurable disturbances can be directly rejected by an appropriate extension of the control law. The solution of this rejection problem can be obtained in a similar manner to that of Problem 1.5. The regulator structure that corresponds to disturbance rejection may be added to either the proportional or, preferably, the proportional-plus-integral control structure. The control law obtained this way may be somewhat more complicated than that which would result by including the disturbances action into the initial model of the system.

The Figure 1.3 represents the proportional-plus-integral control structure with reference model and direct disturbance rejection. The block-diagram extends that in Fig. 1.2 with the regulator terms corresponding to disturbances.

The separate, successive treatment of the regulation, tracking, and direct disturbance rejection properties is very attractive from a practical point of view. The regulation problem is first solved in the control scheme design process, by setting the dynamics of the closed-loop system by an appropriate selection of the weighting matrices Q and R. Then, the control structures for tracking and measurable disturbance rejection, whose matrices are computed in terms of the results of the regulation problem, are

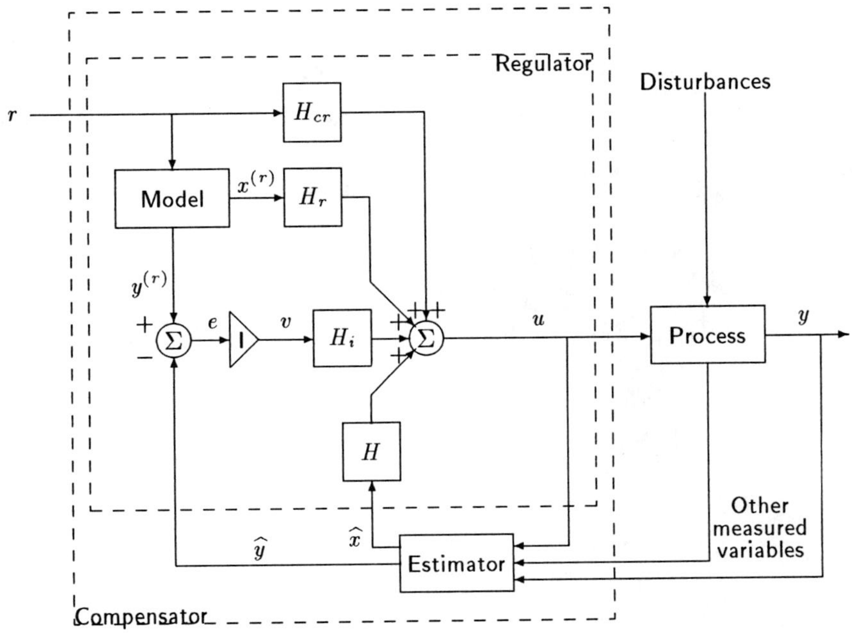

Fig. 1.2 PI control structure with reference model; see text for description.

successively added. By reducing the problem dimension and the number of "degrees of freedom" at each stage of the design process, various tests for selecting the most adequate values for the design parameters can economically be performed. Some results obtained by solving Problems 1.4–1.6 for a numerical example are given in § 1.2.2. A more general treatment of these problems is included in Section 1.3.

Problem 1.7 (Reducing the dimension) For certain practical applications, it is acceptable to solve a reduced order algebraic matrix Riccati equation. For instance, let assume that the matrix A is of the form

$$A = \begin{bmatrix} A_{11} & 0 \\ A_{21} & A_{22} \end{bmatrix},$$

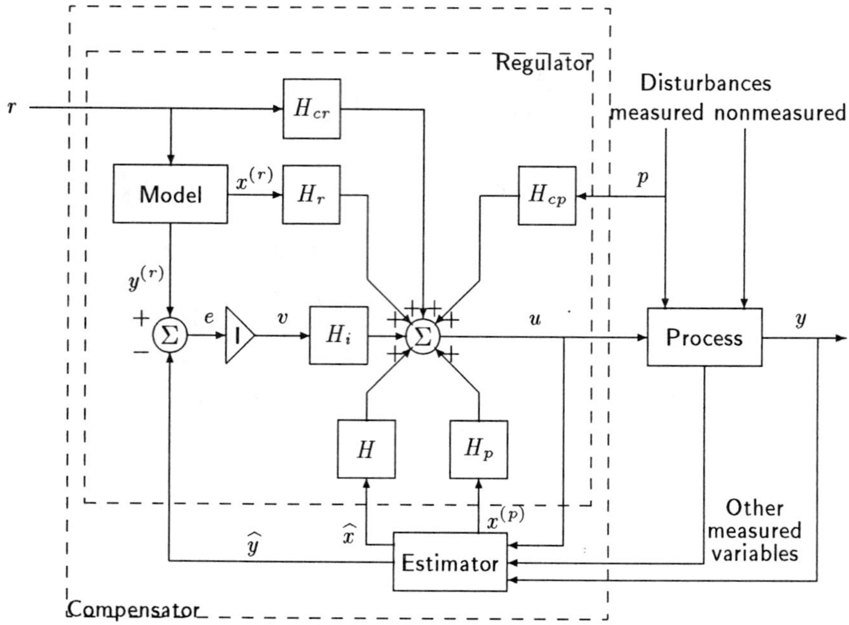

Fig. 1.3 PI control structure with reference model and disturbance rejection; see text for description.

where $\lambda(A_{11}) \cap \lambda(A_{22}) = \emptyset$, and $A_{22} \in \mathbb{R}^{p \times p}$ has stable eigenvalues, which should not be modified by feedback. In this case, it is reasonable that the modes associated with A_{22} be not weighted in the quadratic performance index, hence Q can be chosen as

$$Q = \begin{bmatrix} Q_{11} & 0 \\ 0 & 0 \end{bmatrix}, \qquad Q_{11} \in \mathbb{R}^{(n-p)\times(n-p)}.$$

The reduced optimal problem of order $n - p$ is then solved, resulting in an optimal stabilizing matrix $H_{11} \in \mathbb{R}^{m\times(n-p)}$. The optimal stabilizing matrix for the problem of order n is given by $H = \begin{bmatrix} H_{11} & 0 \end{bmatrix}$. The method is productive, since the value $n - p$ is generally small in practice. Particularly, the method is recommended for systems obtained by discretizing continuous-time systems with time delays. If matrix A has not the form presented above, it is still possible to apply a suitable similarity transforma-

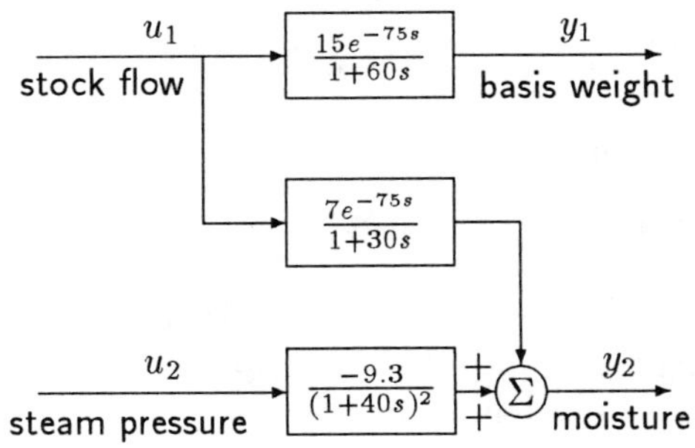

Fig. 1.4 Simplified model of a paper machine; see text for description.

tion, which brings A to that form. For instance, one can use the transpose of the ordered upper real Schur form.

There are many other problems reducible to standard problem. Such a reduction is not always desirable. For instance, the optimal control problem for *descriptor systems*, defined by

$$E\delta(x(t)) = Ax(t) + Bu(t), \tag{1.71}$$

with $E \in \mathbb{R}^{n\times n}$, and δ operator defined in the paragraph titled "Structural properties" (Subsection 1.1.1), is reducible to a standard problem for a system with $\bar{A} := E^{-1}A$ and $\bar{B} := E^{-1}B$, when E is nonsingular; but this transformation should not be used when E is nearly singular. Special algorithms have been proposed for such a case; see, for example, (Mehrmann, 1991).

1.2.2 Numerical example

To illustrate the usefulness of dynamical optimization algorithms, let us consider a simplified model of a paper machine (Foulard, Gentil and Sandraz, 1977; Sandraz, 1973), represented in Fig. 1.4. The physical process includes two output (controlled) variables, the dry basis weight and moisture, and two control variables, the stock flow (of pulp) and steam pressure in the dryer section. The model is valid around a nominal operation regime, defined by the following values or ranges:

- dry basis weight: 72 g/m^2;
- moisture: 6 %;

- stock flow: 4–5 m^3/h;
- steam pressure in the dryer: 0,1–0,25 bar;
- paper machine speed: 600 m/min.

The stock flow influences the two output variables, and so the control problem cannot be properly solved having two independent single-input single-output (monovariable) loops. There is also a dead time, represented by the numerator factor $\exp(-75s)$, on the input-output channels from u_1 to the outputs. This dead time is mainly due to a delay in transport of the pulp from the headbox to the output of the dryer. There are quite large time constants, given by the s coefficients in the transfer functions denominators. Consequently, the process is hardly controlled using conventional control.

Using a sampling period of $\Delta T = 15$ seconds and passing from the transfer functions representation to a state-space representation, we obtain a model with dimensions $n = 9$, $m = l = 2$, defined by the following matrices:

$$A = \begin{bmatrix} 0 & 0 & 0 & 0 & 0 & 0 & 0 & 0 & 0 \\ 1 & 0 & 0 & 0 & 0 & 0 & 0 & 0 & 0 \\ 0 & 1 & 0 & 0 & 0 & 0 & 0 & 0 & 0 \\ 0 & 0 & 1 & 0 & 0 & 0 & 0 & 0 & 0 \\ 0 & 0 & 0 & 1 & 0 & 0 & 0 & 0 & 0 \\ 0 & 0 & 0 & 0 & .222 & .778 & 0 & 0 & 0 \\ 0 & 0 & 0 & 0 & .4 & 0 & .6 & 0 & 0 \\ 0 & 0 & 0 & 0 & 0 & 0 & 0 & 1.372 & -.47 \\ 0 & 0 & 0 & 0 & 0 & 0 & 0 & 1 & 0 \end{bmatrix},$$

$$B = \begin{bmatrix} 1 & 0 & 0 & 0 & 0 & 0 & 0 & 0 & 0 \\ 0 & 0 & 0 & 0 & 0 & 0 & 0 & .098 & 0 \end{bmatrix}^T,$$

$$C = \begin{bmatrix} 0 & 0 & 0 & 0 & 0 & 15 & 0 & 0 & 0 \\ 0 & 0 & 0 & 0 & 0 & 0 & 7 & -5.357 & -3.943 \end{bmatrix}.$$

The process poles (the eigenvalues of A) are: .6616, .7104, .6, .778, and 0 (five multiple poles in the origin). Hence, the open-loop system is stable, and A matrix is singular. The system (A, B, C) is completely reachable and completely reconstructible.[8]

We may ask the automatic control system to satisfy some performance requirements, such as:

- the transient regime, after the action of an output disturbance, or a change of the reference values, be less than $30\Delta T$;

[8] A brute-force test is to find the rank of the controllability and observability matrices, defined by (1.43) and (1.49), respectively.

- overshoots less than 5 % of the nominal values;
- output deviates less than 5 % of the nominal values.

Such performances can be achieved by solving Problem 1.4 with properly selected weighting matrices in the cost functional. This may require some experimentation. Choosing,

$$\widetilde{Q} = \begin{bmatrix} .5 & 0 \\ 0 & 5 \end{bmatrix}, \qquad \widehat{Q} = \begin{bmatrix} .5 & 0 \\ 0 & 5 \end{bmatrix}, \qquad R = \begin{bmatrix} 400 & 0 \\ 0 & 700 \end{bmatrix},$$

and using any algorithm for solving discrete-time algebraic matrix Riccati equations for singular $\bar{A}$ matrices (see Problem 1.4), the following optimal regulator matrices of the proportional-plus-integral control structure are obtained (the values being rounded to four digits for the fractionar part):

$$H = \begin{bmatrix} -.4263 & -.4280 & -.4264 & -.4270 & -.4306 \\ .1094 & .1838 & .2615 & .3297 & .3721 \end{bmatrix.$$
$$\begin{matrix} -1.7014 & -.1445 & .8358 & -.3652 \\ -.2574 & 1.1185 & -4.0160 & 1.3067 \end{bmatrix},$$

$$H_i = \begin{bmatrix} .0271 & .0076 \\ .0041 & -.0680 \end{bmatrix}, \qquad H_a = \begin{bmatrix} .3088 & .0504 \\ .0446 & -.3989 \end{bmatrix}.$$

The Figures 1.5, and 1.6 show the input and output trajectories computed using these regulator matrices, for the following simulated actions:

1. additive disturbance on the dry basis weight: -5 g/m^2 (Fig. 1.5);
2. reference change for the dry basis weight: $+5$ g/m^2 (Fig. 1.6).

Even better results have been obtained for similar actions on the second output, so these results have been omitted. The trajectories have been represented using the same unit on the ordinate axis. The origin on this axis has been translated to the nominal operation point, since only deviates from that point are studied. Note that the regulation requirements are satisfied, but the tracking requirements are not met: there are large overshoots, and large variations of the input values. The undesired effects are caused by the large time constants and time delays of the process.

The tracking behaviour is then improved by using a reference model depicted in Fig. 1.7. Note that the reference model incorporates the process time constants and delays, but the input-output channels are decoupled. The rationale for such a selection is that the model should mimic the essential dynamics of the process, but avoid the influence of a reference input on more than one output. Indeed, when the reference for an output changes, ideally the other output should not be disturbed.

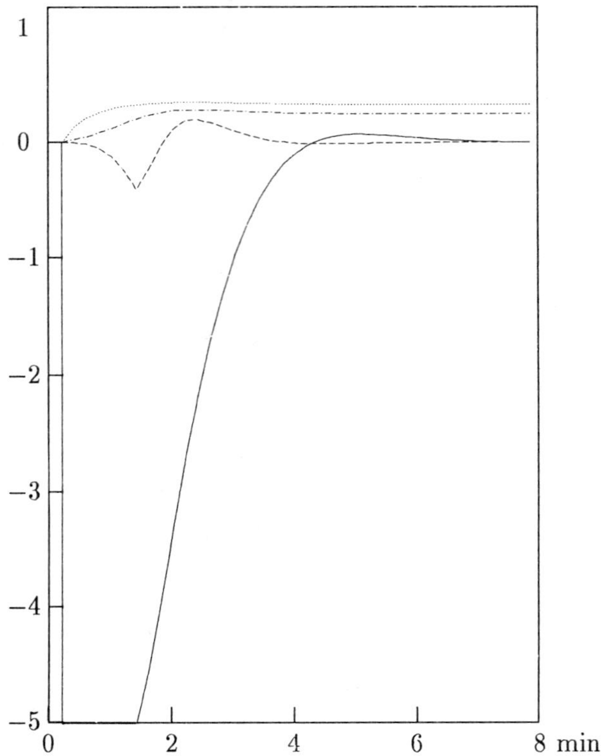

Fig. 1.5 Input and output trajectories for $\Delta y_1 = -5$, using PI control. Trajectories: — basis weight; - - moisture; ··· flow rate; -·- steam pressure. A common unit is used on the vertical axis.

The reference model in Fig. 1.7 has a state-space description given by the following matrices:

$$A_r = \begin{bmatrix} 0 & 0 & 0 & 0 & 0 & 0 & 0 & 0 \\ 1 & 0 & 0 & 0 & 0 & 0 & 0 & 0 \\ 0 & 1 & 0 & 0 & 0 & 0 & 0 & 0 \\ 0 & 0 & 1 & 0 & 0 & 0 & 0 & 0 \\ 0 & 0 & 0 & 1 & 0 & 0 & 0 & 0 \\ 0 & 0 & 0 & 0 & .222 & .778 & 0 & 0 \\ 0 & 0 & 0 & 0 & 0 & 0 & 1.372 & -.47 \\ 0 & 0 & 0 & 0 & 0 & 0 & 1 & 0 \end{bmatrix},$$

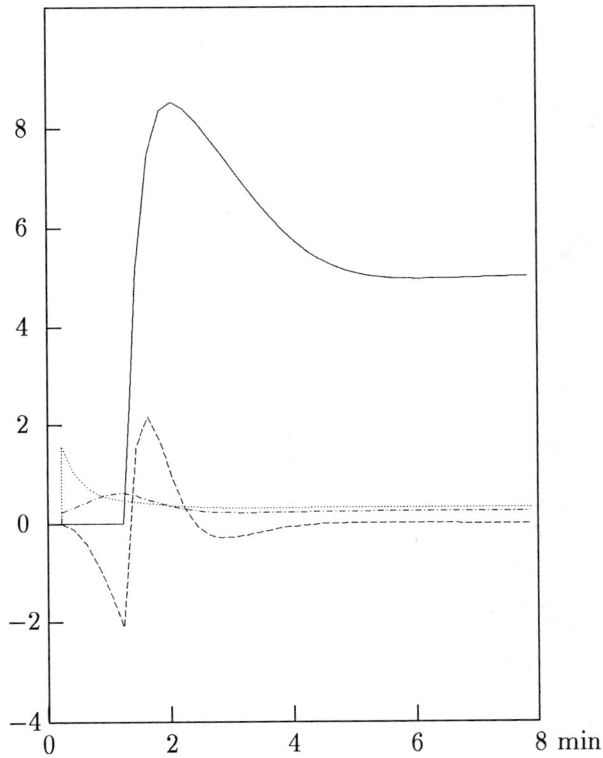

Fig. 1.6 Input and output trajectories for $\Delta r_1 = 5$, using PI control. Trajectories: — basis weight; -- moisture; ··· flow rate; -·- steam pressure. A common unit is used on the vertical axis.

$$
\begin{aligned}
B_r &= \begin{bmatrix} 1 & 0 & 0 & 0 & 0 & 0 & 0 & 0 \\ 0 & 0 & 0 & 0 & 0 & 0 & .098 & 0 \end{bmatrix}^T, \\
C_r &= \begin{bmatrix} 0 & 0 & 0 & 0 & 0 & 1 & 0 & 0 \\ 0 & 0 & 0 & 0 & 0 & 0 & .576 & .424 \end{bmatrix}.
\end{aligned}
$$

Obviously, the reference model is asymptotically stable, the A_r matrix having the same eigenvalues as A, except for .6, which is missing. On solving Problem 1.5, the reference tracking control structure matrices are found as follows:

$$
H_r = \begin{bmatrix} .0204 & .0219 & .0231 & .0240 & .0246 & .1135 & .0901 & -.0394 \\ .0031 & .0033 & .0035 & .0036 & .0037 & .0171 & -.4319 & .1406 \end{bmatrix},
$$

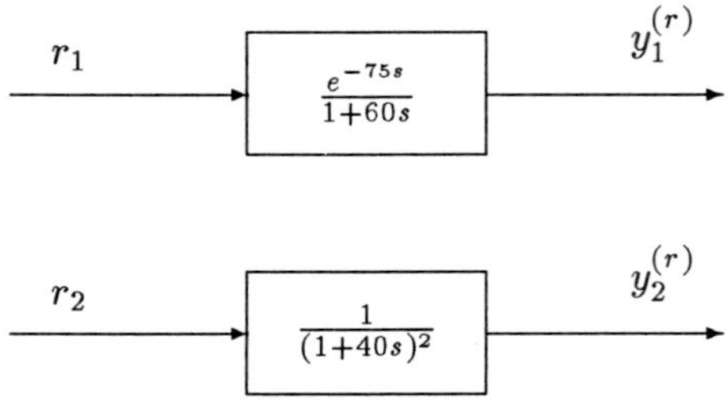

Fig. 1.7 Reference model for paper machine; see text for description.

$$H_{cr} = \begin{bmatrix} .0811 & -.0003 \\ .0104 & -.1076 \end{bmatrix}.$$

Fig. 1.8 shows the trajectories obtained when the same simulated action 2, as in Fig. 1.6, takes place. It becomes apparent that the use of the reference model upgrades the quality of the transient regime to reference changes: the settling time is much shorter, the response has no overshoot (aperiodic aspect), the effects of interactions between an input and other outputs are much lower, and the input magnitudes are drastically reduced, if compared to the proportional-plus-integral control structure.

We have also designed the control structure for direct rejection of a measurable disturbance, that acts through the transfer function

$$\frac{y_p(s)}{p(s)} = \frac{-30e^{-30s}}{1+70s}$$

on the output y_2. The corresponding state-space description is given by:

$$A_p = \begin{bmatrix} 0 & 0 & 0 \\ 1 & 0 & 0 \\ 0 & .192 & .808 \end{bmatrix}, \quad B_p = \begin{bmatrix} 1 \\ 0 \\ 0 \end{bmatrix}, \quad C_p = \begin{bmatrix} 0 & 0 & 0 \\ 0 & 0 & -30 \end{bmatrix},$$

and the associated regulator matrices of the direct disturbance rejection control structure, computed by solving Problem 1.6, are:

$$H_p = \begin{bmatrix} .2436 & .2386 & 1.229 \\ -1.0840 & -1.2820 & -7.437 \end{bmatrix}, \quad H_{cp} = \begin{bmatrix} -.1796 \\ -2.1730 \end{bmatrix}.$$

The input and output trajectories when a step of $-0,07$ in disturbance takes place, are illustrated in Fig. 1.9. For comparison purposes, the trajectories for the PI control structure without, as well as with direct rejection, are all

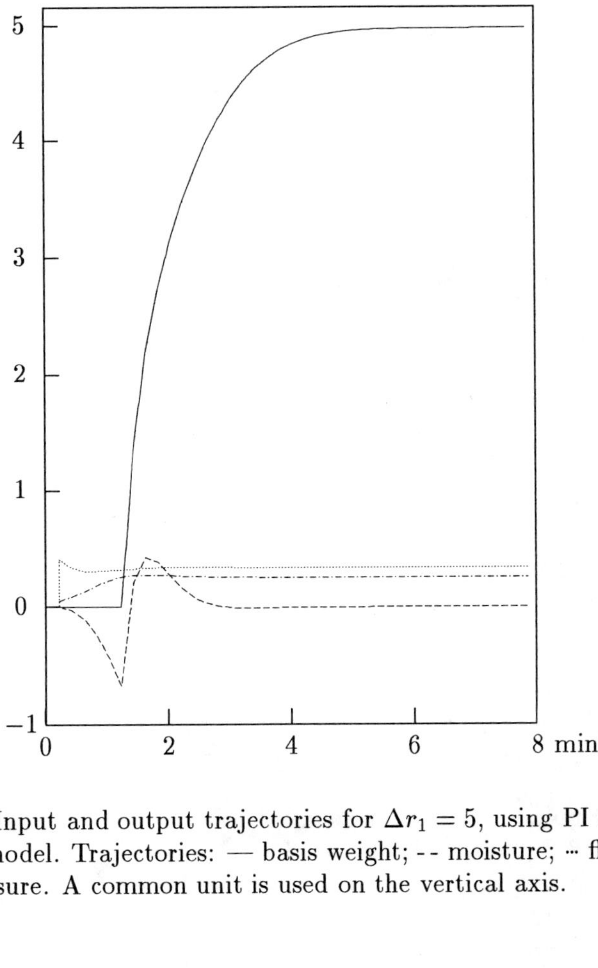

Fig. 1.8 Input and output trajectories for $\Delta r_1 = 5$, using PI control and reference model. Trajectories: — basis weight; -- moisture; ··· flow rate; -·- steam pressure. A common unit is used on the vertical axis.

represented in Fig. 1.9. Note that a remarkable improvement results when using the direct disturbance rejection control structure.

The MATLAB code listed below[9] can be used in computing the input and output trajectories for a discrete-time process controlled by one of the various control strategies mentioned above, for different simulation conditions.

```
function [u,y] = dsicon(ns,n,m,l,nr,np,mp,A,B,C,H,HI,HA,..
                        AR,BR,CR,HR,HCR,AP,BP,CP,HP,HCP,..
                        r,py,p);
```

[9] MATLAB is a registered trademark of The MathWorks, Inc., MA.

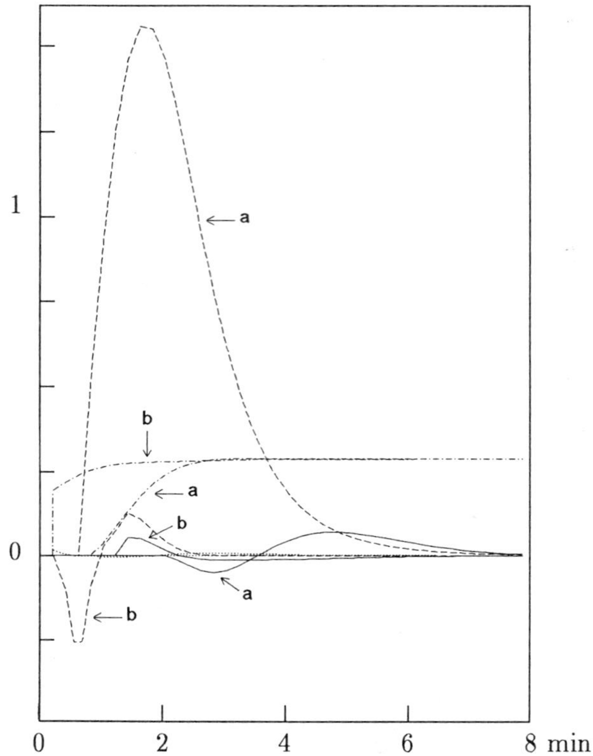

Fig. 1.9 Input and output trajectories for $p = -0.07$, using PI control and reference model: (a) - without disturbance model; (b) - with disturbance model. Trajectories: — basis weight; - - moisture; ··· flow rate; -·- steam pressure. A common unit is used on the vertical axis.

```
% DSICON Discrete-time controlled system simulation.
%        System model:
%          x(t+1) = A*x(t) + B*u(t),
%           y(t)  = C*x(t) + py,  or
%           y(t)  = C*x(t) + yp(t),
%            where: x n-vector, u m-vector, y l-vector.
%        Disturbance model:
%          xp(t+1) = AP*xp(t) + BP*p,
%           yp(t)  = CP*xp(t),
%            where:  xp np-vector, p mp-vector, yp l-vector.
```

```
%        Reference model:
%          xr(t+1) = AR*xr(t) + BR*r,
%           yr(t)  = CR*xr(t),
%            where:  xr nr-vector, r l-vector, yr l-vector.
%        Integrator:
%          v(t+1) = v(t) + r - y(t), or
%          v(t+1) = v(t) + yr(t) - y(t),
%            where: v l-vector.
%        Regulator model:
%           u(t)  = H*x(t) + HI*v(t) + HA*r,
%                            (if nr = 0 and np = 0, or
%                                nr = 0 and np < 0);
%           u(t)  = H*x(t) + HI*v(t) + HR*xr(t) + HCR*r,
%                            (if nr ~= 0 and np = 0);
%           u(t)  = H*x(t) + HI*v(t) + HA*r +
%                          + HP*xp(t) + HCP*p,
%                            (if nr = 0 and np ~= 0);
%           u(t)  = H*x(t) + HI*v(t) + HR*xr(t) + HCR*r +
%                          + HP*xp(t) + HCP*p,
%                            (if nr ~= 0 and np ~= 0).
%        Parameter ns is the number of time steps.
%        Only u(t) and y(t) are stored in arrays u(:,t)
%        and y(:,t), respectively, t = 1:ns.


u = [];       y = [];
x = zeros(n,1);      v = zeros(l,1);
npp = abs(np);
                          % PI regulator without reference and
                          %                 disturbance models
if ~nr & ~np,
  for t=1:ns,
    u = [u H*x+HI*v+HA*r];      y = [y C*x+py];
    v = v + r - y(:,t);         x = A*x + B*u(:,t);
  end
                          % PI regulator with reference model
elseif nr & ~np,
  xr = zeros(nr,1);
  for t=1:ns,
    u = [u H*x+HI*v+HR*xr+HCR*r];      y = [y C*x+py];
    v = v + CR*xr - y(:,t);            x = A*x + B*u(:,t);
    xr = AR*xr + BR*r;
```

```
  end
                    % PI regulator with disturbance model
                    %          without direct rejection
elseif ~nr & np < 0,
  xp = zeros(npp,1);
  for t=1:ns,
    u = [u H*x+HI*v+HA*r];      y = [y C*x+CP*xp];
    v = v + r - y(:,t);         x = A*x + B*u(:,t);
    xp = AP*xp + BP*p;
  end
                    % PI regulator with disturbance model
                    %             and direct rejection
elseif ~nr & np,
  xp = zeros(npp,1);
  for t=1:ns,
    u = [u H*x+HI*v+HA*r+HP*xp+HCP*p];
    y = [y C*x+CP*xp];
    v = v + r - y(:,t);      x = A*x + B*u(:,t);
    xp = AP*xp + BP*p;
  end
                    % PI regulator with reference and
                    %             disturbance models
                    %             and direct rejection
else
  xr = zeros(nr,1);     xp = zeros(npp,1);
  for t=1:ns,
    u = [u H*x+HI*v+HR*xr+HCR*r+HP*xp+HCP*p];
    y = [y C*x+CP*xp];
    v = v + CR*xr - y(:,t);      x  = A*x + B*u(:,t);
    xr = AR*xr + BR*r;           xp = AP*xp + BP*p;
  end
end
```

1.3 Robust Control

This section gives an overview of the main topics in the newly developed robust control theory, that generalizes the traditional control approach. Details and related topics can be found, for instance, in (Başar and Bernhard, 1991; Doyle et al., 1989; Doyle and Stein, 1981; Francis, 1987; Francis and Doyle, 1987; Kwakernaak, 1993; Murciano, 1990; Safonov, Laub and Hartmann, 1981; Stein and Athans, 1987). Some issues related to performance specification for robust analysis and synthesis are first dealt with, and the H_∞ optimization problem is stated. The internal stability problem, the standard robust control problem, as well as some of its special cases are discussed. Finally, the H_2 optimal problem is stated and an algorithm for its solution is described.

The general methodology of robust control synthesis allows the performance specifications be defined as constraints to be satisfied by the closed-loop system. Since performances are usually expressed in the frequency domain, the specific notations used are summarized here.

A linear constant continuous-time system is represented by either a transfer function matrix $G(s)$, or its state-space description

$$[A, B, C, D] := \begin{bmatrix} A & B \\ C & D \end{bmatrix},$$

where s is the complex Laplace variable; so, $G(s) = C(sI - A)^{-1}B + D$, and these two descriptions are considered to be equivalent. A rational function in s with real coefficients is $\Re$-*rational.* $G^\star$ denotes the complex conjugate transpose of G, that is, $G^\star = G(-s)^T$. G is strictly proper if $D := G(+\infty)$ is zero, and G is proper if D is finite. $\Re L_2$ denotes the space of strictly proper transfer function matrices without any poles on the imaginary axis; $\Re L_\infty$ denotes the space of proper transfer function matrices without any poles on the imaginary axis. H_2 denotes the space of stable and strictly proper transfer function matrices; H_∞ denotes the space of stable and proper transfer function matrices. $||G||_2$ is the H_2 norm,

$$||G||_2^2 = \frac{1}{2\pi}\int_{-\infty}^{+\infty} \mathrm{tr}[G(j\omega)^\star G(j\omega)]\, d\omega;$$

$||G||_\infty$ is the L_∞ norm,

$$||G||_\infty = \sup\{\bar{\sigma}(G(j\omega)) \ : \ \omega > 0\},$$

where ω is the frequency variable, $j := \sqrt{-1}$, and $\bar{\sigma}(\cdot)$ and $\underline{\sigma}(\cdot)$ are the largest and smallest singular values, respectively.

1.3.1 Performance specification and the H_∞ optimization problem

Consider the linear multivariable closed-loop system[10] in Fig. 1.10, where $G(s)$ and $K(s)$ are transfer function matrices of the process and regulator,

[10] A dynamical system with one input and one output, is called *monovariable system*; otherwise, it is a *multivariable system.*

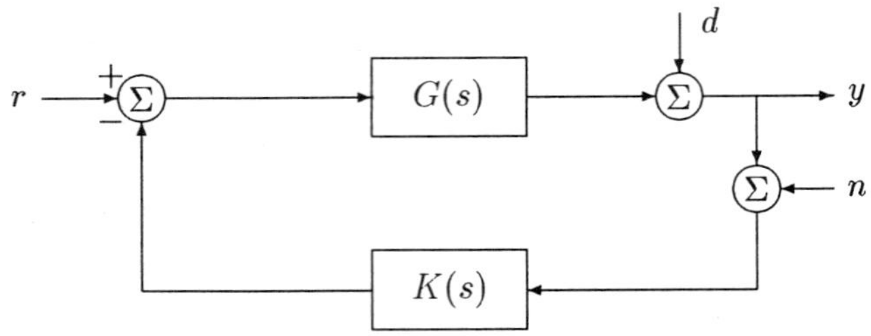

Fig. 1.10 Closed-loop control system configuration; see text for description.

respectively. The $r(s)$ signal is a reference input, $y(s)$ is the output, $d(s)$ is a disturbance input, and $n(s)$ is a measurement noise. (The signals $r(s)$, $y(s)$, $d(s)$ and $n(s)$ are vectors.) The open-loop transfer function matrix is $L(s) := G(s)K(s)$. Define the *sensitivity matrix*, $S(s)$, and *complementary sensitivity matrix*, $T(s)$:

$$S(s) = [I + L(s)]^{-1}, \qquad T(s) = L(s)[I + L(s)]^{-1}. \tag{1.72}$$

These matrices play an important role in the assessment of linear systems robustness. The system output can be decomposed as

$$y(s) = y_r(s) + y_d(s) + y_n(s),$$

where the three terms are associated with the inputs r, d and n:

$$y_r(s) = S(s)G(s)r(s), \quad y_d(s) = S(s)d(s), \quad y_n(s) = -T(s)n(s). \tag{1.73}$$

The following assumptions are usually made

$$\|d(j\omega)\| \le M_d(\omega), \qquad \|y_d(j\omega)\| \le M_{y_d}(\omega), \tag{1.74}$$

$$\|n(j\omega)\| \le M_n(\omega), \qquad \|y_n(j\omega)\| \le M_{y_n}(\omega), \tag{1.75}$$

where M_d, M_n, M_{y_d}, M_{y_n} are small gain bounds.

An insight into the robustness properties is caught by considering a monovariable system. If the system in Fig. 1.10 is stable, then the response to disturbances of frequency ω, $d(t) = e^{j\omega t}$, is determined by the sensitivity function, $y_d(t) = S(j\omega)e^{j\omega t}$, hence the disturbance effect can diminish by imposing the condition $|S(j\omega)| \ll 1$. Similarly, since the system response to the noise, $n(t) = e^{j\omega t}$, is governed by $y_n(t) = -T(j\omega)e^{j\omega t}$, it can diminish by imposing $|T(j\omega)| \ll 1$.

The robustness of control systems also means reduced effects of modelling errors on parameter variations. These errors generally appear at high frequencies. The modelling uncertainties can be described by the model $G'(s) = (I + \Delta_M(s))G(s)$ (see Fig. 1.11), where it is assumed that $\Delta_M(s)$ belongs to the domain

$$\mathcal{D}(s) := \{\Delta_M(s) \; : \; \Delta_M(s) \text{ stable}, \quad |\Delta_M(j\omega)| \leq M(\omega)\}. \tag{1.76}$$

One sufficient condition so that all linear systems $G'(s)$ defined on $\mathcal{D}(s)$ be stable is: $G(s)$ should be stable and the complementary sensitivity function should satisfy $|T(j\omega)| \leq M_T(\omega) < 1/M(\omega)$.

Generally, an uncertain system is reasonably well modelled at high frequencies if a bound $M(\omega)$ that increases with ω is used. Therefore, the complementary sensitivity function should have a small gain to bear a robust stability under uncertainties at high frequencies.

The performance specifications can then be written as:

$$|S(j\omega)| \leq M_S(\omega), \qquad |T(j\omega)| \leq M_T(\omega). \tag{1.77}$$

On the other hand, the sensitivity and complementary sensitivity matrices satisfy the relation $S(j\omega) + T(j\omega) \equiv I$, hence $|S(j\omega)|$ and $|T(j\omega)|$ cannot be simultaneously reduced at the same frequency. A compromise should be made between the robustness properties relative to the sensitivity reduction, quantified by $|S(j\omega)|$, and those relative to the modelling of high frequencies of uncertain systems, quantified by $|T(j\omega)|$.

In practice, uncertainties and measurement noises become large at high frequencies; in addition, it is generally desired to reject disturbances and reduce the sensitivity in the low-frequency range. Consequently, the abovementioned compromise is solved by requesting that $M_S(\omega)$ be small at low frequencies and $M_T(\omega)$ be small at high frequencies.

On the other hand, there is a close connection between the transfer function of an open-loop system and its closed-loop robustness properties, namely (for monovariable systems):

$$\begin{aligned} |L(j\omega)| \gg 1 \quad &\Longleftrightarrow \quad |S(j\omega)| \ll 1 \text{ and } T(j\omega) \approx 1, \\ |L(j\omega)| \ll 1 \quad &\Longleftrightarrow \quad |T(j\omega)| \ll 1 \text{ and } S(j\omega) \approx 1, \\ \left.\begin{array}{r} |L(j\omega)| \approx 1 \\ \arg L(j\omega) \approx \pm 180^\circ \end{array}\right\} \quad &\Longleftrightarrow \quad |S(j\omega)| \gg 1 \text{ and } |T(j\omega)| \gg 1. \end{aligned} \tag{1.78}$$

The approximations (1.78) support the following *practical rules* for designing robust control systems:

1. If the open-loop gain $|L(j\omega)|$ is large, then the closed-loop system has a small sensitivity, and hence, good disturbance rejection properties. But the noise directly occurs at the system output, since $y_n(t) \approx -n(t)$, where $n(t) = e^{j\omega t}$.

2. If the open-loop gain $|L(j\omega)|$ is small, then the noise response is reduced, but the sensitivity is large.

3. For the frequency range where the open-loop gain (in decibels) is close to zero ($|L(j\omega)| \approx 1$), the system phase, $\arg L(j\omega)$, should stay far from $\pm 180°$, to ensure a good stability bound and prevent from a magnification of perturbations and noise.

Therefore, the open-loop gain should be large at low frequencies, to reject disturbances and reduce sensitivity, and small at high frequencies, to guarantee a robust stability. At intermediate frequencies, the system phase should be far from $\pm 180°$, to prevent too large values of $|S|$ and $|T|$. Clearly, the gain and phase should also satisfy the Nyquist stability criterion.

The properties above are generalized to multivariable systems by replacing the absolute value by the largest singular value, defined by

$$\bar{\sigma}(M(j\omega)) := \max\{\|M(j\omega)u\| \ : \ \|u\| = 1\}. \tag{1.79}$$

The robustness specifications are then written as:

$$\bar{\sigma}(S(j\omega)) \leq M_S, \qquad \bar{\sigma}(T(j\omega)) \leq M_T. \tag{1.80}$$

The sensitivity bound should satisfy $M_S \ll 1$ at frequencies for which disturbance rejection and sensitivity reduction are desired. Similarly, the complementary sensitivity bound should satisfy $M_T \ll 1$ at frequencies for which noise rejection and robust stability are desired. At each frequency, there is a compromise between the control properties governed by S and those governed by T. Some approximations similar to (1.78) are used to assess the robustness properties of multivariable systems:

$$\begin{aligned} \bar{\sigma}(L(j\omega)) \gg 1 &\iff \bar{\sigma}(S(j\omega)) \ll 1 \text{ and } T(j\omega) \approx I, \\ \bar{\sigma}(L(j\omega)) \ll 1 &\iff \bar{\sigma}(T(j\omega)) \ll 1 \text{ and } S(j\omega) \approx I. \end{aligned} \tag{1.81}$$

The rules and comments presented for monovariable systems are easily generalized, using these approximations.

One objective of robustness analysis is to find measures of the stability bounds for multivariable systems.[11] Consider the multivariable system in Fig. 1.11, where $G(s)$ and $K(s)$ model the process and regulator, respectively, and $\Delta_M(s)$ and $\Delta_A(s)$ are transfer function matrices of multiplicative and additive perturbations, respectively. The *multiplicative (additive) stability bound* of the system is defined as the "size" of the smallest stable matrix $\Delta_M(s)$ ($\Delta_A(s)$, respectively), that destabilizes the control system in

[11] Such bounds are well-known for monovariable systems.

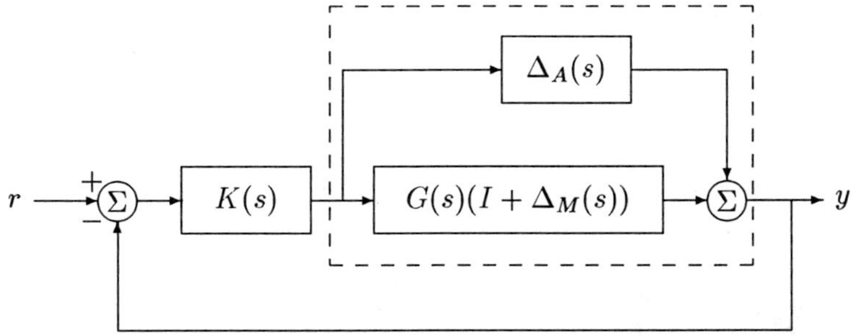

Fig. 1.11 Closed-loop system with multiplicative and additive perturbations; see text for description.

Fig. 1.11 with $\Delta_A = 0$ ($\Delta_M = 0$, respectively). The "size" of a perturbation matrix is measured by the largest singular values, $\bar{\sigma}(\Delta(j\omega))$.

If the nominal system ($\Delta_A = 0$, $\Delta_M = 0$) is stable, then the size of the smallest stable matrix $\Delta_M(s)$, for which the control system with $\Delta_A = 0$ becomes unstable, is

$$\bar{\sigma}(\Delta_M(j\omega)) = 1/\bar{\sigma}(T(j\omega)). \tag{1.82}$$

Similarly, the size of the smallest stable matrix $\Delta_A(s)$, for which the control system with $\Delta_M = 0$ becomes unstable, is

$$\bar{\sigma}(\Delta_A(j\omega)) = 1/\bar{\sigma}(R(j\omega)), \tag{1.83}$$

where $R(s) := K(s)[I + L(s)]^{-1}$. This result helps to specify stability bounds for multivariable systems by inequalities for singular values,

$$\bar{\sigma}(R(j\omega)) \leq |W_1^{-1}(j\omega)|, \qquad \bar{\sigma}(T(j\omega)) \leq |W_2^{-1}(j\omega)|, \tag{1.84}$$

where $|W_1(j\omega)|$ and $|W_2(j\omega)|$ are the sizes of the largest additive and multiplicative perturbation matrices, respectively.

H_∞ optimization problem

The conditions for ensuring robust stability, a good stability bound, and disturbance attenuation are specified by some weighting functions W_i, $i = 1{:}\,3$, as follows:

$$\bar{\sigma}(S(j\omega)) \leq |W_1^{-1}(j\omega)|, \qquad \omega \leq \underline{\omega}_c,$$

$$\bar{\sigma}(R(j\omega)) \leq |W_2^{-1}(j\omega)|,$$

$$\bar{\sigma}(T(j\omega)) \le |W_3^{-1}(j\omega)|, \qquad \omega \ge \bar{\omega}_c,$$

where $(\underline{\omega}_c, \bar{\omega}_c)$ are frequencies corresponding to gain factors of 0 dB for W_1^{-1} and W_3^{-1}. The weighting matrices W_1, W_2 and W_3 weight the error $e = r - y$, control (command) u and output y, respectively.

The following inequalities for singular values (which can be generalized for any number of matrices) hold

$$\frac{1}{\sqrt{2}}\bar{\sigma}\left(\begin{bmatrix} A \\ B \end{bmatrix}\right) \le \max\{\bar{\sigma}(A), \bar{\sigma}(B)\} \le \bar{\sigma}\left(\begin{bmatrix} A \\ B \end{bmatrix}\right). \tag{1.85}$$

Therefore, the above-mentioned conditions can equivalently be expressed (up to a factor of 2) as $||H||_\infty \le 1$, where

$$H := \begin{bmatrix} W_1 S \\ W_2 R \\ W_3 T \end{bmatrix}. \tag{1.86}$$

Hence, one has to construct a stabilizing regulator that should satisfy $||H||_\infty \le 1$.

The system response to a disturbance d and noise n should sometimes be minimized without taking the properties of the matrix $R(s)$ into account. In this case, a stabilizing regulator should be determined so that

$$\left\| \begin{bmatrix} W_1 S \\ W_3(I - S) \end{bmatrix} \right\|_\infty \le 1. \tag{1.87}$$

In particular, minimizing the effect of disturbances can be written as

$$\min ||WS||_\infty, \tag{1.88}$$

where W is a weighting matrix that weights the sensitivity matrix. The problem (1.88) is thus an H_∞ optimization problem.

1.3.2 Standard robust control problem

One contribution of the robust control theory is in establishing a general unifying framework for all classic automatic control problems. This subsection presents the standard robust control problem: finding a stabilizing regulator that minimizes the H_α norm, $\alpha \in \{2, \infty\}$, of the transfer function matrix from an external (disturbance) input w to a controlled output z (see Fig. 1.13 below). Some problems reducible to the standard problem are also dealt with.

Let us first state the internal stability problem—a prerequisite for the standard problem.

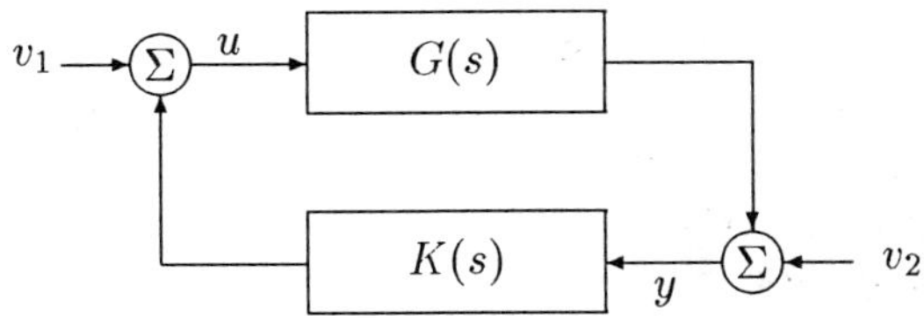

Fig. 1.12 Internal stability configuration

Consider the linear multivariable closed-loop system in Fig. 1.12. The regulator $K(s)$ *internally stabilizes* the linear system $G(s)$ if and only if the four transfer function matrices from v_1, and v_2 to u, and y, are stable. In particular, K stabilizes G (in the traditional sense) since the transfer function matrix from v_1 to y (when $v_2 = 0$) is stable.

For any proper linear system $G(s)$, there are eight *stable* transfer function matrices satisfying the following equations, called the *doubly coprime factorization of* G:

$$G = NM^{-1} = \widetilde{M}^{-1}\widetilde{N}, \qquad \begin{bmatrix} \widetilde{X} & -\widetilde{Y} \\ -\widetilde{N} & \widetilde{M} \end{bmatrix} \begin{bmatrix} M & Y \\ N & X \end{bmatrix} = I. \tag{1.89}$$

It can be shown (Francis, 1987) that the set of all proper $\Re$-rational matrices $K(s)$ stabilizing $G(s)$ is parameterized by the formulas

$$K = (Y - MQ)(X - NQ)^{-1} = (\widetilde{X} - Q\widetilde{N})^{-1}(\widetilde{Y} - Q\widetilde{M}), \tag{1.90}$$

where Q is a proper and stable $\Re$-rational parameter ($Q \in H_\infty$). Given a transfer function matrix P partitioned into

$$P(s) = \begin{bmatrix} P_{11} & P_{12} \\ P_{21} & P_{22} \end{bmatrix}, \tag{1.91}$$

the *linear fractional operator* $\mathcal{F}_l(P, X)$ is defined by

$$\mathcal{F}_l(P, X) = P_{11} + P_{12}X(I - P_{22}X)^{-1}P_{21}.$$

Then, from (1.89)–(1.90), K is parameterized by:

$$K = K_0 + \widetilde{X}^{-1}Q(I + X^{-1}NQ)^{-1}X^{-1} = \mathcal{F}_l(J, Q), \tag{1.92}$$

where $K_0 = \widetilde{X}^{-1}\widetilde{Y}$, Q is proper and stable parameter, and

$$J = \begin{bmatrix} K_0 & \widetilde{X}^{-1} \\ X^{-1} & -X^{-1}N \end{bmatrix}.$$

In particular, the transfer function matrix $K_0 = \mathcal{F}_l(J, 0)$ corresponds to the LQG compensator (regulator plus estimator). By means of a stable transfer function matrix Q, formula (1.92) generates all stabilizing regulators.

It can be proven that any stabilization procedure is equivalent to adding a stable dynamical system to the process and then using a LQG compensator to stabilize the extended system. In terms of state variables, the following result is obtained:

Theorem 1.19 *$K(s)$ internally stabilizes $G(s) := [A, B, C, D]$ if one can construct a system $[A_e, B_e, C_e, D_e]$, where*

$$A_e := \begin{bmatrix} A & 0 \\ 0 & A_a \end{bmatrix}, \quad B_e := \begin{bmatrix} B \\ 0 \end{bmatrix}, \quad C_e := \begin{bmatrix} C & 0 \end{bmatrix}, \quad D_e := D,$$

and A_a is stable, so that K has the form:

$$K(s) = [A_e - B_e F_e - H_e C_e + H_e D_e F_e, H_e, -F_e, 0],$$

where $A_e - B_e F_e$ and $A_e - H_e C_e$ are stable.

Consider now the linear multivariable closed-loop system in Fig. 1.13, where w, u, z, and y are vector signals; w is the external input (usually corresponding to a disturbance or to a measurement noise), u is the command, z is the controlled output (usually, the tracking error), and y is the measured output. The transfer function matrices $P(s)$ and $K(s)$ are $\Re$-rational and proper, by assumption; P represents a "generalized" system (often containing the controlled system), and $K(s)$ stands for the regulator. Partition P as in (1.91). From Fig. 1.13 it follows that

$$z = P_{11} w + P_{12} u, \qquad y = P_{21} w + P_{22} u, \qquad u = K y. \tag{1.93}$$

Hence, using the $\mathcal{F}_l(P, X)$ operator, we obtain

$$z = [P_{11} + P_{12} K (I - P_{22} K)^{-1} P_{21}] w = \mathcal{F}_l(P, K) w.$$

Let T_{zw} denote the transfer function matrix from w to z for the closed-loop system. Let $P(s) := [A, B, C, D]$ be a minimal realization of P.[12] On partitioning the B, C, and D matrices according to the partitions of input and output vectors,

$$B = \begin{bmatrix} B_1 & B_2 \end{bmatrix}, \qquad C = \begin{bmatrix} C_1 \\ C_2 \end{bmatrix}, \qquad D = \begin{bmatrix} D_{11} & D_{12} \\ D_{21} & D_{22} \end{bmatrix},$$

[12]A *minimal realization of a transfer function matrix* is a state-space model having minimal order and the same input-output behaviour.

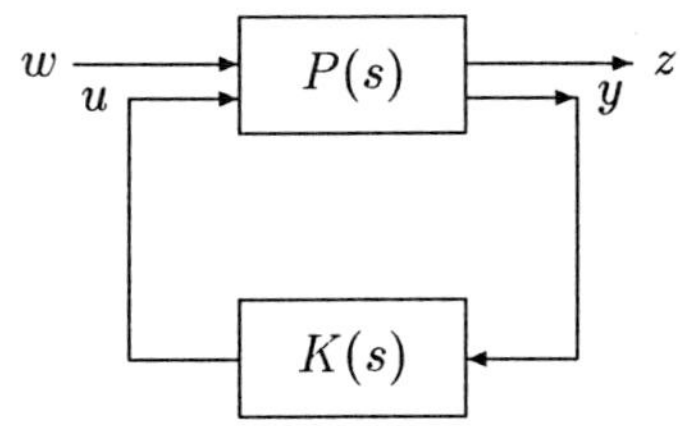

Fig. 1.13 Standard system configuration; see text for description.

it follows that P satisfies the equations:

$$\begin{aligned} \dot{x} &= Ax + B_1 w + B_2 u, \\ z &= C_1 x + D_{11} w + D_{12} u, \\ y &= C_2 x + D_{21} w + D_{22} u. \end{aligned} \tag{1.94}$$

The *standard robust control problem* consists in finding a transfer function matrix $\bar{K}(s)$ that internally stabilizes P and has one of the following properties:

$$\|\mathcal{F}_l(P, \bar{K})\|_{\alpha \in \{2,\infty\}} \leq \gamma \qquad (\gamma > 0),$$
$$\bar{K} = \arg \min_{K(s)} \|\mathcal{F}_l(P, K)\|_\alpha.$$

Remark 1.10 It can be shown that the $K(s)$ regulator internally stabilizes $P(s)$ if and only if $K(s)$ internally stabilizes $P_{22}(s)$. Consequently, the parametrization of the regulators that stabilize $P(s)$ is a special (possibly non-unique) parametrization of the regulators that stabilize P_{22}. Analogously, $K(s)$ internally stabilizes the variant of the standard system in Fig. 1.14 if and only if all the nine closed-loop transfer function matrices from w, v_1, and v_2 to z, u, and y are stable. Therefore, $K(s)$ stabilizes both P_{22} ($v_2 = 0$, $w = 0$), and the transfer function matrix from w to z ($v_1 = 0$, $v_2 = 0$), for which the H_2 or H_∞ norm should be minimized. □

Three problems reducible to the standard robust control problem are now discussed.

Problem 1.8 (Reference tracking) Consider a linear multivariable system $G(s)$ in Fig. 1.15, whose output v must follow the reference signal r. The regulator is defined by the transfer function matrices C_1 and C_2. The reference r belongs to a signal class of the form

$$\{r : r = Ww, W \in H_2, \|w\|_2 \leq 1\},$$

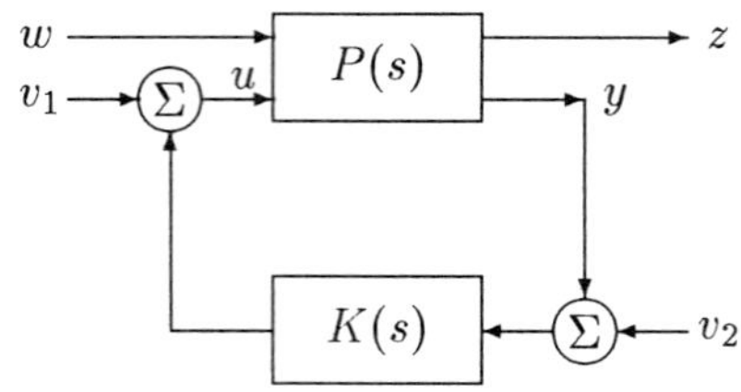

Fig. 1.14 Standard system with additional inputs; see text for description.

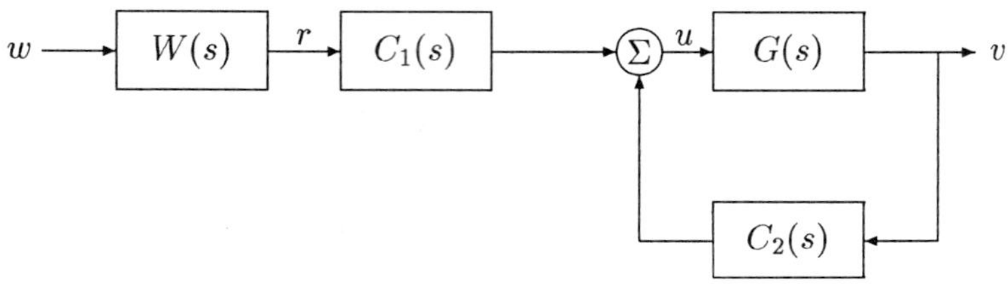

Fig. 1.15 Tracking system; see text for description.

where W is a given transfer function matrix.

The tracking error being $r - v$, the cost functional is chosen as

$$J = (\|r - v\|_2^2 + \|\rho u\|_2^2)^{1/2}. \tag{1.95}$$

The term ρu ($\rho > 0$) makes a proper optimal regulator exist. Note that J is the H_2 norm of the signal $z := [(r-v)^T, \rho u^T]^T$. Hence, one must determine the optimal regulators C_1 and C_2 that minimize the following tracking criterion

$$\sup\{\, \|z\|_2 \; : \; W \in H_2, \|w\|_2 \le 1\};$$

this is equivalent to minimizing the H_∞ norm of the transfer function matrix from w to z in the closed-loop system. The standard problem is obtained by defining:

$$y := \begin{bmatrix} r \\ v \end{bmatrix}, \qquad K := \begin{bmatrix} C_1 & C_2 \end{bmatrix}, \qquad P := \begin{bmatrix} P_{11} & P_{12} \\ P_{21} & P_{22} \end{bmatrix},$$

$$P_{11} := \begin{bmatrix} W \\ 0 \end{bmatrix}, \quad P_{12} := \begin{bmatrix} -G \\ \rho I \end{bmatrix}, \quad P_{21} := \begin{bmatrix} W \\ 0 \end{bmatrix}, \quad P_{22} := \begin{bmatrix} 0 \\ G \end{bmatrix}.$$

It can be shown that the signals z, y, and u satisfy the equations (1.93). Therefore, the problem is to determine the optimal regulator $K^*(s)$ that stabilizes the extended system P, and minimizes $\|\mathcal{F}_l(P, K)\|_\infty$.

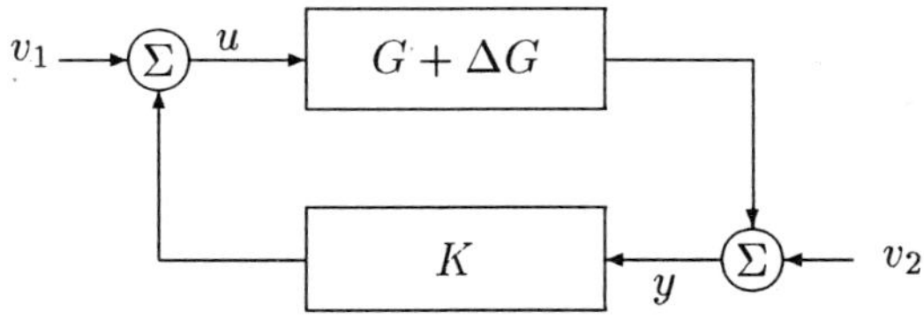

Fig. 1.16 System with modelling errors

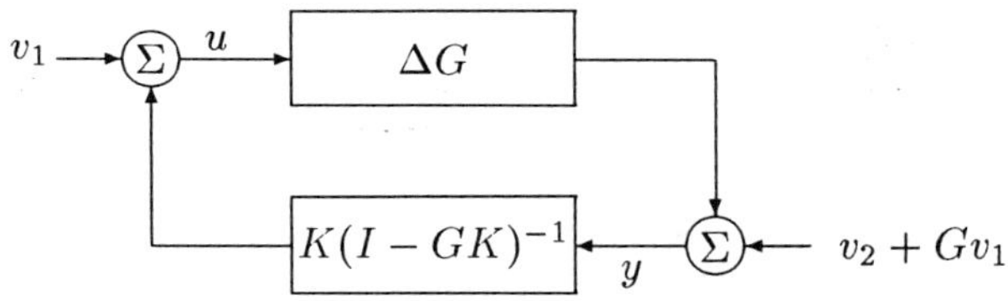

Fig. 1.17 The transformed system

Problem 1.9 (Robust stabilization) Given $G(s)$, a transfer function matrix representing a linear system, and $R(s)$, a complex rational stable scalar-valued function, a neighbourhood of G is defined,

$$V_G := \{\widetilde{G} = G + \Delta G \ : \ ||\Delta G(j\omega)|| < |R(j\omega)|, \quad 0 \leq \omega \leq \infty\},$$

where the transfer function matrices $G + \Delta G$ have the same number of unstable poles (with $\Re(s) \geq 0$). ΔG denotes unknown perturbations, generally due to unmodelled states or parameter variations.

The robust stabilization problem is to find a proper regulator $K(s)$ that internally stabilizes all the systems $\widetilde{G} \in V_G$. The closed-loop system is represented in the diagram in Fig. 1.16, equivalent to that in Fig. 1.17.

Using the small gain theorem (Desoer and Vidyasagar, 1975), the system in Fig. 1.17 is internally stable if

$$||\Delta G K(I - GK)^{-1}||_\infty < 1. \tag{1.96}$$

As $||\Delta G|| < |R|$, it suffices to check that $||RK(I - GK)^{-1}||_\infty \leq 1$. It is then possible to construct the standard problem so that the closed-loop transfer function matrix from w to z be $RK(I - GK)^{-1}$. This is obtained for

$$P := \begin{bmatrix} 0 & RI \\ I & G \end{bmatrix},$$

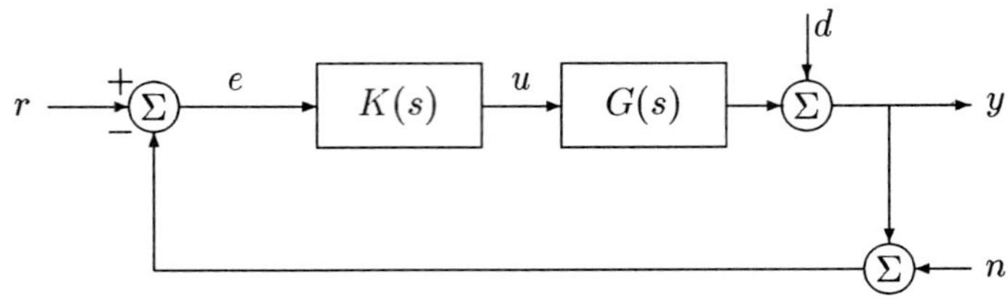

Fig. 1.18 Control system with disturbances

where the definition of the $\mathcal{F}_l(P, K)$ operator was used. Hence, the solution is the regulator K which internally stabilizes the extended system P and makes sure that the H_∞ norm of the transfer function matrix from w to z is less than unity.

Problem 1.10 (Sensitivity reduction) Consider the closed-loop multivariable system depicted in Fig. 1.18. Recall from § 1.3.1 that the closed-loop transfer function matrices $S(s)$, $T(s)$, and $R(s)$ ($L(s) := G(s)K(s)$, $S(s) = [I + L(s)]^{-1}$, $T(s) = I - S(s)$, $R(s) := K(s)S(s)$) play an important role in robustness analysis of multivariable systems. Specifically, the singular values of S determine the disturbance rejection properties, while the singular values of T and R give stability bounds for multivariable control systems.

Let W_1, W_2, and W_3 be the transfer function matrices specifying the robustness, which weight the error e, control u, and output y, respectively. The robust regulator is obtained by solving the suboptimal H_∞ problem, $||H||_\infty < 1$, where H is defined by the relation (1.86). An extended system $P(s)$ is constructed so that $||T_{zw}||_\infty = ||H||_\infty$:

$$P(s) = \left[\begin{array}{c|c} W_1 & -W_1G \\ 0 & W_2 \\ 0 & W_3G \\ \hline I & -G \end{array}\right]$$

If $G(s)$, $W_1(s)$, $W_2(s)$, and $W_3(s)$ have the realizations:

$$\begin{aligned} G(s) &:= [A_G, B_G, C_G, D_G], \\ W_i(s) &:= [A_{W_i}, B_{W_i}, C_{W_i}, D_{W_i}], \qquad i = 1, 2, 3, \end{aligned}$$

then a realization for $P(s)$ is

$$P(s) := \left[\begin{array}{c|cc} A & B_1 & B_2 \\ \hline C_1 & D_{11} & D_{12} \\ C_2 & D_{21} & D_{22} \end{array}\right]$$

$$= \left[\begin{array}{cccc|c|c} A_G & 0 & 0 & 0 & 0 & B_G \\ -B_{W_1}C_G & A_{W_1} & 0 & 0 & B_{W_1} & -B_{W_1}D_G \\ 0 & 0 & A_{W_2} & 0 & 0 & B_{W_2} \\ B_{W_3}C_G & 0 & 0 & A_{W_3} & 0 & B_{W_3}D_G \\ \hline -D_{W_1}C_G & C_{W_1} & 0 & 0 & D_{W_1} & -D_{W_1}D_G \\ 0 & 0 & C_{W_2} & 0 & 0 & D_{W_2} \\ D_{W_3}C_G & 0 & 0 & C_{W_3} & 0 & D_{W_3}D_G \\ \hline -C_G & 0 & 0 & 0 & I & -D_G \end{array}\right].$$

Remark 1.11 If the extended system is used for obtaining the H_∞ control, the weighting matrices W_1, W_2, and W_3 must be chosen so that the matrix D_{12} of the $P(s)$ realization has full column rank. This is quite possible by choosing $W_2 = \varepsilon I$, where ε is a real non-zero number. □

Numerical solution of the standard problem is a difficult task. One approach involves iterative computations and the solution of two algebraic matrix Riccati equations at each iteration.

1.3.3 H_2 optimal control

This subsection briefly presents the design of the robust H_2 optimal control; this is an LQG command for which the control problem is not described by a system of linear equations and by a quadratic criterion, but by a standard problem (§ 1.3.2). Thus, the H_2 control has the advantage of satisfying the given robustness requirements; in terms of the LQG command, this corresponds to a judicious and well-motivated choice of the weighting matrices of the quadratic criterion.

Let $G(s) := [A_G, B_G, C_G, D_G]$ be a linear multivariable system, for which robustness requirements are specified. The corresponding standard problem is constructed (Fig. 1.13).

Problem 1.11 ($\mathbf{H_2}$ optimal control) The H_2 optimal control problem consists in finding an optimal compensator $K^*(s)$ that internally stabilizes $P(s)$ and minimizes the H_2 norm of the closed-loop transfer function matrix from w to z, that is $||T_{zw}||_2$.

Using state variables, the standard system $P(s)$ of the H_2 problem satisfies the equations (1.94) with $D_{11} = 0$,[13] hence,

$$\begin{aligned} \dot{x} &= Ax + B_1 w + B_2 u, \\ z &= C_1 x + D_{12} u, \\ y &= C_2 x + D_{21} w + D_{22} u. \end{aligned} \tag{1.97}$$

Assuming w *Gaussian white noise*, it follows, $||T_{zw}||_2 = ||T_{zw} w||_2 = ||z||_2$. (The same notation is used for the time variables and their s-transforms.) But

$$||z||_2^2 = \frac{1}{2\pi} \int_{-\infty}^{+\infty} \operatorname{tr}[z(j\omega)^\star z(j\omega)]\, d\omega,$$

so that, using Parseval's relation, we find

$$||T_{zw}||_2^2 = \lim_{T\to\infty} \mathcal{E}\left(\frac{1}{T} \int_0^T ||z(t)||_2^2\, dt \right). \tag{1.98}$$

Replacing z, it follows,

$$||T_{zw}||_2^2 = \lim_{T\to\infty} \mathcal{E}\left(\frac{1}{T} \int_0^T \begin{bmatrix} x^T & u^T \end{bmatrix} \begin{bmatrix} C_1^T C_1 & C_1^T D_{12} \\ D_{12}^T C_1 & D_{12}^T D_{12} \end{bmatrix} \begin{bmatrix} x \\ u \end{bmatrix} dt \right). \tag{1.99}$$

According to the equations (1.97) and (1.99), a statement of the LQG control is so obtained. However, the weighting matrices of the performance index, as well as the covariance matrices of the white noise processes, are not determined in an arbitrary manner (as for the LQG case), but they are implicitly determined by the robustness specifications.

We shall now describe an algorithm for the solution of the H_2 optimal control problem. Assume that the matrix pairs (A, B_2) and (C_2, A) are stabilizable and detectable, respectively. Note that the optimal problem defined by the performance index (1.99) is not standard (see § 1.1.2), because it contains the coupling between the state and command. But it is equivalent to a standard problem with the matrices A and Q redefined as (see Problem 1.1 in § 1.2.1):

$$\bar{A} = A - BR^{-1}S^T, \qquad \bar{Q} = Q - SR^{-1}S^T,$$

where S is a coupling matrix. Moreover, the optimal control problem with performance index defined by (1.99) is equivalent to a "normalized" problem, in which the corresponding matrix R is the identity matrix. This

[13] D_{11} must be zero to guarantee a finite value to the performance index of the optimal problem. (To this end, for the sensitivity problem, it is required that $D_{W_1} = 0$.)

normalization can be achieved through either the eigenvalue decomposition of R as, for instance, in the program packages ROBUST (Chiang and Safonov, 1988) or BASILE (Murciano, 1990), or the singular value decomposition (SVD) (see, for example, (Golub and Van Loan, 1983)). In the latter case, omitting the indices (for convenience), if $D = U\Sigma V^T$ is the SVD of D (hence, U and V are orthogonal matrices), then $V\Sigma^{1/2}V^T$ is the square root of D^TD (denoted $(D^TD)^{1/2}$), since,

$$[(D^TD)^{1/2}]^2 = (V\Sigma^{1/2}V^T)^2 = V\Sigma V^T = D^TD.$$

Therefore, only the singular values and right singular vectors of D are needed. This normalization is usually performed before the reduction of the optimal problem to the standard form. By the way, two optimal problems must be solved (for the system (1.97) and the performance index (1.99)), one for regulation, and the other for state estimation. Only the most important computational steps will be presented in the sequel.

Procedure $\mathbf{H_2}$-control

/* 1. [Normalization.] Compute: */

$$N_1 := (D_{12}^T D_{12})^{-1/2}, \quad B_2 \leftarrow B_2 N_1,$$

$$N_2 := (D_{21} D_{21}^T)^{-1/2}, \quad C_2 \leftarrow N_2 C_2,$$

$$D_{12} \leftarrow D_{12} N_1, \quad D_{21} \leftarrow N_2 D_{21}, \quad D_{22} \leftarrow N_2 D_{22} N_1.$$

/* 2. [Reduction to the standard form.] Compute: */

$$\begin{array}{lll} S_1 = C_1^T D_{12}, & A_X := A - B_2 S_1^T, & Q_X := C_1^T C_1 - S_1 S_1^T, \\ S_2 = B_1 D_{21}^T, & A_Y := A - S_2 C_2, & Q_Y := B_1 B_1^T - S_2 S_2^T. \end{array}$$

/* 3. [Solution of Riccati equations.] Solve in X and Y, respectively: */

$$\begin{array}{rcl} A_X^T X + X A_X - X B_2 B_2^T X + Q_X & = & 0, \\ A_Y Y + Y A_Y^T - Y C_2^T C_2 Y + Q_Y & = & 0. \end{array}$$

/* 4. [Preliminary computation for optimal compensator.] Compute: */

$$F := B_2^T X + D_{12}^T C_1, \quad A_F := A - B_2 F, \quad C_F := C_1 - D_{12} F,$$

$$L := Y C_2^T + B_1 D_{21}^T, \quad A_L := A - L C_2, \quad B_L := B_1 - L D_{21},$$

$$\widehat{A} := A - B_2 F - L C_2 + L D_{22} F, \quad F \leftarrow N_1 F, \quad L \leftarrow L N_2.$$

/* 5. [Realization of unique optimal compensator $K^*(s)$.] Construct: */

$$K^*(s) := \left[\begin{array}{c|c} \widehat{A} & L \\ \hline -F & 0 \end{array} \right].$$

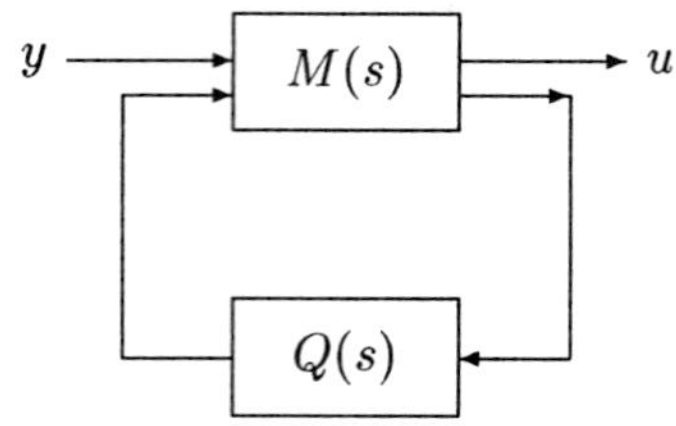

Fig. 1.19 Parametrization of admissible compensators; see text for description.

The optimal compensator satisfies

$$\begin{aligned} \min ||\mathcal{F}_l(P, K^*)||^2 &= \min ||T_{zw}||_2^2 = ||G_c B_1||_2^2 + ||F G_f||_2^2 \\ &= ||G_c L||_2^2 + ||C_1 G_f||_2^2, \end{aligned} \tag{1.100}$$

where

$$G_c(s) := \left[\begin{array}{cc} A_F & I \\ C_F & 0 \end{array} \right], \qquad G_f(s) := \left[\begin{array}{cc} A_L & B_L \\ I & 0 \end{array} \right], \tag{1.101}$$

On the other hand, it can be shown (Doyle et al., 1989) that the set of admissible compensators so that $||T_{zw}||_2 < \gamma$, coincides with the set of all transfer function matrices from y to u of the extended system in Fig. 1.19, defined by the parametrization

$$K_{ad}(s) := \mathcal{F}_l(M(s), Q(s)), \tag{1.102}$$

where $M(s)$ has the realization

$$M(s) := \left[\begin{array}{c|cc} \widehat{A} & L & B_2 \\ \hline -F & 0 & I \\ -C_2 & I & 0 \end{array} \right], \tag{1.103}$$

and $Q(s)$ is a strictly proper and stable parameter so that

$$||Q||_2^2 < \gamma^2 - (||G_c B_1||_2^2 + ||F G_f||_2^2).$$

Note that if $Q = 0$ we find $K^* = \mathcal{F}_l(M, 0)$.

1.4 Basic Numerical Linear Algebra Algorithms and Software

The previous sections have presented the main topics and results in linear-quadratic optimization, and related robust control domain. The rest of this book aims to show how the various problems described are actually solved. This section deals with basic numerical linear algebra algorithms of special interest in the linear-quadratic optimization, and the associated software support. They include, for instance, the computation and application of standard plane rotations, and solving systems of linear algebraic equations (in the usual or least squares sense). Such "low-level" algorithms are used as building blocks for solving more difficult problems, like Riccati equations, often being called by other "high-level" algorithms. On the other hand, the discussion of some numerical linear algebra algorithms offers the opportunity of presenting the specific style of describing algorithms, used in this book. Some relevant numerical issues (numerical stability, conditioning, etc.) are also commented.

1.4.1 Algorithms and programs

Using an intuitive definition, an *algorithm* is a finite succession of well-defined operations, associated with a problem class, which, applied to some *input data*, deliver *results* (representing the solution of the real problem). Here, *operations* means the arithmetical and logical operations which can be executed on a computer, as well as references to the algorithms defined before. It must be emphasized that results are always to be viewed as elements of some *finite* dimensional vector.

There are two main categories of algorithms. *Direct algorithms* find the solution of a given problem after completing a *finite*, fixed number of *steps* or *stages*. *Iterative algorithms* theoretically require to perform an *infinite* number of steps, the solution being obtained as a limit of a sequence. In practice, the execution of iterative algorithms ends when a certain *termination criterion* for convergence or non-convergence is met.

Algorithms are implemented on computers as *programs*, *subprograms*, or *procedures*, which are sequences of *statements* in a certain *programming language*. Most algorithms for solving linear-quadratic optimization problems are implemented in the FORTRAN language. Advantage was taken of the existing (public-domain) numerical linear algebra subroutine libraries: BLAS—Basic Linear Algebra Subprograms—described in (Lawson et al., 1979); LINPACK—for solving linear algebraic equations and least squares problems (Dongarra et al., 1979); and EISPACK—for solving algebraic eigenproblems (Garbow et al., 1977; Smith et al., 1976). These collections have proven to perform efficient computations on conventional and vector computer architectures; but on modern high-performance computers often do they obtain only a small fraction of the peak performance.

The newly developed library LAPACK (Anderson et al., 1992) has been specially designed to supersede LINPACK and EISPACK, by achieving much greater efficiency on vector computers, high-performance "superscalar" workstations and shared memory multiprocessors. LAPACK also extends the functional capabilities, and uses some improved or new algorithms. The LAPACK approach to combining efficiency with portability[14] (on a wide range of machines) is to use the BLAS as a low-level interface to different computer architectures. Beside the original *Level 1* BLAS, performing vector operations, *Level 2* BLAS (Dongarra et al., 1988)—for matrix-vector operations, and *Level 3* BLAS (Dongarra et al., 1990)—for matrix-matrix operations, are highly used. Level 2, and especially Level 3 BLAS are in fact responsible for the efficient use of parallelism.

Level 1 BLAS contains subprograms which implement elementary vector operations, including scalar products, vector norms (or equivalents), and generation and application of standard or modified plane rotations.

Level 2 BLAS includes subprograms performing the following class of matrix-vector operations:

$$\begin{aligned} y &\leftarrow \alpha \mathrm{op}(A)x + \beta y, \\ y &\leftarrow \mathrm{op}(T)x, \\ y &\leftarrow (\mathrm{op}(T))^{-1}x, \\ A &\leftarrow \alpha x \mathrm{op}(y) + A, \\ A &\leftarrow \alpha x \mathrm{op}(y) + \mathrm{op}(\alpha x \mathrm{op}(y)) + A, \end{aligned} \tag{1.104}$$

where α and β are scalars, x and y are vectors, A is a general matrix (sometimes rectangular), T is a triangular matrix, and $\mathrm{op}(M)$ denotes either M, its transpose, or its conjugate transpose. (For the last two operations, $\mathrm{op}(M)$ cannot mean M.)

Similarly, Level 3 BLAS includes subprograms performing the following class of matrix-matrix operations:

$$\begin{aligned} C &\leftarrow \alpha \mathrm{op}(A)\mathrm{op}(B) + \beta C, \\ C &\leftarrow \alpha A \mathrm{op}(A) + \beta C, \\ C &\leftarrow \alpha \mathrm{op}(A) A + \beta C, \\ C &\leftarrow \alpha A \mathrm{op}(B) + \mathrm{op}(\alpha A \mathrm{op}(B)) + \beta C, \end{aligned} \tag{1.105}$$

[14] *Portability* is meant as the possibility to transport the source programs, with minimal changes, from the machine which they were written for, to another machine.

$$
\begin{aligned}
C &\leftarrow \alpha \mathrm{op}(A)B + \mathrm{op}(\alpha \mathrm{op}(A)B) + \beta C, \\
B &\leftarrow \alpha \mathrm{op}(T)B, \qquad B \leftarrow \alpha B \mathrm{op}(T), \\
B &\leftarrow \alpha(\mathrm{op}(T))^{-1}B, \quad B \leftarrow \alpha B(\mathrm{op}(T))^{-1},
\end{aligned}
$$

where A, B, and C are matrices, and other notations have been defined above. There are variants for various kinds of matrices, including the case when A or B are symmetric or Hermitian.

During the last decade, the scientific and engineering computations have benefited from the availability of various interactive program packages, enabling simple and direct control of the entire computational process. Very popular has the MATLAB package become, which was originally devoted to linear algebra calculations (Moler, 1982), and based on LINPACK and EISPACK collections. The basic entity is a matrix, but its dimension is implicit. Algorithms expressed in MATLAB are very similar to the mathematical notation, and the programming effort needed is low. MATLAB offers many analytical commands, including commands for graphics in two- or three dimensions, programmable macros, a fast interpreter, and many facilities (Moler et al., 1985). A number of so-called *MATLAB toolboxes*, devoted to some specific problem-areas of large interest, are now widely available. Examples are the toolboxes for control system design, robust control, system identification, signal processing, neural nets, optimization, etc.

1.4.2 Numerical issues

The basic numerical issues to be considered are problem conditioning, and numerical stability of algorithms.

The analysis of conditioning of a problem is generally a difficult mathematical process, intimately connected with the perturbation theory of that problem. Briefly, a problem is said to be *well conditioned* if small changes in its data result in small changes in its solution. Otherwise, the problem is *ill conditioned.* It is important to know the condition of a problem, since the computed results always correspond to a perturbed problem, due to the rounding-errors produced by the floating-point representation and arithmetic of computers. The conditioning has been analyzed for many problems, particularly in numerical linear algebra, and computable or estimable *condition numbers* have been obtained. For instance, a condition number (with respect to inversion) of a nonsingular matrix A is $\mathrm{cond}(A) := \|A\|\,\|A^{-1}\|$, where $\|\cdot\|$ is any matrix norm consistent to a vector norm. Let $Ax = b$ be a linear algebraic system with A nonsingular. A perturbation δb in the right-hand side, b, produces a perturbation δx in the solution x, where $A\delta x = \delta b$. Using the properties of norms, it follows that $||A|| \, ||x|| \geq ||b||$ and $||\delta x|| \leq ||A^{-1}|| \, ||\delta b||$, so that

$$||\delta x||/||x|| \leq \text{cond}(A)||\delta b||/||b||,$$

hence cond(A) is an upper bound on the ratio of the relative errors in the solution and of relative errors in the right-hand side. It can also be shown (see, for instance, (Golub and Van Loan, 1983; Stewart, 1973)) that a perturbation δA in A causes a perturbation δx in x satisfying the inequality

$$||\delta x||/||x|| \leq \text{cond}(A)||\delta A||/||A||/(1 - \text{cond}(A)||\delta A||/||A||).$$

If cond(A) is large, the relative errors could be significant and the problem is ill conditioned. When A is singular, then cond$(A) = \infty$; when A is close to a singular matrix, then cond(A) is large. The LINPACK and LAPACK packages include subroutines for estimating the condition numbers for various types of matrices and linear algebra problems.

In the same informal way, an algorithm is said to be *numerically stable* if the computed result is (near) the exact solution of a slightly perturbed problem. That means that such an algorithm does not induce more sensitivity on data changes than the inherent sensitivity of the problem. If the problem is well conditioned, then the result computed by a numerically stable algorithm is near the solution; if the problem is ill conditioned, the result could be far from solution, but no algorithm could guarantee an accurate result. On the other hand, an unstable algorithm could give completely erroneous results even for well-conditioned problems. Numerical stability has been proven for many algorithms for solving linear algebra problems (linear algebraic equations, linear least squares, algebraic eigenproblems). For instance, the algorithm for Cholesky factorization of a positive definite matrix is numerically stable; the algorithm for LU factorization with partial pivoting is not numerically stable, but *practically* stable, in the sense that problems for which it fails to give acceptable results, occur very rarely in practice.

Details on rounding-error analysis, conditioning, and numerical stability can be found, for instance, in (Dahlquist and Björk, 1974; Forsythe, Malcolm and Moler, 1977; Golub and Van Loan, 1983; Stewart, 1973; Wilkinson, 1963, 1965).

Let us now briefly discuss the conditioning of the algebraic matrix Riccati equations. One says that a Riccati equation (1.55) or (1.59) is *ill-possed* if the assumptions (i) and (iii) in § 1.1.2 (Theorems 1.16 and 1.17) are not valid, and it is *nearly ill-possed* if a relatively "small" perturbation in the data makes it ill-possed. Here, "small" depends on the accuracy of the data and on the desired accuracy for the solution. The nearly ill-possed problems are often ill conditioned. In a general statement, *the ill-conditioned algebraic matrix Riccati equations are associated with nearly unstabilizable systems and nearly unstable closed-loop system matrices.* The near unstabilizability of a system (A, B) may be interpreted in either a parametric sense, or a control energy sense, i.e. with a nearly singular reachability Gramian, defined, for variable continuous-time systems, as

$$\int_{t_0}^{T} \Phi(\tau, t_0) B(\tau) B(\tau)^T \Phi(\tau, t_0)^T \, d\tau$$

(where $\Phi(\tau, t_0)$ is the transition matrix) and, similarly, for discrete-time systems.

Let $X \neq 0$ be the solution of an equation (1.55) or (1.59), and $\widetilde{X}$ the solution obtained when replacing A, B, Q, and R by the matrices $\widetilde{A}$, $\widetilde{B}$, $\widetilde{Q}$, and $\widetilde{R}$, respectively. If $||A - \widetilde{A}||$, $||B - \widetilde{B}||$, $||Q - \widetilde{Q}||$, and $||R - \widetilde{R}||$, are less than a sufficiently small $\varepsilon > 0$, then $\widetilde{X}$ exists and one can define the *condition number*

$$\kappa(A, B, Q, R) = \lim_{\varepsilon \to 0} ||X - \widetilde{X}|| / ||\widetilde{X}||.$$

Generally, relative errors of size $0(\varepsilon)$ in the data produce relative errors of size $\kappa(A, B, Q, R)\varepsilon$ in the solution. If $\kappa(A, B, Q, R)$ is large, the corresponding algebraic matrix Riccati equation is ill conditioned, and small perturbations in the data may cause large changes in the solution. For the equation (1.55), in (Bunse-Gerstner, Byers and Mehrmann, 1989) it is estimated that

$$\frac{1}{4}\kappa(A, Q, F) \leq (||\Omega^{-1}|| \, ||Q|| + ||\Theta|| \, ||A|| + ||\Pi|| \, ||F||)/||X|| \leq 9\kappa(A, Q, F),$$

where $F := BR^{-1}B^T$, and Ω, Θ and Π are linear operators on $\mathbb{R}^{n \times n}$, defined by

$$\begin{aligned} \Omega(Z) &= (A - FX)Z + Z(A - FX), \\ \Theta(Z) &= \Omega^{-1}(Z^T X + XZ), \\ \Pi(Z) &= \Omega^{-1}(XZX). \end{aligned}$$

Although the fraction above, indicating the size of $\kappa(A, Q, F)$, is a complicated expression, it may be evaluated (at least for small n and m) using, for instance, some routines from the EISPACK or LAPACK packages. Unfortunately, this fraction explicitly depends on X, and this is unavoidable. Note that equation (1.55) is ill conditioned if $||\Omega^{-1}||$ is large, i.e. if the optimal closed-loop system is nearly unstable. (Take into account that, using (1.54), $A - FX = A + BH$.)

Another numerical issue is related to efficiency of computations. To measure efficiency, the *standard floating-point operation*, called *flop*, is frequently used. A *flop* is usually defined as the time needed to evaluate the FORTRAN expression

```
T = A(I,J) + B*A(I,K)
```

so that it includes an addition, a multiplication and few memory references. We shall use *flop* in the sense of a similar, generalized operation $a_{ij} + ba_{ik}$.

For maximal efficiency, some algorithms make use of some *machine-dependent parameters*, like the *relative machine precision*, ε_M, the largest representable number, the least positive representable number, etc. The preferred solution (including in the LAPACK package) is to obtain such parameters using some special routines.

1.4.3 Basic algorithms

Let us first recall several definitions from numerical linear algebra. A matrix $A \in \mathbb{R}^{n\times n}$ is *upper (lower) triangular* if $a_{ij} = 0$, for $i > j$ $(i < j)$; if A is (upper or lower) triangular and $a_{ii} = 1$, $i = 1{:}\,n$, then A is *unit (upper* or *lower) triangular.* A is *diagonal* if it is both upper and lower triangular, and A is an *identity matrix* (I_n) if, in addition, $a_{ii} = 1$, $i = 1{:}\,n$. A *permutation matrix* is obtained by interchanging two rows or columns of I_n. A matrix having non-zero elements on the diagonal and on the two adjacent codiagonals only is called *tridiagonal*, and a triangular tridiagonal matrix is (upper or lower) *bidiagonal.* These definitions also apply in the complex case, and some of them can be extended to the rectangular case, $A \in \mathbb{R}^{m\times n}$.

Let $A \in \mathbb{R}^{m\times n}$. A *submatrix* S of A is a matrix formed by the elements of A found on the intersection of rows $i_1, \ldots, i_k$ $(1 \le i_1 < \cdots < i_k \le m)$ and columns $j_1, \ldots, j_l$ $(1 \le j_1 < \cdots < j_l \le n)$. If $l = k$ and $i_p = j_p$, for $p = 1{:}\,k$, then S is a *principal submatrix*, and if $i_p = p$, for $p = 1{:}\,k$, and $j_r = r$, for $r = 1{:}\,l$, then S is a *leading submatrix.* The notation $A(p{:}\,q, r{:}\,s)$, or simply $a_{p:q,r:s}$, denotes the submatrix defined by the rows $p, p+1, \ldots, q$ and columns $r, r+1, \ldots, s$. The first form coincides with the MATLAB convention for indices.

The usual mathematical notation, possibly extended with some specific symbols, is often used for presentation of algorithms. Sometimes, algorithms are expressed in a special language, called *pseudo-code*, which is similar to high-level programming languages. Many algorithms are described using the natural language and their description is often accompanied by listings of FORTRAN subprograms or ALGOL procedures (Dongarra et al., 1979; Forsythe, Malcolm and Moler, 1977; Garbow et al., 1977; Press et all, 1992; Smith et al., 1976; Wilkinson and Reinsch, 1971).

In this book, algorithms are presented in a simple and concise informal language, similar to the MATLAB notation, which facilitates computer implementations. For instance, $[x; y]$ is equivalent to $(x^T, y^T)^T$, and the notation $i = n{:}-p{:}\,q$ means that the index i successively takes the values n, $n-p$, ..., q. However, we write $y_{:,i}$, instead of $y_{:,i:i}$, for convenience. Mathematical formulas are used rather directly, but are always followed by a ";" sign. The arrow sign $\leftarrow$ is used in statements like, "$a \leftarrow x + a$;" to indicate that the value of the left-hand side variable is overwritten by the value of the expression on the right-hand side, containing that variable. It is not used in a statement defining an algorithm, where the left-hand side variables indicate the output, and they can also be input variables. The two-arrows sign $\leftrightarrow$ denotes an interchange of the values of two variables, that is, "$a \leftrightarrow x$;" is equivalent to "$t = a$; $a = x$; $x = t$;".

Calling statements of previously defined algorithms can be used, and recursive algorithms can simply be defined. The usual IF THEN, IF THEN ELSE, FOR, and WHILE constructions are available. The scope for some of these constructions is not marked by a corresponding END statement when it is obvious from context, or from indentation. Comments of the form "/* Comment */" may be inserted in on any line, but usually they appear flushed to the right, to improve the clarity.

Except for well-motivated instances, matrix dimensions are not specified in the argument list of an algorithm, for convenience. (If needed, calling statements like `size(` A `);`, or `ord(` A `);`, can be used for obtaining the dimensions or order of a matrix A, respectively.) Optional arguments are also possible. Sometimes, for flexibility reasons, the same algorithm can be used in several different manners, as the remarks below indicate. On the other hand, several variants of an algorithm can be used in the same way. It is tacitly assumed that the high-level algorithms possess a certain *context-sensitive* ability to select the most appropriate low-level algorithmic variants for the problem at hand. This implies, for instance, the selection of that algorithm that fully exploits the structure of the problem. Such "intelligent" algorithms are actually implemented in the so-called "expert-driver" routines, as those included in the LAPACK package. In this way, only algorithmic templates are actually used in the main text of the book, rather than descriptions of some source codes. This approach offers flexibility and ensures generality and independence of a specific computer language.

The above comments are now illustrated by several basic algorithms, which will be often used by other high-level algorithms. Real versions of various algorithms will generally be considered, since complex versions often follow easily, by replacing the transpose matrices by conjugate transpose matrices.

Algorithm rot : $(c,s) = \texttt{rot}(\ a,b\ )$

/* Compute a standard plane rotation, */

/* $G := \begin{bmatrix} c & s \\ -s & c \end{bmatrix}$, $c^2+s^2=1$, so that $G\begin{bmatrix} a \\ b \end{bmatrix} = \begin{bmatrix} r \\ 0 \end{bmatrix}$. */

/* 1. Compute r, c, and s. */
IF $|a| > |b|$, THEN $r = a(1+(b/a)^2)^{1/2}$; ELSE $r = b(1+(a/b)^2)^{1/2}$;
IF $r = 0$, THEN $c = 1$; $s = 0$; ELSE $c = a/r$; $s = b/r$;
$a = r$; /* 2. Return r in a. */

Remark 1.12 Algorithm `rot` can also be used as $G = \texttt{rot}(\ a,b\ )$. Note that the input value of a is overwritten with the computed r, but b is not actually zeroed. Some implementations (including BLAS) overwrite b by a value enabling the stable reconstruction of c and s, which need not be stored; the actual implementation could be more sophisticated, to confer good numerical properties. There is also a "square-root free" variant, generating a *modified plane rotation* (see (Lawson et al., 1979), for details). When $a, b \in \mathbb{C}$, a variant of the algorithm above is often used, that computes

$$G = \begin{bmatrix} c & s \\ s & -c \end{bmatrix}, \quad c \in \mathbb{R},\ s \in \mathbb{C},\ c^2 + |s|^2 = 1;$$

this variant will be called `crot`. □

A *plane rotation in the* (i,j) *plane* $(i<j)$ is an $n \times n$ matrix,

$$G(i,j) = I_n + (c-1)(e_i e_i^T + e_j e_j^T) + s(e_i e_j^T - e_j e_i^T), \tag{1.106}$$

where c and s define a standard plane rotation.

Example 1.1 Let $A \in \mathbb{R}^{(n+1)\times n}$ be a matrix whose leading $n \times n$ submatrix is upper triangular.

Algorithm upptri : $A = \texttt{upptri}(\ A\ )$

/* Reduce A to the upper triangular form. */

FOR $i = 1{:}\,n$,

$G = \texttt{rot}(\ a_{ii}, a_{n+1,i}\ )$; $\begin{bmatrix} a_{i,i+1:n} \\ a_{n+1,i+1:n} \end{bmatrix} \leftarrow G \begin{bmatrix} a_{i,i+1:n} \\ a_{n+1,i+1:n} \end{bmatrix}$;

Note that the n transformations computed to annihilate the elements of the last row preserved the upper triangular structure of the given matrix A.

These elements were not actually zeroed, but the leading $n \times n$ submatrix was updated. When calling Algorithm `upptri`, the output may not be specified, i.e. `upptri(` A `);` is a valid calling statement. □

A plane rotation is an *orthogonal matrix*, that is its columns are othonormal. (Two real vectors, q and r, are *othonormal* if $q^T r = 0$ and $\|q\|_2 = \|r\|_2 = 1$.) Hence, G in (1.106) satisfies $G^T G = GG^T = I_n$. If Q_1 and Q_2 are orthogonal and have compatible dimensions, then $Q_1 Q_2$ is also orthogonal. A matrix $Q \in \mathbb{C}^{n \times n}$ is *unitary* if $Q^H Q = I_n$. Orthogonal and unitary matrices have nice numerical properties: are perfectly conditioned, i.e. their condition number (in the Euclidean norm) is one (see § 1.4.2), and preserve the Euclidean and Frobenius matrix norms, i.e. $\|AQ\|_p = \|QA\|_p = \|A\|_p$, for $p \in \{2, F\}$.

A matrix of the form $U = I_n - 2uu^T$, where $u^T u = 1$, $u \in \mathbb{R}^n$ is called *Householder transformation* or *elementary reflector*. It is easy to prove that U is symmetric ($U = U^T$), orthogonal, and involutory ($U^2 = I_n$).[15] If $u^T u \neq 0$, the normalization $u/\|u\|_2$ can be avoided, if considering that

$$U = I_n - 2uu^T/u^T u =: I_n - \beta uu^T, \quad \beta := 2/u^T u. \tag{1.107}$$

By pre- or postmultiplying a matrix $A \in \mathbb{R}^{n \times n}$ with a suitably chosen Householder transformation, it is possible to annihilate simultaneously several elements in a column or row of A, respectively. Let $x \in \mathbb{R}^n$, $x \neq 0$, be given and choose a Householder transformation U so that $Ux = -re_1$, $r \in \mathbb{R}$. Hence, $r = \pm\|x\|_2$, since $|r| = \|Ux\|_2 = \|x\|_2$. If $x = -re_1$, then $U = I_n$. If $x \neq -re_1$, then U defined by (1.107) using $u = x + re_1$ satisfies $Ux = -re_1$. In this case, $u^T u = 2r(x_1 + r)$. For avoiding the possible cancellation error when evaluating $u_1 = x_1 + r$ and $\beta = 1/(r(x_1 + r))$, if $|r| \approx |x_1|$, r is chosen as $r = \mathrm{sign}(x_1)\|x\|_2$. Note that such a choice admits $x = -re_1$ only when $x = 0$. These comments motivate the following algorithm, typically used for $n = 2$, or $n = 3$.

Algorithm Hous_n : $(v, w) =$ `Hous_n(` ξ, y `)`

/* Compute a modified Householder transformation, $U := I_n + vw^T$, */
/* $v = -u/r$, $w = u/(\xi + r)$, so that $Ux = -re_1$, for $x = (\xi, y^T)^T$. */

IF $\|y\|_\infty = 0$, THEN
  $v = 0$; $w = 0$;
ELSE
  $s = \max\{\xi, \|y\|_\infty\}$; $w = [\xi;\ y]/s$; $\widetilde{r} = \mathrm{sign}(\xi)\|w\|_2$; $r = \widetilde{r}s$;
  $v_1 = -(w_1 + \widetilde{r})/\widetilde{r}$; $v_{2:n} = -w_{2:n}/\widetilde{r}$; $w_1 = 1$; $w_{2:n} = v_{2:n}/v_1$; $\xi = -r$;
END IF

[15] Unitary Householder transformations are defined similarly.

Remark 1.13 Note that ξ has been modified. The structure of x vector, $x = (\xi, y^T)^T$, has been explicitly used in the argument list for higher flexibility. When the ξ variable and y elements are stored in equally-spaced memory locations, then Algorithm `Hous_n` can be called as $(v, w) =$ `Hous_n(` x `)`. Such a call can always be used, but written as $(v, w) =$`Hous_n(` $(x_1; x_{2:n})$ `)`. In addition, the notation $U =$ `Hous_n(` x `)` denotes that $U = I_n + vw^T$ is the Householder transformation generated by Algorithm `Hous_n`. Such comments apply to all algorithms which follow. We can also write `Hous_3` instead of `Hous_n`, for $n = 3$. Moreover, $(v, w) =$ `Hous(` n, x `)` can be used as a synonym. □

Algorithm `Hous_n` requires $4n - 1$ multiplications/divisions, n additions and one square root. The computed v and w vectors satisfy the relation $v^T w = -2$.

For large n, it is inconvenient to store two vectors, v and w. Since, generally, x is no longer needed after u has been computed, u can be overwritten on x. Only the first element of x must be modified, since $u_i = x_i$, for $i = 2{:}\, n$.

Algorithm Hous : $(r, \beta) =$ `Hous(` x `)`

/* Compute a Householder transformation, $U := I_n - \beta uu^T$, so that */
/* $Ux = -re_1$ (u overwrites x). */

$s = \|x\|_\infty$;
IF $s \neq 0$, THEN $r = \mathrm{sign}(x_1)s\|x/s\|_2$; $x_1 \leftarrow x_1 + r$; $\beta = 1/(rx_1)$;

Algorithm `Hous` can also be called as $U =$ `Hous(` x `)`, or $(U, r) =$ `Hous(` x `)`. Since it needs $2n + 3$ multiplications/divisions, n additions and one square root, it is more efficient than Algorithm `Hous_n` for $n > 2$, and for large n is about twice faster. Taking also into account the cost of applying the computed Householder transformation to a given matrix, Algorithm `Hous` is preferable for $n > 3$.

Example 1.2 Let $A \in \mathbb{R}^{(n+p)\times n}$ be a given matrix, whose leading $n \times n$ submatrix is upper triangular.

Algorithm hupptri : $A =$ `hupptri(` A `)`

/* Reduce A to the upper triangular form. */

FOR $i = 1{:}\, n$,
  $(U, r) =$ `Hous(` $[\, a_{ii};\, a_{n+1:n+p,i}\,]$ `)`; $A \leftarrow UA$; $a_{ii} = -r$;

For convenience, this algorithm does not explicitly show how the Householder transformations can be efficiently applied. This will be made apparent in Algorithm `QR` below. □

Let $A \in \mathbb{R}^{m \times n}$, with $m \geq n$. There is an orthogonal matrix Q, so that

$$Q^T A = \begin{bmatrix} R \\ 0 \end{bmatrix}, \quad R \in \mathbb{R}^{n \times n} \text{ upper triangular.} \tag{1.108}$$

If Q is partitioned $Q = (Q_1, Q_2)$, where Q_1 has n columns, we obtain the *QR factorization*, $A = Q_1 R$. This says that if $\operatorname{rank}(A) = n$, the columns of Q_1 form an orthonormal basis for the linear subspace generated by columns of A, called the *column* (or *range*) *space of* A, and denoted $\mathcal{R}(A)$,

$$\mathcal{R}(A) := \{y \in \mathbb{R}^m : y = \sum_{i=1}^{n} \alpha_i a_{:,i}, \quad \alpha_i \in \mathbb{R}\}.$$

The rank of A can be defined as $\operatorname{rank}(A) = \dim(\mathcal{R}(A))$. The orthogonal projections on $\mathcal{R}(A)$ and its orthogonal complement (the *null space of* A),

$$\mathcal{N}(A) := \mathcal{R}(A)^{\perp} = \{z \in \mathbb{R}^n : Az = 0\},$$

are defined, respectively, by

$$P_A = Q_1 Q_1^T, \qquad P_A^{\perp} = Q_2 Q_2^T. \tag{1.109}$$

Indeed, any m-vector y can be uniquely written as $y = QQ^T y = P_A y + P_A^{\perp} y$, where $P_A y \in \mathcal{R}(A)$, and $(P_A y)^T (P_A^{\perp} y) = 0$.

Consider the system of linear algebraic equations $Ax = b$, and assume $\operatorname{rank}(A) = n$. It follows that

$$\|Ax - b\|_2 = \|Q^T(Ax - b)\|_2 = \left\| \begin{bmatrix} R \\ 0 \end{bmatrix} x - Q^T b \right\|_2,$$

and this norm is minimized when $Rx = Q_1^T b$, where R is upper triangular and nonsingular. The residual sum of squares is given by $\|Q_2^T b\|_2^2$.

Algorithm QR : $(R, Q, q) = \texttt{QR}(\ A\ )$

```
/* Compute the QR factorization of A ∈ IR^{m×n}, m ≥ n: A = QR. */
FOR i = 1:n,
        /* 1. Compute the Householder transformation U_i for column i. */
    (r, β) = Hous( a_{i:m,i} );
                                        /* 2. Apply U_i to a_{i:m,i+1:n}. */
    FOR j = i + 1:n,
        τ = −β(a_{i:m,i}^T a_{i:m,j});   IF τ ≠ 0, THEN a_{i:m,j} ← a_{i:m,j} + τ a_{i:m,i};
    q_i = a_{ii};   a_{ii} = −r;         /* 3. Save (u_i)_1 and set a_{ii}. */
```

Remark 1.14 Algorithm `QR` returns R in the upper triangular part of A. (We could add $R =$ `triu(` A `);`, where `triu` simply extracts the upper triangular part of a matrix.) It does not explicitly form Q, $Q = U_1 U_2 \cdots U_n$, but the information based on which Q can be reconstructed is stored in the lower triangular part of A and in the n-vector q. This factored form of Q can be easily used when applying Q to another matrix. Obviously, valid calling statements are $(A, Q, q) =$ `QR(` A `)`, or $(A, q) =$ `QR(` A `)`, or even `QR(` A `)`, when Q is not needed. Algorithm `QR` can also be used when $m < n$; in this case, it gives $A = Q[R \;\; S]$, where R is again upper triangular. An alternative is the LQ factorization, defined by

$$A = [L \;\; 0]Q = [L \;\; 0] \begin{bmatrix} Q_1 \\ Q_2 \end{bmatrix} = LQ_1,$$

where L is lower triangular, and Q is orthogonal. The corresponding algorithm is $(L, Q, q) =$ `LQ(` A `)`. LAPACK library includes routines which compute the QR and LQ factorizations, pre- or post-multiply another matrix by Q or Q^T, or generate all og sr part of Q. There is also a variant for the QR factorization with column pivoting, useful when $\operatorname{rank}(A) < \min(m, n)$; the "free" column (i.e., that must not be frozen at one of the leading positions) of largest norm is interchanged with the ith column at the ith stage of the algorithm. □

Many algorithms require, as a basic computational step, the solution of a system of linear algebraic equations. Consider the system $Ax = b$, where $A \in \mathbb{R}^{n \times n}$ and $\operatorname{rank}(A) = n$. In this case, A is *nonsingular*, and the unique solution of the system is $x = A^{-1}b$, where A^{-1}, the *inverse* of A satisfies $A^{-1}A = AA^{-1} = I_n$. If $\operatorname{rank}(A) < n$, then A is *singular*; one necessary and sufficient condition for singularity is that $\det(A) = 0$. As $\det(T) = \prod_{i=1}^{n} t_{ii}$ for a triangular matrix T, it follows that T is nonsingular if and only if $t_{ii} \neq 0$, for all $i \in \{1, \ldots, n\}$; the inverse of T is also triangular.

For any matrix $A \in \mathbb{R}^{n \times n}$, there is an upper triangular matrix U and a matrix L, the product of unit lower triangular and permutation matrices, so that $A = LU$. This is the *LU factorization* of A, and it exists even if A is singular, but in this case the U factor is singular, since $\det(A) = \det(L)\det(U)$. The relation $A = LU$ is equivalent to $L^{-1}A = U$, that is A is "triangularized"; this process is also known as *Gaussian elimination.* The matrix L^{-1} can be written as

$$L^{-1} = L_{n-1}P_{n-1} \cdots L_2 P_2 L_1 P_1, \tag{1.110}$$

where each P_k is a permutation matrix, and each L_k is an *elementary eliminator*, that differs from I_n in its subdiagonal elements of the kth column, hence

$$L_k = I_n + m_k e_k^T, \quad m_k = (0, \ldots, m_{k+1,k}, \ldots, m_{nk})^T. \tag{1.111}$$

The elements m_{ik}, called *(negative) multipliers*, are chosen to annihilate the last $n-k$ components of the kth column of the currently transformed matrix. Permutation P_k is chosen to ensure numerical stability, by requiring that $|m_{ik}| \le 1$, $i = k+1\!:n$. The transformations used in (1.110) are called *stabilized elementary transformations.* The LU factorization is computed by directly applying the permutation and elimination operations on A. The multipliers are stored below the diagonal, in place of the annihilated elements, and the indices specifying the permutations are stored in an additional vector p.

Algorithm LU : $(A, p) = \texttt{LU}(\ A\ )$

/* Compute the LU factorization of $A \in \mathbb{R}^{n\times n}$, */
/* $A = LU$, U upper triangular. */

FOR $k = 1\!:n-1$,
  /* 1. Annihilate the subdiagonal elements in the kth column. */
  $l = \min\{\arg\max_{\,i=k:n} |a_{ik}|\}$; $p_k = l$;
  IF $a_{lk} \ne 0$, THEN
    $a_{kk} \leftrightarrow a_{lk}$; $a_{k+1:n,k} \leftarrow -a_{k+1:n,k}/a_{kk}$;
    /* 2. Apply transformation to columns $k+1$ to n. */
   FOR $j = k+1\!:n$,
    $a_{kj} \leftrightarrow a_{lj}$; $a_{k+1:n,j} \leftarrow a_{k+1:n,j} + a_{k+1:n,k}a_{kj}$;
  END IF
$p_n = n$;

The element a_{lk}, interchanged with a_{kk}, is the *pivot* at step k, and this strategy for pivot selection is called *partial pivoting.*

If the U factor of the LU factorization of A is nonsingular, then the linear system $Ax = b$ can be solved in two stages, by solving two triangular systems, $Ly = b$ and $Ux = y$. The first stage is called *forward elimination*, and the second stage, *backward substitution.*

Algorithm LUsol : $b = \texttt{LUsol}(\ A, p, b\ )$

/* Solve $Ax = b$ using the factorization computed by Algorithm LU. */

  /* 1. Forward elimination: $b \leftarrow L^{-1}b$. */
FOR $k = 1\!:n-1$,
  $b_k \leftrightarrow b_{p_k}$; $b_{k+1:n} \leftarrow b_{k+1:n} + a_{k+1:n,k}b_k$;

/* 2. Backward substitution: $x := b \leftarrow U^{-1}b$. */
FOR $k = n\!:\!-1\!:\!1$,
  $b_k \leftarrow b_k / a_{kk}$; $b_{1:k-1} \leftarrow b_{1:k-1} - a_{1:k-1,k} b_k$;

A similar algorithm can solve $A^T x = b$ using the LU factorization of A. The following algorithm combines the algorithms **LU** and **LUsol** above.

Algorithm solve : $B =$ **solve**(A, B)

/* Solve matrix equation $AX = B$ ($B \in \mathbb{R}^{n\times m}$), using the LU factorization of A. */

(A, p) = **LU**(A);
FOR $k = 1\!:\!m$,
  $b_{:,k}$ = **LUsol**($A, p, b_{:,k}$);

Remark 1.15 Clearly, Algorithm **solve** can be called as **solve**(A, B);, the result being obtained in the parameter B. We will use, from time to time, statements like x = **LUsol**(A, p, b), or X = **solve**(A, B), when the result needs be copied in another variable. □

Similar algorithms can be written for special matrix structures, like symmetric, band, band symmetric, etc. With few exceptions, we will use the same algorithms for all such structures, but assume that all structure of the problem is implicitly exploited for improving the efficiency and the numerical properties. One exception is the following algorithm, applied to positive definite matrices. A matrix A is *positive definite* if it is symmetric ($A = A^T$), and the quadratic form $x^T Ax$ is positive for any real non-zero vector x. In this case, there is a unique upper triangular matrix R with positive diagonal elements, so that $A = R^T R$. This is called the *Cholesky factorization* of A, and can be used in a similar way to the LU factorization. The matrix R is the *Cholesky factor* of A. For obtaining R, we can write

$$\begin{bmatrix} A_{11} & a_k & A_{13} \\ \cdot & a_{kk} & A_{23} \\ \cdot & \cdot & A_{33} \end{bmatrix} = \begin{bmatrix} R_{11}^T & 0 & 0 \\ r_k^T & r_{kk} & 0 \\ R_{13}^T & R_{23}^T & R_{33}^T \end{bmatrix} \begin{bmatrix} R_{11} & r_k & R_{13} \\ 0 & r_{kk} & R_{23} \\ 0 & 0 & R_{33} \end{bmatrix},$$

(where A_{23} and R_{23} have one row), and equating the coefficients of the kth column, we obtain

$$a_k = R_{11}^T r_k, \qquad a_{kk} = r_k^T r_k + r_{kk}^2.$$

Hence, if R_{11} has already been computed, r_k and r_{kk} can be obtained from the relations:

$$R_{11}^T r_k = a_k, \qquad r_{kk}^2 = a_{kk} - r_k^T r_k.$$

Algorithm Chol : $\boxed{A = \texttt{Chol}(\ A\ )}$

/* Compute the Cholesky factorization of $A \in \mathbb{R}^{n\times n}$, */
/* $A = R^T R$, R upper triangular, overwritten on A. */

FOR $k = 1{:}\, n$,
  FOR $j = 1{:}\, k-1$,
    $a_{jk} \leftarrow (a_{jk} - a^T_{1:j-1,j}a_{1:j-1,k})/a_{jj}$;
  $a_{kk} \leftarrow (a_{kk} - a^T_{1:k-1,k}a_{1:k-1,k})^{1/2}$;

In an implementation, the sign of the quantity in parantheses must be checked before computing the square root; if negative, then the given matrix is not positive definite. The last scalar product can efficiently be evaluated during the inner FOR loop. This loop actually solves the system $R_{11}^T r_k = a_k$, where R_{11} is upper triangular, and R_{11} and r_k are stored in A. Therefore, it is possible to use a call to a Level 2 BLAS routine for triangular matrices (see (1.104)):

```
      CALL DTRSV( 'Upper', 'Transpose', 'Non-unit', k-1,
     $            A, LDA, A(1,k), 1 )
```

where `LDA` is the declared number of rows of the two-dimensional array `A`. Such an implementation can provide near peak performance on some computers, like CRAY Y-MP, but it can be far from the optimum for other modern architectures. For increasing the performance in the latter case too, a variant called *block Cholesky factorization* should be used. This variant can be derived using only submatrices, or *blocks* of the original matrix:

$$\begin{bmatrix} A_{11} & A_{12} & A_{13} \\ . & A_{22} & A_{23} \\ . & . & A_{33} \end{bmatrix} = \begin{bmatrix} R_{11}^T & 0 & 0 \\ R_{12}^T & R_{22}^T & 0 \\ R_{13}^T & R_{23}^T & R_{33}^T \end{bmatrix} \begin{bmatrix} R_{11} & R_{12} & R_{13} \\ 0 & R_{22} & R_{23} \\ 0 & 0 & R_{33} \end{bmatrix}.$$

Proceeding as above on the second block-column, we obtain

$$A_{12} = R_{11}^T R_{12}, \qquad A_{22} = R_{12}^T R_{12} + R_{22}^T R_{22}.$$

Hence, if R_{11} has already been computed, R_{12} and R_{22} can be obtained from the relations:

$$R_{11}^T R_{12} = A_{12}, \qquad R_{22}^T R_{22} = A_{22} - R_{12}^T R_{12}.$$

Algorithm CholB : $\boxed{A = \texttt{CholB}(\ A, p\ )}$

/* Compute the block Cholesky factorization of $A \in \mathbb{R}^{n\times n}$, */
/* using blocks of order p. */

FOR $k = 1{:}\, p{:}\, n$,

$l = \min(k + p - 1, n);$

/* 1. Compute R_{12}. */

$a_{1:k-1,k:l} =$ `solve`($a^T_{1:k-1,1:k-1}, a_{1:k-1,k:l}$);

/* 2. Update A_{22}. */

$a_{k:l,k:l} \leftarrow a_{k:l,k:l} - a^T_{1:k-1,k:l} a_{1:k-1,k:l};$

/* 3. Compute R_{22}. */

$a_{k:l,k:l} =$ `Chol`($a_{k:l,k:l}$);

A *recursive algorithm*, containing calls to itself, could somewhat simplify the presentation. When computing R_{12}, at stage 1, it is assumed that `solve` algorithm exploits the fact that its first parameter is the transpose of an upper triangular matrix. A Level 3 BLAS routine (`DTRSM`) can be used (see (1.105)). Stage 2 involves a symmetric updating of a symmetric matrix, achievable through the use of another Level 3 BLAS routine (`DSYRK`). The first two stages are not executed when $k = 1$.

The LAPACK implementation of Algorithm `CholB` actually uses a variant that is more efficient on many modern computer architectures; it is based on matrix-matrix multiplications, rather than on solving triangular systems of equations. Similar block algorithms are also implemented for other factorizations. The block QR factorization algorithm is based on a special representation of products of Householder transformations, $H_1 H_2 \cdots H_p = I_n - VTV^T$, where V is an $n \times p$ matrix whose columns are the individual vectors $u_1, u_2, \ldots, u_p$, associated with $H_1, H_2, \ldots, H_p$, respectively, and T is a $p \times p$ upper triangular matrix.

Unblocked algorithms will generally be described in this book, since they are always needed. (For instance, such algorithms are called when matrix dimensions are less than certain threshold values of the block dimension, p, like 32 or 64, or when there is not enough memory for calling block variants.) Obviously, LAPACK block algorithms can be used for basic linear algebra computations.

1.5 Overview of Algorithms

The first three sections of this chapter show how important powerful tools are for solving Riccati equations. Many algorithms have purposely been proposed in the last decades as Riccati solvers. We can distinguish two important categories: iterative algorithms and direct algorithms. Some algorithms in the first category determine the stationary solution of the differential or difference Riccati equations associated with (1.52), and (1.53), or (1.56), and (1.57) (see also (1.1), and (1.2), or their discrete-time correspondents, (1.18), and (1.19)). This class includes the "doubling algorithms" (see, for instance, (Anderson, 1978)), which have quadratic convergence rate. The algorithms that use the matrix sign function (Barraud, 1979; Byers, 1987) may also be interpreted as doubling algorithms. Another class of iterative algorithms is formed by Newton-type algorithms (Caines and Mayne, 1970, 1971; Hewer, 1971; Kleinman, 1968). Initialized by a stabilizing matrix H, these algorithms solve a sequence of Lyapunov equations. The convergence rate of the iterative process is also quadratic. Although the amount of computations per iteration is larger than that for doubling algorithms (about $12n^3$, compared to $8n^3$ *flops*), Newton algorithms are to be preferred, due to their better numerical properties (accuracy, reliability, stability).

The category of "direct" algorithms includes the algorithms that determine a basis for the stable invariant subspace of a matrix of order $2n$ (for instance, G in (1.60) or (1.61)), associated with the optimal problem of interest; the solution of the algebraic matrix Riccati equation is then obtained by solving a linear matrix equation. In the classical approach (Potter, 1966; Vaughan, 1969, 1970), the basis was formed by the eigenvectors and principal vectors corresponding to the stable eigenvalues. This technique is not to be recommended, due to possible numerical difficulties related to the computation of the eigenvectors corresponding to (near) multiple eigenvalues. (See, for instance, (Golub and Van Loan, 1983; Stewart, 1973; Wilkinson, 1965).) An elegant approach consists in using the Schur vectors (Laub, 1979; Pappas, Laub and Sandell, 1980; Van Dooren, 1981). It should be emphasized that all these approaches employ the *iterative* QR or QZ algorithms (Moler and Stewart, 1973), to determine a basis of the stable subspace. More efficient variants have lately been proposed (Bunse-Gerstner and Mehrmann, 1986; Bunse-Gerstner, Mehrmann and Watkins, 1989). They exploit the special structure of the matrix (or matrices) which a special version of the QR or QZ algorithm is applied to.

It is not possible to state the absolute superiority of a Riccati solver over others, because, as we shall see, each algorithm has both advantages and disadvantages. The application of one or another algorithm depends on the problem concerned, on its nature and conditioning. To our knowledge, proofs of the numerical stability of the Riccati solvers are still lacking; most of the algorithms to be presented consist in numerically stable stages. Orthogonal transformations, real Schur form and (numerically) orthogonal bases for the invariant subspaces of interest are often used in this book. It is possible to solve some problems with dense matrices of order of tens or hundreds of state variables.

Some of the most powerful and remarkable algorithms currently available for solving algebraic matrix Riccati equation (1.55) or (1.59), and for determining the optimal stabilizing matrix H in (1.54) or (1.58), will be described in the following chapters of this book. The next paragraphs of this section briefly discuss some special algorithms for solving Riccati equations. They are important not only for historical or pedagogical reasons, but also for solving related optimization problems (for instance, for time-variant systems). Special attention is paid to the discrete-time case, since it is more appealing from an implementation viewpoint, and more appropriate for dealing with time-variant systems, and optimal estimation problems.

1.5.1 Direct iteration algorithms

The so-called *direct iteration algorithms* compute the stationary solution of the associated difference Riccati equation (for $k = 0{:}\, k_{max} - 1$):[16]

$$X_{k+1} = A^T X_k A - A^T X_k B (R + B^T X_k B)^{-1} B^T X_k A + Q, \qquad (1.112)$$

for a given $X_0 \geq 0$, usually chosen as $X_0 = \alpha^2 I_n$, $\alpha \neq 0$. The iteration (1.112) also holds for time-variant systems. For the time-invariant case, the basic result is (Dorato and Levis, 1971):

Theorem 1.20 *If the (A,B) pair is stabilizable, then the matrix sequence in (1.112) has a limiting solution X which satisfies the algebraic matrix Riccati equation.*

The direct implementation of the recursion (1.112) is not numerically appealing. Better results are obtained when using the mathematically equivalent formulation (see Theorem 1.2 in § 1.1.1)

$$\begin{aligned} X_{k+1} &= A_k^T X_k A_k + H_k^T R H_k + Q, \\ H_k &:= -(R + B^T X_k B)^{-1} B^T X_k A, \quad A_k := A + B H_k. \end{aligned} \qquad (1.113)$$

Computations are summarized in the folowing algorithm.

Algorithm Ricdit : $X = \texttt{Ricdit}(\, A, B, Q, R, X_0, k_{max}, \varepsilon \,)$

/* Solve (1.59) iterating on (1.113). */

$X = X_0$;

/* 1. Iterate. */

FOR $k = 1{:}\, k_{max}$,

/* 2. Compute H_{k-1} in H and A_{k-1} in $\widehat{A}$ using X_{k-1}. */

$H = -(R + B^T X B)^{-1} B^T X A; \quad \widehat{A} = A + BH;$

[16] Time dependency has been replaced by an index notation, which is more convenient for discrete-time systems.

/* 3. Compute X_k in $\widehat{X}$. */

$\widehat{X} = \widehat{A}^T X \widehat{A} + H^T R H + Q;$

/* 4. Check convergence. */

$e = \|\widehat{X} - X\|; \quad X = \widehat{X};$
IF $e \leq \varepsilon \|X\|$, THEN STOP;

/* 5. No convergence. */

"No convergence"; STOP;

Remark 1.16 Note that Algorithm **Ricdit** also computes an approximation of the optimal gain matrix. This matrix could be specified as output data by writing the defining statement of the algorithm as

Algorithm Ricdit : $(X, H, k) =$ **Ricdit(** $A, B, Q, R, X_0, k_{max}, \varepsilon$ **)**

/* Solve ... */

The number of iterations performed has also been included. For instance, the following call:

$(X, H, k) =$ **Ricdit(** $A, B, Q, R, 100I_{15}, 50, \varepsilon_M$ **)**

tries to solve the problem in 50 iterations with an accuracy equal to relative machine precision, ε_M, and with initial value $X_0 = 100I_{15}$. □

Algorithm **Ricdit** can easily be translated into a MATLAB code, as follows:

```
function [X,H,k] = Ricdit( A, B, Q, R, X_0, kmax, epsil )
%Ricdit  Iterative solution of discrete-time algebraic
%        matrix Riccati equation

if nargin <= 6, epsil = eps; end
X = X_0;
for k = 1 : kmax,
  H = -inv(R + B'*X*B)*B'*X*A;       A_hat = A + B*H;
  X_hat = A_hat'*X*A_hat + H'*R*H + Q;
  e = norm(X_hat - X);      X = X_hat;
  if e <= epsil*norm(X), quit;
end
disp('Warning: Maximum number of iterations exceeded');
```

In the code above, the relative machine precision is used for the ε parameter in Algorithm **Ricdit**, if the **epsil** parameter is omitted. The parameters X_0 and $kmax$ could be dealt with similarly, providing default values in the code.

Algorithm **Ricdit** can be applied on-line, for instance, to solve some simple adaptive control problems. For example, Alag and Kaufman (1977) made use of such an algorithm to update the stabilizing gain matrix, when the A and B matrices get changed. The algorithm is initialized on each iteration by the result of the preceding one, and so, the convergence is obtained in only two-three iterations.

Algorithm **Ricdit** requires about $1.5n^3 + 2.5nm(n+m) + m^3/6$ *flops* per iteration, taking into account the symmetry of X matrices and other intermediate results. It needs $5n^2 + 2nm + m^2$ memory locations for A, B, Q, R, H, $\widehat{A}$, X, and $\widehat{X}$. (The matrix X_0 could be overwritten by X.) If the symmetric matrices Q, R, X, and $\widehat{X}$ are stored in a packed form (see, for instance, (Dongarra et al., 1979)), then $1.5n(n-1)+0.5m(m-1)$ locations can be saved.

Algorithm **Ricdit** can easily be modified to solve optimal control problems for time-variant systems and/or with finite final time. In the first case, indices should also be added to the A, B, Q, R, and H matrices. In the second case, only H should be indexed. The iteration should be performed in backward time, and the H_k matrices should be stored, in order to be subsequently applied on-line. The convergence test should be omitted in the time-variant case.

Algorithm **Ricdit** has a slow, linear convergence rate and its numerical properties are unsatisfactory. There is no numerical guarantee that the positive (semi)-definite character of X_k is preserved.

An iterative algorithm to prevent the altering of positiveness of X has been proposed, for instance, in (Barraud, 1974). The idea, met with in the theory of sequential estimation (see also (Bierman, 1977)) is to iterate on the square root of X. The equation (1.112) can be re-written as follows:

$$\widetilde{X}_k = X_k - X_k B(R + B^T X_k B)^{-1} B^T X_k, \tag{1.114}$$

$$X_{k+1} = A^T \widetilde{X}_k A + Q. \tag{1.115}$$

Factoring $Q = UU^T$, $R = V^T V$, $X_k = S_k S_k^T$, $\widetilde{X}_k = \widetilde{S}_k \widetilde{S}_k^T$, where U, V^T, S_k, and $\widetilde{S}_k$ are lower triangular matrices, we obtain

$$\begin{aligned} \widetilde{S}_k \widetilde{S}_k^T &= S_k S_k^T - S_k S_k^T B(V^T V + B^T S_k S_k^T B)^{-1} B^T S_k S_k^T \\ &= S_k S_k^T - S_k S_k^T Y(I_m + Y^T S_k S_k^T Y)^{-1} Y^T S_k S_k^T, \end{aligned}$$

where $Y := BV^{-1}$, hence Y is the unique solution of the linear system $YV = B$, with V nonsingular.

Denoting $Z = S_k^T Y$, we have

$$I_m + Y^T S_k S_k^T Y = I_m + Z^T Z =: a,$$

and then

$$\widetilde{S}_k = S_k(I_n - Za^{-1}Z^T)^{1/2}.$$

For the scalar input case ($m = 1$), it follows that

$$I_n - a^{-1}ZZ^T = (I_n - \alpha ZZ^T)^2,$$

where α is a root of the quadratic equation $(Z^TZ)\alpha^2 - 2\alpha + a^{-1} = 0$, hence $\alpha = 1/(a \pm a^{1/2})$; the root $\alpha = 1/(a + a^{1/2})$ is chosen in order to avoid possible loss of accuracy due to the cancellation of significant digits. Therefore,

$$\widetilde{S}_k = S_k - S_kZZ^T/(a + a^{1/2}).$$

The technique can be extended to the multi-input case, by entering Y column by column, taking into account the triangular structure of S_k. (Another approach is to use Lemma 1.1 in § 1.5.2.) Relation (1.115) is now re-written as

$$S_{k+1}S_{k+1}^T = A^T\widetilde{S}_k\widetilde{S}_k^TA + UU^T = \begin{bmatrix} A^T\widetilde{S}_k & U \end{bmatrix} T^TT \begin{bmatrix} \widetilde{S}_k^TA \\ U^T \end{bmatrix},$$

where matrix T is orthogonal and chosen so that S_{k+1} be lower triangular.

The following algorithm is obtained.

Algorithm Ricditsr : $X =$ `Ricditsr(` $A, B, Q, R, S_0, k_{max}, \varepsilon$ `)`

/* Solve (1.59) iterating on the square root of X_k in (1.114)–(1.115). */

$S = S_0$;

/* 1. Factor $R = V^TV$, $Q = U^TU$, V, U upper triangular.[17] */

`Chol(` R `);` `Chol(` Q `);`

/* 2. Solve in Y, $YR = B$.[18] */

$Y = BR^{-1}$;

/* 3. Iterate. */

FOR $k = 1\colon k_{max}$,

/* 4. Compute $\widetilde{S}_k$. */

$\widetilde{S} = S$;

FOR $i = 1\colon m$,

$z = S^Ty_{:,i}$; $a = z^Tz + 1$; $\widetilde{S} \leftarrow \widetilde{S} - \widetilde{S}zz^T/(a + a^{1/2})$;

---

[17] If Q is not positive definite, Cholesky factorization with pivoting can be used.

[18] The inverse should not be computed on an implementation; a statement like $X =$ `solve(` R^T, B^T `);` $Y = X^T$; is more appropriate. Note that the Level 3 BLAS includes a routine for solving $YR = B$, where R is upper triangular.

/* 5. Compute S_{k+1} in $\widehat{S}$ using Householder transformations, H_i. */

$\widehat{S} = \begin{bmatrix} \widetilde{S}^T A \\ Q \end{bmatrix}$;

FOR $i = 1{:}\,n$,

$(H_i, r) =$ `Hous`($\widehat{S}_{i:n+i,i}$); $\widehat{S} \leftarrow H_i \widehat{S}$; $\widehat{s}_{ii} = -r$;

/* 6. Check convergence. */

$\widehat{S} \leftarrow \widehat{S}^T_{1:n,1:n}$; $e = \|\widehat{S} - S\|$; $S = \widehat{S}$;

IF $e \le \varepsilon \|S\|$, THEN $X = SS^T$; STOP;

/* 7. No convergence. */

"No convergence"; STOP;

The algorithm requires about $2.7n^3 + 3n^2 m$ *flops* per iteration, and additional $(n^3 + m^3)/6 + m^2 n$ *flops* for initialization. It needs $5n^2 + nm + m^2$ memory locations for data and working arrays (assuming overwriting whenever possible).

The convergence rate does not improve over Algorithm **Ricdit**, but the positivity of X is ensured. The conditioning of computations is better than that for Algorithm **Ricdit**, since $\text{cond}(X) = \text{cond}(S)^2$.

1.5.2 Doubling algorithms

The direct iteration algorithms should not be used for solving the associated algebraic matrix Riccati equation, since they have linear convergence rate. More efficient will the so-called *doubling algorithms* or Newton's algorithms (see Chapter 2) be, since they have quadratic convergence.

The doubling algorithms do not compute every term X_k of the sequence defined in (1.112), but only the terms X_{2^k}, hence the time-interval between two successively generated elements doubles on each iteration. To obtain $X_{2^{k+1}}$ from X_{2^k}, appropriate updating relations are used. Such relations are obtained, for instance, from the Hamiltonian formulation of the optimal problem.

It can be shown (Anderson, 1978) that X_{k+1} in (1.112) can be re-written as

$$X_{k+1} = (\bar{G}_{21} + \bar{G}_{22} X_k)(\bar{G}_{11} + \bar{G}_{12} X_k)^{-1}, \tag{1.116}$$

where $\bar{G}_{ij}$ are $n \times n$ submatrices of the matrix

$$\bar{G} = \begin{bmatrix} A^{-1} & A^{-1}F \\ QA^{-1} & A^T + QA^{-1}F \end{bmatrix} =: \begin{bmatrix} \bar{G}_{11} & \bar{G}_{12} \\ \bar{G}_{21} & \bar{G}_{22} \end{bmatrix}, \tag{1.117}$$

where $F := BR^{-1}B^T$. Note that $\bar{G}$ is just the inverse of the matrix G in (1.61), and we assumed that A matrix is nonsingular.[19]

[19] If A is singular, then the same results are obtained by a taking-limit argument.

Consider now the linear difference matrix equation

$$\begin{bmatrix} U_{k+1} \\ V_{k+1} \end{bmatrix} = \begin{bmatrix} \bar{G}_{11} & \bar{G}_{12} \\ \bar{G}_{21} & \bar{G}_{22} \end{bmatrix} \begin{bmatrix} U_k \\ V_k \end{bmatrix}. \tag{1.118}$$

If U_k and V_k satisfy the relation $V_k U_k^{-1} = X_k$, then $V_p U_p^{-1} = X_p$, for any $p \geq k$. Assume that $U_0 = I_n$, $V_0 = 0$ (hence $X_0 = 0$), and let Φ be the transition matrix for (1.118). Then,

$$\begin{bmatrix} U_{2^k} \\ V_{2^k} \end{bmatrix} = \Phi^{2^k} \begin{bmatrix} U_0 \\ V_0 \end{bmatrix} = \Phi^{2^k} \begin{bmatrix} I_n \\ 0 \end{bmatrix}. \tag{1.119}$$

The required updating relations derive by an induction argument from the condition $\Phi^{2^{k+1}} = \Phi^{2^k}\Phi^{2^k}$. These relations are incorporated into the next algorithm.

Algorithm Ricditd : $X =$ `Ricditd`$(A, B, Q, R, k_{max}, \varepsilon)$

/* Solve (1.59) using the doubling formulas based on (1.119). */

/* 1. Start iteration. */

$a = A; \quad b = BR^{-1}B^T; \quad c = Q;$

/* 2. Iterate. */

FOR $k = 1{:}\, k_{max}$,

/* 3. Update a, b, and c. */

$d = (I_n + bc)^{-1}; \quad \widehat{a} = ada; \quad \widehat{b} = b + adba^T; \quad \widehat{c} = c + a^T cda;$
$a = \widehat{a}; \quad b = \widehat{b};$

/* 4. Check convergence. */

$e = \|\widehat{c} - c\|; \quad c = \widehat{c};$
IF $e \leq \varepsilon\|c\|$, THEN $X = c$; STOP;

/* 5. No convergence. */

"No convergence"; STOP;

From a numerical point of view, Algorithm `Ricditd` is comparable to Algorithm `Ricdit`, so it has the disadvantages of numerical instability and possible ill conditioning of computations. From the efficiency point of view, it is clearly superior. Assuming that Algorithm `Ricdit` converges in $k_f = 2^k$ iterations, then Algorithm `Ricditd` converges in k iterations. About $9n^3$ *flops* are required per iteration, and additional $m^3/6 + n^2m + nm^2/2$ *flops* for initialization. For $m \geq n/5$, Algorithm `Ricditd` is more efficient than Algorithm `Ricdit` for $k_f \geq 24$. Algorithm `Ricditd` needs $7n^2 + nm + m^2$ memory locations for data and working arrays, assuming overwriting whenever possible (for instance, a and c can overwrite A and Q, respectively).

Note that a, b, and c in the above algorithm depend on k. (These matrix variables could be written as a_k, b_k, and c_k, respectively, and the "hat" variables are a_{k+1}, b_{k+1}, and c_{k+1}, respectively.) This dependence has been suppressed, not only for convenience, but also for a rational implementation uses overwriting of variables, to conserve the memory space. When b_k and c_k are nonsingular, then these matrices may be put in a convenient factored form. If $c_0 > 0$, it follows that $c_k > 0$ for $k \geq 0$, and one can use a square root form of Algorithm **Ricditd**, similar to that of Algorithm **Riccditsq**. Denoting

$$b_k = F_k F_k^T, \qquad c_k = G_k^T G_k, \qquad X_k = G_k F_k,$$

where F_k and G_k are upper triangular matrices, the updating relations become (Sidhu and Bierman, 1977):

$$a_{k+1} = a_k G_k^{-1}(I_n + X_k X_k^T)^{-1} G_k a_k, \tag{1.120}$$

$$F_{k+1}F_{k+1}^T = F_k F_k^T + (a_k F_k)(I_n + X_k^T X_k)^{-1}(a_k F_k)^T, \tag{1.121}$$

$$G_{k+1}^T G_{k+1} = G_k^T G_k + (G_k a_k)^T (I_n + X_k X_k^T)^{-1}(G_k a_k). \tag{1.122}$$

The matrix inversions in (1.120)–(1.122) can be avoided on using the following easily provable lemma.

Lemma 1.1 *Let X be an $n \times p$ matrix, and let S be an orthogonal matrix, so that $\begin{bmatrix} I_n & X \end{bmatrix} S = \begin{bmatrix} U & 0 \end{bmatrix}$, where $U \in \mathbb{R}^{n\times n}$ is upper triangular. Then,*

$$(I_n + XX^T)^{-1} = S_{11}S_{11}^T, \qquad (I_p + X^T X)^{-1} = S_{22}S_{22}^T,$$

where S_{ii}, $i = 1{:}\,2$, are the diagonal blocks of S of order n, and p, respectively.

The following "square-root" algorithm is obtained.

Algorithm Ricditds : $X =$ `Ricditds(` $A, B, Q, R, k_{max}, \varepsilon$ `)`

/* Solve (1.59) using the square-root doubling formulas. */

/* 1. Start iteration: F and G upper triangular. */

$a = A;\quad F = (BR^{-1}B^T)^{1/2};\quad G = (Q^{1/2})^T;\quad X = GF;$

/* 2. Iterate. */

FOR $k = 1{:}\,k_{max}$,

/* 3. Find S orthogonal, L lower triangular. */

$(L, S) =$ `LQ(` $\begin{bmatrix} I_n & X \end{bmatrix}$ `);` $\quad Y = S_{1:n,1:n} G a;$

/* 4. Update a. */

$\widehat{a} = aG^{-1}S_{1:n,1:n}^T Y;$

/* 5. Update F, using S_F orthogonal; $\widehat{F}$ upper triangular. */

$L = \texttt{LQ}(\ \left[\begin{array}{cc} F & aFS^T_{n+1:2n,n+1:2n} \end{array}\right]\ ); \quad \widehat{F} = L^T;$

/* 6. Update G, using S_G orthogonal; $\widehat{G}$ upper triangular. */

$\widehat{G} = \texttt{QR}(\ \left[\begin{array}{c} G \\ Y \end{array}\right]\ );$

/* 7. Check convergence. */

$\widehat{X} = \widehat{G}\widehat{F}; \quad e = \|\widehat{X} - X\|; \quad G = \widehat{G}; \quad F = \widehat{F}; \quad X = \widehat{X};$

IF $e \le \varepsilon\|X\|$, THEN STOP;

/* 8. No convergence. */

"No convergence"; STOP;

Note that the orthogonal matrices S_F and S_G are only implicitly defined, and need not be stored.

Algorithm **Ricditds** may be applied if Q is a positive semi-definite matrix. Essentially it has the same convergence rate as Algorithm **Ricditd**, but the numerical performances are higher, comparable to those of Algorithm **Ricditsq**. About $7n^3$ *flops* are required per iteration (taking into account the triangular structure of G, and F). Algorithm **Ricditd** needs $7n^2$ memory locations for data and working arrays (assuming overwriting whenever possible).

1.5.3 Algorithms for sequential state estimation

The optimal linear discrete-time state estimation problem (filtering or prediction) has been briefly discussed in § 1.1.1. The solution of this problem for the time-variant case and/or finite final time is given by the sequential Kalman estimator. For the time-invariant infinite final time case, the solution is the asymptotic estimator, which can be determined by solving the corresponding discrete-time algebraic matrix Riccati equation.

Due to its simplicity and versatility, the sequential Kalman estimator has become a fundamental tool in the analysis and solution of linear estimation problems. Impressive numerical experience has drawn attention to the fact that not careful implementation of the recursive relations can generate divergence phenomena (Bierman, 1977), characterized by a discrepancy between the numerical and the predicted analytical results. There are many reasons for divergence, such as: the use of incorrect *a priori* statistics or incorrect models (unmodelled dynamics), nonlinearities, and rounding-errors in computations.

Many approaches have been made for avoiding divergence phenomena. For instance, for reducing the effects of imperfect modelling, matrix A is partitioned in (Mahalanabis, 1972) as

$$A = \begin{bmatrix} A_{11} & A_{12} \\ 0 & A_{22} \end{bmatrix},$$

where A_{11} and A_{12} are exactly known, and A_{22} is unknown, but its elements have bounded variations around some known nominal values, represented by the matrix A_{22}^0. The C matrix is assumed to be exactly known. Partitioning the state vector $x = \begin{bmatrix} x_1^T & x_2^T \end{bmatrix}^T$, according to the partition of A, a filter for x_2 is computed (using A_{22}^0) so that the effect of A_{22} variations should be minimized. The filter for x_1 is determined in the standard way, but including inter-covariances and the x_2 estimate.

The asymptotic filter (in the time-invariant case) produces biased estimates if the process is incorrectly modelled. This drawback can be removed by using error integrators, see (Lee and Athans, 1979). This is the dual problem of the control with zero steady-state error (Problem 1.4 in Subsection 1.2.1).

The Kalman filter can prove very sensitive to rounding-errors. Numerical difficulties have been encountered, for instance, for problems with (nearly) perfect measurements and (nearly) singular covariance matrices. The rounding-errors can alter the non-negative definiteness character of covariance matrices. Various *ad hoc* techniques have been advanced to improve the accuracy and preserve the positiveness and symmetry of the computed covariance. They include, for instance, alteration of the covariance elements, to ensure such properties, or use more stable calculations (like the expression $(A - KC)X(A - KC)^T + KRK^T + Q$, instead of the first or second formulas in (1.32)).

A better alternative is the so-called *square-root filter*. This approach implicitly preserves the symmetry and non-negative definiteness. Moreover, the square-root algorithms can conveniently be used on process computers with reduced word-length, and can easily be implemented on parallel computer architectures (see, for instance, (Morf et al., 1979a, 1979b)).

Some basic square-root algorithms applied to either the covariance matrix or its inverse, the *information matrix*, are briefly discussed below. The covariance-based algorithms are recommended for applications involving frequent estimates, scalar measurements and (nearly) perfect knowledge of the process model. The information-based algorithms are ideally suited for problems involving large data sets, unfrequent estimates, and possible lack of *a priori* knowledge.

For convenience, we start with the important problem of parametric estimation, and then we add the dynamics and process noise to estimate the state.

Let the measurement equation be written as

$$y = Cx + e, \tag{1.123}$$

where $y \in \mathbb{R}^l$, $x \in \mathbb{R}^n$, and e is a Gaussian process with $\mathcal{E}\{e\} = 0$, $\mathcal{E}\{ee^T\} = I_l$.[20] Assume the *a priori* information $\left[\, \widetilde{S}_0 \;\; \widetilde{y}_0 \,\right]$ available, corresponding to the equation

$$\widetilde{y}_0 = \widetilde{S}_0 x + \widetilde{e}_0,$$

where $\widetilde{S}_0$ is a square matrix, and $\mathcal{E}\{\widetilde{e}_0\} = 0$, $\mathcal{E}\{\widetilde{e}_0\widetilde{e}_0^T\} = I_l$. The information $\left[\, \widetilde{S}_0 \;\; \widetilde{y}_0 \,\right]$ can be computed from the initial covariance matrix, $\widetilde{X}_0$, and initial state estimate, $\widetilde{x}_0$, using the relations

$$\widetilde{X}_0^{-1} = \widetilde{S}_0\widetilde{S}_0^T, \qquad \widetilde{y}_0 = \widetilde{S}_0\widetilde{x}_0.$$

So, $\widetilde{S}_0$ is the square-root of the *a priori* information matrix, $\widetilde{X}_0^{-1}$. When *a priori* knowledge is not available, simply set $\widetilde{S}_0 = 0$, $\widetilde{y}_0 = 0$. Without loss of generality, $\widetilde{S}_0$ may be assumed as upper triangular.

The following algorithm computes the least squares estimate of the parameter vector x and its information matrix.

Algorithm Pardinf : $(\widehat{S}, \widehat{y}, \widehat{x}) = \texttt{Pardinf(}\; \widetilde{S}_0, \widetilde{y}_0, C, y \;\texttt{)}$

/* Estimate x in (1.123) using square-root information factorization. */

/* 1. Process the measurements ($\widehat{S}$ upper triangular). */

$$(\widehat{S}, Q) = \texttt{QR(}\left[\begin{array}{c} \widetilde{S}_0 \\ C \end{array}\right]\texttt{)}; \qquad \left[\begin{array}{c} \widehat{y} \\ r \end{array}\right] = Q^T \left[\begin{array}{c} \widetilde{y}_0 \\ y \end{array}\right]; \qquad \widehat{x} = \widehat{S}^{-1}\widehat{y};$$

The computations can easily be performed using the following MATLAB statements:

```
[l,n] = size(C);
[Q, S_hat] = qr([ S_tilde; C ]);
y_hat = Q'*[ y_tilde; y ];        r = y_hat(n+1:n+l);
y_hat = y_hat(1:n);           S_hat = S_hat(1:n,1:n);
x_hat = inv(S_hat)*y_hat;
X_hat = inv(S_hat)'*inv(S_hat);
```

which also compute the covariance matrix.

The r vector in Algorithm `Pardinf` and MATLAB code contains the residuals of the least squares approximation. If $\widetilde{S}_0$ is upper triangular, it

[20] If e is a correlated noise with $\mathcal{E}\{ee^T\} = R > 0$ and R is well-conditioned, the problem can be restated as above for the normalized variables, $\bar{y} = \bar{C}x + \bar{e}$, to be obtained by premultiplying (1.123) by L^{-1}, where $R = LL^T$.

is possible to exploit this structure using a variant of Algorithm `QR`, that determines the Householder transformations as in Algorithm `hupptri` (see Example 1.2 in § 1.4.3). In an implementation, the arrays $\widetilde{S}_0$ and $\widetilde{y}_0$ can be overwritten by $\widehat{S}$ and $\widehat{y}$, respectively.

Algorithm `Pardinf` is numerically stable and needs about $n^2 l + n^3/6$ *flops* and $(n+l)(n+1)$ memory locations. It can easily be used for processing q measurement data sets,

$$y_k = C_k x + e_k, \qquad k = 1{:}\,q. \tag{1.124}$$

Obviously, a global approach is to apply `Pardinf` on the concatenated data, $y = [\, y_1^T \ y_2^T \ \ldots \ y_q^T \,]^T$, and similarly for C. This could not work for very large data sets. An alternative is to process the data sequentially, in the form

Algorithm Pardinfs : $(\widehat{S}, \widehat{x}) =$ `Pardinfs`$(\ \widetilde{S}_0, \widetilde{y}_0\ )$

/* Estimate x in (1.124) using `Pardinf` sequentially. */

/* 1. Initialize the calculations. */

$\widehat{S} = \widetilde{S}_0; \qquad \widehat{y} = \widetilde{y}_0;$

/* 2. Process the measurements sequentially ($\widehat{S}$ upper triangular). */

FOR $k = 1{:}\,q$,

`Read`$(\ C_k, y_k\ ); \quad (\widehat{S}, \widehat{y}, \widehat{x}) =$ `Pardinf`$(\ \widehat{S}, \widehat{y}, C_k, y_k\ );$

The algorithm above assumes that each new data set is read from external media at each time step k (see the FOR loop). The use of as large data blocks as possible (depending on the available memory space) is highly recommended. It is possible to exploit the upper triangular structure of the matrix $\widehat{S}$. The algorithm can be modified so as to deliver parameter estimates and covariances for specified values of k. In addition, the residual sum of squares can also be obtained easily, using r in Algorithm `Pardinf`, that contains the residuals.

These comments point to the modifications needed for including dynamics and noise. Assume that x changes according to the equation

$$x_{k+1} = A_k x_k + G_k v_k, \tag{1.125}$$

where A_k is assumed nonsingular and $v_k \in \mathbb{R}^p$ is a Gaussian process with $\mathcal{E}\{v_k\} = \bar{v}_k$, $\mathcal{E}\{(v_k - \bar{v}_k)(v_j - \bar{v}_j)^T\} = Q_k \delta_{kj}$, $Q_k > 0$, and δ_{kj} denotes the Kronecker symbol (δ_{kj} equals one if $k = j$ and zero, otherwise.) The problem is to find an estimate of x_{q+1} using the q measurement data sets

$$y_k = C_k x_k + e_k, \qquad k = 1{:}\,q.$$

The solution technique consists in getting rid of the influence of variables x_k, v_k, for $k = 1{:}\, q$, using some orthogonal transformations. This technique can also be extended for the *smoothing problem*, when the estimate x_r, $1 \le r \le q$, is required.

A priori information, if available, is represented by the data equations

$$\widetilde{y}_0 = \widetilde{S}_0 x_0 + \widetilde{e}_0, \qquad y_k^v = S_k^v v_k + e_k^v, \qquad Q_k = (S_k^v)^{-1}(S_k^v)^{-T},$$

where $\widetilde{e}_0$ and e_k^v have zero mean and unit covariance.

Algorithm xdinf : $\boxed{(\widehat{S}, \widehat{x}) = \texttt{xdinf}(\ \widetilde{S}_0, \widetilde{y}_0\ )}$

/* Estimate x_{q+1} sequentially. */

/* 1. Initialize the calculations. */

$\widehat{S} = \widetilde{S}_0; \qquad \widehat{y} = \widetilde{y}_0;$

FOR $k = 1{:}\, q$,

/* 2. Process the measurements ($\widetilde{S}$ upper triangular). */

$\texttt{Read}(\ S_k^v, y_k^v, A_k, G_k, C_k, y_k\ ); \quad (\widetilde{S}, \widetilde{y}) = \texttt{Pardinf}(\ \widehat{S}, \widehat{y}, C_k, y_k\ );$

/* 3. Process the dynamics data ($\widehat{S}$ upper triangular). */

$\widetilde{S} \leftarrow \widetilde{S} A_k^{-1};$

$$\left(\begin{bmatrix} \widetilde{S}_k^v & \widetilde{S}_k^x \\ 0 & \widehat{S} \end{bmatrix}, Q \right) = \texttt{QR}\left(\begin{bmatrix} S_k^v & 0 \\ -\widetilde{S} G_k & \widetilde{S} \end{bmatrix} \right); \quad \begin{bmatrix} \widehat{y}_k^v \\ \widehat{y} \end{bmatrix} = Q^T \begin{bmatrix} y_k^v \\ \widetilde{y} \end{bmatrix};$$

/* 4. Set final estimate. */

$\widehat{x} = \widehat{S}^{-1}\widehat{y};$

Algorithm `xdinf` needs about $2n^3 + n^2 l$ *flops* (for $p = 1$) and $(n+l+1)(n+1) + (n+p+1)(n+p)$ memory locations. In practice, the S_k^v matrices can be assumed as diagonal. Indeed, if the initial matrix Q_k is not diagonal, the problem can be diagonalized, while noting that $G_k v_k = G_k(S_k^v)^{-1}(S_k^v v_k)$ and the vector $S_k^v v_k$ has identity as covariance matrix. The algorithm can be extended to correlated and non-zero mean noise.

Consider now factorization algorithms applied to the covariance matrix, X. We will assume that $X > 0$ and that the measurements are processed one by one. Let $X = UDU^T$, where U is unit upper triangular and D is diagonal, the so-called *U-D factorization* (Bierman, 1976, 1977; Thornton and Bierman, 1978). Consider the scalar measurement equation

$$y = c^T x + e, \tag{1.126}$$

where $c \in \mathbb{R}^n$, and e is a Gaussian variable with $\mathcal{E}\{e\} = 0$, $\mathcal{E}\{e^2\} = \sigma$. Assume that *a priori* information is provided by the initial estimate $\widetilde{x}$ and covariance $\widetilde{X}$, with $\widetilde{X} = \widetilde{U}\widetilde{D}\widetilde{U}^T$, where $\widetilde{U}$ is unit upper triangular and $\widetilde{D} = \text{diag}(\widetilde{d}_1, \widetilde{d}_2, \ldots \widetilde{d}_n)$. Let $\widetilde{U}_i := \widetilde{U}_{:,i}$, $i = 1{:}\,n$.[21] Using the matrix inversion lemma (Golub and Van Loan, 1983; Goodwin and Payne, 1977), the updating equation for X after a new measurement is processed, can be written as

$$X = \widetilde{X} - \widetilde{X}cc^T\widetilde{X}^T/(\sigma + c^T\widetilde{X}c). \tag{1.127}$$

Denoting,

$$f = \widetilde{U}^Tc, \quad v = \widetilde{D}f, \quad K = \widetilde{U}v, \quad b = 1/(\sigma + c^T\widetilde{X}c),$$

then the factored form of (1.127) is

$$UDU^T = \widetilde{U}(\widetilde{D} - b(\widetilde{D}\widetilde{U}^Tc)(\widetilde{D}\widetilde{U}^Tc)^T)\widetilde{U}^T = \widetilde{U}(\widetilde{D} - bvv^T)\widetilde{U}^T.$$

Therefore, if $\bar{U}$ and $\bar{D}$ are the U-D factors of the matrix $\widetilde{D} - bvv^T$ (i.e., $\bar{U}\bar{D}\bar{U}^T = \widetilde{D} - bvv^T$), we can identify $U = \widetilde{U}\bar{U}$ and $D = \bar{D}$, since $\widetilde{U}$ and $\bar{U}$ are upper triangular matrices. Hence, updating the U-D factorization of $\widetilde{X}$ is reducible to a special case of updating matrix factorization after a rank-one change. The factors $\bar{U}$ and $\bar{D}$ can be determined by specializing the algorithm of Thornton and Bierman (1978, Appendix II), retrieved in a different manner in (Sima, 1986), namely, for $j = n{:}\,-1{:}\,1$,

$$\bar{d}_j = \widetilde{d}_j - b^j v_j^2, \quad b^{j-1} = b^j\widetilde{d}_j/\bar{d}_j, \quad \bar{u}_{ij} = -b^j v_j v_i/\bar{d}_j, \ (i < j),$$

and $\bar{u}_{jj} = 1$, $b^n = b$. Unfortunately, an algorithm based on these relations raises numerical difficulties. From the first two equalities, it follows that

$$1/b^{j-1} = 1/b^j - v_j^2/\widetilde{d}_j,$$

so that, denoting $a^j = 1/b^j$, we obtain the following stable recursive relations

$$a^j = a^{j-1} + v_j^2/\widetilde{d}_j, \quad \bar{d}_j = \widetilde{d}_j a^{j-1}/a^j, \quad \bar{u}_{ij} = -v_j v_i/(a^{j-1}\widetilde{d}_j),$$

and we choose

$$a^0 = 1/b - \sum_{j=1}^{n} v_j^2/\widetilde{d}_j,$$

so that $a^n = 1/b$. Such updating formulas are also derived in (Gill and Murray, 1977).

Taking now the structure of the matrix $\bar{U}$ into account, namely

$$\bar{U}_j := \bar{U}_{:,j} = (s^j v_1, s^j v_2, \ldots, s^j v_{j-1}, 1, 0, \ldots, 0)^T,$$

[21] Similar notation is used for other matrices.

with $s^j := -v_j/(a^{j-1}\widetilde{d}_j)$, then

$$U_j = \widetilde{U}\bar{U}_j = s^j\widetilde{U}(v_1, v_2, \ldots, v_{j-1}, 1/s^j, 0, \ldots, 0)^T = \widetilde{U}_j + s^j K_{j-1},$$

where

$$K_{j-1} := \sum_{i=1}^{j-1} v_i \widetilde{U}_i.$$

The following algorithm uses the formulas above, but the calculations are performed within the same arrays containing the data, to reduce the memory requirements. We have also suppressed the "tilde" on U and D, for convenience.

Algorithm Pardcov : $(x, K) =$ `Pardcov(` $U, D, \widetilde{x}, c, y, \sigma$ `)`

/* Estimate x in (1.126) using the U-D covariance factorization. */

/* 1. Initialize the calculations. */

$f = U^T c; \quad a = \sigma; \quad K = 0;$

/* 2. Update U and D. */

FOR $j = 1{:}\,n$,
  $v = d_j f_j; \quad k_j = v; \quad \widehat{a} = a + v f_j; \quad d_j \leftarrow d_j a/\widehat{a};$
  $w = u_{1:j-1,j} - (f_j/a)k_{1:j-1}; \quad k_{1:j-1} \leftarrow k_{1:j-1} + v u_{1:j-1,j};$
  $u_{1:j-1,j} = w;$

/* 3. Set parameter estimate x. */

$x = \widetilde{x} + K(y - c^T\widetilde{x})/\widehat{a};$

For $j = n$, Algorithm `Pardcov` obtains $\widehat{a} = 1/b = \sigma + c^T\widetilde{X}c$ and the returned K is the Kalman gain vector, $K = \widetilde{U}\widetilde{D}\widetilde{U}^T c = \widetilde{X}c$. The updated values for the input matrices $\widetilde{U}$ and $\widetilde{D}$ are also returned to the input/output array variables U and D, respectively. Only $1.5n^2 + 6.5n$ *flops* and $n^2 + 5n$ memory locations are required. The numerical properties are comparable to those of the square-root algorithms, but no square-roots are needed here.

To process p scalar measurements, Algorithm `Pardcov` should be applied p times.

Setting $S = UD^{1/2}$ in the above relations, the following square-root algorithm can easily be obtained.

Algorithm Pardcovs : $(x, K) =$ `Pardcovs(` $S, \widetilde{x}, c, y, \sigma$ `)`

/* Estimate x in (1.126) using the SS^T covariance factorization. */

/* 1. Initialize the calculations. */

$f = S^T c; \quad a = \sigma; \quad K = 0;$

/* 2. Update S. */

FOR $j = 1{:}\,n$,
  $k_j = f_j s_{jj}$; $\widehat{a} = a + f_j^2$; $d = (a/\widehat{a})^{1/2}$;
  $w = d(s_{1:j-1,j} - (f_j/a)k_{1:j-1})$; $k_{1:j-1} \leftarrow k_{1:j-1} + f_j s_{1:j-1,j}$;
  $s_{1:j-1,j} = w$; $s_{jj} \leftarrow d s_{jj}$;

/* 3. Set parameter estimate x. */

$x = \widetilde{x} + K(y - c^T\widetilde{x})/\widehat{a}$;

Algorithm **Pardcovs** requires $2n^2 + 7n$ *flops*. Similar algorithms have been proposed, for instance, by Clarke (1980), Clarke and Gawthrop (1981), Peterka (1975).

To deal with dynamics and process noise, Algorithm **Pardcov** must be extended by the calculations needed for covariance updating. If $\widehat{x}_k$ and $\widehat{X}_k = \widehat{U}\widehat{D}\widehat{U}^T$ are parameter estimates and their covariance matrix after the last measurement processed at time k, then the updated estimate and covariance at time $k+1$ are given by:

$$\begin{aligned} \widetilde{x}_{k+1} &= A_k\widehat{x}_k + G_k\widetilde{v}_k, \\ \widetilde{X}_{k+1} &= A_k\widehat{X}_kA_k^T + G_kQ_kG_k^T. \end{aligned} \tag{1.128}$$

The indices will, for convenience, be suppressed in the sequel. The numerical problem is now to find the U-D factors $\widetilde{U}$ and $\widetilde{D}$, so that $\widetilde{X} = \widetilde{U}\widetilde{D}\widetilde{U}^T$. The calculation of $\widetilde{X}$ using (1.128) and factoring the result is obviously not advisable. The algorithm below uses a modified Gram-Schmidt triangularization procedure, of which numerical accuracy is comparable to Householder triangularization.

Assume that Q is a diagonal matrix. (If Q is not diagonal, then it is factored as $Q = \bar{U}\bar{D}\bar{U}^T$ and G and Q are replaced by $G\bar{U}$ and $\bar{D}$, respectively, so that $GQG^T = (G\bar{U})\bar{D}(G\bar{U})^T$.) Relation (1.128) can be re-written as

$$\widetilde{X} = \big[\,A\widehat{U}\;\; G\,\big]\mathrm{diag}(\widehat{D}, Q)\big[\,A\widehat{U}\;\; G\,\big]^T =: WDW^T,$$

where W has n rows and $n+p$ columns,

$$\begin{aligned} W &= \big[\,A\widehat{U}\;\; G\,\big] =: \big[\,w_1\;\; w_2\;\; \ldots\;\; w_n\,\big]^T, \\ D &= \mathrm{diag}(\widehat{D}, Q) = \mathrm{diag}(d_1, \ldots, d_{n+p}). \end{aligned}$$

Using a well-known result on weighted Gram-Schmidt orthogonalization (see (Golub and Van Loan, 1983)), it follows that if $\{w_j, j = 1{:}\,n\}$ are linearly independent $(n+p)$-vectors and if $\{d_j, j = 1{:}\,n+p\}$ are positive scalars, then the vectors s_i, $i = n{:}\,-1{:}\,1$, defined by the relations

$$s_n = w_n; \qquad s_j = w_j - \sum_{i=j+1}^{n} \frac{w_j^T D s_i}{s_i^T D s_i} s_i, \quad j = n-1{:}\,-1{:}\,1, \tag{1.129}$$

satisfy $s_j \neq 0$, $j = 1{:}\,n$, and $s_k^T D s_j = 0$, for $k \neq j$.

The algorithm below uses a modification of relations (1.129), that improves the numerical accuracy. It is assumed that $\{w_j, j = 1{:}\,n\}$ are linearly independent, and $D > 0$.

Algorithm xcovGS : $x =$ `xcovGS`$(\ \widehat{U}, \widehat{D}, \widehat{x}, \widetilde{v}, A, G, Q\ )$

/* Estimate $\widetilde{x}_{k+1}$ in (1.128) using Gram-Schmidt orthogonalization. */

/* 1. Construct $W = \begin{bmatrix} w_1 & w_2 & \dots & w_n \end{bmatrix}^T$, $D = \text{diag}(d_1, \dots, d_{n+p})$. */

$W = \begin{bmatrix} A\widehat{U} & G \end{bmatrix}$; $\quad D = \text{diag}(\widehat{D}, Q)$;

/* 2. Update $\widehat{U}$, $\widehat{D}$, and W. */

FOR $j = n{:} -1{:}\, 2$,

$\quad \widehat{d}_j = w_{j,:} D w_{j,:}^T$; $\quad \widehat{u}_{1:j-1,j} = w_{1:j-1,:} D w_{j,:}^T / \widehat{d}_j$;

$\quad w_{1:j-1,:}^T \leftarrow w_{1:j-1,:}^T - \widehat{u}_{1:j-1,j} w_{j,:}^T$;

/* 3. Set $\widehat{d}_1$ and estimate $\widetilde{x}_{k+1}$. */

$\widehat{d}_1 = w_{1,:} D w_{1,:}^T$; $\quad x = A\widehat{x} + G\widetilde{v}$;

On return, the matrices $\widehat{U}$, $\widehat{D}$ contain the updated ("tilde") quantities, and x contains $\widetilde{x}_{k+1}$. The rows of W finally contain the orthogonal s_k vectors mentioned above. Algorithm `xcovGS` requires about $1.5n^3$ *flops*, and $(n+p)(n+1) + n^2$ memory locations. It must be initialized by Algorithm `Pardcov`, for finding the U-D factors $\widehat{U}$ and $\widehat{D}$ from measurements, which are processed one by one.

1.5.4 Defect correction

Our overview of algorithms for solving linear-quadratic optimization problems concludes by briefly presenting a *defect correction* technique, that allows efficient and accurate solutions of algebraic matrix Riccati equation to obtain. This technique is based on a property of equations (1.55) or (1.59), namely that, for any approximate solution $\widehat{X}$, the error $E := X - \widehat{X}$ satisfies a Riccati equation of the same type. Let $Y = Y^T$ be the *residual of the computed solution* $\widehat{X}$, that is

$$Y = A^T\widehat{X} + \widehat{X}A - \widehat{X}BR^{-1}B^T\widehat{X} + Q, \tag{1.130}$$

$$Y = A^T\widehat{X}A - \widehat{X} - A^T\widehat{X}B(R + B^T\widehat{X}B)^{-1}B^T\widehat{X}A + Q. \tag{1.131}$$

By replacing $X = \widehat{X} + E$ in the equations (1.55) or (1.59), respectively, one obtains (Mehrmann and Tan, 1988),

$$A_c^T E + E A_c - E B R^{-1} B^T E + Y = 0, \tag{1.132}$$

$$E = A_c^T E A_c - A_c^T E B(\widehat{R} + B^T E B)^{-1} B^T E A_c + Y, \tag{1.133}$$

where $\widehat{R} := R + B^T \widehat{X} B$, and $A_c := A + B\widehat{H}$ is the state matrix of the closed-loop system, $\widehat{H}$ being computed from the relations (1.54) and (1.58) corresponding to $\widehat{X}$ (instead of X). Therefore, the correction equations (1.132) or (1.133) have the same form as the equations (1.55) or (1.59), respectively. In addition, the stabilizability and detectability properties of the initial system are inherited by the system corresponding to (1.132) and (1.133). Consequently, one can devise a general procedure for solving the algebraic matrix Riccati equations, that, in the first stage, determines by *any* method an approximate solution $\widehat{X}$, and in the second stage iteratively corrects this approximation. Therefore, the *defect correction procedure* is the following.

Procedure RicCorr

/* 1. [Initialization by an approximate solution.] */
Solve (1.55) (or (1.59)), obtaining $\widehat{X}$.
Conv = False.

/* 2. [Iteration for defect correction.] */
WHILE Conv = False, DO
  Compute Y by (1.130) (or (1.131)).
  Solve (1.132) (or (1.133) in E.
  $\widehat{X} \leftarrow \widehat{X} + E$.
  IF $\|Y\|/\|\widehat{X}\| \le c_1\varepsilon_M$ or $\|E\|/\|\widehat{X}\| \le c_2\varepsilon_M$, THEN Conv = True.
END WHILE

This procedure obtains a solution $\widehat{X}$ and an estimate of its error, E. If possibly different algorithms used for solving the equations (1.55) or (1.59) and (1.132) or (1.133), respectively, are enough accurate to get the first significant decimal digits of E correct, then $\{\widehat{X}\}$ tends to X. Theoretically, the procedure converges in one iteration. In practice, several iterations might be necessary. When the iterative process converges, the accuracy of the computed residual Y decreases, so that, eventually, the significant decimal digits of the current E are no longer correct. Even if this limit accuracy has been acquired, usually $\|E\|$ does not overestimate $\|X - \widehat{X}\|$ by a factor greater than 10. Often, one or two iterations of the procedure will do to achieve the limit accuracy. Very good results can thus be obtained, having even "poor" algorithms for solving algebraic matrix Riccati equations.

Two termination criteria are used in Procedure `RicCorr`. The first criterion is based on the relative residual of the current approximate solution $\widehat{X}$, and the second one uses the relative error in $\widehat{X}$. The scalars c_1 and c_2 are usually chosen as linear functions of n, like $10n$, or $100n$, and ε_M is the relative machine precision. An upper bound on the number of iterations in the WHILE loop should be used, to guarantee the termination for ill-conditioned Riccati equations. Numerical results obtained using `RicCorr` are described in (Sima, 1993).

The defect correction procedure for solving algebraic matrix Riccati equations is similar to the iterative refinement technique, used for many years for improving approximate solutions of simultaneous linear algebraic equations (see, for instance, (Golub and Van Loan, 1983)). Consider the linear system $Ax = b$, where $A \in \mathbb{R}^{n \times n}$, and suppose that matrix A has been factored and the factors have been used to compute an initial x. Each iteration of the refinement process computes the *residual*, $r(x) := b - Ax$, and determines a correction z by solving $Az = r(x)$, using the previously computed factors of A. Most implementations calculate $r(x)$ in higher precision than the working precision of the calculations. Recently, Turner and Walker (1992) have considered an abstract improvement algorithm that generalizes the iterative refinement, and efficiently computes high accuracy solutions, making use of relatively few high accuracy computations. Such an algorithm has been used in the cited reference for solving a discretized linear elliptic boundary value problem, and for computing a step of an inexact Newton method, using finite differences, for a discretized nonlinear elliptic boundary value problem.

References

Alag, G., and Kaufman, H. (1977). An implementable digital adaptive flight controller design using stabilized single stage algorithms, *IEEE Trans. Automat. Control*, *AC-22*(*5*): 780–788.

Anderson, B. D. O. (1978). Second-order convergent algorithms for the steady-state Riccati equation, *Int. J. Control*, *28*(*2*): 295–306.

Anderson, B. D. O., and Moore, J. B. (1971). *Linear Optimal Control*, Prentice-Hall, Englewood Cliffs, New Jersey.

Anderson, B. D. O., and Moore, J. B. (1979). *Optimal Filtering*, Prentice-Hall, Englewood Cliffs, New Jersey.

Anderson, E., Bai, Z., Bischof, C., Demmel, J., Dongarra, J., Du Croz, J., Greenbaum, A., Hammarling, S., McKenney, A., Ostrouchov, S., and Sorensen, D. (1992). *LAPACK Users' Guide*, SIAM, Philadelphia.

Åström, K. J. (1970). *Introduction to Stochastic Control Theory*, Academic Press, New York.

Athans, M., and Falb, P. L. (1966). *Optimal Control*, McGraw-Hill, New York.

Barnett, S. (1975). *Introduction to Mathematical Control Theory*, Clarendon Press, Oxford.

Barraud, A. Y. (1974). Sur la résolution numérique des équations de Riccati—cas discret, *R.A.I.R.O. Revue Française d'Automatique, Informatique et de Recherche Opérationnelle*, (*J-3*): 75–95.

Barraud, A. Y. (1979). Investigations autour de la fonction signe d'une matrice. Application à l'équation de Riccati, *R.A.I.R.O. Automatique/Systems Analysis and Control*, *13*(*4*): 335–368.

Başar, T., and Bernhard, P. (1991). *H^∞-Optimal Control and Related Minimax Design Problems: A Dynamic Game Approach*, Systems and Control: Foundations and Applications, Birkhäuser, Boston, MA.

Bellman, R. (1957). *Dynamic Programming*, Princeton University Press, Princeton, New Jersey.

Bellman, R., and Dreyfus, S. (1962). *Applied Dynamic Programming*, Princeton University Press, Princeton, New Jersey.

Bierman, G. J. (1976). Measurement updating using the U-D factorization, *Automatica*, *12*: 375–382.

Bierman, G. J. (1977). *Factorization Methods for Discrete Sequential Estimation*, Academic Press, New York.

Bittanti, S., Laub, A. J., and Willems, J. C., eds. (1991). *The Riccati Equation 1990*, Academic Press, New York.

Brocket, R. A. (1970). *Finite Dimensional Linear Systems*, Wiley, New York.

Bucy, R. S. (1975). Structural stability for the Riccati equation, *SIAM J. Control*, *13*: 749–753.

Bunse-Gerstner, A., Byers, R., and Mehrmann, V. (1989). "Numerical methods for algebraic Riccati equations", Lecture Notes of the Workshop on "The Riccati Equation in Control, Systems, and Signals", Como, Italy (S. Bittanti, ed.), Pitagora Editrice Bologna, pp. 107–115.

Bunse-Gerstner, A., and Mehrmann, V. (1986). A symplectic QR like algorithm for the solution of the real algebraic Riccati equation, *IEEE Trans. Automat. Control*, *AC-31*(*12*): 1104–1113.

Bunse-Gerstner, A., Mehrmann, V., and Watkins, D. (1989). An SR algorithm for Hamiltonian matrices, based on Gaussian elimination, *Methods of Oper. Res.*, *58*: 15–26.

Byers, R. (1987). Solving the algebraic Riccati equation with the matrix sign function, *Lin. Alg. Appl.*, *85*: 267–279.

Caines, P. E., and Mayne, D. Q. (1970). On the discrete time matrix Riccati equation of optimal control, *Int. J. Control*, *12*(*5*): 785–794.

Caines, P. E., and Mayne, D. Q. (1971). On the discrete time matrix Riccati equation of optimal control—A correction, *Int. J. Control*, *14*(*1*): 205–207.

Casti, J. L. (1977). *Dynamical Systems and Their Applications: Linear Theory*, Academic Press, New York, San Francisco, London.

Chiang, R. Y., and Safonov, M. G. (1988). *Robust-Control Toolbox User's Guide*, The MathWorks, Inc., South Natick, MA.

Chung, K. L. (1974). *A Course in Probability Theory*, Academic Press, New York, second edition.

Clarke, D. W. (1980). Some implementation considerations of self-tuning controllers, *Numerical Techniques for Stochastic Systems* (F. Archetti and M. Cugiani, eds.), North-Holland Publishing Co., New York.

Clarke, D. W., and Gawthrop, P. J. (1981). Implementation and application of microprocessors-based self-tuners, *Automatica*, *17*: 233–244.

Clements, D. J., and Anderson, B. D. O. (1977). Linear-quadratic discrete-time control and constant directions, *Automatica*, *13*: 255–264.

Comte, G., Perdon, A. M., and Wyman, B., eds. (1991). *New Trends in Systems Theory*, Birkhäuser, Boston, MA.

Dahlquist, G., and Björck, A. (1974). *Numerical Methods*, Prentice-Hall, Inc., Englewood Cliffs, New Jersey.

Desoer, C. A., and Vidyasagar, M. (1975). *Feedback Systems: Input-Output Properties*, Academic Press, New York.

Dongarra, J. J., Bunch, J. R., Moler, C. B., and Stewart, G. W. (1979). *LINPACK Users' Guide*, SIAM, Philadelphia.

Dongarra, J. J., Du Croz, J., Duff, I. S., and Hammarling, S. (1990). Algorithm 679: A set of Level 3 Basic Linear Algebra Subprograms, *ACM Trans. Math. Softw.*, *16*: 1–17, 18–28.

Dongarra, J. J., Du Croz, J., Hammarling, S., and Hanson, R. J. (1988). Algorithm 656: An extended set of Fortran Basic Linear Algebra Subprograms, *ACM Trans. Math. Softw.*, *14*: 1–17, 18–32.

Dorato, P., and Levis, A. H. (1971). Optimal linear regulators: The discrete-time case, *IEEE Trans. Automat. Control*, *AC–16*(*6*): 613–620.

Doyle, J. C., Glover, K., Khargoneckar, P. P., and Francis, B. A. (1989). State-space solutions to standard H_2 and H_∞ control problems, *IEEE Trans. Automat. Control*, *AC–34*(*4*): 831–847.

Doyle, J. C., and Stein, G. (1981). Multivariable feedback design: Concepts for a classical/modern synthesis, *IEEE Trans. Automat. Control*, *AC–26*(*1*): 4–16.

Forsythe, G. E., Malcolm, M. A., and Moler, C. B. (1977). *Computer Methods for Mathematical Computations*, Prentice-Hall, Englewood Cliffs, New Jersey.

Foulard, C., Gentil, S., and Sandraz, J. P. (1977). *Commande et Régulation par Calculateur Numérique: De la Théorie aux Applications*, Eyrolles, Paris.

Francis, B. A. (1987). *A Course in H_∞ Control Theory*, volume 88 of *Lect. Notes in Control and Information Sciences*, Springer-Verlag, New York.

Francis, B. A., and Doyle, J. C. (1987). Linear control theory with an H_∞ optimality criterion, *SIAM J. Control & Optimiz.*, *25*: 815–844.

Garbow, B. S., Boyle, J. M., Dongarra, J. J., and Moler, C. B. (1977). *Matrix Eigensystem Routines—EISPACK Guide Extension*, volume 51 of *Lect. Notes in Computer Science*, Springer-Verlag, Berlin.

Gill, P. E., and Murray, W. (1977). Modification of matrix factorizations after a rank-one change, *The State of the Art in Numerical Analysis* (D. Jacobs, ed.), Academic Press, London, pp. 55–83.

Gill, P. E., Murray, W., and Wright, M. H. (1981). *Practical Optimization*, Academic Press, New York.

Golub, G. H., and Van Loan, C. F. (1983). *Matrix Computations*, The Johns Hopkins University Press, Baltimore, Maryland.

Goodwin, G. C., and Payne, R. L. (1977). *Dynamic System Identification: Experiment Design and Data Analysis*, volume 136 of *Math. in Sci. & Engineering*, Academic Press, New York.

Hewer, G. A. (1971). An iterative technique for the computation of the steady state gains for the discrete optimal regulator, *IEEE Trans. Automat. Control, AC–16(4)*: 382–384.

Kailath, T. (1980). *Linear Systems*, Prentice-Hall, Englewood Cliffs, New Jersey.

Kalman, R. E., Falb, P. L., and Arbib, M. A. (1969). *Topics in Mathematical System Theory*, McGraw-Hill, New York.

Kleinman, D. L. (1968). On an iterative technique for Riccati equation computations, *IEEE Trans. Automat. Control, AC–13*: 114–115.

Kučera, V. (1972). A contribution to matrix quadratic equations, *IEEE Trans. Automat. Control, AC–17(3)*: 344–347.

Kwakernaak, H. (1993). Robust control and H_∞-optimization—Tutorial paper, *Automatica, 29(2)*: 255–273.

Kwakernaak, H., and Sivan, R. (1972). *Linear Optimal Control Systems*, Wiley-Interscience, New York.

Lancaster, P., Ran, A. C. M., and Rodman, L. (1986). Hermitian solutions of the discrete algebraic Riccati equation, *Int. J. Control, 44*: 777–802.

Lancaster, P., Ran, A. C. M., and Rodman, L. (1987). An existence and monotonicity theorem for the discrete algebraic matrix Riccati equation, *Lin. and Multil. Alg., 20*: 353–361.

Lancaster, P., and Rodman, L. (1980). Existence and uniqueness theorems for the algebraic Riccati equation, *Int. J. Control, 32*: 285–309.

Laub, A. J. (1979). A Schur method for solving algebraic Riccati equations, *IEEE Trans. Automat. Control, AC–24(6)*: 913–921.

Lawson, C. L., Hanson, R. J., Kincaid, D. R., and Krogh, F. T. (1979). Basic Linear Algebra Subprograms for Fortran usage, *ACM Trans. Math. Softw., 5(3)*: 308–323.

Lee, W.-H., and Athans, M. (1979). The discrete-time compensated Kalman filter, *Int. J. Control, 29(2)*: 293–311.

Ljung, L. (1987). *System Identification: Theory for the User*, Prentice-Hall, Englewood Cliffs, New Jersey.

Ljung, L., and Söderström, T. (1983). *Theory and Practice of Recursive Identification*, MIT, Cambridge.

Luenberger, D. G. (1969). *Optimization by Vector Space Methods*, John Wiley and Sons, Inc., New York.

Mahalanabis, A. K. (1972). On minimizing the divergence in discrete filters, *IEEE Trans. Automat. Control*, *AC–17*(*2*): 239–240.

Martensson, K. (1971). On the matrix Riccati equation, *Info. Sci.*, *3*: 17–49.

Mehrmann, V. (1991). *The Autonomous Linear Quadratic Control Problem. Theory and Numerical Solution*, volume 163 of *Lect. Notes in Control and Information Sciences* (M. Thoma and A. Wyner, eds.), Springer-Verlag, Berlin.

Mehrmann, V., and Tan, E. (1988). Defect correction methods for the solution of algebraic Riccati equations, *IEEE Trans. Automat. Control*, *AC–33*(*7*): 695–698.

Mendel, J. M., and Gieseking, D. L. (1971). Bibliography on the Linear-Quadratic-Gaussian problem, *IEEE Trans. Automat. Control*, *AC–16*(*6*): 847–869.

Moler, C. B. (1982). "Demonstration of a matrix laboratory", Numerical Analysis, Proc. Cocoyoc, Mexico, 1981 (J. P. Hennart, ed.), volume 909 of *Lect. Notes in Math.*, Springer-Verlag, Berlin.

Moler, C. B., Little, J. N., Bangert, S. N., and Kleiman, S. (1985). *PC-MATLAB User's Guide*, The MathWorks, Inc., Natick, MA.

Moler, C. B., and Stewart, G. W. (1973). An algorithm for generalized matrix eigenvalue problems, *SIAM J. Numer. Anal.*, *10*(*2*): 241–256.

Molinari, B. P. (1973a). The stabilizing solution of the algebraic Riccati equation, *SIAM J. Control*, *11*(*2*): 262–271.

Molinari, B. P. (1973b). Equivalence relations for the algebraic Riccati equation, *SIAM J. Control*, *11*(*2*): 272–285.

Molinari, B. P. (1977). The time-invariant linear-quadratic optimal control problem, *Automatica*, *13*: 347–357.

Morf, M., Dobbins, J. R., Friedlander, B., and Kailath, T. (1979a). Square-root algorithms for parallel processing in optimal estimation, *Automatica*, *15*: 299–306.

Morf, M., Dobbins, J. R., Friedlander, B., and Kailath, T. (1979b). Reply to the discussion of "Square-root algorithms for parallel processing in optimal estimation", *Automatica, 15*: 499.

Murciano, A. (1990). BASILE — La commande robuste. Version 0.0, Technical report, INRIA.

Nashed, Z. M., ed. (1976). *Generalized Inverses and Applications*, Academic Press, New York.

Pappas, T., Laub, A. J., and Sandell, N. R. (1980). On the numerical solution of the discrete-time algebraic Riccati equation, *IEEE Trans. Automat. Control, AC-25(4)*: 631–641.

Payne, H. J., and Silverman, L. M. (1973). On the discrete time algebraic Riccati equation, *IEEE Trans. Automat. Control, AC-18*: 226–234.

Peterka, V. (1975). A square-root filter for a real time multivariable regression, *Kybernetika, 11*: 53.

Petkov, P. H., Christov, N. D., and Konstantinov, M. M. (1991). *Computational Methods for Linear Control Systems*, Prentice-Hall International Ltd., U. K.

Pontryagin, L. S., Boltyanskii, V., Gamkrelidze, R., and Mishenko, E. (1962). *The Mathematical Theory of Optimal Processes*, Interscience, New York.

Potter, J. E. (1966). Matrix quadratic solutions, *SIAM J. Appl. Math., 14*: 496–500.

Press, W. H., Teukolsky, S. A., Wetterling, W. T., and Flannery, B. P. (1992). *Numerical Recipes. The Art of Scientific Computing*, Cambridge University Press, New York, second edition.

Rubio, J. E. (1971). *The Theory of Linear Systems*, Academic Press, New York.

Safonov, M. G., Laub, A. J., and Hartmann, G. L. (1981). Feedback properties of multivariable systems: The role and use of the return difference matrix, *IEEE Trans. Automat. Control, AC-26(1)*: 47–65.

Sage, A. P., and White, C. C., III. (1977). *Optimum Systems Control*, Prentice-Hall, Inc., Englewood Cliffs, New Jersey, second edition.

Sandraz, J. P. (1973). *Identification et Commande Multidimensionnelle d'une Unité Pilote de Fabrication de Papier*, Thèse de docteur ingenieur, Université Scientifique et Medicale, Institut Polytechnique de Grenoble, July 1973.

Shinners, S. M. (1992). *Modern Control System Theory and Design*, John Wiley and Sons, Chichester, UK.

Sidhu, G. S., and Bierman, G. J. (1977). Integration-free interval doubling for Riccati equation solutions, *IEEE Trans. Automat. Control*, *AC-22*(*5*): 831–834.

Sima, V. (1986). "Factorization techniques in discrete time adaptive control", A Bridge between Control Science and Technology (J. Gertler and L. Keviczky, eds.), volume 2, Pergamon Press, Oxford.

Sima, V. (1993). "Iterative improvement of Riccati equations solutions", Preprints of the 9-th International Conference on "Control Systems and Computer Science (CSCS 9)", May 25–28, 1993, volume 1, Politehnica University of Bucharest, Faculty of Control and Computers, Bucharest, pp. 6–14.

Smith, B. T., Boyle, J. M., Dongarra, J. J., Garbow, B. S., Ikebe, Y., Klema, V. C., and Moler, C. B. (1976). *Matrix Eigensystem Routines—EISPACK Guide*, volume 6 of *Lect. Notes in Computer Science*, Springer-Verlag, New York, second edition.

Söderström, T., and Stoica, P. (1989). *System Identification*, Prentice-Hall, London.

Stein, G., and Athans, M. (1987). The LQG/LTR procedure for multivariable feedback control design, *IEEE Trans. Automat. Control*, *AC-32*: 105–114.

Stewart, G. W. (1973). *Introduction to Matrix Computations*, Academic Press, New York.

Thornton, C. L., and Bierman, G. J. (1978). Filtering and error analysis via the UDU^T covariance factorization, *IEEE Trans. Automat. Control*, *AC-23*: 901–907.

Turner, K., and Walker, H. F. (1992). Efficient high accuracy solutions with GMRES(m), *SIAM J. Sci. Stat. Comput.*, *13*: 815–825.

Van Dooren, P. (1981). A generalized eigenvalue approach for solving Riccati equations, *SIAM J. Sci. Stat. Comput.*, *2*: 121–135.

Van Overschee, P., and De Moor, B. (1994). N4SID: Two subspace algorithms for the identification of combined deterministic-stochastic systems, *Automatica, 30(1)*.

Vaughan, D. R. (1969). A negative exponential solution for the matrix Riccati equation, *IEEE Trans. Automat. Control, AC-14*: 72–75.

Vaughan, D. R. (1970). A nonrecursive algebraic solution for the discrete Riccati equation, *IEEE Trans. Automat. Control, AC-15*(*5*): 597–599.

Verhaegen, M. (1993a). Subspace model identification. Part 3: Analysis of the ordinary output-error state-space model identification algorithm, *Int. J. Control, 58*(*3*): 555–586.

Verhaegen, M. (1993b). Application of a subspace model identification technique to identify LTI systems operating in closed-loop, *Automatica, 29*(*4*): 1027–1040.

Verhaegen, M. (1994). Identification of the deterministic part of MIMO state space models given in innovations form from input-output data, *Automatica, 30*(*1*): 61–74.

Verhaegen, M., and Dewilde, P. (1992a). Subspace model identification. Part 1: The output-error state-space model identification class of algorithms, *Int. J. Control, 56*(*5*): 1187–1210.

Verhaegen, M., and Dewilde, P. (1992b). Subspace model identification. Part 2: Analysis of the elementary output-error state-space model identification algorithm, *Int. J. Control, 56*(*5*): 1211–1241.

Wilkinson, J. H. (1963). *Rounding Errors in Algebraic Processes*, Prentice-Hall, Englewood Cliffs, New Jersey.

Wilkinson, J. H. (1965). *The Algebraic Eigenvalue Problem*, Oxford University Press (Clarendon), London.

Wilkinson, J. H., and Reinsch, C., eds. (1971). *Handbook for Automatic Computation. Vol. II: Linear Algebra*, Springer-Verlag, Berlin.

Willems, J. C. (1971). Least squares stationary optimal control and the algebraic Riccati equation, *IEEE Trans. Automat. Control, AC-16*: 621–634.

Willems, J. C. (1974). On the existence of a nonpositive solution to the Riccati equation, *IEEE Trans. Automat. Control, AC-19*: 592–593.

Wonham, W. M. (1968). On a matrix equation of stochastic control, *SIAM J. Control & Optimiz.*, *6*: 681–697.

Wonham, W. M. (1979). *Linear Multivariable Control: A Geometric Approach*, Springer-Verlag, New York, second edition.

2

Newton Algorithms

This chapter deals with *Newton-type algorithms* for solving algebraic matrix Riccati equations for both continuous-time and discrete-time systems. Properly initialized (by stabilizing matrices), Newton algorithms are sometimes to be preferred to other algorithms, due to their fast, quadratic convergence rate, and good numerical properties (stability, reliability, accuracy). The main computational task of each iteration is to solve a linear matrix equation of Lyapunov type. A section of this chapter discusses several algorithms for solving Lyapunov equations; both general *Bartels-Stewart algorithms*, and particular *Hammarling's algorithms*, for stable non-negative definite Lyapunov equations, are included. Another section presents numerically sound *stabilization algorithms* for linear dynamical systems, needed for initializing Newton algorithms. All these algorithms, including Newton solvers, are based on the *real Schur form* (RSF) of a matrix. The computation of RSF and orthogonal bases for invariant subspaces, needed in the present and subsequent chapters, is addressed in an ample section. After all prerequisites have been described, algorithmic templates for Newton solvers are given for continuous-time and discrete-time systems. They use previously defined basic algorithms. Techniques for efficiency improvement, and computational issues are also presented.

2.1 Basic Theory

This chapter investigates Newton's algorithms for the numerical solution of continuous-time and discrete-time algebraic matrix Riccati equations, defined by (1.55) and (1.59) in § 1.1.2, re-written here for convenience

$$A^T X + XA - XBR^{-1}B^T X + Q = 0, \tag{2.1}$$

$$X = A^T XA - A^T XB(R + B^T XB)^{-1} B^T XA + Q, \tag{2.2}$$

where $A \in \mathbb{R}^{n\times n}$, $B \in \mathbb{R}^{n\times m}$, $Q \in \mathbb{R}^{n\times n}$, and $R \in \mathbb{R}^{m\times m}$, are given matrices. Under certain conditions (see, for instance, Theorems 1.16, 1.17, and 1.18 in § 1.1.2), there is a unique symmetric positive semi-definite solution, $X \in \mathbb{R}^{n\times n}$, for each of these Riccati equations. Newton's algorithms compute this solution and also give the optimal gain matrix, defined by

$$\begin{aligned} H &:= -R^{-1}B^T X, &(2.3)\\ H &:= -(R + B^T XB)^{-1} B^T XA, &(2.4)\end{aligned}$$

respectively. The basic theoretical results are stated in the next two subsections.

2.1.1 Newton's method

The essence of Newton's method consists in obtaining the solution of the Riccati equation (2.1) or (2.2), respectively, as a limit of a sequence of solutions of some *Lyapunov equations*. The kth iteration, $k = 1, 2, \ldots$, of the iterative Newton's process covers the following stages:

(i) Compute $A_k := A + BH_k$.

(ii) Solve the Lyapunov equation in (2.5) or (2.6), respectively,

$$\begin{aligned} A_k^T X_k + X_k A_k + Q_k &= 0, &(2.5)\\ X_k &= A_k^T X_k A_k + Q_k, &(2.6)\end{aligned}$$

where $Q_k := H_k^T RH_k + Q$.

(iii) Compute H_{k+1} using (2.7) or (2.8), respectively,

$$\begin{aligned} H_{k+1} &= -R^{-1}B^T X_k, &(2.7)\\ H_{k+1} &= -(R + B^T X_k B)^{-1} B^T X_k A. &(2.8)\end{aligned}$$

The iterative process must be initialized by a stabilizing matrix H_1, i.e. A_1 is stable, hence $\lambda(A_1) \subset \mathbb{C}^-$. Such a matrix can be computed using the algorithms presented at Section 2.4. With such an initialization, the corresponding Lyapunov equation at stage (ii) has a unique solution $X_1 \geq 0$. The matrix H_2, computed at stage (iii), is also a stabilizing matrix and so it is possible to begin the next iteration. All generated matrices A_k are stable, and the matrices X_k are symmetric positive semi-definite.

Theorem 2.1 (Kleinman, 1968; Hewer, 1971) *If $Q > 0$ and H_1 is a stabilizing matrix, then the sequence $\{X_k\}$ of solutions of Lyapunov equations (2.5) or (2.6), respectively, has the following properties:*

(i) $0 \leq X \leq \cdots \leq X_{k+1} \leq X_k$, $k = 1, 2, \ldots$, *and* $\lim_{k\to\infty} X_k = X$;

(ii) the convergence rate of the process is quadratic, that is

$$||X - X_{k+1}|| \leq c||X - X_k||^2, \qquad k = 1, 2, \ldots,$$

where c is a constant independent of X.

We shall indicate the main steps of the proof, by specializing the proof in (Mehrmann, 1991), attributed to Elsner. The continuous-time case only is discussed, since similar arguments apply to the discrete-time case as well. Using (2.7), and defining $F := BR^{-1}B^T$, equation (2.5) gets equivalent to

$$(A - FX_{k-1})^T X_k + X_k(A - FX_{k-1}) = -Q - X_{k-1}FX_{k-1}, \tag{2.9}$$

which can be re-written as

$$\begin{aligned}(A - FX_k)^T X_k + X_k(A - FX_k) = -Q - X_kFX_k \\ - (X_k - X_{k-1})F(X_k - X_{k-1}),\end{aligned} \tag{2.10}$$

or, combining (2.9) and (2.10), it follows

$$\begin{aligned}(A - FX_k)^T(X_k - X_{k+1}) + (X_k - X_{k+1})(A - FX_k) = \\ - (X_{k-1} - X_k)F(X_{k-1} - X_k), \qquad k = 2, 3, \ldots.\end{aligned} \tag{2.11}$$

By Lyapunov's theorem (Theorem 1.6, § 1.1.1), we inductively obtain that $X_k \geq 0$ (from (2.9)), $A - FX_k$ is stable (from (2.10)), and $X_k - X_{k+1} \geq 0$ (from (2.11)). Thus, $\{X_k\}$ is a non-increasing sequence, bounded from below, hence it is convergent, and $X := \lim_{k\to\infty} X_k$ is positive semi-definite. From (2.10), we have

$$(A - FX)^T X + X(A - FX) = -Q - XFX, \tag{2.12}$$

so that $X \geq 0$ solves (2.1). Using (2.12), we obtain

$$\begin{aligned}(A - FX_k)^T X + X(A - FX_k) = -Q - XFX + \\ (X - X_k)FX + XF(X - X_k),\end{aligned} \tag{2.13}$$

and, subtracting (2.10) from (2.13), after simple manipulations, we get

$$\begin{aligned}(A - FX)^T(X - X_k) + (X - X_k)(A - FX) = \\ (X - X_k)F(X - X_k) + (X_k - X_{k-1})F(X_k - X_{k-1}).\end{aligned} \tag{2.14}$$

Denoting $\widetilde{F}_k := (X_k - X_{k-1})F(X_k - X_{k-1})$, it follows that

$$\begin{aligned} 0 &\leq X_k - X \leq \int_0^\infty e^{t(A-FX)^T} \widetilde{F}_k e^{t(A-FX)^T}\, dt \\ &\leq \|X_k - X_{k-1}\|^2 \int_0^\infty e^{t(A-FX)} F e^{t(A-FX)}\, dt \\ &=: \|X_k - X_{k-1}\|^2 Y. \end{aligned} \tag{2.15}$$

This implies

$$\|X_k - X\| \leq \|Y\|\,\|X_k - X_{k-1}\|^2 =: c\|X_k - X_{k-1}\|^2 \leq c\|X - X_{k-1}\|^2, \tag{2.16}$$

because $X_{k-1} - X \geq X_{k-1} - X_k \geq 0$.

Remark 2.1 The proof in (Mehrmann, 1991) applies to more general algebraic Riccati equations for complex descriptor systems defined by (1.71) in § 1.2.1

$$E\delta(x(t)) = Ax(t) + Bu(t),$$

where E is $n \times n$, and δ operator is defined in § 1.1.1 (see (1.37)); the cost functional includes coupling terms with weighting matrices S and S^H (see Problem 1.1 in § 1.2.1). The corresponding algebraic Riccati equations are

$$\begin{aligned}&A^H XE + E^H XA \\ &-(B^H XE + S^H)^H R^{-1}(B^H XE + S^H) + Q = 0,\end{aligned} \tag{2.17}$$

$$\begin{aligned}&E^H XE = A^H XA \\ &-(B^H XA + S^H)^H (R + B^H XB)^{-1}(B^H XA + S^H) + Q.\end{aligned} \tag{2.18}$$

In this case, Newton iteration uses A_k as before, with

$$H_{k+1} = -R^{-1}(B^H X_k E + S^H), \tag{2.19}$$

$$H_{k+1} = -(R + B^H X_k B)^{-1}(B^H X_k A + S^H), \tag{2.20}$$

and X_k is the solution of one of the following Lyapunov equations

$$A_k^H X_k E + E^H X_k A_k + Q_k = 0, \tag{2.21}$$

$$E^H X_k E = A_k^H X_k A_k + Q_k, \tag{2.22}$$

where $Q_k := H_k^H R H_k + Q - S H_k - H_k^H S^H$. If the output y is weighted in the cost functional, instead of x, then Q and S are replaced by $C^H Q C$ and $C^H S$, respectively (see Problem 1.3 in § 1.2.1). The result proven in (Mehrmann, 1991) assumes that E is nonsingular, so that the problem is transformed into a standard one. □

Note that Theorem 2.1 assumes that matrix Q is positive definite. In practice, Newton's method could still converge even if Q is positive semidefinite. The other requirement in the theorem statement, namely, that H_1 is a stabilizing matrix, is an essential one. Such a matrix exists if the pair (A, B) is stabilizable (Wonham, 1979). A necessary and sufficient condition for stabilizability is stated in Theorem 1.11 (§ 1.1.1). In the following subsection we shall present several results in the construction of a stabilizing matrix; specific algorithms will be given in Section 2.4.

2.1.2 Stabilization methods

The first two results refer to continuous-time systems.

Theorem 2.2 (Kleinman, 1970) *Let the continuous-time system defined by the pair (A, B) be stabilizable. Then*

$$H = -B^T W(\Delta)^{\dagger}, \tag{2.23}$$

is a stabilizing matrix, where

$$W(\Delta) := \int_0^{\Delta} e^{-At} B B^T e^{-A^T t}\, dt, \qquad \Delta > 0. \tag{2.24}$$

If the pair (A, B) is controllable, then the pseudoinverse $W(\Delta)^{\dagger}$ can be replaced by the usual inverse $W(\Delta)^{-1}$.

Theorem 2.3 (Armstrong, 1975) *Let the continuous-time system defined by the pair (A, B) be stabilizable. Then*

$$H = -B^T Z^{\dagger}, \tag{2.25}$$

is a stabilizing matrix, where $Z = Z^T \geq 0$ satisfies

$$(A + \beta I_n) Z + Z (A + \beta I_n)^T = 2 B B^T, \tag{2.26}$$

with $\beta > \|A\|$. If the pair (A, B) is controllable, then $Z > 0$, and $Z^{\dagger}$ can be replaced by Z^{-1}.

Instead of proving these results, we shall illustrate the connection between them, as established in (Varga, 1981a). The matrix $W(\Delta)$ in (2.24) can be evaluated as (see (Van Loan, 1978))

$$W(\Delta) = E_2(\Delta)^T G_1(\Delta), \tag{2.27}$$

where $E_2(\Delta)$ and $G_1(\Delta)$ are blocks of the matrix exponential

$$\exp\left(\begin{bmatrix} A & BB^T \\ 0 & -A^T \end{bmatrix}\Delta\right) = \exp\begin{bmatrix} E_1(\Delta) & G_1(\Delta) \\ 0 & E_2(\Delta) \end{bmatrix}, \tag{2.28}$$

with

$$\begin{aligned} E_1(\Delta) &= e^{A\Delta}, \quad E_2(\Delta) = e^{-A^T\Delta}, \\ G_1(\Delta) &= \int_0^\Delta e^{A(\Delta-t)}BB^T e^{-A^T t}\,dt. \end{aligned} \tag{2.29}$$

We replace the given matrix pair (A, B) by a modified one, $(A+\beta I_n, B)$, where $\beta > \|A\|$. Such a choice of β guarantees that all eigenvalues of $A+\beta I_n$ are unstable, i.e. $\Re(\lambda_i) \in \mathbb{C}^+$. We can simplify the computation of the matrix exponential in (2.28), by transforming the matrix

$$F := \begin{bmatrix} A+\beta I_n & BB^T \\ 0 & -(A+\beta I_n)^T \end{bmatrix}$$

into a block-diagonal form. Given the chosen β, the following Lyapunov equation

$$(A+\beta I_n)Z + Z(A+\beta I_n)^T = BB^T, \tag{2.30}$$

has a unique solution Z (see Theorem 1.6 in § 1.1.1 and the preamble of Section 2.3). Defining

$$V = \begin{bmatrix} I_n & Z \\ 0 & I_n \end{bmatrix}, \quad \text{hence} \quad V^{-1} = \begin{bmatrix} I_n & -Z \\ 0 & I_n \end{bmatrix},$$

we obtain,

$$VFV^{-1} = \begin{bmatrix} A+\beta I_n & 0 \\ 0 & -(A+\beta I_n)^T \end{bmatrix}.$$

Using a property of the matrix exponential, $\exp(VFV^{-1}) = V\exp(F)V^{-1}$, it follows

$$\exp(F\Delta) = V^{-1}\exp\left(\begin{bmatrix} A+\beta I_n & 0 \\ 0 & -(A+\beta I_n)^T \end{bmatrix}\Delta\right)V, \tag{2.31}$$

hence, specializing (2.29) and (2.27), we have

$$
\begin{aligned}
E_1(\Delta) &= e^{(A+\beta I_n)\Delta}, \qquad E_2(\Delta) = e^{-(A+\beta I_n)^T\Delta}, \\
G_1(\Delta) &= E_1(\Delta)Z - ZE_2(\Delta), \\
W(\Delta) &= E_2(\Delta)^T G_1(\Delta) = Z - E_2(\Delta)^T Z E_2(\Delta) \\
&= Z - e^{-(A+\beta I_n)\Delta} Z e^{-(A+\beta I_n)^T\Delta}.
\end{aligned}
\tag{2.32}
$$

As the matrix $-(A+\beta I_n)$ is stable, we can take the limit in (2.32) for $\Delta \to \infty$, thus obtaining $W(\infty) = Z$, while with (2.23), the stabilizing matrix becomes

$$H = -B^T Z^\dagger. \tag{2.33}$$

This is the same expression as in (2.25), but derived from Kleinman's result on. The only difference is that the right-hand side of the Lyapunov equation that gives Z is BB^T in (2.30), and $2BB^T$ in (2.26). But the proof in (Armstrong, 1975) is still valid if the term $2BB^T$ is replaced by γBB^T, where $0 < \gamma \leq 2$. The value $\gamma = 2$ determines a minimum norm of H.

Now, the main results for discrete-time systems are stated.

Theorem 2.4 (Kleinman, 1974) *Let the discrete-time system defined by the pair (A, B) be stabilizable, with A nonsingular. Then*

$$H = -B^T[(A^T)^N W_N^\dagger A^N]A = -(I_m + B^T V_N B)^{-1} B^T V_N A, \tag{2.34}$$

is a stabilizing matrix, where $N \geq n$, and W_N, V_N are defined by

$$W_N := \sum_{i=0}^{N} A^i BB^T (A^T)^i, \qquad V_N := (A^T)^N W_{N-1}^\dagger A^N. \tag{2.35}$$

If the pair (A, B) is controllable, then the pseudoinverse $W_N^\dagger$ can be replaced by the usual inverse W_N^{-1}.

Theorem 2.5 (Armstrong and Rublein, 1976) *Let the discrete-time system defined by the pair (A, B) be stabilizable, and assume that*

$$A = \begin{bmatrix} A_{11} & A_{12} \\ 0 & A_{22} \end{bmatrix}, \qquad B = \begin{bmatrix} B_1 \\ 0 \end{bmatrix}, \tag{2.36}$$

with A_{11} nonsingular and A_{22} (asymptotically) stable. Then

$$H = -B^T(Z + BB^T)^\dagger A, \tag{2.37}$$

is a stabilizing matrix, where $Z = Z^T \geq 0$ satisfies

$$AZA^T = \beta^2 Z + 2BB^T, \tag{2.38}$$

with $0 < \beta < \min\{1, \min\{|\lambda_i| : \lambda_i \in \lambda(A), \lambda_i \neq 0\}\}$. *If* (A, B) *is controllable, and* A *is nonsingular, then* $Z > 0$, *and* $Z^{\dagger}$ *can be replaced by* Z^{-1}.

Remark 2.2 The representation of (A, B) in (2.36), with A_{22} asymptotically stable, can be obtained from a general pair using a nonsingular transformation T (see (1.47) and Theorem 1.11 in § 1.1.1). □

The connection between the results stated in Theorems 2.4 and 2.5 is established in (Varga, 1981a) in a manner similar to that for continuous-time systems.

A somewhat more general result than in Theorem 2.4 is proven in (Kleinman, 1974); namely, the stabilizing matrix is

$$H = -RB^T[(A^T)^N W_N^{\dagger} A^N]A = -(R + B^T V_N B)^{-1} B^T V_N A,$$

where R is any positive definite matrix, and the factor BB^T in the definition of W_N in (2.35) is replaced by $BR^{-1}B^T$. Moreover, the results in Theorem 2.4 are still true for $N \geq n - \text{rank}(B) + 1$. There are many other results concerning the stabilization of linear systems; see, for instance, (Kwon and Pearson, 1975). As we are mainly interested in reliable and efficient algorithms for solving Riccati equations and related problems, we shall not discuss other theoretical results here. Good algorithms for stabilization will be described in Section 2.4. It is worth mentioning that some of the theoretical requirements (like nonsingularity of A in discrete-time case[1]) are not explicitly made on an implementation. On the other hand, some theoretical results, although constructive, are useless from a numerical point of view.

Another approach to obtaining an initial stabilizing matrix is the *pole assignment* approach, described, for instance, in (Byers and Nash, 1989; Kautsky, Nichols and Van Dooren, 1985; Miminis and Paige, 1982, 1988; Petkov, Christov and Konstantinov, 1984; Petkov, Konstantinov and Christov, 1985; Varga, 1981b). This approach determines a matrix H so that the closed-loop matrix $A + BH$ has desired eigenvalues. Obviously, eigenvalues of the controllable part of the system can only be assigned. Pole assignment is actually an *inverse eigenvalue problem*, because the eigenvalues are known (see, for instance, (Miminis and Paige, 1982), for the single-input case). Byers and Nash (1989) minimize the condition number of the eigenvector matrix—a nonlinear optimization problem—and propose reliable derivative evaluations and carefully selected algorithmic parameters for fast convergence. As pole assignment accomplishes more than actually

[1] Results have been reached (Kwon and Person, 1975) without the nonsingularity assumption, but they are more complicated.

needed for solving linear-quadratic optimization problems, and as it can deliver stabilizing matrices with quite large entries, we will not pursue this track anymore.

The discussion in this section points out to the main prerequisites in developing powerful algorithms based on Newton's method. These prerequisites will be elaborated in the next three sections. The results and algorithms to be presented will often be used in the remaining part of the book. Beside standard material, included for self-containedness and reference purposes, some very recent algorithmic developments, incorporated in the LAPACK package (Anderson et al., 1992), will also be addressed.

2.2 Computation of Real Schur Form and Invariant Subspaces

The main computational tool described in this section is the QR algorithm. When applied to a matrix A reduced to the upper Hessenberg form, this algorithm computes, remarkably efficient and reliable, the real Schur form (RSF) of A. This form reveals the eigenvalues of A, and the transformation matrix involved in the calculations gives bases for certain invariant subspaces of special interest in many applications, including the solution of linear and quadratic matrix equations.

2.2.1 Basic definitions and properties

The natural framework within which to deal with algebraic eigenproblems is the linear space $\mathbb{C}^n$, since real matrices can also have complex eigenvalues. So, in this section we shall generally consider complex matrices. Let us first summarize some numerical linear algebra definitions and results, which do matter in developing efficient algorithms for solving linear-quadratic optimization problems. Details and full proofs can be found, for instance, in (Dahlquist and Björk, 1974; Golub and Van Loan, 1983; Nash, 1979; Stewart, 1973; Stoer and Bulirsch, 1980; Wilkinson, 1965).

Let $A \in \mathbb{C}^{n \times n}$. A vector $x \in \mathbb{C}^n$, $x \neq 0$, is a *(right) eigenvector* of A, corresponding to the *eigenvalue* $\lambda \in \mathbb{C}$, if $Ax = \lambda x$. The collection of eigenvalues of A is denoted by $\lambda(A)$, and is called the *spectrum* of A.

If x is an eigenvector of A corresponding to λ, then αx is also an eigenvector, for any $\alpha \neq 0$, $\alpha \in \mathbb{C}$. More generally, if $x_1, x_2, \ldots, x_k$, are eigenvectors of A corresponding to λ, then any non-zero linear combination of these eigenvectors is an eigenvector of A corresponding to λ. As λ is an eigenvalue of A if and only if $A - \lambda I_n$ is singular, it follows that the eigenvalues of A satisfy the *characteristic equation* of A, $\det(\lambda I_n - A) = 0$, i.e. they are the roots of the *characteristic polynomial* of A, $\det(\lambda I_n - A)$. Therefore, $A \in \mathbb{C}^{n\times n}$ has exactly n complex eigenvalues, some of which may be repeated. If an eigenvalue λ is repeated k times, λ is said to be a *multiple eigenvalue with (algebraic) multiplicity* k; if $k = 1$, then λ is a *simple eigenvalue.* As the characteristic polynomial of a real matrix $A \in \mathbb{R}^{n\times n}$ has real coefficients, it follows that complex eigenvalues of A occur in complex conjugate pairs. It is easy to see that $\lambda(A^T) = \lambda(A)$, and $\lambda(A^H) = \{\bar{\lambda} : \lambda \in \lambda(A)\}$. Clearly, if $\lambda \in \lambda(A)$, then there is a vector $y \neq 0$ so that $A^H y = \bar{\lambda} y$; since $y^H A = \lambda y^H$, y is called a *left eigenvector* of A corresponding to λ.

If $\lambda \in \lambda(A)$ with eigenvector x, then, by definition, $\lambda = x^H Ax/x^H x$, the right-hand side being the *Rayleigh quotient.* It follows that the eigenvalues of a Hermitian matrix are real, and those of a positive (semi)-definite matrix are positive (non-negative, respectively).

If λ is an eigenvalue of $A \in \mathbb{C}^{n\times n}$, with eigenvector x, and $\alpha, \mu \in \mathbb{C}$, then the matrices αA, $A - \mu I_n$, and A^{-1} (if A is nonsingular) have an eigenvalue $\alpha\lambda$, $\lambda - \mu$, and $1/\lambda$, respectively, with eigenvector x. Other useful properties of $\lambda_i \in \lambda(A)$, $i = 1\colon n$, are:

$$\mathrm{tr}(A) := \sum_{i=1}^{n} a_{ii} = \sum_{i=1}^{n} \lambda_i; \qquad \det(A) = \prod_{i=1}^{n} \lambda_i.$$

The last property shows that A is singular if and only if there is at least one index i so that $\lambda_i = 0$. Moreover, if $A, B^H \in \mathbb{C}^{m\times n}$, $m \geq n$, then $\lambda(AB) = \lambda(BA) \cup \{0, 0, \ldots, 0\}$, where the last set has $m - n$ elements. If U is a unitary matrix, and $\lambda \in \lambda(U)$, then $|\lambda| = 1$, because $\|Ux\|_2 = \|x\|_2$.

By definition, an eigenvector of A defines an one-dimensional subspace, which is invariant on a premultiplication by A. More generally, a subspace $\mathcal{S} \subset \mathbb{C}^n$ is A-*invariant* if and only if $Ax \in \mathcal{S}$ for any $x \in \mathcal{S}$. Note that if $AX = XB$, where $B \in \mathbb{C}^{k\times k}$, and $X \in \mathbb{C}^{n\times k}$, then the subspace spanned by the columns of X, $\mathcal{R}(X)$, is A-invariant, because any $x \in \mathcal{R}(X)$ is a linear combination of the columns of X, and likewise is Ax_i $(= Xb_i)$, for any column x_i of X. Moreover, if $By = \lambda y$, then $A(Xy) = \lambda(Xy)$. Therefore, if X has full column rank, then $AX = XB$ implies $\lambda(B) \subset \lambda(A)$. If X is square and nonsingular, it follows that $\lambda(A) = \lambda(B)$, where $B = X^{-1}AX$.

Two matrices $A, B \in \mathbb{C}^{n\times n}$ are *similar* if there is a nonsingular matrix P, such that $B = P^{-1}AP$. Using the results above, if λ is an eigenvalue of A, with eigenvector x, then λ is an eigenvalue of $P^{-1}AP$, with eigenvector $P^{-1}x$. By a special choice of P, as a product of some *similarity transformations*, matrix A can be reduced to a simpler form, which the eigenvalues can directly be obtained from. (Note that merely pre- or postmultiplication of a matrix by another matrix does not preserve the eigenvalues; in particular, a permutation of the rows of a matrix can change the eigenvalues.)

One of the reduced forms of special importance is the *quasi-triangular form* (QTF), in which the matrix is block upper triangular,[2] and each diagonal block has order at most two. If the 2×2 diagonal blocks of a QTF matrix $A \in \mathbb{R}^{n \times n}$ correspond to conjugate eigenvalues only, the matrix is said to be in a *real Schur form* (RSF), also known as *Wintner-Murnaghan canonical form.* If each 2×2 diagonal block of A (in RSF) has its diagonal elements equal and off-diagonal elements of opposite sign, then A is in a *canonical real Schur form.* This form is used, for instance, by the LAPACK routines (Anderson et al., 1992). As a block triangular matrix T is singular if and only if one of its diagonal blocks, say T_{ii}, is singular, it follows that

$$\lambda(T) = \bigcup_{i=1}^{q} \lambda(T_{ii}),$$

where q is the number of diagonal blocks. In particular, the eigenvalues of a triangular matrix are just its diagonal elements.

The set of eigenvectors corresponding to an eigenvalue λ of an $n \times n$ matrix A, together with the null vector, form a linear subspace, $\mathcal{X}_\lambda$, defined by $\mathcal{X}_\lambda := \mathcal{N}(A - \lambda I_n)$. Clearly, $\dim(\mathcal{X}_\lambda) = n - \operatorname{rank}(A - \lambda I_n)$. Let k be the multiplicity of λ. Sometimes, for $k > 1$, $\dim(\mathcal{X}_\lambda) < k$; for example, an *elementary Jordan block*, $J_\lambda^{(k)} \in \mathbb{C}^{k \times k}$, which is an upper bidiagonal matrix with elements equal to λ on the diagonal and 1 on the first superdiagonal, has $\dim(\mathcal{N}(J_\lambda^{(k)} - \lambda I_k)) = 1$, because $\operatorname{rank}(J_\lambda^{(k)} - \lambda I_k) = k - 1$.

If y and x are left and right eigenvectors, respectively, corresponding to eigenvalues λ_1 and λ_2 of A, with $\lambda_1 \neq \lambda_2$, then $y^H x = 0$. But if x and y correspond to the same simple eigenvalue λ, then it can be shown that $y^H x \neq 0$. This may not hold when λ has multiplicity $k > 1$; for instance, if $A = J_\lambda^{(k)}$, then any such eigenvectors x and y are scalar multiples of e_1 and e_k, respectively.

A square matrix A is *defective* if it has an eigenvalue λ with multiplicity k, with less than k associated linearly independent eigenvectors. Matrix $J_\lambda^{(k)}$ is defective for $k > 1$. The maximum number of linearly independent eigenvectors corresponding to λ is $\dim(\mathcal{N}(A - \lambda I_n))$; this number is called the *geometric multiplicity* of λ. If the algebraic multiplicity of λ is greater than its geometric multiplicity, λ is a *defective* eigenvalue. On the other hand, if $\lambda_i \in \lambda(A)$, $i = 1{:}k$, and $\lambda_i \neq \lambda_j$, $i \neq j$, then the corresponding eigenvectors, $x_1, x_2, \ldots, x_k$, are linearly independent. A matrix which is not defective is called *nondefective.*

[2] A matrix is *block upper (lower) triangular* if it is an upper (lower) triangular matrix with matrix elements, and square matrices on its diagonal.

The *standard eigenvalue problem* for a matrix $A \in \mathbb{C}^{n\times n}$ asks for determining its eigenvalues and corresponding eigenvectors.

Numerical computation of eigenvalues as roots of the characteristic polynomial is not advisable, since severe loss of accuracy can occur. For instance, consider

$$A = \begin{bmatrix} 1 & \varepsilon \\ \varepsilon & 1 \end{bmatrix}, \qquad \text{with } \lambda_1 = 1+\varepsilon, \quad \lambda_2 = 1-\varepsilon,$$

and the characteristic polynomial $\lambda^2 - 2\lambda + (1-\varepsilon^2)$. Assume that $\varepsilon^2 < \varepsilon_u < \varepsilon < 1$, where ε_u is the *roundoff unit* of the computer used for calculations, i.e. the smallest positive representable number so that the floating-point representation of $1+\varepsilon_u$, $\mathrm{fl}(1+\varepsilon_u)$, is greater than 1 (but $\mathrm{fl}(1+\nu) = 1$, for $\nu < \varepsilon_u$). It follows that $\mathrm{fl}(1-\varepsilon^2) = 1$, so that the computed roots will be $\widehat{\lambda}_1 = \widehat{\lambda}_2 = 1$. Therefore, half of the working precision can be lost.

The difficulty illustrated above becomes more severe for large eigenproblems. For this reason, sophisticated iterative algorithms operating on A, without resorting to characteristic polynomial, are used for the numerical solution. To gain in efficiency, matrix A has to be initially reduced to simpler forms. General similarity transformations, $P^{-1}AP$, have the disadvantage of matrix P turning to be ill conditioned. A remedy will be to restrict the transformations to a class of well-conditioned matrices: unitary matrices (or orthogonal matrices, for the real case); such matrices are easily inverted ($U^{-1} = U^H$), and are perfectly conditioned ($\mathrm{cond}_2(U) = 1$). Moreover, from a theoretical point of view, unitary matrices naturally yield backward error analyses, in the sense that a perturbation in the result can be explained by a perturbation of the *same* size in the original problem; indeed, assuming that the calculation of U^HAU introduces an error F in the result, if we denote $E = UFU^H$, then $U^H(A+E)U = U^HAU + F$, and $\|E\|_2 = \|F\|_2$.

Let x be an eigenvector of $A \in \mathbb{C}^{n\times n}$, corresponding to an eigenvalue λ. Assume that x has been normalized so that $\|x\|_2 = 1$. Clearly, one can find a matrix $U \in \mathbb{C}^{n\times(n-1)}$ so that $[\,x \;\; U\,]$ is unitary. As $Ax = \lambda x$, $x^Hx = 1$, and $U^Hx = 0$, it follows

$$[\,x \;\; U\,]^H A [\,x \;\; U\,] = \begin{bmatrix} x^H \\ U^H \end{bmatrix} [\,\lambda x \;\; AU\,] = \begin{bmatrix} \lambda & x^HAU \\ 0 & U^HAU \end{bmatrix}, \tag{2.39}$$

that is a block upper triangular matrix, having as eigenvalues the scalar λ and the eigenvalues of the matrix $C := U^HAU$, of order $n-1$. So, the problem order has been reduced by one, by *deflating* the *known* eigenvalue λ, based on the use of a unitary similarity transformation. Note that the matrix $[\,x \;\; U\,]$ can be determined as an elementary reflector (see § 1.4.3).

Indeed, assuming that y is an eigenvector corresponding to λ, and that the elementary reflector R has been determined so that $Ry = \pm\|y\|_2 e_1$, it follows that the first column of R is an eigenvector of unit length.

In practice, an approximate eigenvector x will be used, so that $g := U^H Ax \neq 0$. For reducing the problem order, the elements of g must be zeroed, but this corresponds to introducing an error of size $\|g\|_2$ in the original matrix. Hence, it is important to reach conditions under which $\|g\|_2$ is small. Let x be an approximate eigenvector corresponding to the approximate eigenvalue μ. The accuracies of x and μ can be assessed using $\|r\|_2$, where $r := Ax - \mu x$ is the residual. (If x and μ are exact, then $r = 0$.) The minimization of $\|r\|_2^2 = \|Ax - \mu x\|_2^2$ is a linear least squares problem with unknown μ determined by the normal equation $(x^H x)\mu = x^H Ax$. Hence, $\|r\|_2$ is minimal when $\mu = x^H Ax / x^H x$, the Rayleigh quotient of x; that value of μ is actually the estimate for λ obtained from (2.39), by setting g to zero. But it is easy to prove that the error induced by setting $g = 0$ in the deflation procedure has the same size as the smallest residual which can be obtained from the approximate eigenvector x, i.e., $\|g\|_2 = \|Ax - (x^H Ax)x\|_2$. Standard results for systems of linear algebraic equations (see, for instance, (Forsythe, Malcolm and Moler, 1977)) show that this residual can be very small, even when x is a poor approximation.

Efficient algorithms for solving eigenproblems are based on the following result (Golub and Van Loan, 1983; Stewart, 1973; Wilkinson, 1965):

Theorem 2.6 (Schur decomposition) *If $A \in \mathbb{C}^{n\times n}$ has eigenvalues λ_1, λ_2, ..., λ_n (with multiplicities counted), then there is a unitary matrix U, so that $T := U^H AU$ is upper triangular with diagonal elements λ_1, λ_2, ..., λ_n, in that order.*

The *Schur decomposition* is $A = UTU^H$. Denoting $T = D + N$, where $D = \mathrm{diag}(\lambda_1, \ldots, \lambda_n)$, and N is a strictly upper triangular matrix, and defining $u_i := u_{:,i}$, then relation $AU = UT$ gives

$$Au_k = \lambda_k u_k + \sum_{i=1}^{k-1} n_{ik} u_i, \qquad k = 1\colon n,$$

i.e. the subspaces $\mathcal{S}_k := \mathcal{R}(u_1, \ldots, u_k)$, spanned by u_1, ..., u_k, are A-invariant, for $k = 1\colon n$. The vectors u_i are called *Schur vectors*. In addition, as $U_k := \begin{bmatrix} u_1 & \ldots & u_k \end{bmatrix}$ satisfies $\lambda(U_k^H AU_k) = \{\lambda_1, \ldots, \lambda_k\}$, and the eigenvalues can be arbitrarily ordered, it follows that there is at least one invariant subspace associated with each subset of k eigenvalues.

For real matrices it is cheaper to restrict the similarity transformations to the class of orthogonal matrices.

Theorem 2.7 (Real Schur decomposition) *If $A \in \mathbb{R}^{n\times n}$, then there is an orthogonal matrix U, so that U^TAU is in quasi-triangular form; in addition, U can be chosen so that U^TAU is in real Schur form, and its diagonal blocks appear in any desired order.*

From $U^HAU = T = \mathrm{diag}(\lambda_1, \ldots, \lambda_n) + N$, it follows that

$$\|N\|_F = \|A\|_F - \sum_{i=1}^{n} |\lambda_i|^2,$$

so $\|N\|_F$ is independent of U. There is a class of matrices, called *normal* or *diagonalizable* matrices, for which $N = 0$. (This class includes Hermitian and symmetric matrices, and all matrices with simple eigenvalues only.) But if $\|N\|_F \neq 0$, to make T "more diagonal", i.e. to continue the reduction beyond the (quasi-)triangular form, it is necessary to use non-unitary similarity transformations. Of much importance, particularly in theory, is the *Jordan canonical form* (JCF),

$$J = \mathrm{diag}(J_{\lambda_1}^{(m_1)}, J_{\lambda_2}^{(m_2)}, \ldots, J_{\lambda_r}^{(m_r)}), \tag{2.40}$$

where $J_\lambda^{(k)}$ denotes an elementary Jordan block defined above. For any matrix $A \in \mathbb{C}^{n\times n}$, there are some unique numbers $\lambda_k \in \lambda(A)$, and unique positive integers m_k, $k = 1{:}r$ (with $m_1 + \cdots + m_r = n$), so that A is similar to J in (2.40). The JCF is unique, not considering the order of the Jordan blocks. It is worth mentioning that the same eigenvalue may appear in several blocks (of equal or different dimension). In this case, the JCF is said *derogatory*. A general matrix is *derogatory* if its JCF is derogatory. If the JCF of a matrix has a single Jordan block for each eigenvalue, then that matrix is *nonderogatory*.

Let X be a matrix that reduces A to its associated JCF, and define $x_i := x_{:,i}$. From $X^{-1}AX = \mathrm{diag}(J_{\lambda_1}^{(m_1)}, J_{\lambda_2}^{(m_2)}, \ldots)$, it follows that x_1, x_2, $\ldots$, x_{m_1} satisfy

$$\begin{aligned} Ax_1 &= \lambda_1 x_1, \\ Ax_{i+1} &= \lambda_1 x_{i+1} + x_i, \qquad i = 1{:}m_1 - 1. \end{aligned} \tag{2.41}$$

The other eigenvectors satisfy similar *chain relations*. Note that

$$x_j, \quad j \in \{1, 1+m_1, 1+m_1+m_2, \ldots, 1+m_1+\cdots+m_{r-1}\}$$

are eigenvectors of A. The vectors x_i, $i = 1{:}n$, are called *generalized eigenvectors* or *principal vectors* of A.

Unfortunately, the numerical computation of JCF of a defective matrix A is a very difficult task. It can be shown that the set of $n \times n$ nondefective matrices is dense in $\mathbb{C}^{n\times n}$, and hence small perturbations in a defective matrix may drastically alter its JCF. An associated difficulty is that a "nearly" defective matrix may have an ill-conditioned matrix of eigenvectors. Clusters of eigenvalues whose eigenvectors are nearly linearly dependent are often encountered in numerical computations.

Let us now discuss the *conditioning* of eigenvalues and eigenvectors. Generally, if λ is a defective eigenvalue, then perturbations of order ε in A can produce perturbations of size $0(\varepsilon^{1/p})$ in λ, where $p \geq 2$. It can be shown (see, for instance, (Stewart, 1973)), that if x and y are right and left eigenvectors of A corresponding to a simple eigenvalue λ, and $x^H x = y^H x = 1$, then $\|y\|_2$ is a *condition number* for λ. It measures the absolute sensitivity of λ to perturbations in the elements of A. Likewise, if $x^H x = y^H y = 1$, then a condition number for λ is $1/|y^H x|$; if that value is large, then A is relatively close to a matrix with multiple eigenvalue λ. Note that the condition numbers are not invariant with respect to diagonal similarity transformations; in practice, relevant values correspond to a transformation $D^{-1}AD$, where D is a diagonal matrix chosen so that $\|D^{-1}AD\|_2$ is minimal. The *balancing procedure* included in the packages EISPACK (Smith et al., 1976) and LAPACK (Anderson et al., 1992) is usually suitable for this aim; it includes a further preprocessing step, called *permutation*, that reorders the rows and columns of the matrix to make it more nearly an upper triangular matrix (see § 2.2.2, for more details). Clearly, eigenvalues of A and $U^H AU$, for unitary U, have the same condition numbers.

Condition numbers for eigenvectors have also been derived. It can be shown (see (Stewart, 1973)) that a lower bound on a condition number for an eigenvector x, corresponding to a simple eigenvalue λ, is the reciprocal of the *separation* of $\lambda =: \lambda_1$ from the other eigenvalues of A,

$$\text{sep} = \min_{i=2:n} \{\, |\lambda - \lambda_i| \,\}.$$

Hence, if λ is near other eigenvalue(s) of A, then the corresponding eigenvector is ill conditioned.

The perturbation theory for multiple eigenvalues is quite complicate. A multiple eigenvalue may or may not be ill conditioned, depending on several factors. Condition numbers like those mentioned above can also be defined for clusters of eigenvalues. It is worth mentioning that an invariant subspace associated with a cluster of eigenvalues may be insensitive to perturbations in matrix elements, if that cluster is well separated from the other eigenvalues. For this reason, it is advisable to use Schur vectors for determining bases for such subspaces, instead of eigenvectors (and principal vectors), which could be ill conditioned even for matrices with well-separated eigenvalues.

Theoretical and practical details concerning ill-conditioned eigensystems, and the computation of the Jordan canonical form, can be found, for instance, in (Golub and Wilkinson, 1976; Golub and Van Loan, 1983).

The LAPACK package includes routines which estimate tight lower bounds on reciprocal condition numbers for individual eigenvalues and eigenvectors, and for clusters of eigenvalues and their associated invariant subspaces. For instance, the bounds for eigenvalues will not underestimate the true values by factors larger than $n^{1/2}$. The calculations are done for the eigenvalues collected in the leading principal submatrix of order k of RSF of T, T_{11}, where

$$T = \begin{bmatrix} T_{11} & T_{12} \\ 0 & T_{22} \end{bmatrix}.$$

The reciprocal condition number of the average of the eigenvalues of T_{11} is computed as $1/(1 + \|R\|_F^2)^{1/2}$, where R is given by the solution of the Sylvester equation

$$T_{11}R - RT_{22} = T_{12}. \tag{2.42}$$

(It can be shown that

$$P = \begin{bmatrix} I_k & R \\ 0 & 0 \end{bmatrix}$$

is the projector on the invariant subspace associated with T_{11}, and the true reciprocal condition number desired is $1/(1 + \|R\|_2^2)^{1/2}$. Note that, with (2.42), $PT = TP$.) Similarly, the reciprocal condition number of the invariant subspace spanned by the first k Schur vectors is the *separation* of the matrices T_{11} and T_{22}, defined as the smallest singular value of the linear transformation $\mathcal{T}$ in (2.42) from X to $T_{11}X - XT_{22}$ (Varah, 1979):

$$\text{sep}(T_{11}, T_{22}) = \min_{X \neq 0} \{ \|T_{11}X - XT_{22}\|_F / \|X\|_F \}.$$

This value is estimated using the condition estimator in (Higham, 1988), based on calculations of $\mathcal{T}X$ and $\mathcal{T}^T X$, i.e. the solution of an equation like (2.42), and of a similar equation with T_{11} and T_{22} replaced by their transposes, respectively. It can be shown that $\text{sep}(T_{11}, T_{22})$ can be very small for non-normal matrices. An example is the class of *Frank matrices*, defined by $a_{ij} = 0$; $a_{ij} = n+1-\max(i,j)$, $j \geq i-1$. For $n = 12$, and $k \geq 8$, sep is of order of 10^{-6} (Varah, 1979). Small values of sep indicate possible numerical difficulties when using the corresponding invariant subspace. Another, even more interesting, example is given below.

Example 2.1 Consider the matrix G in (1.60), § 1.1.2, re-written here

$$G = \begin{bmatrix} A & -F \\ -Q & -A^T \end{bmatrix},$$

Table 2.1 Reciprocal condition numbers for the cluster of stable eigenvalues and associated G-invariant subspace in Example 2.1.

ε	s	sep
1	0.7405	$4.4946 \cdot 10^{-1}$
10^{-1}	0.9621	$9.9125 \cdot 10^{-3}$
10^{-2}	0.9996	$9.9991 \cdot 10^{-5}$
10^{-3}	1.0000	$1.0000 \cdot 10^{-6}$
10^{-4}	1.0000	$1.0000 \cdot 10^{-8}$
10^{-5}	1.0000	$9.9999 \cdot 10^{-11}$
10^{-6}	1.0000	$9.9912 \cdot 10^{-13}$
10^{-7}	0.9877	$1.0093 \cdot 10^{-14}$
10^{-8}	0.9410	$7.3815 \cdot 10^{-16}$

where A, F, and Q are defined by (Ammar, Benner and Laub, 1993; Arnold and Laub, 1984):

$$A = \begin{bmatrix} -\varepsilon & 1 & 0 & 0 \\ -1 & -\varepsilon & 0 & 0 \\ 0 & 0 & \varepsilon & 1 \\ 0 & 0 & -1 & \varepsilon \end{bmatrix}, \quad F = Q = \begin{bmatrix} 1 \\ 1 \\ 1 \\ 1 \end{bmatrix} \begin{bmatrix} 1 & 1 & 1 & 1 \end{bmatrix}.$$

Matrix G has four stable (with negative real parts) and four unstable eigenvalues. For $\varepsilon \to 0$, a pair of complex conjugate eigenvalues approaches the imaginary axis, hence G tends to a matrix which is not dichotomic. (The underlying continuous-time system approaches an unstabilizable one.) Table 2.1 gives the estimated reciprocal condition numbers for the cluster of stable eigenvalues, s, and for the associated G-invariant subspace, sep, computed in double precision arithmetic using LAPACK routines on an IBM PC-compatible computer. It can be seen that s is about one for all values of ε in the range $(10^{-8}, 1)$, so the cluster of stable eigenvalues is well conditioned, but sep decreases steadily and is about the order of the relative machine precision, ε_M, for $\varepsilon = 10^{-8}$. The stable eigenvalues tend to -3.7321, -0.2679, and $-.5\varepsilon^2 \pm j$. Actually, for $\varepsilon \leq 10^{-1}$, the value of sep is a very tight measure of the distance between the set of stable eigenvalues and the set of unstable eigenvalues of G. The standard Riccati solvers will have difficulties with this example for small values of ε, and the value of sep could explain their behaviour. □

2.2.2 Preprocessing algorithms

Let us now consider the practical computational algorithms for solving standard eigenproblems for general matrices. The main algorithmic tool is the iterative *QR algorithm*, that reduces a matrix A to a quasi-triangular form by unitary similarity transformations. As QR algorithm requires a prohibitive effort if applied on a full matrix (about $2n^3$ *flops* per iteration), A should first be reduced to a simpler form (the upper Hessenberg form), using unitary similarity transformations. This reduction is usually preceded by a matrix *balancing algorithm*, which equilibrates the row and column 1-norms (as specified below), and isolates, using appropriate permutations, the leading and trailing upper triangular submatrices of A, if possible. Specifically, let $A \in \mathbb{R}^{n\times n}$ be given, and let P be a product of permutation matrices, so that

$$A \leftarrow P^T AP = \begin{bmatrix} T & X & Y \\ 0 & B & Z \\ 0 & 0 & R \end{bmatrix} \begin{matrix} \} & k-1 \\ \} & l-k+1\ , \\ \} & n-l \end{matrix} \qquad (2.43)$$

where T and R are upper triangular matrices, and B is a square matrix so that $B - \text{diag}(B)$ has no zero rows and columns, $B = a_{k:l,k:l}$. The diagonal elements of T and R are the eigenvalues of A which have been isolated. The remaining eigenvalues of A are the eigenvalues of B. The balancing algorithm iteratively applies diagonal similarity transformations, $B \leftarrow D^{-1}BD$, using integer powers of the machine basis, or of ten—in LAPACK—as scaling factors. This process makes the 1-norms of each row of B and its corresponding column nearly equal, and may reduce $\|B\|_1$ and improve the accuracy of the computed eigenvalues and/or eigenvectors. (This is possible because the available algorithms for solving eigenproblems usually produce results with errors of order $\varepsilon_M\|A\|_F$, and reducing $\|B\|_1$ may reduce $\|A\|_F$.) The transformed matrix A is

$$A \leftarrow \begin{bmatrix} T & XD & Y \\ 0 & D^{-1}BD & D^{-1}Z \\ 0 & 0 & R \end{bmatrix}.$$

We do not detail this algorithm, but refer to (Parlett and Reinsch, 1969) or to its implementations in EISPACK (Smith et al., 1976) or LAPACK (Anderson et al., 1992) packages. Note that LAPACK routines for computing eigenvalues and real Schur form do not employ equilibration, but merely permutation. The balancing algorithm can be defined by the following statement

$(A, d, k, l) =$ `balanc(` A `);`

where d contains the scaling factors and permutation indices, and k, and l are the indices defining the submatrix B in (2.43).

An $n \times n$ matrix A is in *upper Hessenberg form* (UHF) if $a_{ij} = 0$, for $i > j+1$, $j = 1{:}n-2$. A general matrix $A \in \mathbb{R}^{n\times n}$ can be reduced to UHF using $n-2$ orthogonal similarity transformations. Let $A_1 = A$. At step i of the algorithm, matrix A_1 has already been reduced to A_i, which is upper Hessenberg in its first $i-1$ columns. Matrix A_{i+1} is obtained from A_i by computing

$$A_{i+1} = U_i A_i U_i, \qquad U_i := I_n - \beta_i u_i u_i^T, \tag{2.44}$$

where U_i is a Householder transformation chosen to annihilate the elements $i+2{:}n$, in the ith column of A_i. Note that the transformation in (2.44) preserves the zeroes already introduced in the first $i-1$ columns.

Clearly, after $n-2$ steps one obtains the upper Hessenberg matrix A_{n-1}, similar to A. If A is a symmetric matrix, then A_{n-1} is symmetric tridiagonal. The Householder transformations U_i can be computed using Algorithm `Hous` in § 1.4.3, and can be applied as in Algorithm `QR`, or using Level 2 BLAS operations (1.104), as shown below. A template algorithm is now given, which assumes a previous application of balancing algorithm. (If balancing has not been used, k must be set to 1 and l to n.)

Algorithm Hess : $(A, Q, \tau) =$ `Hess`$(\ A, k, l\ )$

/* Reduce a matrix $A \in \mathbb{R}^{n\times n}$ to upper Hessenberg form, */
/* using an orthogonal similarity transformation: $A \leftarrow Q^T A Q$, */
/* assuming $a_{1:k-1,1:k-1}$ and $a_{l+1:n,l+1:n}$ already upper triangular. */

$Q = I_n;\quad \tau_{1:k-1} = 0;\quad \tau_{l-1:n-1} = 0;$
FOR $i = k{:}l-2$,

/* 1. Compute U_i to annihilate $a_{i+2:l,i}$ and update Q. */
$(U, r) =$ `Hous`$(\ a_{i+1:l,i}\ )$; $\quad q_{:,i+1:l} \leftarrow q_{:,i+1:l} U$;

/* 2. Apply U_i to $a_{1:l,i+1:l}$ from the right. */
$a_{1:l,i+1:l} \leftarrow a_{1:l,i+1:l} U$;

/* 3. Apply U_i to $a_{i+1:l,i+1:n}$ from the left. */
$a_{i+1:l,i+1:n} \leftarrow U a_{i+1:l,i+1:n}$; $\quad \tau_i = a_{i+1,i}$; $\quad a_{i+1,i} = -r$;

Remark 2.3 Algorithm `Hess` returns the Hessenberg matrix in the upper triangular part and the elements of the first subdiagonal of A (in the columns k to l). For convenience, it also computes the transformation matrix Q, $Q = U_k U_{k+1} \cdots U_{l-2}$. When k and/or l are not specified, it is assumed that $k = 1$ and/or $l = n$. Valid calling statements are $(A, Q) =$ `Hess`$(\ A\ )$, or $(A, \tau) =$ `Hess`$(\ A\ )$, when Q is not needed. In the latter case, we will assume that Q is not explicitly formed. However, the information based on which Q can be reconstructed is stored in the lower triangular part of A, under the first subdiagonal, and in the $(n-1)$-vector τ, containing (in positions $k, \ldots, l-2$) the scalar factors of the elementary reflectors. This case is actually implemented in the EISPACK and LAPACK packages, which include separate routines for accumulating Q or applying it to other matrices. A similar algorithm is used for a complex matrix A, and unitary matrix Q. □

The implementation of Algorithm **Hous**, used in **Hess**, could somewhat differ from that suggested in § 1.4.3. For instance, the EISPACK implementation uses the 1-norm for scaling the input vector x, instead of the infinity-norm. The LAPACK implementation is more sophisticated. It uses the 2-norm, but to avoid underflows, hence inaccurate calculations of $\|x\|_2$ and x_1, x is rescaled and $\|x\|_2$ and x_1 are recomputed. Moreover, it finally scales the computed Householder vector u so that its first element is $u_1 = 1$. Specifically, let $U = I_m - \beta uu^T$ be a Householder transformation, where (see § 1.4.3)

$$u_1 = x_1 + r, \quad \beta = 1/(r(x_1 + r)), \quad r = \text{sign}(x_1)\|x\|_2.$$

We obtain $U = I_m - \widetilde{\beta}\widetilde{u}\widetilde{u}^T$, where $\widetilde{u} := u/(x_1 + r)$, and $\widetilde{\beta} := \beta(x_1 + r)^2$. Therefore,

$$\widetilde{u}_1 = 1, \quad \widetilde{u}_{2:m} = x_{2:m}/(x_1 + r), \quad \widetilde{\beta} = 1 + x_1/r.$$

It follows that $1 \le \widetilde{\beta} \le 2$. The first element of $\widetilde{u}$ is not stored, so x_1 can take its updated value, $-r$. We will now suppress the tilde sign, for convenience. Obviously, LAPACK Householder transformations can replace the modified Householder transformations, generated by Algorithm **Hous_m**, and can be economically applied to given matrices. For instance, when applied from the left, $C \leftarrow UC$ (if $\beta \neq 0$), one has first to compute $w := C^T u$, and then update $C \leftarrow C - \beta uw^T$, using Level 2 BLAS operations (see (1.104) in § 1.4.3). The fact that $u_1 = 1$ is not exploited. The product CU is computed similarly.

Algorithm **Hess** requires about $5n^3/3$ *flops*, assuming $k = 1$, $l = n$. If the applied Householder transformations are accumulated, the additional effort needed is $2n^3/3$ *flops* (for $k = 1, l = n$). The algorithm is numerically stable, the computed matrix $\widehat{A}$ being orthogonally similar to $A + E$, where

$$\|E\|_F \le \gamma n_b^2 \varepsilon_M \|A\|_F, \qquad n_b := l - k + 1, \tag{2.45}$$

and γ is a constant of order of unity. The factor n_b^2 in (2.45) can be replaced by n_b, if the scalar products involved are accumulated in double precision.

In addition, the orthogonal matrix Q, that transforms $A+E$ into $\widehat{A}$, is just the product of the orthogonal matrices U_k, U_{k+1}, ..., U_{l-2}, where U_i corresponds to the exact transformation for the computed A_i. It follows that if the information stored in the lower part of A and in τ is used to generate the matrix product $U_k U_{k+1} \cdots U_{l-2}$, then the computed result is very close to the matrix Q that reduces $A+E$.

Let us briefly discuss the *block algorithm* implemented in LAPACK. Let p be the block size and set $\widetilde{U} = I_n$. The block algorithm, call it `HessB`, processes "simultaneously" the columns from i to $q := i + \min(p, l-i) - 1$ of A, for $i = k{:}\,p{:}\,l-1-\widetilde{n}$, where $\widetilde{n}$ denotes a dimension parameter indicating when to switch from blocked to unblocked algorithm. For each value of i, `HessB` first reduces the columns $i{:}\,q$ of the currently transformed matrix actually by using the same Householder transformations as the unblocked algorithm does for that block of columns, but returning the matrices $V \in \mathbb{R}^{n\times p}$ and $T \in \mathbb{R}^{p\times p}$ of the current block reflector $U := U_1 U_2 \cdots U_p = I_n - VTV^T$, and the matrix $Y := AVT$, $Y \in \mathbb{R}^{n\times p}$ (Schreiber and Van Loan, 1989). The columns of V are the individual vectors associated with the Householder transformation matrices U_1, U_2, ..., U_p, and T is an upper triangular matrix. Level 2 BLAS and few Level 1 BLAS operations are involved in obtaining T and Y. For instance, assuming that $j-1$ Householder transformations have already been processed, and the jth transformation has been computed, the jth column of T results from the following calculations

$$\begin{array}{lcl} t_{1:j-1,j} & = & -\beta_j v_{j:l,1:j-1}^T v_{j:l,j}; \\ t_{1:j-1,j} & \leftarrow & t_{1:j-1,1:j-1} t_{1:j-1,j}; \qquad t_{jj} = \beta_j. \end{array}$$

The intermediate result computed in the first relation can also be used to obtain the jth column of Y.

Clearly, extra effort is needed for computing the elements of the matrices T and Y, but this is partly compensated for by greater efficiency in applying the block form. Indeed, the reduction of the columns $i{:}\,q$ of A to upper Hessenberg form is equivalent to the following calculations

$$U^T A U = (I - VT^T V^T) A (I - VTV^T) = (I - VT^T V^T)(A - YV^T).$$

Therefore, the block reflector U is applied to $a_{1:l,q:l}$ from the right, by computing $A \leftarrow A - YV^T$, a Level 3 BLAS operation. Then, it is still to apply the block reflector U^T to $a_{i+1:l,q:n}$ from the left, by computing $A \leftarrow A - VT^T V^T A$. This involves Level 2 BLAS and few Level 1 BLAS operations. Indeed, consider that the block transformation $I - VT^T V^T$ is to be applied to a vector b, where

$$V = \begin{bmatrix} V_1 \\ V_2 \end{bmatrix}, \qquad b = \begin{bmatrix} b_1 \\ b_2 \end{bmatrix},$$

V_1 and b_1 having $i-1$ rows, and V_1 being unit lower triangular. The computations can be accomplished as follows

$$w := V_1^T b_1 + V_2^T b_2; \quad w \leftarrow T^T w; \quad b_2 \leftarrow b_2 - V_2 w; \quad b_1 \leftarrow b_1 - V_1 w.$$

The transformations used for reducing the current block of columns are similarly accumulated, giving $\widetilde{U} \leftarrow \widetilde{U}U$. After the loop indexed by i has been completed, the unblocked algorithm is called to reduce the rest of the matrix:

$$(A, Q, \tau) = \texttt{Hess}(\ A, i, l\ ); \quad Q \leftarrow \widetilde{U}Q;.$$

The logic for determining the actual block size is quite complicate. For optimal performance, **HessB** needs extra np memory locations for workspace, where p is the optimal block size, and a working array of dimension (65,64) for storing T. (The maximal block size is 64.) The optimal block size is determined as $\min\{64, b\}$, where b is a machine-dependent value returned by an "environmental" routine, called ILAENV. The resulted value should be at least 2, but less than $l-k+1$, to use the block algorithm. If these constraints are satisfied, then the point, $\widetilde{n}$, when to switch from blocked to unblocked algorithm, is determined using again ILAENV (last block is always handled by unblocked algorithm). If there is no enough workspace to use the optimal p, the minimal value of p is determined, and p is then reduced to the maximal value possible with the available workspace, or use of the unblocked algorithm is forced. A similar or simpler logic is implemented for other LAPACK block algorithms.

2.2.3 The QR algorithm

The iterative QR algorithm is the most widely used general-purpose algorithm for reducing a given matrix to (quasi-)triangular form by unitary similarity transformations. Using it in conjunction with a preliminary reduction to the upper Hessenberg form, this algorithm has proven to be one of the most efficient methods for solving eigenproblems with dense matrices.

Let $A \in \mathbb{C}^{n\times n}$. The QR algorithm is based on the fact that if $A = QR$, and $B = RQ$, where Q is a unitary matrix, and R is upper triangular, then $B = RQ = Q^H AQ$ is unitarily similar to A. By repeated application of this result, one can obtain a sequence of matrices, A_s, unitarily similar to $A_1 := A$, defined by

$$A_s = Q_s R_s, \qquad A_{s+1} = R_s Q_s = Q_s^H A_s Q_s, \tag{2.46}$$

that tends, under certain conditions, to an upper triangular matrix. For instance, if $\lambda_i \in \lambda(A)$, $i = 1{:}\,n$, satisfy $|\lambda_1| > |\lambda_2| > \cdots > |\lambda_n|$, and if the matrix Y, having as the ith row the left eigenvector of A corresponding to λ_i, $i = 1{:}\,n$, has an LU factorization, then the subdiagonal elements of A_s approach zero, and

$$a_{ii}^{(s)} \to \lambda_i, \quad \text{as } s \to \infty, \quad i = 1{:}\,n.$$

This and other similar results explain some of the observed numerical properties of the QR algorithm. The result suggests that the QR algorithm will present the tendency to determining the eigenvalues in a *decreasing order* along the diagonal. When A has p eigenvalues with equal moduli, the limit matrix is generally not triangular, but has a $p \times p$ diagonal block, with those p values as eigenvalues.

The convergence rate of the process (2.46) is often unsatisfactory. (Usually, the subdiagonal element $a_{ij}^{(s)}$ asymptotically behaves like $\gamma_{ij}(\lambda_i/\lambda_j)^s$, where γ_{ij} is a constant.) In order to increase the convergence, one may use, at each step s, the matrix $A_s - \delta_s I_n$, where δ_s is a properly chosen *origin shift*. When the algorithm converges, δ_s tends to an eigenvalue of A. If $A_s - \delta_s I_n$ is nonsingular, then it has an essentially unique QR factorization (see § 1.4.3). The relations (2.46) are replaced by

$$\begin{aligned} A_s - \delta_s I_n &= Q_s R_s, \\ A_{s+1} &= R_s Q_s + \delta_s I_n = Q_s^H A_s Q_s. \end{aligned} \tag{2.47}$$

From the first relation, it follows that $Q_s = (A_s - \delta_s I_n)^{-H} R_s^H$. Denoting by $q_i^{(s)}$ the ith column of Q_s, we obtain

$$q_n^{(s)} = Q_s e_n = \bar{r}_{nn}(A_s^H - \bar{\delta}_s I_n)^{-1} e_n, \tag{2.48}$$

as $R_s^H e_n = \bar{r}_{nn} e_n$; hence $q_n^{(s)}$ is an approximate eigenvector of A_s^H, obtained after having applied one step of *inverse iteration* (see, for instance, (Golub and Van Loan, 1983; Stewart, 1973; Wilkinson, 1965)) with shift $\bar{\delta}_s$ to the vector e_n. If A_s is partitioned in the form

$$A_s = \begin{bmatrix} C_s & h_s \\ g_s^H & a_{nn}^{(s)} \end{bmatrix}, \tag{2.49}$$

and A_{s+1} is done the same, it follows that $\|g_s\|_2$ and $\|g_{s+1}\|_2$ are the sizes of the minimal residuals for e_n and $q_n^{(s)}$, respectively, regarded as approximate eigenvectors of A_s^H. If $\bar{\delta}_s$ is nearly an eigenvalue of A^H, then $q_n^{(s)}$ will be an approximate eigenvector much more accurate than e_n, and hence $\|g_{s+1}\|_2 < \|g_s\|_2$. In this way, even if δ_s is only an approximate eigenvalue of A, one can expect the vectors g_s (hence the off-diagonal elements of the last row in A_s) tend quickly to zero. A natural choice for $\bar{\delta}_s$ in (2.48) is the Rayleigh quotient, $e_n^T A e_n$, hence $\delta_s = a_{nn}^{(s)}$. In this case, the quantities $||g_s||_2$ are the sizes of the residuals corresponding to the Rayleigh quotient iteration, and if this process does converge, it is known that g_s will usually approach zero quadratically (or cubically, for Hermitian matrices). Once g_s is small enough, we may neglect it, and the process can restart with the smaller submatrix C_s in (2.49).

It can also be shown that the QR algorithm is an extension of the *power method*, so that, if it converges to an eigenvalue, it is going to find out other eigenvalues too, hence all subdiagonal elements of A_s can approach zero somewhat more slowly than $a_{n-1,n}^{(s)}$. Indeed, denoting $\widetilde{Q}_s = Q_1 Q_2 \cdots Q_s$ and $\widetilde{R}_s = R_s R_{s-1} \cdots R_1$, it follows that $A_{s+1} = \widetilde{Q}_s^H A \widetilde{Q}_s$, and, as $\widetilde{Q}_s$ is unitary,

$$A_{s+1} - \delta_s I_n = \widetilde{Q}_s^H (A - \delta_s I_n) \widetilde{Q}_s.$$

Combining with the second relation (2.47),

$$R_s = (A_{s+1} - \delta_s I_n) Q_s^H = \widetilde{Q}_s^H (A - \delta_s I_n) \widetilde{Q}_{s-1},$$

and postmultiplying by $\widetilde{R}_{s-1}$, one derives that

$$\widetilde{Q}_s \widetilde{R}_s = (A - \delta_s I_n) \widetilde{Q}_{s-1} \widetilde{R}_{s-1}.$$

Hence, by induction, we have

$$\widetilde{Q}_s \widetilde{R}_s = (A - \delta_s I_n)(A - \delta_{s-1} I_n) \cdots (A - \delta_1 I_n). \tag{2.50}$$

The significance of this result is made clear in the case of unshifted QR algorithm. Indeed, if $\delta_1 = \delta_2 = \cdots = \delta_s = 0$, then (2.50) becomes $A^s = \widetilde{Q}_s \widetilde{R}_s$, hence $\widetilde{Q}_s \widetilde{R}_s$ is the QR factorization of A^s. In particular, as $\widetilde{R}_s e_1 = \widetilde{r}_{11}^{(s)} e_1$, the first column of $\widetilde{Q}_s$ is given by $\widetilde{r}_{11}^{(s)} \widetilde{q}_1^{(s)} = \widetilde{Q}_s \widetilde{R}_s e_1 = A^s e_1$, so that $\widetilde{q}_1^{(s)}$ is just the vector that would be obtained after s iterations of the power method applied to e_1. If A has a dominant eigenvalue λ_1, and the first component of the corresponding eigenvector is non-zero, then $\widetilde{q}_1^{(s)}$ approaches that eigenvector. If we partition (with some other notation than in (2.49)),

$$A_s = \begin{bmatrix} a_{11}^{(s)} & h_s^H \\ g_s & C_s \end{bmatrix},$$

it follows similarly that $||g_{s+1}||_2$ is the size of the minimal residual of $\widetilde{q}_1^{(s)}$, regarded as an approximate eigenvector of A. As $\widetilde{q}_1^{(s)}$ approaches an eigenvector of A, g_s should approach zero. Hence the subdiagonal elements of the first column of A tend to zero. The convergence rate depends on the ratio $|\lambda_2/\lambda_1|$, where it is assumed that $|\lambda_1| > |\lambda_2| \geq \cdots \geq |\lambda_n|$.

In practice, the reduction of the subdiagonal elements may not be large in any iteration, but, after several iterations, it becomes significant. If A is in upper Hessenberg form, and a subdiagonal element, $a^{(s)}_{i+1,i}$, has become negligible, then the problem can be decomposed into two smaller subproblems. Even if this does not happen, the element $a^{(s)}_{ii}$ may become a better and better approximate eigenvalue, so that at the time when it is used as a shift, the convergence should be very fast.

If A is in upper Hessenberg form (UHF), then all iterates A_s of the QR algorithm will be in UHF, and the computations can be reformulated so that each iteration costs $0(n^2)$ *flops* only. Indeed, let $A \in \mathbb{R}^{n\times n}$ be in UHF. The QR factorization of A can be performed using a sequence of $n-1$ plane rotations P_i, $i = 1{:}\,n-1$ (see § 1.4.3), where each rotation P_i, in the plane $(i, i+1)$, annihilates the element $(i+1, i)$, $i = 1{:}\,n-1$, so that $P_{n-1}\cdots P_2P_1A = R$ is upper triangular. Hence $A = QR$, where $Q = P_1^TP_2^T\cdots P_{n-1}^T$, or, equivalently, $Q^TA = R$. Therefore, for obtaining the next iterate, one should compute $RQ = RP_1^TP_2^T\cdots P_{n-1}^T$. By the nature of P_1, the product RP_1^T differs from R in its first two rows and columns at most, and it can have the $(2,1)$ element non-zero. In general, postmultiplication of $RP_1^TP_2^T\cdots P_{j-1}^T$ by P_j^T affects the jth and $(j+1)$th columns only, and introduces a single non-zero element in the position $(j+1, j)$. So, RQ is finally in UHF.

Actually, pre- and postmultiplications get intermixed. Once P_2P_1A has been computed, the first two columns are never used in the remaining steps of the triangularization process. As postmultiplication by P_1^T can change the first two columns only, this can be done before the reduction proceeds. In general, after j rotations have been applied, the reduced matrix is $P_j\cdots P_2P_1AP_1^TP_2^T\cdots P_{j-1}^T$, and, postmultiplying it by P_j^T, let us obtain an upper Hessenberg matrix similar to A.

We have described the computations in a typical QR iteration, called a *QR step*, and performed in the following algorithm, which also incorporates shifts.

Algorithm eQRs : $(A, c, s) = \texttt{eQRs}(\ A, h\ )$

/* Apply a QR step with shift h to matrix $A \in \mathbb{R}^{n\times n}$ in UHF. */

$a_{11} \leftarrow a_{11} - h$;
FOR $i = 1{:}\,n$,

  IF $i < n$, THEN
    $(c_i, s_i) = \texttt{rot}(\ a_{ii}, a_{i+1,i}\ )$;   /* 1. Find P_i. */

$a_{i+1,i} = 0; \quad a_{i+1,i+1} \leftarrow a_{i+1,i+1} - h;$

$a_{i:i+1,i+1:n} \leftarrow \begin{bmatrix} c_i & s_i \\ -s_i & c_i \end{bmatrix} a_{i:i+1,i+1:n};$ /* 2. Premultiply by P_i. */

END IF
IF $i > 1$, THEN

/* 3. Postmultiply by P_{i-1}^T. */

$a_{1:i,i-1:i} \leftarrow a_{1:i,i-1:i} \begin{bmatrix} c_{i-1} & -s_{i-1} \\ s_{i-1} & c_{i-1} \end{bmatrix};$

$a_{i-1,i-1} \leftarrow a_{i-1,i-1} + h;$

END IF
END FOR

$a_{nn} \leftarrow a_{nn} + h;$

Remark 2.4 The plane rotations used by the algorithm are stored in the $(n-1)$-vectors c and s. These values can be used for accumulating the transformations applied. If no accumulation is needed, or Algorithm **eQRs** is modified to perform it inside, then only the values c_{i-1} and s_{i-1} from the previous step should be stored. Since the shift h is actually subtracted (and finally added to the diagonal elements), we say that **eQRs** is "with explicit shifts", or simply, *explicit*. There is also an "implicit" variant of the QR step algorithm, to be discussed later on. □

Algorithm **eQRs** involves about $4n^2$ *flops*. So, if an average of r QR steps should determine an eigenvalue, the explicit QR algorithm will require $4rn^3$ *flops*. The value of r is generally quite small. In practice, one can obtain a higher efficiency by repeated application of Algorithm **eQRs** with different shifts $h_1, h_2, \ldots$. If $h_s = a_{nn}^{(s)}$ is taken as a shift, we may expect that ultimately $a_{n,n-1}^{(s)}$ approaches zero at least quadratically. Once $a_{n,n-1}^{(s)}$ is small enough, it may be zeroed, so that $\lambda_n := a_{nn}^{(s)}$ is an approximate eigenvalue of A, and the remaining eigenvalues are those of the leading principal submatrix of order $n-1$ in A_s. Therefore, the algorithm can be applied to this smaller submatrix. In general, if m eigenvalues have been computed, one should only work with the leading principal submatrix of order $n-m$, hence a significant increase in efficiency can be obtained. Moreover, according to the previously mentioned results, one may expect that other subdiagonal elements approach zero as well. Although this convergence will be relatively slow, for large n one of these elements may effectively become zero after a number of iterations. In this case, the matrix splits into the form

$$A_s = \begin{bmatrix} B & X & Y \\ 0 & C & Z \\ 0 & 0 & U \end{bmatrix},$$

where U is upper triangular, and B and C are in UHF. Hence, one may apply the QR algorithm to the matrix C, and after having determined its eigenvalues, one proceeds similarly on with B. If C lies between rows and columns l and $n-m$ of A_s, then a QR iteration restricted to C is simply obtained by replacing in Algorithm `eQRs` the loop indices 1 and n by l and $n-m$, respectively. Nevertheless, for computing eigenvectors or Schur form, the rotations applied on rows (columns) of C should also be applied on Z (and X). The same technique is used to process the results returned by `balanc` and `Hess` algorithms.

A natural criterion for deciding on an element $a_{l,l-1}^{(s)}$ as being *negligible*, hence "small" enough to be zeroed, would be $|a_{l,l-1}^{(s)}| \leq \varepsilon_M \|A\|$, where $\|\cdot\|$ is an easily calculable norm (for instance, $\|\cdot\|_\infty$ or $\|\cdot\|_1$). As A is unitarily similar to A_s, the result of zeroing $a_{l,l-1}^{(s)}$ corresponds to a perturbation of order $0(\varepsilon_M \|A\|)$ or smaller in A, which is of the order of errors made for representing A on a computer. But this criterion is not suitable for matrices having norms of consecutive rows (or columns) with very different orders of size, and not strictly decreasing, because some eigenvalues may then be smaller than $\varepsilon_M \|A\|$. If, for instance, $a_{nn}^{(s)}$ tends to such an eigenvalue, the criterion may lead to annihilation of $a_{n,n-1}^{(s)}$ even if $|a_{n,n-1}^{(s)}| > |a_{nn}^{(s)}|$. Therefore, the subdiagonal elements of A_s should be compared to the neighbouring elements, using criteria of the form

$$\begin{aligned} |a_{l,l-1}^{(s)}| &\leq \varepsilon_M \min\{\, |a_{l-1,l-1}^{(s)}|, |a_{ll}^{(s)}|\,\}, \quad \text{or} \\ |a_{l,l-1}^{(s)}| &\leq \varepsilon_M (|a_{l-1,l-1}^{(s)}| + |a_{ll}^{(s)}|). \end{aligned} \tag{2.51}$$

The first criterion is too stringent. Both have the disadvantage of possibly asking for extra iterations, but very fast convergence of the QR algorithm makes the extra effort usually insignificant, and entirely motivated by the resulting increase in reliability. The LAPACK routines use a refined version of the second criterion; if the sum in parantheses is zero, then the 1-norm of the current trailing principal submatrix is used instead.

Algorithm `eQRs` is numerically stable. The matrix obtained by applying it to A, with shift h, is unitarily similar to $A+E$, where

$$\|E\|_F \leq \phi(n)\varepsilon_M \max\{\, \|A\|_F, |h|\,\}, \tag{2.52}$$

and $\phi(n)$ is a slowly increasing function of n. If the elements of A are reasonably balanced, the well-conditioned eigenvalues of A will not be much perturbed. The result (2.52) is still good for matrices with elements which decrease along the diagonal, because small eigenvalues will then be determined first, and $|h|$ in (2.52) may increase only thereafter.

The issues above lay the basis for a general algorithm for finding all eigenvalues of a matrix. Such an algorithm is used, for instance, in the EISPACK package to solve eigenproblems for symmetric matrices. For general matrices, however, Algorithm **eQRs** cannot effectively be used with real arithmetic; this is a reasonable requirement when $A \in \mathbb{R}^{n\times n}$, because A can then be reduced to quasi-triangular form using (real) orthogonal similarity transformations (Theorem 2.7 in § 2.2.1). In such a case, it would be useful that $h \in \mathbb{R}$, but the shift $h = a_{nn}$ cannot approximate a complex eigenvalue. As fast convergence of the QR algorithm requires that good approximation of an eigenvalue be used as a shift, it follows that **eQRs** cannot efficiently compute the complex eigenvalues.

Consider now the *implicit* QR algorithm, which does not actually subtract the shift from the diagonal elements. An upper Hessenberg matrix $B \in \mathbb{R}^{n\times n}$ is called *unreduced* if all its subdiagonal elements are nonzero, i.e. $b_{i+1,i} \neq 0$, $i = 1{:}\,n-1$. The QR algorithm exclusively works on unreduced matrices, but a reduced matrix may be split into unreduced matrices.

Unreduced Hessenberg matrices have the following important property (Golub and Van Loan, 1983; Stewart, 1973; Wilkinson, 1965):

Theorem 2.8 *Let A, B, $Q \in \mathbb{C}^{n\times n}$, where matrix Q is unitary, and B is an unreduced Hessenberg matrix with* positive *subdiagonal elements. If $B = Q^H AQ$, then both B and Q are uniquely determined by the first column of Q (or, equivalently, the first row of Q^H).*

Indeed, let $q_i := Q_{:,i}$, $i = 1{:}\,n$, and assume that we have determined q_1, q_2, $\ldots$, q_k, and the first $k-1$ columns of B. As q_1 is given, it suffices to show how q_{k+1} and the kth column of B are computed. From $QB = AQ$, taking the structure of B into account, we have

$$b_{k+1,k}q_{k+1} + b_{kk}q_k + \cdots + b_{1k}q_1 = Aq_k.$$

Premultiplying by q_i^H, as Q is unitary, it follows that $b_{ik} = q_i^H Aq_k$, for $i = 1{:}\,k$. Since $b_{k+1,k} \neq 0$, q_{k+1} is given by

$$q_{k+1} = (Aq_k - \sum_{i=1}^{k} b_{ik}q_i)/b_{k+1,k}.$$

The normalization condition, $\|q_{k+1}\|_2 = 1$, determines a unique $b_{k+1,k} > 0$.

Note that the positivity condition on $b_{k+1,k}$ has only been required for guaranteeing the uniqueness. Otherwise, $b_{k+1,k}$ and q_{k+1} are determined up to a constant factor with unit modulus.

Assume now that a QR step with shift h has been applied on an unreduced upper Hessenberg matrix A, resulting in a matrix B, $B = Q^H AQ$, where Q is the unitary part of the QR factorization of A. We prove that B may alternately be obtained using the following procedure:

1. Find a unitary matrix P^H whose first column coincides with that of the matrix Q.

2. Apply Algorithm **Hess** (§ 2.2.2) (with $k = 1$, $l = n$) for reducing PAP^H to a matrix B' in upper Hessenberg form.

Indeed, note that $B' = Q'^H AQ'$, with $Q'^H = U_{n-2}U_{n-3}\cdots U_1 P$, where U_i, $i = 1{:}\,n-2$, are the Householder transformations used by **Hess**. But, due to their special form, the premultiplication of a matrix by U_i does not modify the first row of that matrix. So, Q'^H has the first row identical to that of P and, by construction, to that of Q^H. Hence, using Theorem 2.8, it follows that $Q = Q'$ and $B = B'$.

The effect of the shift h is incorporated into the matrix P. In order to determine P, one has to compute the first column of Q. But $A - hI_n = QR$, where R is upper triangular. Hence, the first column of Q is a multiple of the first column of $A - hI_n$,

$$r_{11}q_1 = QRe_1 = (A - hI_n)e_1 = (a_{11} - h, a_{21}, 0, \ldots, 0)^T =: a.$$

Choosing P so that $Pa = \pm\|a\|_2 e_1$, hence $P^H e_1 = \pm a/\|a\|_2$, it follows that the first column of P^H is the same as that of Q, not considering the sign. If A is real and in upper Hessenberg form, we may choose P as a plane rotation in the plane $(1, 2)$, which zeroes the second element in a.

We will now show how stage 2 of the procedure above can efficiently be carried out, taking the special structure of matrix PAP^H into account.

The implicit QR algorithm with origin shifts for a matrix $A \in \mathbb{R}^{n\times n}$ is defined by the relations (2.47), where $A_1 = A$. It can be proven that if all shifts are zero ($\delta_s \equiv 0$), then A_s tends to a matrix in the quasi-triangular form, i.e. for which $a^{(s)}_{i+1,i}a^{(s)}_{i+2,i+1} = 0$, $i = 1{:}\,n-2$, so that all eigenvalues are either isolated on the diagonal, or are those of some 2×2 diagonal submatrices. For fast convergence it is essential to use shifts, and each shift should ultimately be close to an eigenvalue of A (hence of A_s, $s = 1, 2, \ldots$). But some eigenvalues may be complex, and if a complex shift δ_s is used in (2.47), then A_{s+1} will generally be a complex matrix. This difficulty can theoretically be overcome by applying (2.47) again with the shift $\bar{\delta}_s$, so that A_{s+2} should be real. Although the QR step is numerically stable, in that it preserves the eigenvalues, the strategy just described may produce in practice a matrix A_{s+2} with significant imaginary components; merely neglecting them could bring in unacceptable errors. Fortunately, the difficulty can be stepped over by simultaneously performing two QR steps, with shifts δ_s and δ_{s+1}, either both real or complex conjugate. From (2.47), we obtain

$$A_{s+2} = Q_{s+1}^H Q_s^H A_s Q_s Q_{s+1}, \qquad \text{or} \quad Q_s Q_{s+1} A_{s+2} = A_s Q_s Q_{s+1},$$
$$(Q_s Q_{s+1})(R_{s+1} R_s) = (A_s - \delta_s I_n)(A_s - \delta_{s+1} I_n) =: M.$$

(The last equality follows from

$$Q_{s+1} R_{s+1} = A_{s+1} - \delta_{s+1} I_n = Q_s^H A_s Q_s - \delta_{s+1} I_n,$$

by pre- and postmultiplication by Q_s and $Q_s^H(A_s - \delta_s I_n) = R_s$, respectively.) Denoting $Q := Q_s Q_{s+1}$ and $R := R_{s+1} R_s$, where Q is unitary and R upper triangular, we deduce that $QR = M$, i.e. Q and R perform the QR factorization of M; in addition, $QA_{s+2} = A_s Q$. Without loss of generality, we may assume that the Hessenberg matrix A_{s+2} is unreduced. According to Theorem 2.8, if from A_s one determines (by any method) an upper Hessenberg matrix H, so that $A_s \widetilde{Q} = \widetilde{Q} H$ (or $\widetilde{Q}^H A_s \widetilde{Q} = H$), where $\widetilde{Q}$ is unitary, and its first column coincides with that of Q, then $\widetilde{Q} = Q$ and $H = A_{s+2}$. If $\delta_s, \delta_{s+1} \in \mathbb{R}$, or $\delta_{s+1} = \bar{\delta}_s \in \mathbb{C}$, matrix M is real, and, as $QR = M$, it follows that Q is real, and so orthogonal. But any real $n \times n$ matrix can be triangularized by successive premultiplication by $n-1$ real Householder transformations, $U_1, U_2, \ldots, U_{n-1}$, where $U_j = I_n - 2u_j u_j^T$, and u_j has its first $j-1$ elements zero, $\|u_j\|_2 = 1$ (see the preamble to Algorithm **QR** in § 1.4.3). Hence we may take $Q^T = U_{n-1} \cdots U_2 U_1$, so that the first column of $Q = U_1 U_2 \cdots U_{n-1}$ coincides with the first column (or row) of U_1. But U_1 is determined by the first column of M, that has the form $(p, q, r, 0, \ldots, 0)^T \in \mathbb{R}^n$, where

$$\begin{aligned} p &= a_{11}^2 - a_{11}(\delta_s + \delta_{s+1}) + \delta_s \delta_{s+1} + a_{12} a_{21}, \\ q &= a_{21}(a_{11} + a_{22} - \delta_s - \delta_{s+1}), \\ r &= a_{32} a_{21}. \end{aligned} \tag{2.53}$$

(In (2.53), and in the following relations, a_{ij} denotes $a_{ij}^{(s)}$.) For this reason, the vector u_1, defining U_1, has three non-zero elements at most (the first ones), and hence the matrix $U_1 A_s U_1$ has a "bump" of extra (possibly) non-zero elements (if compared to an upper Hessenberg form), in positions (3,1), (4,1), and (4,2). If $U_1 A_s U_1$ is then reduced to the matrix H, in UHF, from all above it follows that $H = A_{s+2}$. This reduction can be performed using Algorithm **Hess**, but, to gain in efficiency, advantage can be taken of the special structure of $U_1 A_s U_1$ and of the matrices generated by **Hess**. Indeed, if $\widetilde{U}_k$, $k = 1{:}n-2$, are the elementary reflectors used by **Hess**, it is easily seen that the vectors $\widetilde{u}_k$, defining $\widetilde{U}_k$, have three non-zero elements at most, and the successive similarity transformations will move the bump along the diagonal. The extra non-zero $(n, n-2)$ element of the matrix $\widetilde{U}_{n-3}\cdots\widetilde{U}_1 U_1 A_s U_1 \widetilde{U}_1 \cdots \widetilde{U}_{n-3}$ can be zeroed using a plane rotation in the plane $(n-1, n)$. The transformations $\widetilde{U}_k$, $k = 1\colon n-3$, have the form $\widetilde{U}_k = \text{block-diag}(I_k, R_k, I_{n-k-3})$, where R_k is a 3×3 reflector, chosen to annihilate the elements $(k+2, k)$ and $(k+3, k)$ of the matrix $\widetilde{U}_{k-1}\cdots\widetilde{U}_1 U_1 A_s U_1 \widetilde{U}_1 \cdots \widetilde{U}_{k-1}$. Clearly, U_1 has the same nature as $\widetilde{U}_k$, and the operations involved are similar. The Householder transformations R_k can be determined by Algorithm `Hous_3` (§ 1.4.3). Such a choice allows their efficient application, as $R_k = I_3 + vw^T$, where $v, w \in \mathbb{R}^3$, and $w_1 = 1$.

So, a step of the implicit QR algorithm with double shift, briefly called a *double QR step*, has been defined. With the usual choice of shifts, δ_s and δ_{s+1} are the eigenvalues of the lowermost current submatrix $a_{i-1:i,i-1:i}$ ($i = n$, initially). Therefore,

$$\begin{aligned}\delta_s + \delta_{s+1} &= a_{i-1,i-1} + a_{ii},\\ \delta_s\delta_{s+1} &= a_{i-1,i-1}a_{ii} - a_{i-1,i}a_{i,i-1}.\end{aligned} \tag{2.54}$$

Denoting $a_{10} = p/a_{21}$, $a_{20} = q/a_{21}$, and $a_{30} = r/a_{21}$, and using (2.54) in (2.53), it follows that

$$\begin{aligned}a_{10} &= ((a_{ii} - a_{11})(a_{i-1,i-1} - a_{11}) - a_{i,i-1}a_{i-1,i})/a_{21} + a_{12},\\ a_{20} &= a_{22} - a_{11} - (a_{ii} - a_{11}) - (a_{i-1,i-1} - a_{11}),\\ a_{30} &= a_{32}.\end{aligned} \tag{2.55}$$

As only the direction of the vector $(p, q, r, 0, \ldots, 0)^T$ is important, the factor a_{21} can be omitted in p, q, and r, and one may work with a_{10}, a_{20}, and a_{30}.

More generally, the following algorithm performs an implicit double QR step on a diagonal principal submatrix of A and can apply the transformations on the entire matrix. (This is useful for computing RSF of A.)

Algorithm idQRs : $(A, U) = \texttt{idQRs}(\ A, a_{10}, a_{20}, a_{30}, U, l, m, i, \underline{i}, \bar{\imath}\ )$

```
/* Apply an implicit double QR step to the submatrix a_{m:i,m:i} */
/* of A ∈ IR^{n×n} in UHF (m ≥ l), with parameters a_10, a_20, a_30; */
/* apply the transformations to the rows i_:i and columns m:ī */
/* (i_ ≤ m, ī ≥ i); update the transformation matrix U. */

IF  m > l,  THEN  a_{m,m-1} ← −a_{m,m-1};
FOR k = m − 1: i − 3,
    IF  k = m − 1,  THEN  x = a_{1:3,0};  ELSE  x = a_{k+1:k+3,k};
                /* 1. Find a modified Householder transformation U_k. */
    (v, w) = Hous_3( x );
    IF  k ≥ m,  THEN  a_{k+1,k} = x_1;   a_{k+2:k+3,k} = 0;
```

FOR $j = k+1 : \bar{\imath}$,

/* 2. Premultiply by U_k. */

$t = w^T a_{k+1:k+3,j}$; $a_{k+1:k+3,j} \leftarrow a_{k+1:k+3,j} + tv$;

FOR $j = \underline{i} : \min(k+4, i)$,

/* 3. Postmultiply by U_k. */

$t = a_{j,k+1:k+3}v$; $a_{j,k+1:k+3} \leftarrow a_{j,k+1:k+3} + tw^T$;

$u_{:,k+1:k+3} \leftarrow u_{:,k+1:k+3}(I_3 + vw^T)$;

END FOR

/* 4. Find and apply the final plane rotation. */

$(c, s) = \mathtt{rot}(\ a_{i-1,i-2}, a_{i,i-2}\ )$;

$$a_{i-1:i,i-1:i} \leftarrow \begin{bmatrix} c & s \\ -s & c \end{bmatrix} a_{i-1:i,i-1:i};$$

$$a_{\underline{i}:i,i-1:i} \leftarrow a_{\underline{i}:i,i-1:i} \begin{bmatrix} c & -s \\ s & c \end{bmatrix}; \quad u_{:,i-1:i} \leftarrow u_{:,i-1:i} \begin{bmatrix} c & -s \\ s & c \end{bmatrix};$$

For the moment, we may assume that $l = m = \underline{i} = 1$, and $i = \bar{\imath} = n$. In this case, Algorithm **idQRs** requires about $5n^2$ *flops* (without accumulating the transformations in U). The matrix produced is exactly similar to a matrix $A + E$, where E satisfies $\|E\|_F \le \phi(n)\varepsilon_M \|A\|_F$, and the function $\phi(n)$ grows slower than n^2.

Algorithm **idQRs** can iteratively be applied to find all eigenvalues of a matrix. Using the shifts implicitly defined by (2.54), the convergence is usually very fast and, at some s, either $a_{i,i-1}$ or $a_{i-1,i-2}$ becomes negligible. In the first case, a_{ii} may be considered as eigenvalue, and, eliminating the ith row and column, a Hessenberg matrix of order $i-1$ is obtained. Likewise, in the second case, the eigenvalues of the 2×2 current trailing submatrix of A_s are eigenvalues of A, and the problem reduces to a similar one, of order $i-2$. Obviously, the final matrix cannot generally be in a triangular form, but only in QTF.

The process described above may not converge in some special cases. For instance, for the matrix defined by $a_{j+1,j} = 1$, $j = 1 : n-1$, $a_{1n} = 1$, the remaining elements being zero, the shifts will be zero, and A is orthogonal, and invariant with respect to the unshifted QR algorithm. Nevertheless, if Algorithm **idQRs** with arbitrary shifts is applied, the convergence will be very fast. For this reason, if the convergence has not been obtained for an eigenvalue or a pair of eigenvalues of the block $(i-1:i, i-1:i)$, the usual shifts are replaced, after 10 and 20 iterations, by some "exceptional" shifts, defined by

$$\begin{aligned} \delta_s + \delta_{s+1} &= 1.5(|a_{i,i-1}| + |a_{i-1,i-2}|), \\ \delta_s \delta_{s+1} &= (|a_{i,i-1}| + |a_{i-1,i-2}|)^2. \end{aligned} \tag{2.56}$$

Table 2.2 Sample of the numbers of QR iterations performed for solving eigenproblems of various order.

Order n	Iterations	Order n	Iterations
4	6	30	54
8	14	40	60
12	17	60	115
16	17	64	85
20	35	80	112

Current implementations use $\delta_s + \delta_{s+1} = 2\alpha$ and $\delta_s \delta_{s+1} = \alpha^2 - w$, where $\alpha = 0.75s$, $s := |a_{i,i-1}| + |a_{i-1,i-2}|$, and hence $w = -0.4375s^2$. The maximum number of iterations for an eigenvalue (or a pair of eigenvalues) is 30. (The LAPACK routines allow for a total of $30n$ iterations, not for 30 iterations for each eigenvalue.) In practice, the average number of QR iterations per eigenvalue, r, does not exceed 2, and often $r = 1.7$. Hence the QR algorithm requires about $5rn^3$ *flops*. If the transformations used are accumulated, extra $2.5rn^3$ *flops* are needed. This accumulation could be an option of Algorithm `idQRs` (and of other related algorithms), because it is not needed when only eigenvalues are required. Table 2.2 indicates typical performances for problems of various order, encountered when solving Riccati equations.

After each double QR step, the subdiagonal elements are checked for negligibility, using the second criterion (2.51). If there are negligible elements, then the current submatrix is split into two (or more) Hessenberg submatrices. Let l be the largest index so that $a_{l,l-1}$ is negligible, and $l = 1$, if no such element exists. Then the iteration can continue for the submatrix $a_{l:i,l:i}$. After l has been found, one checks if the submatrix with indices $l: i-2$ has two consecutive subdiagonal elements whose product is negligible; in such a case, it is possible to work with a still smaller submatrix, beginning in the row and column m, where m satisfies

$$|a_{m,m-1}|\,(|q_m|+|r_m|) \le \varepsilon_M\,|p_m|\,(|a_{m-1,m-1}|+|a_{mm}|+|a_{m+1,m+1}|), \quad (2.57)$$

with

$$\begin{aligned} p_m &= a_{mm}^2 - a_{mm}(\delta_s + \delta_{s+1}) + \delta_s\delta_{s+1} + a_{m,m+1}a_{m+1,m}, \\ q_m &= a_{m+1,m}(a_{mm} + a_{m+1,m+1} - \delta_s - \delta_{s+1}), \\ r_m &= a_{m+2,m+1}a_{m+1,m}. \end{aligned} \quad (2.58)$$

In fact, if the Householder transformation $\widetilde{U}_m$, defined by p_m, q_m, and r_m in (2.58), were applied, it would generate non-zero elements in positions

$(m+1, m-1)$ and $(m+2, m-1)$. The required condition is that these elements should be negligible. In order to obtain a satisfactory criterion, without being too complicate, approximations of these elements (which are valid when they are negligible) are used. The condition (2.57) has been obtained from such considerations. Since the relations (2.58) have the same form as (2.53), searching for subdiagonal elements with negligible product may be combined with finding the transformation at step 1 of Algorithm **idQRs**, computed by **Hous_3**. If p_m, q_m, and r_m in (2.58) (hence their moduli in (2.57)) are divided by $a_{m+1,m}$, if $a_{m+1,m} \neq 0$, the computing formulas have the form (2.55), so that they should simultaneously provide the values corresponding to the incorporation of the shifts δ_s and δ_{s+1}. In addition, the values p_m, q_m, and r_m (and those to be determined in the double QR step) are scaled by $|p_m|+|q_m|+|r_m|$. Therefore, the QR iteration uses only the transformations $\widetilde{U}_m$, $\widetilde{U}_{m+1}$, ..., $\widetilde{U}_{i-3}$, and these are only applied to the submatrix in rows $m{:}\,i$ and columns $m-1{:}\,i$. If Schur form T is desired, the transformations should be applied on all rows and columns, as previously specified with Algorithm **eQRs**.

In an implementation, the final plane rotation in Algorithm **idQRs** is replaced by a modified Householder transformation defined by **Hous_2** (see § 1.4.3). Step 4 is omitted, and an extra cycle of the FOR loop is performed. The LAPACK routines use a modification of Algorithm **Hous**, described after the statement of Algorithm **Hess** in § 2.2.2.

The eigenvalues of the 2×2 diagonal submatrices can be computed by the usual relations. For instance, consider that

$$a_{i-1:i,i-1:i} = \begin{bmatrix} \alpha & \beta \\ \gamma & \delta \end{bmatrix}, \tag{2.59}$$

is the lowermost current submatrix, lying in the rows and columns $i-1$ and i. Denote $w := \beta\gamma$, $p := (\alpha - \delta)/2$, and let λ_{i-1} and λ_i be the eigenvalues of this submatrix. As $\lambda_{i-1} + \lambda_i = \alpha + \delta$ and $\lambda_{i-1}\lambda_i = \alpha\delta - w$, it follows that

$$\{\, \lambda_{i-1}, \lambda_i \,\} = \{\, \delta + p \pm (p^2 + w)^{1/2} \,\}. \tag{2.60}$$

Complex conjugate roots are obtained if $p^2 + w < 0$. Otherwise, denoting $z := p + \operatorname{sign}(p)(p^2 + w)^{1/2}$, then $\lambda_{i-1} = \delta + z$, and $\lambda_i = \lambda_{i-1}$, if $z = 0$, or $\lambda_i = \delta - w/z$, if $z \neq 0$. The LAPACK routines use special formulas, for the canonical real Schur form, which we shall discuss in § 2.2.4.

The template algorithm below calls Algorithm **Schur2**, completely described in § 2.2.4, for obtaining eigenvalues of the 2×2 diagonal submatrices. Algorithm **Schur2** is defined by the following statement:

$(\Theta, c, s, \Lambda) =$ **Schur2**(Θ);

where $\Theta \in \mathbb{R}^{2\times 2}$ is given. We will assume, for the moment, that `Schur2` sets $c = 1$, $s = 0$, and computes in Λ the eigenvalues of Θ, using, for instance, the formulas (2.60) above, when Θ is the matrix in (2.59). We may also consider that the last two input parameters of Algorithm `Schur` below are set as follows: $\underline{l} = 1$, $\underline{n} = n$.

Algorithm Schur : $(A, U, \Lambda) =$ `Schur(` $A, U, \underline{l}, \underline{n}$ `)`

$/*$ Reduce $A \in \mathbb{R}^{n\times n}$ in upper Hessenberg form to RSF, $T = U^T AU$, $*/$
$/*$ using implicit double-shift QR algorithm, compute eigenvalues Λ, $*/$
$/*$ and update the transformation matrix U, $*/$
$/*$ assuming $a_{1:\underline{l}-1,1:\underline{l}-1}$ and $a_{\underline{n}+1:n,\underline{n}+1:n}$ already reduced. $*/$

$/*$ 1. Start main loop: i decreases from $\underline{n}$ to $\underline{l}$ in steps of 1 or 2. $*/$
$\underline{i} = 1$; $\bar{i} = n$; $itn = 30(\underline{n} - \underline{l} + 1)$; $i = \underline{n}$;
WHILE $i \geq \underline{l}$, DO
$l = \underline{l}$; $/*$ 2. Perform QR iterations on $a_{\underline{l}:i,\underline{l}:i}$ until trailing $*/$
$/*$ submatrix of order 1 or 2 splits off. $*/$
FOR $itr = 0{:}\, itn$,
$/*$ 3. Look for a single small subdiagonal element. $*/$
FOR $k = i{:}\, -1{:}\, l+1$,
IF $|a_{k,k-1}| \leq \varepsilon_M(|a_{k-1,k-1}| + |a_{kk}|)$, THEN EXIT LOOP;
$l = k$; IF $l > \underline{l}$, THEN $a_{l,l-1} = 0$; $/*$ $a_{l,l-1}$ negligible. $*/$
IF $l \geq i - 1$, THEN EXIT LOOP; $/*$ $a_{l:i,l:i}$ splits off. $*/$
$/*$ Active submatrix $a_{l:i,l:i}$, where $l = \underline{l}$, or $a_{l,l-1}$ is negligible. $*/$
IF $itr = 10$ OR $itr = 20$, THEN
$/*$ 4. Compute exceptional shifts. $*/$
$s = |a_{i,i-1}| + |a_{i-1,i-2}|$; $\alpha = .75s$; $\delta = \alpha$; $w = -0.4375s^2$;
ELSE
$/*$ 5. Compute usual shifts. $*/$
$\alpha = a_{i-1,i-1}$; $\delta = a_{ii}$; $w = a_{i,i-1}a_{i-1,i}$;
END IF
$/*$ 6. Look for two consecutive small subdiagonal elements. $*/$
FOR $m = i - 2{:}\, -1{:}\, l$,
$a_{10} = ((\alpha - a_{mm})(\delta - a_{mm}) - w)/a_{m+1,m} + a_{m,m+1}$;
$a_{20} = a_{m+1,m+1} - a_{mm} - (\alpha - a_{mm}) - (\delta - a_{mm})$;
$a_{30} = a_{m+2,m+1}$;
IF $m \neq l$, THEN
$t = |a_{m-1,m-1}| + |a_{mm}| + |a_{m+1,m+1}|$;
IF $|a_{m,m-1}|(|a_{20}| + |a_{30}|) \leq \varepsilon_M |a_{10}| t$, THEN EXIT LOOP;
END FOR
$/*$ 7. Apply a double QR step on the matrix $a_{m:i,m:i}$. $*/$

/* If only eigenvalues are desired, set $\underline{i} = m$, $\bar{\imath} = i$. */

$(A, U) =$ **idQRs**($A, a_{10}, a_{20}, a_{30}, U, l, m, i, \underline{i}, \bar{\imath}$);

END FOR

/* 8. No convergence. */

IF $itr > itn$, THEN "No convergence at eigenvalue i"; STOP;

IF $l = i$, THEN $\lambda_i = a_{ii}$; /* Here, l is either i or $i-1$. */

IF $l = i-1$, THEN

$(a_{l:i,l:i}, c, s, \lambda_{l:i}) =$ Schur2($a_{l:i,l:i}$); $a_{l:i,l:i} \leftarrow \begin{bmatrix} c & s \\ -s & c \end{bmatrix} a_{l:i,l:i}$;

$a_{\underline{i}:i-2,l:i} \leftarrow a_{\underline{i}:i-2,l:i} \begin{bmatrix} c & -s \\ s & c \end{bmatrix}$; $u_{:,l:i} \leftarrow u_{:,l:i} \begin{bmatrix} c & -s \\ s & c \end{bmatrix}$;

$itn \leftarrow itn - itr$; $i \leftarrow l - 1$;

END WHILE

Note that Algorithm **Schur** assumes that Schur form is desired, so the transformations used by Algorithm **idQRs** at step 7 are applied to the entire matrix A ($\underline{i} = 1$, $\bar{\imath} = n$). If the input parameter U is set to identity matrix, then the returned U contains the product of the transformations performed by Algorithm **Schur**. If the input U is the matrix produced by Algorithm **Hess**, then the output U contains the product of the transformations which reduced the original matrix to the quasi-triangular form. The calculations of U, and/or Λ, are skipped over if the corresponding parameters are missing. Algorithm **Schur** can take the output of the balancing procedure into account. The implementation of the negligibility test at step 3 could be more sophisticated, as shown in Algorithm **mSchur** below, which use multiple-shift QR iterations (Bai and Demmel, 1989). It also refines the essential calculations in order to avoid overflows and underflows and to increase accuracy and reliability of results. For instance, if κ denotes the smallest representable number whose reciprocal is also representable, also called *safe minimum*, and $\mathcal{K} = 1/\kappa$, then overflow should not occur if $\|A\| \le \mathcal{K}^{1/2}$.

Algorithm **mSchur** has the same function and essentially the same logic as Algorithm **Schur**. The main difference lies in that it can simultaneously apply p shifts, with $p > 2$, using a multiple-shift QR step.[3] Let G be the current trailing submatrix of order $q \ge p$, and S its lowermost submatrix of order p. The eigenvalues w_j of S, $j = 1{:}p$, are found by **Schur**, and are used as shifts. Algorithm **mSchur** computes a vector v, defined as

$$v := (G - w_1 I_q)(G - w_2 I_q) \cdots (G - w_p I_q) e_1,$$

and determines a Householder transformation H that reduces v to e_1. By

[3] We will use such a technique in Chapter 4, for the special case of Hamiltonian matrices.

applying H from left and right to G (in A) a non-zero bump below the subdiagonal is produced. Then, the upper Hessenberg form is restored by suitably defined Householder transformations. The process is similar to that in Algorithm **idQRs**, but transformations with order larger than three are used. The advantage will be that of the calculations being implemented using Level 2 BLAS and Level 1 BLAS operations, which improve efficiency on modern computer architectures. The value of parameter p is machine dependent, the optimum value typically placing between 4 and 16; it cannot be too large, because the effort for computing a multiple-shift QR step becomes quite demanding for high values.

Algorithm mSchur : $(A, U, \Lambda) = \texttt{mSchur}(\ A, U\ )$

```
/* Reduce A ∈ IR^{n×n} in upper Hessenberg form to RSF, T = U^T AU, */
/* using implicit multiple-shift QR algorithm, compute eigenvalues Λ, */
/* and update the transformation matrix U. */

  /* 1. Set to zero the rows and columns below the first subdiagonal. */
FOR j = 1: n − 2,
   a_{j+2:n,j} = 0;
p = min{ n, 15 };   K = 1/κ;   s_min = κn/ε_M;   itn = 30n;   i = n;
  /* 2. Start main loop; each iteration works with an active submatrix */
    /* of order at most p, a_{l:i,l:i}, i = n: −p: 1; l = 1, or a_{l,l−1} negligible. */
WHILE i ≥ 1, DO
   l = 1;              /* 3. Apply multiple-shift QR iterations on a_{1:i,1:i}. */
   FOR itr = 0: itn,
                       /* 4. Look for a single small subdiagonal element. */
     FOR k = i: −1: l + 1,
       t = |a_{k−1,k−1}| + |a_{kk}|;   IF t = 0, THEN t = ||a_{l:i,l:i}||_1;
       IF |a_{k,k−1}| ≤ max{ ε_M t, s_min }, THEN EXIT LOOP;
     l = k;   IF l > 1, THEN a_{l,l−1} = 0;          /* a_{l,l−1} negligible. */
     IF l ≥ i − p + 1, THEN EXIT LOOP;              /* a_{l:i,l:i} splits off. */
                                                   /* Active submatrix is a_{l:i,l:i}. */
     IF itr = 20 OR itr = 30, THEN
       FOR j = i − p + 1: i,
         λ_j = 1.5(|a_{j,j−1}| + |a_{jj}|);   /* 5. Compute exceptional shifts. */
     ELSE
         /* 6. Use eigenvalues of p × p trailing submatrix as shifts.[4] */
       S = a_{i−p+1:i,i−p+1:i};   (S, λ_{i−p+1:i}) = Schur( S );
     END IF
```

[4] If Algorithm Schur failed to converge, unconverged diagonal elements are used as remaining shifts.

```
        /* 7. Form v = (G − w_1 I)(G − w_2 I)···(G − w_p I)e_1, where */
                  /* G := a_{l:i,l:i}, and w_{j−i+p} = λ_j, j = i − p + 1: i. */
  v_{2:p+1} = 0;   v_1 = 1;   k = 1;
  FOR j = i − p + 1: i,
    IF ℑ(λ_j) ≥ 0, THEN
      IF ℑ(λ_j) = 0, THEN
                                                      /* Real shift. */
        ṽ = v_{1:k+1};   v ← a_{l:l+k,l:l+k−1} ṽ − λ_j v;   k ← k + 1;
      ELSE
                                  /* Complex conjugate pair of shifts. */
        ṽ = v_{1:k+1};   ṽ ← a_{l:l+k,l:l+k−1} v − 2ℜ(λ_j) ṽ;
        t = 1/ max{ ||ṽ||_∞, s_min };   ṽ ← ṽt;   t ← t|λ_j|^2;
        v ← a_{l:l+k+1,l:l+k} ṽ + tv;   k ← k + 2;
      END IF
            /* 8. Scale v_{1:k} so that ||v||_∞ = 1. If v = 0, set v = e_1. */
      t = ||v_{1:k}||_∞;
      IF t = 0, THEN
        v_{1:k} = e_1;
      ELSE
        t = max{ t, s_min };   v ← v/t;
      END IF
    END IF
  END FOR
                                  /* 9. Apply multiple-shift QR step. */
  FOR k = l: i − 1,
    q = min{ p+1, i − k + 1 };   IF k > l, THEN v = a_{k:k+q−1,k−1};
              /* 10. Find current Householder transformation H. */
    (H, r) = Hous( v );
    IF k > l, THEN a_{k,k−1} = −r;   a_{k+1:i,k−1} = 0;
              /* 11. Apply H from the left to transform the rows */
                                  /* of the matrix in columns k to n. */
    a_{k:k+q−1,k:n} ← H a_{k:k+q−1,k:n};
          /* 12. Apply H from the right to transform the columns */
                      /* of the matrix in rows 1 to min{k + q, i}. */
    a_{1:min{k+q,i},k:k+q−1} ← a_{1:min{k+q,i},k:k+q−1} H;
            /* 13. Accumulate transformations in the matrix U. */
    u_{:,k:k+q−1} ← u_{:,k:k+q−1} H;
  END FOR
END FOR
```

IF $itr > itn$, THEN "No convergence at eigenvalue i"; STOP;

/* 14. Apply double-shift QR algorithm to a */
/* submatrix of order p or less. */

$(A, U, \Lambda) =$ `Schur(` A, U, l, i `);`

/* 15. Restart the main loop. */

$itn = itn - itr$; $i = l - 1$;
END WHILE

2.2.4 Real Schur form computation and ordering

The matrix obtained by the implicit QR algorithm is in QTF. It can be reduced to the real Schur form (RSF), representing the maximum reduction achievable by real arithmetic, exclusively using orthogonal transformations. In order to find RSF, the 2×2 diagonal blocks corresponding to pairs of real eigenvalues should be triangularized. Suitably chosen plane rotations can be used. Note that the following transformation should be accomplished for the block in position $i - 1$:

$$\begin{bmatrix} c & s \\ -s & c \end{bmatrix} \begin{bmatrix} \alpha & \beta \\ \gamma & \delta \end{bmatrix} \begin{bmatrix} c & -s \\ s & c \end{bmatrix} = \begin{bmatrix} \lambda_{i-1} & * \\ 0 & \lambda_i \end{bmatrix}, \tag{2.61}$$

where the previous notations have been used (see (2.59)), and $*$ denotes an element whose value is meaningless in determining the rotation. By zeroing the (2,1) element, we obtain the equation $c^2\gamma - cs(\alpha - \delta) - s^2\beta = 0$, or, equivalently, $c^2\gamma - 2pcs - s^2\beta = 0$. Hence, the ratio c/s must have the value $c/s = (p \pm (p^2 + w)^{1/2})/\gamma$. So, we may choose c and s so that $c/s = z/\gamma$. Taking $c = z/r$, $s = \gamma/r$, where $r = \gamma^2 + z^2$, the condition $c^2 + s^2 = 1$ holds. Equivalently, c and s define the rotation which annihilates the second element of the vector $(z, \gamma)^T$. Such formulas are implemented in the EISPACK package.

The LAPACK routines use a different algorithm, that also brings the 2×2 diagonal blocks to the canonical real Schur form. Specifically, this algorithm computes the Schur factorization of a real 2×2 general matrix in a standard form:

$$\begin{bmatrix} \alpha & \beta \\ \gamma & \delta \end{bmatrix} = \begin{bmatrix} c & -s \\ s & c \end{bmatrix} \begin{bmatrix} \widetilde{\alpha} & \widetilde{\beta} \\ \widetilde{\gamma} & \widetilde{\delta} \end{bmatrix} \begin{bmatrix} c & s \\ -s & c \end{bmatrix}, \tag{2.62}$$

where either $\widetilde{\gamma} = 0$, so that $\widetilde{\alpha}$ and $\widetilde{\delta}$ are real eigenvalues of the matrix (and $|\widetilde{\alpha}| \geq |\widetilde{\delta}|$), or $\widetilde{\alpha} = \widetilde{\delta}$ and $\widetilde{\beta}\widetilde{\gamma} < 0$, so that $\widetilde{\alpha} \pm (\widetilde{\beta}\widetilde{\gamma})^{1/2}$ are complex conjugate eigenvalues.

If $\gamma = 0$, then $c = 1$, and $s = 0$, and the given values are not modified. If $\beta = 0$, then $c = 0$, and $s = 1$; so, $\widetilde{\alpha} = \delta$, $\widetilde{\delta} = \alpha$, $\widetilde{\beta} = -\gamma$, $\widetilde{\gamma} = 0$. Otherwise, if the given matrix is not in standardized form, a plane rotation is first chosen which makes the diagonal elements equal. Denoting $t := \alpha - \delta$, $\sigma := \beta + \gamma$, from (2.62) with $\widetilde{\delta} = \widetilde{\alpha}$, it follows that c and s should satisfy

$$\begin{aligned} tc + \sigma s + (\widetilde{\beta} + \widetilde{\gamma})s &= 0, \\ \sigma c - ts - (\widetilde{\beta} + \widetilde{\gamma})c &= 0. \end{aligned}$$

These equations give the "stable" formula $c/s = -(\sigma + \text{sign}(\sigma)\tau)/t$, where $\tau := (\sigma^2 + t^2)^{1/2}$. Using $c^2 + s^2 = 1$, the definitions for c and s in Algorithm `Schur2` below can be obtained. Finally, if $\widetilde{\beta}\widetilde{\gamma} > 0$ in the transformed matrix, then another plane rotation, with parameters $\widetilde{c}$ and $\widetilde{s}$, is chosen to annihilate the $(2,1)$ element. The relations for $\widetilde{c}$ and $\widetilde{s}$ follow by specializing those described previously (see (2.61)), for $\alpha = \delta$.

Algorithm Schur2 : $(\alpha, \beta, \gamma, \delta, c, s, \lambda_1, \lambda_2) = \texttt{Schur2}(\ \alpha, \beta, \gamma, \delta\ )$

/* Compute the canonical real Schur form and the eigenvalues */
/* of a real 2×2 general matrix. */

IF $\gamma = 0$, THEN $c = 1$; $s = 0$; STOP;
IF $\beta = 0$, THEN
  $c = 0$; $s = 1$; $\alpha \leftrightarrow \delta$; $\beta = -\gamma$; $\gamma = 0$; STOP;
IF $\alpha = \delta$ AND $\beta\gamma < 0$, THEN STOP;

/* 1. Equate the diagonal elements. */
$t = \alpha - \delta$; $p = t/2$; $\sigma = \beta + \gamma$; $\tau = (\sigma^2 + t^2)^{1/2}$;
$c = ((1 + |\sigma|/\tau)/2)^{1/2}$; $s = -\text{sign}(\sigma)p/(\tau c)$;
$\begin{bmatrix} \alpha & \beta \\ \gamma & \delta \end{bmatrix} \leftarrow \begin{bmatrix} c & s \\ -s & c \end{bmatrix} \begin{bmatrix} \alpha & \beta \\ \gamma & \delta \end{bmatrix} \begin{bmatrix} c & -s \\ s & c \end{bmatrix}$;
$t = (\alpha + \delta)/2$; $\alpha = t$; $\delta = t$;
IF $\gamma \neq 0$, THEN
  IF $\beta \neq 0$, THEN
    IF $\beta\gamma > 0$, THEN
      /* 2. Triangularize a block with real eigenvalues. */
      $p = \text{sign}(\gamma)(\beta\gamma)^{1/2}$; $\tau = 1/(|\beta + \gamma|)^{1/2}$;
      $\widetilde{c} = |\beta|^{1/2}\tau$; $\widetilde{s} = |\gamma|^{1/2}\tau$;
      $\alpha = t + p$; $\delta = t - p$; $\beta \leftarrow \beta - \gamma$; $\gamma = 0$;
      $t = c\widetilde{c} - s\widetilde{s}$; $s \leftarrow c\widetilde{s} + s\widetilde{c}$; $c = t$;
    END IF
  ELSE
    $\beta = -\gamma$; $\gamma = 0$; $t = c$; $c = -s$; $s = t$;
  END IF
END IF

/* 3. Set eigenvalues (real, if $\gamma = 0$). */
$\tau = (|\beta\gamma|)^{1/2}$; $\lambda_1 = \alpha + j\tau$; $\lambda_2 = \delta - j\tau$;

For convenience, the four parameters α, β, γ, and δ, in the defining statement of Algorithm `Schur2` can be replaced by the name of the corresponding matrix. No doubt, the transformations just described are useless if only eigenvalues are to be found, but they are essential in the computation of Schur vectors, and eigenvectors.

Assume that matrix A has been reduced to UHF using Algorithm `Hess` (or its block correspondent) in § 2.2.2, and that the transformations performed during this process have been accumulated, resulting in an orthogonal matrix U. If we proceed on by accumulating in U all column transformations determined by the QR algorithm, including the rotations used to triangularize the 2×2 blocks with real eigenvalues (as in Algorithm `(m)Schur`), then an orthogonal matrix U is finally obtained, which accomplishes the reduction of the original matrix A to RSF, i.e. $U^T AU = T$, where T is in RSF.

The reduction of a matrix in UHF to RSF requires about $5rn^3$ *flops* (including accumulation of transformations), where r is the average number of QR iterations per eigenvalue. Adding the effort for initial reduction to UHF, a total of about $(2 + 5r)n^3$ *flops* is obtained. A practical approximation will be $10n^3$ *flops*.

We shall now discuss how orthogonal bases for invariant subspaces are computed. The vectors $v_1, v_2, \ldots, v_q$ generate an *A-invariant subspace* of order q, $\mathcal{S}$, if $Av_i \in \mathcal{S}$, for $i = 1{:}q$. Clearly, if A is similar to J in JCF from (2.40), then each eigenvector x_j generates an A-invariant subspace of order 1, because, using such relations as (2.41),

$$Ax_j = \lambda_l x_j, \qquad j = 1 + \sum_{k=0}^{l-1} m_k, \quad l = 1{:}r,$$

where $m_0 = 0$, and m_k, $k = 1{:}r$, are the orders of elementary Jordan blocks (or multiplicities of the eigenvalues of A, if A is nonderogatory). Moreover, from (2.41), it follows that the principal vectors x_i, corresponding to the Jordan block $J_{\lambda_l}^{(m_l)}$, generate an A-invariant subspace $\mathcal{S}_l$ of order m_l, associated with the eigenvalue λ_l. One can similarly say about the invariant subspaces associated with a set of eigenvalues of A. For some applications it is necessary to determine orthogonal bases for these subspaces. Such bases can theoretically be obtained by orthogonalizing the principal vectors (using, for example, the Gram-Schmidt process). In practice, this approach is unsatisfactory, because the principal vectors can be nearly linearly dependent (Golub and Wilkinson, 1976). Orthogonal bases for invariant subspaces associated with a set of eigenvalues can reliably be computed using the reduction to RSF.

Let $A \in \mathbb{R}^{n\times n}$, and $T = U^T AU$, where U is an orthogonal matrix which reduces A to T, in RSF. For convenience, we shall first assume that T is upper triangular. Let us consider the eigenvalues in the first s positions along the diagonal of T. We can write

$$U^T AU = \begin{bmatrix} T_{11} & T_{12} \\ 0 & T_{22} \end{bmatrix}, \qquad T_{11} \in \mathbb{R}^{s\times s}, \qquad AU_s = U_s T_{11}, \qquad (2.63)$$

where U_s consists in the first s columns of U, $U_s := U_{:,1:s}$. (The relation (2.63) is still true if T_{11} contains 2×2 blocks, provided that such a block does not begin at position s.) The last relation shows that the first s columns of U form an orthogonal basis for the A-invariant subspace associated with the s eigenvalues of T_{11}. Indeed, any column u_i of U_s, $i = 1{:}\,s$, satisfies $Au_i = U_s t_i$, where t_i denotes the ith column of T_{11}, hence Au_i is a linear combination of the orthogonal columns of U_s.

There is no direct method for finding an orthogonal basis for an A-invariant subspace associated with a set of eigenvalues, λ_i, which are not in the leading positions of T. But, given an arbitrary collection of λ_i, it is possible to transform T into an upper triangular matrix $\widetilde{T}$, having the eigenvalues λ_i in the leading positions, by means of some orthogonal similarity transformations. Hence, there is an orthogonal matrix Y, so that

$$Y^T TY = \widetilde{T}. \qquad (2.64)$$

Matrix Y can be obtained as a product of some plane rotations, each one exchanging adjacent 1×1 diagonal blocks. Indeed, let

$$T = \begin{bmatrix} p & q \\ 0 & r \end{bmatrix}.$$

We show that there is a plane rotation R so that,

$$R\begin{bmatrix} p & q \\ 0 & r \end{bmatrix}R^T = \begin{bmatrix} r & q \\ 0 & p \end{bmatrix}.$$

As

$$R\begin{bmatrix} p & q \\ 0 & r \end{bmatrix}R^T = R\left(\begin{bmatrix} 0 & q \\ 0 & r-p \end{bmatrix} + pI_2\right)R^T,$$

it is enough to prove that there is an R so that

$$R\begin{bmatrix} 0 & q \\ 0 & r-q \end{bmatrix}R^T = \begin{bmatrix} r-p & q \\ 0 & 0 \end{bmatrix}, \qquad \text{or}$$

$$\begin{bmatrix} c & s \\ -s & c \end{bmatrix}\begin{bmatrix} 0 & q \\ 0 & r-q \end{bmatrix} = \begin{bmatrix} r-p & q \\ 0 & 0 \end{bmatrix}\begin{bmatrix} c & s \\ -s & c \end{bmatrix},$$

from which it follows the condition $(r-p)c - qs = 0$, hence R is defined by

$$c = q/\rho, \quad s = (r-p)/\rho, \quad \rho := ((r-p)^2 + q^2)^{1/2}.$$

The exchange can also be performed in the complex case, replacing q by $\bar{q}$ in the transformed matrix. When $T \in \mathbb{R}^{n\times n}$ has 2×2 diagonal blocks, corresponding to complex conjugate eigenvalues, an analogous result is valid, for which a complex pair is always dealt with as the spectrum of a real 2×2 block. There are algorithms which exchange 1×1 by 2×2 (or 2×2 by 1×1) and 2×2 by 2×2 adjacent blocks (Stewart, 1976). In the latter case, some extra QR steps can be taken, until $t_{j+n_2,j+n_2-1}$ may be regarded as negligible, where n_1 and n_2 are orders of the first and second diagonal block, and j is the original beginning position of the first block. Note that the 2×2 blocks do not remain invariant, but only their eigenvalues do. Details about such an algorithm are given in a more general case in Section 3.2 of Chapter 3.

An alternate approach is based on solving a special Sylvester equation like (2.42), where T_{11}, T_{22}, and T_{12}, are adjacent blocks of T, and the first two should be swapped. Powerful algorithms for computing the solution of Sylvester equations are presented in § 2.3.1. In order to avoid overflows, the solution is scaled; actually, one solves $T_{11}X - XT_{22} = sT_{12}$ for X, where s is the returned scaling factor. When $n_1 = n_2 = 2$, the swapping algorithm generates and applies two elementary reflectors H_1 and H_2 so that

$$H_2H_1 \begin{bmatrix} -x_{11} & -x_{12} \\ -x_{21} & -x_{22} \\ s & 0 \\ 0 & s \end{bmatrix} = \begin{bmatrix} * & * \\ 0 & * \\ 0 & 0 \\ 0 & 0 \end{bmatrix}, \tag{2.65}$$

where x_{ij}, $i,j = 1{:}2$, are the elements of X.

As the transformed matrix T could be far from Schur form, the transformations are first provisionally applied to a working matrix D, and one tests if the swap can reliably be performed. If not, the blocks are not swapped, T is unchanged, and a warning is issued. Otherwise, the transformations are also applied to T. A detailed algorithm is listed below; it uses Algorithm `Sylsc2` described in § 2.3.1.

Algorithm swap2 : $(T,U) = \texttt{swap2}(\, T, j, n_1, n_2, U\,)$

/* Swap adjacent diagonal blocks T_{11} and T_{22} of T in canonical RSF, */
/* $T_{11} = t_{j:j_1,j:j_1}$, $T_{22} = t_{j_1+1:j_2,j_1+1:j_2}$, $j_1 := j + n_1 - 1$, $j_2 := j_1 + n_2$, */
/* $n_1 \le 2$, $n_2 \le 2$, and update U. */

IF $n_1 = 1$ AND $n_2 = 1$, THEN

/* 1. Swap two 1×1 blocks. */

$p = t_{jj}$; $q = t_{j,j+1}$; $r = t_{j+1,j+1}$; $(c, s) = \texttt{rot}(\ q, r - p\ )$;

$t_{j:j+1,j+2:n} \leftarrow \begin{bmatrix} c & s \\ -s & c \end{bmatrix} t_{j:j+1,j+2:n}$;

$t_{1:j-1,j:j+1} \leftarrow t_{1:j-1,j:j+1} \begin{bmatrix} c & -s \\ s & c \end{bmatrix}$; $t_{jj} = r$; $t_{j+1,j+1} = p$;

$u_{:,j:j+1} \leftarrow u_{:,j:j+1} \begin{bmatrix} c & -s \\ s & c \end{bmatrix}$; /* 2. Accumulate transformation. */

ELSE

/* Swapping involves at least one 2×2 block. */

$j_1 := j + n_1 - 1$; $j_2 := j_1 + n_2$; $D = t_{j:j_2,j:j_2}$;

$b = \max\{\ 10\varepsilon_M \|D\|_a, \kappa/\varepsilon_M\ \}$;

/* 3. Solve $T_{11}X - XT_{22} = sT_{12}$ for X. */

$(X, s) = \texttt{Sylsc2}(\ t_{j:j_1,j:j_1}, -t_{j_1+1:j_2,j_1+1:j_2}, t_{j:j_1,j_1+1:j_2}\ )$;

/* 4. Swap the adjacent diagonal blocks. */

IF $n_1 = 1$ AND $n_2 = 2$, THEN

/* 5. Generate reflector H so that $(s, x_{11}, x_{12})H = (0, 0, *)$. */

$(v, w) = \texttt{Hous\_3}(\ x_{12}, s, x_{11}\ )$; $\tau = t_{jj}$;

$D \leftarrow (I_3 + vw^T)D(I_3 + vw^T)$; /* 6. Apply H provisionally. */

IF $\max\{\ |d_{31}|, |d_{32}|, |d_{33} - \tau|\ \} > b$, THEN

"Swap rejected"; STOP;

/* 7. Apply transformation to T and U. */

$t_{j:j+2,j:n} \leftarrow (I_3 + vw^T)t_{j:j+2,j:n}$;

$t_{1:j+1,j:j+2} \leftarrow t_{1:j+1,j:j+2}(I_3 + vw^T)$;

$t_{j+2,j:j+1} = 0$; $t_{j+2,j+2} = \tau$;

$u_{:,j:j+2} \leftarrow u_{:,j:j+2}(I_3 + vw^T)$;

ELSE IF $n_1 = 2$ AND $n_2 = 1$, THEN

/* 8. Generate H so that $H(-x_{11}, -x_{21}, s) = (*, 0, 0)$. */

$(v, w) = \texttt{Hous\_3}(\ -x_{11}, -x_{21}, s\ )$; $\tau = t_{j+2,j+2}$;

$D \leftarrow (I_3 + vw^T)D(I_3 + vw^T)$; /* 9. Apply H provisionally. */

IF $\max\{\ |d_{21}|, |d_{31}|, |d_{11} - \tau|\ \} > b$, THEN

"Swap rejected"; STOP;

/* 10. Apply transformation to T and U. */

$t_{j:j+2,j+1:n} \leftarrow (I_3 + vw^T)t_{j:j+2,j+1:n}$;

$t_{1:j+2,j:j+2} \leftarrow t_{1:j+2,j:j+2}(I_3 + vw^T)$;

$t_{j+1:j+2,j} = 0$; $t_{jj} = \tau$;

$u_{:,j:j+2} \leftarrow u_{:,j:j+2}(I_3 + vw^T)$;

ELSE IF $n_1 = 2$ AND $n_2 = 2$, THEN

/* 11. Generate H_1 and H_2 so that (2.65) holds. */

$$(v,w) = \texttt{Hous\_3}(\ -x_{11}, -x_{21}, s\ ); \quad \tau = (-x_{12}, -x_{22}, 0)^T w;$$
$$y = (-x_{22} + \tau v_2, \tau v_3, s)^T; \quad (\tilde{v}, \tilde{w}) = \texttt{Hous\_3}(\ y\ );$$

/* 12. Apply $H_2 H_1$ provisionally. */

$$d_{1:3,:} \leftarrow (I_3 + vw^T) d_{1:3,:}; \quad d_{:,1:3} \leftarrow d_{:,1:3}(I_3 + vw^T);$$
$$d_{2:4,:} \leftarrow (I_3 + \tilde{v}\tilde{w}^T) d_{2:4,:}; \quad d_{:,2:4} \leftarrow d_{:,2:4}(I_3 + \tilde{v}\tilde{w}^T);$$

IF $\max\{\,|d_{31}|, |d_{32}|, |d_{41}|, |d_{42}|\,\} > b$, THEN
"Swap rejected"; STOP;

/* 13. Apply transformation to T and U. */

$$t_{j:j+2,j:n} \leftarrow (I_3 + vw^T) t_{j:j+2,j:n};$$
$$t_{1:j+3,j:j+2} \leftarrow t_{1:j+3,j:j+2}(I_3 + vw^T);$$
$$t_{j+1:j+3,j:n} \leftarrow (I_3 + \tilde{v}\tilde{w}^T) t_{j+1:j+3,j:n};$$
$$t_{1:j+3,j+1:j+3} \leftarrow t_{1:j+3,j+1:j+3}(I_3 + \tilde{v}\tilde{w}^T); \quad t_{j+2:j+3,j:j+1} = 0;$$
$$u_{:,j:j+2} \leftarrow u_{:,j:j+2}(I_3 + vw^T); \quad u_{:,j+1:j+3} \leftarrow u_{:,j+1:j+3}(I_3 + \tilde{v}\tilde{w}^T);$$

END IF

END IF

The new 2×2 blocks T_{11} and/or T_{22}, if any, can be standardized using an algorithm as follows

IF $n_2 = 2$, THEN

/* Standardize new 2×2 block T_{11}. */

$$(t_{j:j+1,j:j+1}, c, s, \lambda_{j:j+1}) = \texttt{Schur2}(\ t_{j:j+1,j:j+1}\ );$$
$$t_{j:j+1,j+2:n} \leftarrow \begin{bmatrix} c & s \\ -s & c \end{bmatrix} t_{j:j+1,j+2:n};$$
$$t_{1:j-1,j:j+1} \leftarrow t_{1:j-1,j:j+1} \begin{bmatrix} c & -s \\ s & c \end{bmatrix};$$
$$u_{:,j:j+1} \leftarrow u_{:,j:j+1} \begin{bmatrix} c & -s \\ s & c \end{bmatrix}; \qquad \text{/* Accumulate transformation. */}$$

END IF

More generally, let Algorithm **swap** be defined by the statement:

$$(T, U) = \texttt{swap}(\ T, i, j, U\ );$$

which reorders the real Schur factorization of a real matrix, $A = UTU^T$, in canonical Schur form, so that the diagonal block of T with row index i is moved to row j ($1 \le i \le n; 1 \le j \le n$). Algorithm **swap** consists of a sequence of interchanges of adjacent blocks using calls to Algorithm **swap2**. If $i < j$ each block is swapped with the next one below; if $i > j$ each block is swapped with the next one above. If two adjacent blocks are too close to swap (the problem is very ill conditioned), then a warning is issued; T may have been partially reordered. (No problems can occur if two 1×1 blocks are swapped.) It is numerically possible that a pair of complex eigenvalues be transformed into two real eigenvalues, i.e. a 2×2 block breaks into two

1×1 blocks; in such a case, each of these blocks is swapped individually. Note that the i and j parameters could also be specified as outputs. For instance, if i pointed to the second row of a 2×2 block, it should be changed to point to the first row; j always points to the first row of the block in its final position (which may differ from the given value by $+1$ or -1).

Algorithm `swap` is used in Algorithm `Schord`, detailed below, that reorders the real Schur factorization of a real matrix, $A = UTU^T$, so that a selected cluster of eigenvalues appears in the leading diagonal blocks of the upper quasi-triangular matrix T, and the leading columns of U form an orthonormal basis of the corresponding A-invariant subspace. If the cluster includes a pair of complex conjugate eigenvalues, then both are automatically selected. Actually, the T matrix is reordered by an orthogonal similarity transformation $Z^T TZ$, and the matrix U of Schur vectors is updated by postmultiplying it by Z. (This updating can be optionally performed.) Clearly, if a complex eigenvalue is very ill conditioned, then its final value may differ significantly from its value before reordering. If some eigenvalues are too close to separate (the problem is very ill conditioned), then reordering of T fails.

Algorithm Schord : $(T, U) = \texttt{Schord}(\ T, select, U\ )$

```
/* Order the real Schur factorization of T ∈ IR^{n×n} in RSF, so that */
/* the eigenvalues λ_i, for which select_i is TRUE, appear in the leading */
/* diagonal blocks; update U. */

k~ = 0;   pair = FALSE;

        /* 1. Collect the selected blocks at the upper-left corner of T. */
FOR k = 1:n,
   IF pair = TRUE, THEN
      pair = FALSE;
   ELSE
      s = select_k;
      IF k < n, THEN
         IF t_{k+1,k} ≠ 0, THEN pair = TRUE;   s = s OR select_{k+1};
      IF s = TRUE, THEN
         k~ ← k~ + 1;   k^ = k;   /* 2. Swap the kth block to position k~. */
         IF k ≠ k~, THEN (T, U) = swap( T, k^, k~, U );
         IF pair = TRUE, THEN k~ ← k~ + 1;
   END IF
END FOR
```

The algorithms for (re)ordering the diagonal blocks of a matrix in RSF are stable in practice. The computing effort invested is not too large, but it obviously depends on the initial and final configuration of blocks. It is estimated that ordering a RSF requires no more than 15 % of the effort needed to obtain RSF.

Exact computations have been assumed in relations (2.63) and (2.64). In practice, the computed T satisfies $T = U^T(A+E)U$ (instead of (2.63)), where U is exactly orthogonal, and $\|E\|/\|A\|$ is a modest multiple of ε_M. In addition, the computed matrix $\widehat{U}$ is very near U, and, in particular, it has orthogonal columns almost to working precision. Moreover, (2.64) should be replaced by $Y^T(T+F)Y =: \widetilde{T}$, where $\|F\|/\|T\|$ (and hence, $\|F\|/\|A\|$) is of order of ε_M. Then, finally, $Y^TU^T(A+E+G)UY = \widetilde{T}$, where $G := UFU^T$ and $\|G\|_2 = \|F\|_2$, and the first s columns of UY form an orthogonal basis for the A-invariant subspace associated with those s selected eigenvalues. If, for instance, the selected λ_i are all eigenvalues of A so that $\Re(\lambda_i) < 0$, where A is the state matrix of a continuous-time system, then an orthogonal basis for the *stable* A-invariant subspace is obtained.

It is worth mentioning that invariant subspaces corresponding to multiple eigenvalues may be insensitive to perturbations, hence the approach presented above is definitely superior to the approach based on principal vectors.

2.3 Solving Sylvester and Lyapunov Equations

In analysis and synthesis of linear dynamical systems, but also in other domains, it is frequently necessary to solve the following linear matrix equations, called *Sylvester equations*:

$$\widetilde{A}\widetilde{X} + \widetilde{X}\widetilde{B} = \widetilde{C}, \qquad \widetilde{X} + \widetilde{A}\widetilde{X}\widetilde{B} = \widetilde{C}, \tag{2.66}$$

where $\widetilde{A} \in \mathbb{R}^{m\times m}$, $\widetilde{B} \in \mathbb{R}^{n\times n}$, and $\widetilde{C} \in \mathbb{R}^{m\times n}$, are given matrices, and $\widetilde{X} \in \mathbb{R}^{m\times n}$ should be computed. The first equation has a unique solution if and only if $\lambda_i + \mu_j \neq 0$, where $\lambda_i \in \lambda(\widetilde{A})$, $i = 1{:}\,m$, and $\mu_j \in \lambda(\widetilde{B})$, $j = 1{:}\,n$. Similarly, the second equation has a unique solution if and only if $\lambda_i\mu_j \neq 1$ (Barnett, 1975).

Important special cases of equations (2.66) are the *Lyapunov equations*,[5]

$$\widetilde{A}^T\widetilde{X} + \widetilde{X}\widetilde{A} = \widetilde{C}, \qquad \widetilde{X} - \widetilde{A}^T\widetilde{X}\widetilde{A} = \widetilde{C}, \tag{2.67}$$

[5]The minus sign in the second equation is used to get the usual equation in practice (see, for instance, (2.6) in § 2.1.1), but there is no difficulty to deal with an equation having a plus sign.

where $\widetilde{A}, \widetilde{C}, \widetilde{X} \in \mathbb{R}^{n\times n}$, and $\widetilde{C} = \widetilde{C}^T$, which are encountered, for example, in stability analyses (see Theorem 1.6 in § 1.1.1), or in the solution of algebraic Riccati equations by Newton's method, as we have seen in this chapter (§ 2.1.1). The two equations in (2.67) are usually referred to as the *continuous-time Lyapunov equation*, and *discrete-time Lyapunov equation*, respectively, since they frequently arise in the analysis of continuous-time and discrete-time dynamical systems, respectively. Similar names are sometimes used for Sylvester equations.

Each of the equations in (2.66) or (2.67) may be re-written and then solved as a system of linear algebraic equations. Nevertheless, this direct approach is inefficient for medium- or large scale problems. For instance, the linear system equivalent to (2.67) has $n(n+1)/2$ equations and unknowns (taking advantage of symmetry), which involve about $n^6/24$ *flops*, and $n^4/4$ memory locations.

If certain convergence conditions hold, then some easily implementable iterative methods may be used (see, for instance, (Hoskins, Meek, and Walton, 1977a, 1977b)). But more attractive are the *transformation methods*, which bring the equations to certain reduced, simpler to solve forms. A very efficient and reliable method, due to Bartels and Stewart (1972), is based on the reduction of $\widetilde{A}$ and $\widetilde{B}$ to real Schur form (RSF), or only to quasi-triangular form (QTF). Comparative evaluations of various methods for solving Sylvester equations and Lyapunov equations are made, for instance, in (Bélanger and McGillivray, 1976; Sima, 1980a).

The next two subsections describe efficient transformation-based solvers for equations (2.66) and (2.67), respectively. Subsection 2.3.3 deals with specific solvers for "stable" Lyapunov equations (2.67), with $\widetilde{C} \leq 0$.

2.3.1 Solving Sylvester equations

The transformation methods are based on the equivalence of two problems. The equations in (2.66) are, respectively, equivalent to

$$AX + XB = C, \qquad X + AXB = C, \tag{2.68}$$

where $A := U^{-1}\widetilde{A}U$, $B := V^{-1}\widetilde{B}V$, $C := U^{-1}\widetilde{C}V$, and $X := U^{-1}\widetilde{X}V$, where U and V are some nonsingular matrices, chosen so that A and B have "simpler" forms. Once X has been computed, by solving one of the equations (2.68), the solution $\widetilde{X}$ of the corresponding equation (2.66) is obtained from $\widetilde{X} = UXV^{-1}$. As the transformation technique described involves the solution of systems of linear algebraic equations having U and V as coefficient matrices, it follows that the computed results will generally have relative errors of order $\varepsilon_M(\mathrm{cond}(U)+\mathrm{cond}(V))$. Ill-conditioned transformation matrices must be avoided. The procedures and algorithms dealt with in the sequel make use of orthogonal, hence perfectly conditioned, matrices.

The computational procedure based on the Bartels-Stewart method is the following.

Procedure SylSch

/* 1. [Reduce $\widetilde{A}^T$ and $\widetilde{B}$ to RSF (or QTF), A and B, and update $\widetilde{C}$.] */

$$A = U^T \widetilde{A}^T U, \qquad B = V^T \widetilde{B} V, \qquad C = U^T \widetilde{C} V. \tag{2.69}$$

/* 2. [Solve the transformed equation.] Solve in X */

$$A^T X + XB = C, \qquad \text{or} \quad X + A^T XB = C. \tag{2.70}$$

/* 3. [Find the solution.] */

$$\widetilde{X} = UXV^T. \tag{2.71}$$

The equations in (2.70) are also of Sylvester type, but simpler to solve, because the columns of X can be determined by a back substitution process, slightly complicated by possible occurrence of the 2×2 blocks on the diagonals of A and B. Consider first the continuous-time case. Denoting $x_j := X_{:,j}$, and $c_j := C_{:,j}$, it follows that, if $b_{l+1,l} = 0$ (for an index $l \in \{1, \ldots, n-1\}$), then

$$(A^T + b_{ll} I_m) x_l = c_l - \sum_{j=1}^{l-1} b_{jl} x_j. \tag{2.72}$$

So, once $x_1, \ldots, x_{l-1}$ are known, x_l can be computed by solving a quasi-triangular system of order m, which involves about $m^2/2$ *flops*, if the right-hand side is known. If $b_{l+1,l} \neq 0$, then, equating the lth and $(l+1)$th columns of the transformed Sylvester equation, we obtain

$$A^T \begin{bmatrix} x_l & x_{l+1} \end{bmatrix} + \begin{bmatrix} x_l & x_{l+1} \end{bmatrix} \begin{bmatrix} b_{ll} & b_{l,l+1} \\ b_{l+1,l} & b_{l+1,l+1} \end{bmatrix} =$$

$$\begin{bmatrix} c_l & c_{l+1} \end{bmatrix} - \sum_{j=1}^{l-1} \begin{bmatrix} b_{jl} x_j & b_{j,l+1} x_j \end{bmatrix}, \tag{2.73}$$

that asks for the simultaneous determination of x_l and x_{l+1}. Actually, by considering the quasi-triangular form of A, the problems in (2.72) and (2.73) reduce to the solution of a set of systems of linear algebraic equations of order at most two, and four, respectively. Indeed, one have to compute successively the elements x_{kl}, or x_{kl} and $x_{k,l+1}$, respectively, if $a_{k+1,k} = 0$

(in A), or the elements x_{kl} and $x_{k+1,l}$, or x_{kl}, $x_{k+1,l}$, $x_{k,l+1}$, and $x_{k+1,l+1}$, respectively, if $a_{k+1,k} \neq 0$, where the index k is initialized by 1 and incremented by 1 or 2, until $k = m$. The already determined elements are used to update the right-hand side in (2.72) or (2.73), respectively, before calculating the elements with row indices k, or k and $k+1$, respectively. For instance, when $a_{k+1,k} \neq 0$ and $b_{l+1,l} \neq 0$, the coefficient matrix of the 4×4 linear system is

$$\begin{bmatrix} a_{kk} + b_{ll} & a_{k+1,k} & b_{l+1,l} & 0 \\ a_{k,k+1} & a_{k+1,k+1} + b_{ll} & 0 & b_{l+1,l} \\ b_{l,l+1} & 0 & a_{kk} + b_{l+1,l+1} & a_{k+1,k} \\ 0 & b_{l,l+1} & a_{k,k+1} & a_{k+1,k+1} + b_{l+1,l+1} \end{bmatrix}, \tag{2.74}$$

and the unknown vector is $\left(x_{kl}, x_{k+1,l}, x_{k,l+1}, x_{k+1,l+1}\right)^T$. Note that the first equation (2.70), hence (2.72) or (2.73), may alternately be written as

$$A_{kk}^T X_{kl} + X_{kl} B_{ll} = C_{kl} - R_{kl}, \quad R_{kl} := \sum_{i=1}^{k-1} A_{ik}^T X_{il} + \sum_{j=1}^{l-1} X_{kj} B_{jl}, \tag{2.75}$$

for $l = 1\colon q$ and $k = 1\colon p$, where A_{kk}, B_{ll}, C_{kl}, and X_{kl} are block submatrices defined by the partitioning induced by the Schur forms of A and B; obviously, p and q are equal to m and n, respectively, if all blocks A_{kk} and B_{ll} have order 1, and are less than those values, otherwise. Let $\widehat{k}$ and $\widehat{l}$ be the orders of A_{kk} and B_{ll}, respectively, which can be 1 or 2. Denote by $V \otimes W$ the *Kronecker product* of matrices V and W, i.e. the block-matrix whose (i,j) element is $v_{ij}W$. It follows that the matrix in (2.74) is just $I_{\widehat{l}} \otimes A_{kk}^T + B_{ll}^T \otimes I_{\widehat{k}}$, and the linear system to be solved is

$$(I_{\widehat{l}} \otimes A_{kk}^T + B_{ll}^T \otimes I_{\widehat{k}})\mathrm{vec}(X_{kl}) = \mathrm{vec}(C_{kl} - R_{kl}), \tag{2.76}$$

where $\mathrm{vec}(M)$ denotes the vector obtained by stacking all columns of matrix M one over the next one. Relation (2.76) holds for any values of $\widehat{k}$ and $\widehat{l}$. In the algorithm below, we use Algorithm `solve` in § 1.4.3 for determining X_{kl}, which is overwritten on C_{kl}.

Algorithm Sylsc : $C =$ `Sylsc`(A, B, C)

/* Solve Sylvester equation $A^T X + XB = C$, with $A \in \mathbb{R}^{m \times m}$ and */
/* $B \in \mathbb{R}^{n \times n}$ in QTF; C is overwritten by X. */

$l = 1$;
WHILE $l \leq n$, DO

$k = 1;\quad \widetilde{l} = l;\quad$ IF $\ l < n$ AND $b_{l+1,l} \neq 0,\ $ THEN $\ \widetilde{l} = l + 1;$
IF $\ l > 1,\ $ THEN /* 1. Update C for contribution of B. */
$c_{:,l:\widetilde{l}} \leftarrow c_{:,l:\widetilde{l}} - c_{:,1:l-1} b_{1:l-1,l:\widetilde{l}};$
WHILE $k \leq m$, DO
$\widetilde{k} = k;\quad$ IF $\ k < m$ AND $a_{k+1,k} \neq 0,\ $ THEN $\ \widetilde{k} = k + 1;$
IF $\ k > 1,\ $ THEN /* 2. Update C for contribution of A. */
$c_{k:\widetilde{k},l:\widetilde{l}} \leftarrow c_{k:\widetilde{k},l:\widetilde{l}} - a^T_{1:k-1,k:\widetilde{k}} c_{1:k-1,l:\widetilde{l}};$
/* 3. Solve reduced-order linear system[6]. */
$c_{k:\widetilde{k},l:\widetilde{l}} = \texttt{solve}(\ I_{\widetilde{l}-l+1} \otimes a^T_{k:\widetilde{k},k:\widetilde{k}} + b^T_{l:\widetilde{l},l:\widetilde{l}} \otimes I_{\widetilde{k}-k+1}, \mathrm{vec}(c_{k:\widetilde{k},l:\widetilde{l}})\ );$
$k = \widetilde{k} + 1;$
END WHILE
$l = \widetilde{l} + 1;$
END WHILE

Procedure `SylSch` is then very easily implemented using Algorithm `Sylsc`.

Algorithm Sylc : $C = \texttt{Sylc}(\ A, B, C\ )$

/* Solve Sylvester equation $AX + XB = C$, $A \in \mathbb{R}^{m \times m}$, $B \in \mathbb{R}^{n \times n}$; */
/* C is overwritten by X. */

$(A, U) = \texttt{Hess}(\ A^T\ );\quad (A, U) = \texttt{mSchur}(\ A, U\ );$
$(B, V) = \texttt{Hess}(\ B\ );\quad (B, V) = \texttt{mSchur}(\ B, V\ );\quad C \leftarrow U^T C V;$
$C = \texttt{Sylsc}(\ A, B, C\ );\quad C \leftarrow U C V^T;$

Clearly, Algorithm `Schur` can be used instead of Algorithm `mSchur`.

Our implementation of Algorithm `Sylsc` works with A^T instead of A, and uses Level 1 BLAS operations of the form $y \leftarrow \alpha x + y$ for updating C, but it is apparent that Level 2 or even Level 3 BLAS operations can also be used. The LAPACK implementation also uses Level 1 BLAS scalar products only (see below).

The linear systems of order 2 or 4 have previously been solved by a variant of Gaussian elimination with partial pivoting, called *Crout factorization.* Highly reliable algorithms are implemented in the LAPACK package, and are used to solve the continuous-time Sylvester equation in (2.70), with A and B in canonical RSF. As these algorithms are also recommended for solving discrete-time Sylvester equations, as well as Lyapunov equations, we shall detail the main algorithm here.

The basic algorithm, which we call `Sylsc2`, defined by

$$(X, s) = \texttt{Sylsc2}(\ \mathrm{op}(A), \sigma\mathrm{op}(B), C\ );$$

[6] It is assumed that Algorithm `solve` unpacks the solution $\mathrm{vec}(c_{k:\widetilde{k},l:\widetilde{l}})$.

solves for the $m \times n$ matrix X, $0 \le m, n \le 2$, the Sylvester equation

$$\mathrm{op}(A)X + \sigma X \mathrm{op}(B) = sC,$$

where A is $m \times m$, B is $n \times n$, C is $m \times n$, $\sigma = \pm 1$, and $\mathrm{op}(M)$ denotes M or M^T; the computed scale factor s is chosen less than or equal to 1 to prevent overflows when determining the solution X. The infinity-norm of the solution is also computed. If A and B have too close eigenvalues, then A or B is perturbed to get a nonsingular equation. For instance, when $m = n = 1$, one computes $\tau = a_{11} + \sigma b_{11}$, and if $|\tau| \le \kappa/\varepsilon_M$ (where κ is the safe minimum, see § 2.2.3), then $\tau = \kappa/\varepsilon_M$ is used, and a warning is issued; usually, $s = 1$, but, if $|c_{11}|\kappa/\varepsilon_M > |\tau|$, then $s = 1/|c_{11}|$; anyhow, $x_{11} = sc_{11}/\tau$.

For the 1 by 2 case (with $\mathrm{op}(B) = B$), the equivalent system matrix is

$$\begin{bmatrix} a_{11} + \sigma b_{11} & \sigma b_{21} \\ \sigma b_{12} & a_{11} + \sigma b_{22} \end{bmatrix} \begin{bmatrix} x_{11} \\ x_{12} \end{bmatrix} = \begin{bmatrix} c_{11} \\ c_{12} \end{bmatrix}.$$

The coefficient matrix should be transposed if $\mathrm{op}(B) = B^T$. The 2 by 1 case is defined similarly. These 2×2 systems, denoted as $Sx = c$, are solved using Gaussian elimination with *complete pivoting*, setting pivots $\widetilde{s}_{jj}$ less than $s_{\min} := \max\{\,\kappa/\varepsilon_M, \varepsilon_M\|A\|_a, \varepsilon_M\|B\|_a\,\}$ to $s_{\min}$. The transformations used are simultaneously applied to the right-hand side, obtaining the transformed c, $\widetilde{c}$. Before the solution elements are set, scaling is applied, if needed; typically, $s = 1$, but if $2\kappa/\varepsilon_M|\widetilde{c}_j| > |\widetilde{s}_{jj}|$, then $s = 1/(2\|\widetilde{c}\|_\infty)$, $\widetilde{c} \leftarrow s\widetilde{c}$, and the solution is computed by the usual formulas.

For the 2 by 2 case, with $\mathrm{op}(A) = A^T$, $\mathrm{op}(B) = B$, the system is

$$\begin{bmatrix} a_{11} & a_{21} \\ a_{12} & a_{22} \end{bmatrix} \begin{bmatrix} x_{11} & x_{12} \\ x_{21} & x_{22} \end{bmatrix} + \sigma \begin{bmatrix} x_{11} & x_{12} \\ x_{21} & x_{22} \end{bmatrix} \begin{bmatrix} b_{11} & b_{12} \\ b_{21} & b_{22} \end{bmatrix} = \begin{bmatrix} c_{11} & c_{12} \\ c_{21} & c_{22} \end{bmatrix},$$

which is equivalent to a 4×4 linear system, $Sx = c$, with S as in (2.74), for $k = l = 1$, σb instead of b, and x and c taken as 4-dimensional vectors. These systems are also solved using complete pivoting, setting pivots less than $s_{\min}$ to $s_{\min}$. Rows and columns are actually interchanged, using Level 1 BLAS operations. The coefficient matrix is triangularized by elementary transformations, which are simultaneously applied to the right-hand side. Let $\widetilde{S}$ and $\widetilde{c}$ be the transformed S and c, respectively. A scaling factor is computed as follows:

$s = 1$; $\quad f := 8\kappa/\varepsilon_M$;
IF $f|\widetilde{c_1}| > |\widetilde{s}_{11}|$ OR $f|\widetilde{c_2}| > |\widetilde{s}_{22}|$ OR $f|\widetilde{c_3}| > |\widetilde{s}_{33}|$ OR $f|\widetilde{c_4}| > |\widetilde{s}_{44}|$,
THEN $\quad s = 1/(8\|\widetilde{c}\|_\infty)$; $\quad \widetilde{c} \leftarrow s\widetilde{c}$;

Then, x is obtained in the usual manner, by back substitution; clearly, the column permutations (equivalent to interchanges of some x elements), are taken into account.

The LAPACK algorithm for Sylvester equations solves the matrix equation:

$$\mathrm{op}(A)X + \sigma X\mathrm{op}(B) = sC, \tag{2.77}$$

where $\mathrm{op}(M)$ denotes M or M^T, $\sigma = \pm 1$, $A \in \mathbb{R}^{m\times m}$, and $B \in \mathbb{R}^{n\times n}$ are both in upper canonical RSF, the right-hand side C and the solution X are $m \times n$, and s is a computed scale factor, set less than or equal to 1 to avoid overflow in X. The solution is obtained in C. If A and B have common or very close eigenvalues, then perturbed values are used to solve the equation (but the matrices A and B keep unchanged). Caution is taken to control overflow in a robust way, on a wide range of machines, including those with large exponent range, like Cray machines (taking the square roots of κ and $\mathcal{K}$, instead of κ and $\mathcal{K} = 1/\kappa$, respectively). Let

$$s_{\min} = \max\{\kappa mn/\varepsilon_M, \varepsilon_M\|A\|_a, \varepsilon_M\|B\|_a\}.$$

The perturbations of A and B are no larger than $s_{\min}$. Parameter $\mathcal{K}$ is used for scaling control.

For solving $AX + \sigma XB = sC$, the (k,l)th block of X is determined using

$$A_{kk}X_{kl} + \sigma X_{kl}B_{ll} = C_{kl} - R_{kl}, \quad R_{kl} := \sum_{i=k+1}^{m} A_{ki}X_{il} + \sigma\sum_{j=1}^{l-1} X_{kj}B_{jl},$$

for $l = 1{:}\,q$ and $k = p{:}\,-1{:}\,1$. The sizes of A_{kk} and B_{ll} are 1 or 2, depending on $a_{k,k-1}$ and $b_{l+1,l}$ equal to or different from 0, respectively.

For $A^TX + \sigma XB = sC$, one solves

$$A_{kk}^TX_{kl} + \sigma X_{kl}B_{ll} = C_{kl} - R_{kl}, \quad R_{kl} := \sum_{i=1}^{k-1} A_{ik}^TX_{il} + \sigma\sum_{j=1}^{l-1} X_{kj}B_{jl},$$

for $l = 1{:}\,q$ and $k = 1{:}\,p$. The sizes of A_{kk} and B_{ll} are 1 or 2, depending on $a_{k+1,k}$ and $b_{l+1,l}$ equal to or different from 0, respectively.

For $A^TX + \sigma XB^T = sC$, one solves

$$A_{kk}^TX_{kl} + \sigma X_{kl}B_{ll}^T = C_{kl} - R_{kl}, \quad R_{kl} := \sum_{i=1}^{k-1} A_{ik}^TX_{il} + \sigma\sum_{j=l+1}^{n} X_{kj}B_{lj}^T,$$

for $l = q{:}\,-1{:}\,1$ and $k = 1{:}\,p$. The sizes of A_{kk} and B_{ll} are 1 or 2, depending on $a_{k+1,k}$ and $b_{l,l-1}$ equal to or different from 0, respectively.

For $AX + \sigma XB^T = sC$, one solves

$$A_{kk}X_{kl} + \sigma X_{kl}B_{ll}^T = C_{kl} - R_{kl}, \quad R_{kl} := \sum_{i=k+1}^{m} A_{ki}X_{il} + \sigma \sum_{j=l+1}^{n} X_{kj}B_{lj}^T,$$

for $l = q{:}-1{:}1$ and $k = p{:}-1{:}1$. The sizes of A_{kk} and B_{ll} are 1 or 2, depending on $a_{k,k-1}$ and $b_{l,l-1}$ equal to or different from 0, respectively.

No matter the specific Sylvester equation to be solved, in each of the four possible cases (depending on the orders of A_{kk} and B_{ll}), the matrix R_{kl} is evaluated using two, four, four, or eight scalar products, respectively; it is then subtracted from C_{kl} in an auxiliary matrix, and the resulting linear system is solved using, for instance, Algorithm `Sylsc2`, described above.[7]

The global scale factor s is initially set to 1. If a local scale factor f different by 1 must be used for the current system, then the entire current C matrix is scaled by that factor: $C \leftarrow fC$, $s \leftarrow fs$. Then C_{kl} is replaced by the current X_{kl}.

The *discrete-time Sylvester equations* have been paid considerably less attention in the literature than the continuous-time case, although they have also various applications (see, for instance, Example 2.3 below). An algorithm similar to `Sylc` can be used for solving the second Sylvester equation in (2.66). At stage 2 of the procedure `SylSch`, the transformed equation, $X + A^T XB = C$, is also reducible to a set of systems of linear algebraic equations of order at most four. For instance, when $a_{k+1,k} \neq 0$ and $b_{l+1,l} \neq 0$, the coefficient matrix of the 4×4 linear system is

$$\begin{bmatrix} a_{kk}b_{ll} + 1 & a_{k+1,k}b_{ll} & a_{kk}b_{l+1,l} & a_{k+1,k}b_{l+1,l} \\ a_{k,k+1}b_{ll} & a_{k+1,k+1}b_{ll} + 1 & a_{k,k+1}b_{l+1,l} & a_{k+1,k+1}b_{l+1,l} \\ a_{kk}b_{l,l+1} & a_{k+1,k}b_{l,l+1} & a_{kk}b_{l+1,l+1} + 1 & a_{k+1,k}b_{l+1,l+1} \\ a_{k,k+1}b_{l,l+1} & a_{k+1,k+1}b_{l,l+1} & a_{k,k+1}b_{l+1,l+1} & a_{k+1,k+1}b_{l+1,l+1} + 1 \end{bmatrix}. \tag{2.78}$$

This matrix can be re-written as $B_{ll}^T \otimes A_{kk}^T + I_{\widehat{k}\widehat{l}}$, where $\widehat{k}$ and $\widehat{l}$ are the orders of A_{kk} and B_{ll}, respectively. This representation also holds for any values of $\widehat{k}$ and $\widehat{l}$.

If the (k,l)th blocks of both terms of the second relation (2.70) are equated, exploiting the structure of A and B, the following relation, adapted

---

[7] Although it has all the functionality needed, LAPACK implementation actually uses `Sylsc2` only when both submatrices A_{kk} and B_{ll} have order 2; the case when one of them has order 1, and the other one has order 2 is dealt with by a special, but similar algorithm, which has $s_{\min}$ as input parameter, while the case when both A_{kk} and B_{ll} have order 1 is solved in-line.

from (Sima, 1980a), is obtained

$$X_{kl} + A_{kk}^T X_{kl} B_{ll} = C_{kl} - \sum_{j=1}^{l} \sum_{i=1}^{k-1} A_{ik}^T X_{ij} B_{jl} - \sum_{j=1}^{l-1} A_{kk}^T X_{kj} B_{jl}. \qquad (2.79)$$

If the submatrices X_{ij}, $i = 1{:}k-1$, $j = 1{:}l$, and X_{kj}, $j = 1{:}l-1$, have already been determined, then equation (2.79) is a linear system of order $\widetilde{kl}$, that can easily be solved in X_{kl}. Hence, the X_{kl} blocks can be computed for fixed k and for l increased from 1 in steps of 1 or 2; index k varies in the same manner. The calculations can be efficiently organized by defining, for each k,

$$W_j := \sum_{i=1}^{k-1} A_{ik}^T X_{ij}, \qquad j = 1{:}l,$$

and $W_j = 0$, for $k = 1$. Clearly, the matrix $\begin{bmatrix} W_1 & W_2 & \ldots \end{bmatrix}$ is either $1 \times n$ or $2 \times n$, and the same working array can be used for it, for any k. Relation (2.79) is equivalent to

$$\begin{aligned} X_{kl} + A_{kk}^T X_{kl} B_{ll} &= C_{kl} - \sum_{j=1}^{l} W_j B_{jl} - \sum_{j=1}^{l-1} A_{kk}^T X_{kj} B_{jl} \\ &= C_{kl} - \sum_{j=1}^{l-1} (W_j + A_{kk}^T X_{kj}) B_{jl} - W_l B_{ll}. \end{aligned}$$

If we update W_j, $W_j \leftarrow W_j + A_{kk}^T X_{kj}$, just after X_{kj} has been computed, then the basic equation is

$$X_{kl} + A_{kk}^T X_{kl} B_{ll} = C_{kl} - \sum_{j=1}^{l} W_j B_{jl}.$$

The following algorithm is obtained.

Algorithm Sylsd : $\boxed{C = \texttt{Sylsd}(\,A, B, C\,)}$

/* Solve Sylvester equation $X + A^T X B = C$, with $A \in \mathbb{R}^{m \times m}$ and */
/* $B \in \mathbb{R}^{n \times n}$ in QTF; C is overwritten by X. */

$k = 1$;
WHILE $k \le m$, DO
  $l = 1$; $\widetilde{k} = k$; IF $k < m$ AND $a_{k+1,k} \neq 0$, THEN $\widetilde{k} = k + 1$;
  WHILE $l \le n$, DO

$\tilde{l} = l;$ IF $l < n$ AND $b_{l+1,l} \neq 0,$ THEN $\tilde{l} = l + 1;$

/* 1. Construct W and update C. */

$w_{1:\tilde{k}-k+1,l:\tilde{l}} = 0;$ IF $k > 1,$ THEN $w_{:,l:\tilde{l}} = a^T_{1:k-1,k:\tilde{k}} c_{1:k-1,l:\tilde{l}};$

$c_{k:\tilde{k},l:\tilde{l}} \leftarrow c_{k:\tilde{k},l:\tilde{l}} - w_{:,1:\tilde{l}} b_{1:\tilde{l},l:\tilde{l}};$

/* 2. Solve reduced-order linear system. */

$c_{k:\tilde{k},l:\tilde{l}} = \texttt{solve}(\ b^T_{l:\tilde{l},l:\tilde{l}} \otimes a^T_{k:\tilde{k},k:\tilde{k}} + I_{(\tilde{k}-k+1)(\tilde{l}-l+1)}, \mathrm{vec}(c_{k:\tilde{k},l:\tilde{l}})\ );$

/* 3. Update W. */

IF $l < n,$ THEN $w_{:,l:\tilde{l}} \leftarrow w_{:,l:\tilde{l}} + a^T_{k:\tilde{k},k:\tilde{k}} c_{k:\tilde{k},l:\tilde{l}};$

$l = \tilde{l} + 1;$

END WHILE

$k = \tilde{k} + 1;$

END WHILE

Procedure `SylSch`, for general matrices, is then very easily implemented in Algorithm `Syld`, which is identical to Algorithm `Sylc`, except that Algorithm `Sylsd` is called instead of `Sylsc`. So, we do not mind it, but only give the defining statement

$C = \texttt{Syld}(\ A, B, C\ );$

Our implementation of Algorithm `Sylsd` uses Level 1 BLAS operations (scalar products) for computing/updating W and updating C. It operates with the transpose of matrix W. Although an $n \times 2$ working array would be needed, to store the current matrix W^T, this was not used, so that both continuous-time and discrete-time codes have the same parameter lists. Instead, we have used the first two columns of the array containing matrix B, in RSF, and the possible non-zero submatrix $b_{1:3,1:2}$ was initially saved in a local 3×2 working array. (This submatrix is finally restored.) It would be useful to develop a new code for `Sylsd`, using the LAPACK approach for solving the reduced-order linear systems.

Our implementation of Algorithm `Sylc` and Algorithm `Syld` transposes the matrix A in RSF, so `Sylsc` and `Sylsd` work with a lower RSF matrix.

The total computing effort for solving a general continuous-time Sylvester equation is about $10(m^3 + n^3) + 2.5(m^2n + mn^2)$ *flops*, provided that the reduction of a matrix of order q to RSF requires $10q^3$ *flops* (including accumulation of transformations). The discrete-time equation needs a bit larger effort.

If $m \geq n$, then it is more advantageous to use the *Hessenberg-Schur method* (Golub, Nash and Van Loan, 1979), which reduces $\widetilde{B}^T$ to RSF, while $\widetilde{A}$ is only reduced to UHF, i.e., $U^T\widetilde{A}U = A$, where A is an upper Hessenberg matrix. The transformed equation, $AX + XB^T = C$, can be solved in a similar manner as described at Procedure `SylSch`.

Procedure SylHes

/* 1. [Reduce $\widetilde{A}$ to UHF, A, and $\widetilde{B}^T$ to RSF, B.] */

$$A = U^T \widetilde{A} U, \qquad B = V^T \widetilde{B}^T V, \qquad C = U^T \widetilde{C} V. \tag{2.80}$$

/* 2. [Solve the transformed equation.] Solve in X */

$$AX + XB^T = C, \qquad \text{or} \quad X + AXB^T = C. \tag{2.81}$$

/* 3. [Find the solution.] */

$$\widetilde{X} = UXV^T. \tag{2.82}$$

Procedure `SylHes` has the same form as `SylSch`. The fact that $\widetilde{A}$ and $\widetilde{B}^T$ are reduced in `SylHes`, and not $\widetilde{A}^T$ and $\widetilde{B}$, as in `SylSch`, is not important. (We could unify the presentation, but the algorithms we implemented use the relations given above.) The essential issue is that `SylHess` reduces A to UHF. This determines the process of solving the reduced equations (2.81) at stage 2. Consider the first equation (2.81). If $b_{l,l-1} = 0$, then the lth column of X, x_l, should satisfy a linear system like in (2.72),

$$(A + b_{ll} I_m) x_l = c_l - \sum_{j=l+1}^{n} b_{lj} x_j. \tag{2.83}$$

where $c_j := C_{:,j}$. So, once $x_n, \ldots, x_{l+1}$ are known, x_l can be computed by solving an upper Hessenberg system of order m. Using Gaussian elimination with partial pivoting, this computation requires about m^2 *flops* (if the right-hand side is known). If $b_{l,l-1} \neq 0$, then equating the $(l-1)$th and lth columns in $AX + XB^T = C$, one obtains,

$$A \begin{bmatrix} x_{l-1} & x_l \end{bmatrix} + \begin{bmatrix} x_{l-1} & x_l \end{bmatrix} \begin{bmatrix} b_{l-1,l-1} & b_{l,l-1} \\ b_{l-1,l} & b_{ll} \end{bmatrix}$$

$$= \begin{bmatrix} c_{l-1} & c_l \end{bmatrix} - \sum_{j=l+1}^{n} \begin{bmatrix} b_{l-1,j} x_j & b_{lj} x_j \end{bmatrix}. \tag{2.84}$$

This relation is a linear system of order $2m$ in x_{l-1} and x_l. By a suitable reordering of equations and variables, the coefficient matrix of this system is almost upper triangular, having only the first two subdiagonals non-zero. Indeed, the initial coefficient matrix is

$$H := \begin{bmatrix} A + b_{l-1,l-1} I_m & b_{l-1,l} I_m \\ b_{l,l-1} I_m & A + b_{ll} I_m \end{bmatrix}, \tag{2.85}$$

or $H = I_2 \otimes A + B_{ll} \otimes I_m$, and, denoting $y_i = x_{i,l-1}$, $y_{m+i} = x_{il}$, $i = 1\!:\!m$, the desired form is obtained by ordering the y components as y_1, y_{m+1}, y_2, y_{m+2}, $\ldots$, y_m, y_{2m}, and reordering the equations similarly. Define

$$\begin{bmatrix} \alpha & \gamma \\ \beta & \delta \end{bmatrix} := \begin{bmatrix} b_{l-1,l-1} & b_{l-1,l} \\ b_{l,l-1} & b_{ll} \end{bmatrix},$$

and let $P = \begin{bmatrix} e_1 & e_3 & \ldots & e_{2m-1} & e_2 & e_4 & \ldots & e_{2m} \end{bmatrix} \in \mathbb{R}^{2m\times 2m}$ be the permutation matrix involved. The linear system $Hy = d$ (where d is the corresponding right-hand side in (2.84)), is equivalent to $PHP(Py) = Pd$. Clearly, the matrix PHP has the following form (for $m = 3$)

$$\begin{bmatrix} a_{11}+\alpha & \gamma & a_{12} & 0 & a_{13} & 0 \\ \beta & a_{11}+\delta & 0 & a_{12} & 0 & a_{13} \\ a_{21} & 0 & a_{22}+\alpha & \gamma & a_{23} & 0 \\ 0 & a_{21} & \beta & a_{22}+\delta & 0 & a_{23} \\ 0 & 0 & a_{32} & 0 & a_{33}+\alpha & \gamma \\ 0 & 0 & 0 & a_{32} & \beta & a_{33}+\delta \end{bmatrix}.$$

Using Gaussian elimination with partial pivoting, the resulting system can be solved in $6m^2$ *flops* (if the right-hand side is known).

A template algorithm is given below.

Algorithm Sylhc : $C = $ `Sylhc(` A, B, C `)`

```
/* Solve Sylvester equation AX + XB^T = C, with A ∈ IR^{m×m} in UHF */
/* and B ∈ IR^{n×n} in QTF; C is overwritten by X. */

l = n;   P = [ e_1 e_3 ... e_{2m-1} e_2 e_4 ... e_{2m} ];
WHILE l ≥ 1, DO
   l~ = l;   IF l > 1 AND b_{l,l-1} ≠ 0, THEN l~ = l - 1;
   IF l < n, THEN
      c_{:,l~:l} ← c_{:,l~:l} - c_{:,l+1:n} b^T_{l~:l,l+1:n};        /* 1. Update C. */
   IF l~ = l, THEN
                 /* 2. Solve an upper Hessenberg system of order m. */
      c_{:,l} = solve( A + b_{ll} I_m, c_{:,l} );
   ELSE
            /* 3. Solve a quasi-upper Hessenberg system of order 2m. */
      d := Pvec(c_{:,l~:l});   d = solve( P(I_2 ⊗ A + b_{l~:l,l~:l} ⊗ I_m)P, d );
      d ← Pd;   c_{:,l~} = d_{1:m};   c_{:,l} = d_{m+1:2m};
   END IF
   l = l~ - 1;
END WHILE
```

The implementation requires an auxiliary working array of dimension about $2m^2 + 5m$ memory locations, to store the coefficient matrix. Note that Procedure **SylSch** does not need extra workspace for coefficient matrix, but needs m^2 locations for storing U. Here, the orthogonal matrix U that reduces A to UHF may be retained in a factored form, below the first subdiagonal of A and in an $(m-1)$-dimensional array (see Algorithm **Hess** in § 2.2.2).

A similar approach is used to solve the discrete-time equation in (2.81). Instead of equation (2.84), we have

$$A \begin{bmatrix} x_{l-1} & x_l \end{bmatrix} \begin{bmatrix} b_{l-1,l-1} & b_{l,l-1} \\ b_{l-1,l} & b_{ll} \end{bmatrix} + \begin{bmatrix} x_{l-1} & x_l \end{bmatrix} = \begin{bmatrix} c_{l-1} & c_l \end{bmatrix} - A \sum_{j=l+1}^{n} \begin{bmatrix} b_{l-1,j}x_j & b_{lj}x_j \end{bmatrix}, \tag{2.86}$$

hence, the coefficient matrix corresponding to (2.85) is $H = B_{ll} \otimes A + I_{2m}$, or

$$H = \begin{bmatrix} b_{l-1,l-1}A + I_m & b_{l-1,l}A \\ b_{l,l-1}A & b_{ll}A + I_m \end{bmatrix}. \tag{2.87}$$

Using the notation above for y_i, y_{m+i}, $i = 1\colon m$, by ordering the y components and the equations as previously mentioned, then we obtain an almost upper triangular coefficient matrix, having only the first three subdiagonals non-zero. Actually, the third subdiagonal also has zero elements in the columns with even index. Indeed, the matrix PHP has the form (for $m = 3$)

$$\begin{bmatrix} a_{11}\alpha + 1 & a_{11}\gamma & a_{12}\alpha & a_{12}\gamma & a_{13}\alpha & a_{13}\gamma \\ a_{11}\beta & a_{11}\delta + 1 & a_{12}\beta & a_{12}\delta & a_{13}\beta & a_{13}\delta \\ a_{21}\alpha & a_{21}\gamma & a_{22}\alpha + 1 & a_{22}\gamma & a_{23}\alpha & a_{23}\gamma \\ a_{21}\beta & a_{21}\delta & a_{22}\beta & a_{22}\delta + 1 & a_{23}\beta & a_{23}\delta \\ 0 & 0 & a_{32}\alpha & a_{32}\gamma & a_{33}\alpha + 1 & a_{33}\gamma \\ 0 & 0 & a_{32}\beta & a_{32}\delta & a_{33}\beta & a_{33}\delta + 1 \end{bmatrix}.$$

Algorithm **Sylhd**, for solving the transformed discrete-time Sylvester equation in (2.81), is similar to Algorithm **Sylhc**.

The total computing effort for solving the continuous-time equation is about $5m^3/3 + 10n^3 + 5m^2n + 2.5mn^2$ *flops*, assuming that B has $n/2$ blocks of order 2 along its diagonal; this is the worst case from the efficiency viewpoint. The ratio between the estimates of computing effort for Hessenberg-Schur and Bartels-Stewart methods is

$$(1 + 3r + 3r^2/2 + 6r^3)/(6 + 3r/2 + 3r^2/2 + 6r^3), \quad \text{where } r := n/m.$$

This shows that, for $m \geq n$, the first method is by 30 % – 70 % more efficient than the second method, depending on m and n. If $m < n$, the transposed problem can efficiently be solved by the Hessenberg-Schur method. Consequently, the Bartels-Stewart method can be preferred only when $\widetilde{A}$ and $\widetilde{B}$ matrices are already in RSF. This is the case, for instance, when computing estimates of the reciprocal condition numbers for eigenvalues or invariant subspaces (see § 2.2.1). Clearly, there is no reason why to use the Hessenberg-Schur method for solving a Lyapunov equation.

The algorithms described above are implemented so that the result X is obtained in the array initially containing the $\widetilde{C}$ matrix. Moreover, the reductions to RSF or UHF can be operated in the arrays initially containing $\widetilde{A}$ and $\widetilde{B}$. Therefore, the Bartels-Stewart method requires $2m^2 + 2n^2 + mn$ memory locations; the Hessenberg-Schur method needs about $3m^2 + 2n^2 + mn + 0(m)$ locations, if the transformation matrix U, which reduces A to UHF, is not accumulated, but it is represented in the factored form. The m^2 extra workspace necessary for the second method is not generally regarded as a serious disadvantage.

In our implementation of the Hessenberg-Schur method, matrix U is stored in the factored form; the corresponding Householder transformations are applied to another matrix, either from the left or from the right, using a special-purpose routine. The construction and solution of the $2m$- or m-order systems of linear algebraic equations take place in separate, but similar code sequences. The solution is obtained by calling special routines, which take the almost triangular structure of the permuted coefficient matrix into account. A compact row-wise storage scheme is employed, which is very advantageous in the present context. Specialized Gaussian elimination with partial pivoting is used. For increasing the efficiency, the row interchanges are not actually performed; only the associated row indices, stored in an integer working array, are interchanged. Level 1 BLAS operations of the form $y \leftarrow \alpha x + y$ are used for updating C, as well as the coefficient matrix, during the Gaussian elimination process. The transpose of the coefficient matrix is actually stored, in order to use column-wise operations as much as possible. The routine for solving the $2m$-order linear systems associated with the discrete-time case is more complicated than that for continuous-time case, because of the presence of non-zero elements in the third subdiagonal (on odd-numbered columns).

The *sensitivity* of the solution of a Sylvester equation is related to the norm of the inverse of the linear operator involved. Consider $AX+XB = C$, with general matrices A and B, and assume that $C \neq 0$. Let

$$(A + E)\widehat{X} + \widehat{X}(B + F) = C + G,$$

be a perturbed equation, where

$$\|E\|_F \le \varepsilon_M \|A\|_F, \quad \|F\|_F \le \varepsilon_M \|B\|_F, \quad \|G\|_F \le \varepsilon_M \|C\|_F,$$
$$\varepsilon_M(\|A\|_F + \|B\|_F)\,\|\phi^{-1}\| \le 1/2, \tag{2.88}$$

with $\phi(Z) := AZ + ZB$ nonsingular. Now, $\|\phi^{-1}\| = \|(I_n \otimes A + B^T \otimes I_m)^{-1}\|_2$, and (Golub, Nash and Van Loan, 1979)

$$\|X - \widehat{X}\|_F / \|X\|_F \le 4\varepsilon_M(\|A\|_F + \|B\|_F)\,\|\phi^{-1}\|.$$

Hence, the computed solution may have relative errors of order $\varepsilon_M\|\phi^{-1}\|$, whatever the solution algorithm will be. If $\|\phi^{-1}\|$ is large, then the continuous-time Sylvester equation is ill conditioned. An estimate for $\|\phi^{-1}\|$ can be computed as described in § 2.2.1. It can be shown that, if

$$\varepsilon(2 + \varepsilon)(\|A\|_F + \|B\|_F)\,\|\phi^{-1}\| \le 1/2,$$

where ε is a small multiple of ε_M, then the solution computed by the algorithms described above satisfy

$$\|X - \widehat{X}\|_F / \|X\|_F \le (9\varepsilon + 2\varepsilon^2)(\|A\|_F + \|B\|_F)\,\|\phi^{-1}\|, \tag{2.89}$$

hence the numerical properties of these algorithms are very close to the ideal ones. Moreover, the residual of computed solution satisfies the inequality

$$\|AX + XB - C\|_F \le (10\varepsilon + 3\varepsilon^2)(\|A\|_F + \|B\|_F)\,\|X\|_F,$$

whose right-hand side does not depend on $\|\phi^{-1}\|$. Hence, the algorithms based on orthogonal reduction to UHF and/or QTF (or RSF) are numerically stable. Other transformation methods, such as those based on a reduction to JCF, are not advisable in general, because the transformation matrices U and V can be ill conditioned, so that large relative errors should result in C and $\widetilde{X}$ being computed from the equations $UC = \widetilde{C}V$ and $\widetilde{X}V = UX$.

All algorithms described can be implemented so that efficient solution of any number of Sylvester equations having the same matrices A and B, but different right-hand side matrices C, is possible. Indeed, the reduction of A and/or B to RSF and/or UHF can be performed as an option. In particular, such a technique is useful for iterative improvement of approximate solutions.

The idea of Bartels-Stewart and Hessenberg-Schur methods can be extended to solving problems of the form $AXL + MXB = C$, using the QZ algorithm of Moler and Stewart (1973), instead of the QR algorithm. For the first method, one has to reduce the matrix pairs (A, M) and (B^T, L^T) to

Schur-triangular forms. For the second method, only the pair (B^T, L^T) should be reduced to a Schur-triangular form, while (A, M) is reduced to Hessenberg-triangular form. See Section 3.2 in Chapter 3 for an overview of the QZ algorithm.

We conclude this subsection by presenting two examples.

Example 2.2 As an application of Sylvester equation solvers, consider the estimation of the reciprocal condition number of a cluster of k eigenvalues of a matrix $T \in \mathbb{R}^{n\times n}$, already in RSF, and having those k eigenvalues in its leading positions. As shown in § 2.2.1 (see (2.42)), we can use the following algorithmic sequence:

$$/* \text{ Solve Sylvester equation for } R\text{: } T_{11}R - RT_{22} = T_{12}. \; */$$
$$W = t_{1:k,k+1:n}; \quad W = \texttt{Sylsc}(\ t^T_{1:k,1:k}, -t_{k+1:n,k+1:n}, W\ );$$
$$/* \text{ Estimate the reciprocal of the condition number. } */$$
$$r = \|W\|_F; \quad \text{IF} \;\; r = 0, \;\; \text{THEN} \;\; s = 1; \;\; \text{ELSE} \;\; s = 1/(1+r^2)^{1/2};$$

An estimate of the reciprocal condition number of the invariant subspace associated with that cluster of eigenvalues, $\text{sep}(T_{11}, T_{22})$ can also be obtained by solving some Sylvester equations (see § 2.2.1). □

Example 2.3 Consider a discrete-time optimal control problem with reference tracking (see Problem 1.5 in § 1.2.1). Then, one of the regulator matrices corresponding to the reference model control structure is just the solution of the Sylvester equation (1.70), re-written here as $X = EXA_r + F$, where A_r is the state matrix of the reference model, and E and F are defined in terms of the solution of a Riccati equation for the proportional or proportional-plus-integral control structure. We take as A_r the matrix used in § 1.2.2, F defined by $F = 0$, $f_{ii} = i$, $i = 1{:}8$ ($F \in \mathbb{R}^{9\times 8}$), and

$$E = \begin{bmatrix} -.425 & 1 & 0 & 0 & 0 & 0 & 0 & .01068 & 0 \\ -.427 & 0 & 1 & 0 & 0 & 0 & 0 & .01803 & 0 \\ -.425 & 0 & 0 & 1 & 0 & 0 & 0 & .02568 & 0 \\ -.426 & 0 & 0 & 0 & 1 & 0 & 0 & .03234 & 0 \\ -.430 & 0 & 0 & 0 & 0 & .222 & .4 & .03646 & 0 \\ -1.700 & 0 & 0 & 0 & 0 & .778 & 0 & -.25090 & 0 \\ -.143 & 0 & 0 & 0 & 0 & 0 & .6 & .10976 & 0 \\ .834 & 0 & 0 & 0 & 0 & 0 & 0 & .97804 & 1 \\ -.366 & 0 & 0 & 0 & 0 & 0 & 0 & -.34162 & 0 \end{bmatrix}.$$

The matrix F has been chosen arbitrarily, for convenience, but E corresponds to a proportional control structure. Rounding to three digits after the decimal point, the solution obtained is

$$X = \begin{bmatrix} 15.331 & .257 & .200 & .156 & .121 & .425 & 2.748 & -1.333 \\ .182 & 14.437 & .340 & .265 & .206 & .722 & 3.790 & -1.742 \\ .074 & .288 & 12.519 & .404 & .314 & 1.102 & 4.519 & -1.920 \\ -.002 & .177 & .368 & 9.582 & .453 & 1.587 & 4.737 & -1.729 \\ -.051 & .099 & .256 & .429 & 5.630 & 2.207 & 4.201 & -.977 \\ -.303 & .252 & .823 & 1.447 & 2.162 & 13.576 & -38.055 & 18.477 \\ -.018 & -.014 & -.011 & -.009 & -.007 & -.023 & 27.442 & -8.589 \\ .317 & .247 & .192 & .149 & .116 & .407 & 20.066 & 3.024 \\ -.178 & -.139 & -.108 & -.084 & -.065 & -.229 & -11.330 & 3.695 \end{bmatrix},$$

and the residual of X is $1.02 \cdot 10^{-13}$. □

2.3.2 Solving Lyapunov equations

As Lyapunov equations are special Sylvester equations, there is no difficulty to adapt the Algorithms `Sylsc` and `Sylsd` to Lyapunov equations. It is, however, possible to take advantage of the symmetry. Let $\widetilde{A} = UAU^T$ be a Schur factorization of $\widetilde{A}$. Procedure SylSch in § 2.3.1 specializes as follows.

Procedure LyaSch

/* 1. [Reduce $\widetilde{A}$ to upper RSF (or QTF), A, and update $\widetilde{C}$.] */

$$A = U^T\widetilde{A}U, \qquad C = U^T\widetilde{C}U. \tag{2.90}$$

/* 2. [Solve the transformed equation.] Solve in X */

$$A^TX + XA = C, \qquad \text{or} \quad X - A^TXA = C. \tag{2.91}$$

/* 3. [Find the solution.] */

$$\widetilde{X} = UXU^T. \tag{2.92}$$

The calculations of C in (2.90) and $\widetilde{X}$ in (2.92) can be performed taking the symmetry into account, as described in (Bartels and Stewart, 1972). Let $\widetilde{C} = T + T^T$, where T is an upper triangular matrix. It follows that,

$$C = U^T(T + T^T)U =: U^TV + V^TU, \qquad V := TU.$$

The product $V = TU$ can be formed, using a Level 3 BLAS operation (see (1.105) in § 1.4.3), in about $0.5n^3$ *flops*, which is half the value corresponding to forming $\widetilde{C}U$. Then, $C = U^TV + V^TU$ is obtained with another Level 3 BLAS operation; only the upper (or lower) triangular part of C has to be actually computed.

The solutions of equations in (2.91) can be computed by specializing the algorithms for Sylvester equations, replacing B by A. Clearly, the calculations are again reducible to solving systems of linear algebraic equations of order at most four; the same basic algorithms as those discussed in § 2.3.1 can be used.

Writing down the first equation in (2.91) per blocks, defined by the block structure of A, the equation for the X_{kl} block is (see (2.75))

$$A_{kk}^T X_{kl} + X_{kl} A_{ll} = C_{kl} - \sum_{i=1}^{k-1} A_{ik}^T X_{il} - \sum_{j=1}^{l-1} X_{kj} A_{jl}. \qquad (2.93)$$

Due to the symmetry of the solution, it is enough to compute the blocks X_{kl} for $l = 1{:}p$ and $k = l{:}p$. Note that for $k = l$, equation (2.93) reduces to a linear system of order at most 3. Otherwise, the calculations are performed as in Algorithm `Sylsc`. Similar remarks apply to the discrete-time case. Therefore, it is possible to write down algorithms similar to `Sylsc` and `Sylsd`.

We shall define even conceptually simpler algorithms. Based on the approach in (Bartels and Stewart, 1972), the matrices in (2.91) may be partitioned as

$$A = \begin{bmatrix} A_{11} & \widehat{A}_{12} \\ 0 & \widehat{A}_{22} \end{bmatrix}, \quad C = \begin{bmatrix} C_{11} & \widehat{C}_{12} \\ \widehat{C}_{12}^T & \widehat{C}_{22} \end{bmatrix}, \quad X = \begin{bmatrix} X_{11} & \widehat{X}_{12} \\ \widehat{X}_{12}^T & \widehat{X}_{22} \end{bmatrix},$$

where A_{11}, C_{11}, and X_{11} have order 1 or 2. The continuous-time Lyapunov equation in (2.91) is equivalent to

$$\begin{aligned} A_{11}^T X_{11} + X_{11} A_{11} &= C_{11}, & (2.94)\\ A_{11}^T \widehat{X}_{12} + \widehat{X}_{12} \widehat{A}_{22} &= \widehat{C}_{12} - X_{11} \widehat{A}_{12}, & (2.95)\\ \widehat{A}_{22}^T \widehat{X}_{22} + \widehat{X}_{22} \widehat{A}_{22} &= \widehat{C}_{22} - \widehat{A}_{12}^T \widehat{X}_{12} - \widehat{X}_{12}^T \widehat{A}_{12}. & (2.96) \end{aligned}$$

Now, equation (2.94) is a Lyapunov equation of order 1 or 2. In the latter case, it can be cast as a linear system of order 3:

$$\begin{bmatrix} a_{11} & a_{21} & 0 \\ a_{12} & a_{11} + a_{22} & a_{21} \\ 0 & a_{12} & a_{22} \end{bmatrix} \begin{bmatrix} x_{11} \\ x_{12} \\ x_{22} \end{bmatrix} = \begin{bmatrix} c_{11}/2 \\ c_{12} \\ c_{22}/2 \end{bmatrix}. \qquad (2.97)$$

Equation (2.95) is a Sylvester equation with A_{11}^T and $\widehat{A}_{22}$ in RSF (or QTF), and equation (2.96) is a Lyapunov equation of order $n-1$ or $n-2$, which is solved in the same manner as the initial equation of order n. So, we can define a *recursive algorithm* for solving continuous-time Lyapunov equations.

Algorithm Lyac : $C =$ `Lyac(` A, C `)`

/* Solve Lyapunov equation $A^T X + XA = C$, $A \in \mathbb{R}^{n \times n}$ in QTF; */

/* C is overwritten by X. */

$\ell = 1$; IF $n > 1$ AND $a_{21} \neq 0$, THEN $\ell = 2$;

/* 1. Solve reduced-order symmetric linear system (2.94). */

$c_{1:\ell,1:\ell} =$ `solve`($I_\ell \otimes a^T_{1:\ell,1:\ell} + a^T_{1:\ell,1:\ell} \otimes I_\ell, \mathrm{vec}(c_{1:\ell,1:\ell})$);

IF $n > \ell$, THEN

/* 2. Solve transformed Sylvester equation (2.95). */

$l = \ell + 1$; $c_{1:\ell,l:n} =$ `Sylsc`($a_{1:\ell,1:\ell}, a_{l:n,l:n}, c_{1:\ell,l:n} - c_{1:\ell,1:\ell}a_{1:\ell,l:n}$);

/* 3. Solve the remaining Lyapunov equation (2.96). */

$c_{l:n,l:n} =$ `Lyac`($a_{l:n,l:n}, c_{l:n,l:n} - a^T_{1:\ell,l:n}c_{1:\ell,l:n} - c^T_{1:\ell,l:n}a_{1:\ell,l:n}$);

Obviously, the calculations for updating C at step 3 can be performed taking advantage of the symmetry, as described above. The total computing effort is about $10n^3 + 3.5n^3 = 13.5n^3$ *flops*.

The recursive algorithm for solving discrete-time Lyapunov equations is given below, without further explanation, based on the following relations—the discrete-time correspondents of (2.94–2.96),

$$X_{11} - A_{11}^T X_{11} A_{11} = C_{11}, \tag{2.98}$$

$$\widehat{X}_{12} - A_{11}^T \widehat{X}_{12} \widehat{A}_{22} = \widehat{C}_{12} + A_{11}^T X_{11} \widehat{A}_{12}, \tag{2.99}$$

$$\widehat{X}_{22} - \widehat{A}_{22}^T \widehat{X}_{22} \widehat{A}_{22} = \widehat{C}_{22} + \widehat{A}_{12}^T X_{11} \widehat{A}_{12} + \widehat{A}_{22}^T \widehat{X}_{12}^T \widehat{A}_{12} + \widehat{A}_{12}^T \widehat{X}_{12} \widehat{A}_{22}, \tag{2.100}$$

and (2.97),

$$\begin{bmatrix} a_{11}^2 - 1 & 2a_{11}a_{21} & a_{21}^2 \\ a_{11}a_{12} & a_{11}a_{22} + a_{12}a_{21} - 1 & a_{21}a_{22} \\ a_{12}^2 & 2a_{12}a_{22} & a_{22}^2 - 1 \end{bmatrix} \begin{bmatrix} x_{11} \\ x_{12} \\ x_{22} \end{bmatrix} = - \begin{bmatrix} c_{11} \\ c_{12} \\ c_{22} \end{bmatrix}. \tag{2.101}$$

The last equation follows from a linear system of order 4, with coefficient matrix $A_{11}^T \otimes A_{11}^T - I_4$, re-written using $x_{12} = x_{21}$.

Algorithm Lyad : $C =$ `Lyad`(A, C)

/* Solve Lyapunov equation $X - A^T XA = C$, $A \in \mathbb{R}^{n\times n}$ in QTF; */
/* C is overwritten by X. */

$\ell = 1$; IF $n > 1$ AND $a_{21} \neq 0$, THEN $\ell = 2$;

/* 1. Solve reduced-order symmetric linear system (2.98). */

$c_{1:\ell,1:\ell} =$ `solve`($a^T_{1:\ell,1:\ell} \otimes a^T_{1:\ell,1:\ell} - I_{2\ell}, -\mathrm{vec}(c_{1:\ell,1:\ell})$);

IF $n > \ell$, THEN

/* 2. Solve transformed Sylvester equation (2.99). */

$l = \ell + 1;$

$c_{1:\ell,l:n} = \texttt{Sylsd}(\ -a_{1:\ell,1:\ell}, a_{l:n,l:n}, c_{1:\ell,l:n} + a^T_{1:\ell,1:\ell} c_{1:\ell,1:\ell} a_{1:\ell,l:n}\ );$

/* 3. Solve the remaining Lyapunov equation (2.100). */

$c_{l:n,l:n} \leftarrow c_{l:n,l:n} + a^T_{1:\ell,l:n} c_{1:\ell,1:\ell} a_{1:\ell,l:n};$

$c_{l:n,l:n} \leftarrow c_{l:n,l:n} + a^T_{l:n,l:n}(c^T_{1:\ell,l:n} a_{1:\ell,l:n}) + (c^T_{1:\ell,l:n} a_{1:\ell,l:n})^T a_{l:n,l:n};$

$c_{l:n,l:n} = \texttt{Lyad}(\ a_{l:n,l:n}, c_{l:n,l:n}\ );$

Details about the standard, non-recursive discrete-time algorithm can be found in (Barraud, 1977; Kitagawa, 1977; Sima, 1980a). The linear system at step 1 is the symmetric system in (2.101). A technique similar to that described for discrete-time Sylvester equation is used at step 3.

2.3.3 Solving stable non-negative definite Lyapunov equations

This section deals with the numerical solution of some continuous-time and discrete-time Lyapunov equations of the form

$$\widetilde{A}^H \widetilde{X} + \widetilde{X}\widetilde{A} = -\widetilde{C}, \qquad \widetilde{A}^H \widetilde{X}\widetilde{A} - \widetilde{X} = -\widetilde{C}, \tag{2.102}$$

where $\widetilde{C}$ is positive semi-definite ($\widetilde{C} = \widetilde{C}^H \geq 0$), $\widetilde{A}$ and $\widetilde{C}$ are given $n \times n$ complex matrices, and $\widetilde{X}$ is to be determined. If $\widetilde{A}$ is a stable matrix, then each equation in (2.102) has a unique Hermitian positive semi-definite solution, $\widetilde{X} = \widetilde{X}^H \geq 0$. Hence, the solution can be factored as $\widetilde{X} = \widetilde{U}^H \widetilde{U}$, where $\widetilde{U}$ is an upper triangular matrix with non-negative (real) diagonal elements—the *Cholesky factor* of $\widetilde{X}$. Many algorithms can advantageously use the factor $\widetilde{U}$ instead of the matrix $\widetilde{X}$. Moreover, if matrix $\widetilde{X}$ is nonsingular, then $\mathrm{cond}_2(\widetilde{X}) = (\mathrm{cond}_2(\widetilde{U}))^2$, so that $\widetilde{X}$ could be more ill conditioned with respect to inversion than $\widetilde{U}$.

The equations in (2.102) can be re-written as

$$\begin{aligned} \widetilde{A}^H(\widetilde{U}^H\widetilde{U}) + (\widetilde{U}^H\widetilde{U})\widetilde{A} &= -\widetilde{B}^H\widetilde{B}, \\ \widetilde{A}^H(\widetilde{U}^H\widetilde{U})\widetilde{A} - \widetilde{U}^H\widetilde{U} &= -\widetilde{B}^H\widetilde{B}, \end{aligned} \tag{2.103}$$

where matrix $\widetilde{B}$ is $m \times n$ ($m \geq n$ or $m < n$). Let $\widetilde{S}$ be a unitary matrix that reduces $\widetilde{A}$ to an upper Schur form, that is $A := \widetilde{S}^H \widetilde{A}\widetilde{S}$ is upper triangular. (Hence, a_{ii} ($i = 1\colon n$) are the eigenvalues of $\widetilde{A}$.) Let

$$\widetilde{B} = \widetilde{Q}\begin{bmatrix} \widetilde{R} \\ 0 \end{bmatrix} \ (\text{if } m \geq n), \quad \text{or } \widetilde{B} = \widetilde{Q}\begin{bmatrix} \widetilde{R} & \widetilde{T} \end{bmatrix} \ (\text{if } m < n),$$

be a QR factorization of $\widetilde{B}$ (Golub and Van Loan, 1983), where $\widetilde{R}$ is an $n \times n$ (or $m \times m$, respectively) upper triangular matrix with real diagonal elements (which could turn positive). Finally, let $\widetilde{R}\widetilde{S} = QR$ (or $[\,\widetilde{R}\;\; \widetilde{T}\,]\widetilde{S} = QR$) be a QR factorization of the matrix product $\widetilde{R}\widetilde{S}$ (or $[\,\widetilde{R}\;\; \widetilde{T}\,]\widetilde{S}$, respectively). The equations in (2.102) are then equivalent to the following equations,

$$\begin{aligned} A^H(U^H U) + (U^H U)A &= -R^H R, \\ A^H(U^H U)A - U^H U &= -R^H R, \end{aligned} \tag{2.104}$$

where U is upper triangular, if U is the Cholesky factor from the QR factorization of $\widetilde{U}\widetilde{S}$. Solving an equation (2.104) in U, the Cholesky factor $\widetilde{U}$ of the solution of original equation (2.102) is obtained from the QR factorization of the matrix product $U\widetilde{S}^H$.

Equations like (2.103), and particularly the corresponding real equations (with $\widetilde{A} \in \mathbb{R}^{n \times n}$, $\widetilde{B} \in \mathbb{R}^{m \times n}$), often arise in various techniques for analysis and design of control systems.

In the sequel, we consider the efficient and numerically stable solution of the complex Lyapunov equations (2.104) and their real counterparts,

$$\begin{aligned} A^T(U^T U) + (U^T U)A = -R^T R, \\ A^T(U^T U)A - U^T U = -R^T R, \end{aligned} \tag{2.105}$$

where A is in an upper real Schur form, and R is upper triangular. The basic method is proposed in (Hammarling, 1982); see also (Varga, 1990) for the discrete-time case.

First, we investigate the continuous-time Lyapunov equation, described by the first equations in (2.104) and (2.105). Partitioning

$$U = \begin{bmatrix} u_{11} & u^H \\ 0 & U_1 \end{bmatrix}, \quad R = \begin{bmatrix} r_{11} & r^H \\ 0 & R_1 \end{bmatrix}, \quad A = \begin{bmatrix} a_{11} & a^H \\ 0 & A_1 \end{bmatrix}, \tag{2.106}$$

where u_{11}, r_{11}, and a_{11} are 1×1 submatrices, from the first equation (2.104) we have

$$\begin{aligned} |u_{11}|^2 &= -|r_{11}|^2/(a_{11} + \bar{a}_{11}), \\ u_{11}(A_1^H + a_{11} I_{n-1})u &= -r_{11} r - |u_{11}|^2 a, \\ A_1^H(U_1^H U_1) + (U_1^H U_1)A_1 &= -R_1^H R_1 - r r^H - (v u^H + u v^H), \end{aligned}$$

where $v := A_1^H u + \bar{u}_{11} a$. Denoting

$$\alpha := \operatorname{sign}(r_{11})(-(a_{11} + \bar{a}_{11}))^{1/2}, \qquad y := r - \alpha u, \tag{2.107}$$

from the above equations, we obtain

$$
\begin{aligned}
u_{11} &= r_{11}/\alpha, && (2.108)\\
(A_1^H + a_{11}I_{n-1})u &= -r\alpha - au_{11}, && (2.109)\\
A_1^H(U_1^H U_1) + (U_1^H U_1)A_1 &= -\widehat{R}^H\widehat{R} := -(R_1^H R_1 + yy^H), && (2.110)
\end{aligned}
$$

where $\widehat{R}$ is the Cholesky factor of the matrix $R_1^H R_1 + yy^H$. Several comments on the equations in (2.108)–(2.110) are useful. First, since A (or $\widetilde{A}$) is a stable matrix, we deduce that $\Re(a_{11}) = \Re(\lambda_1) < 0$, hence α in (2.107) is a real scalar and u_{11} is chosen to be positive. If $r_{11} \neq 0$, then α can be written as $\alpha = r_{11}u_{11}^{-1}$, because $u_{11} \neq 0$; this definition gives the equation (2.109). If $r_{11} = 0$, then the definition of α in (2.107) is valid too, and we obtain $u_{11} = 0$; the relation which delivered (2.109) holds for every u, including the solution of (2.109). The stability assumption about the matrix A ensures that $A_1^H + a_{11}I_{n-1}$ is nonsingular. Finally, it is easy to prove the relation of updating the Cholesky factorization in (2.110), after adding the rank-one matrix yy^H. Note that (2.110) has the same form as the first equation in (2.104), but its order is $n-1$.

The solution procedure is more involved in the real case, due to possible 2×2 diagonal blocks in the real Schur form of the matrix A in the first equation (2.105). So, a_{11}, u_{11}, and r_{11} in (2.106) are either scalars or 2×2 matrices, and a, u and r are either vectors or two-column matrices. Assuming for the moment that u_{11} is nonsingular, and denoting

$$
\alpha := r_{11}u_{11}^{-1}, \qquad y := r - u\alpha^T, \tag{2.111}
$$

the equations (2.108)–(2.110) should be replaced by

$$
\begin{aligned}
a_{11}^T(u_{11}^T u_{11}) + (u_{11}^T u_{11})a_{11} &= -r_{11}^T r_{11}, && (2.112)\\
A_1^T u + u(u_{11}a_{11}u_{11}^{-1}) &= -r\alpha - au_{11}^T, && (2.113)\\
A_1^T(U_1^T U_1) + (U_1^T U_1)A_1 &= -\widehat{R}^T\widehat{R} := -(R_1^T R_1 + yy^T). && (2.114)
\end{aligned}
$$

Clearly, in the scalar case, when a_{11} is an 1×1 matrix, the equations (2.112)–(2.114) are special cases of (2.108)–(2.110). So, we consider that a_{11} is a 2×2 matrix.

The equation (2.112) can be re-written in the form of the first relation (2.104). To this end, the matrix a_{11} is reduced to the upper triangular form (in $\mathbb{C}$) using a complex plane rotation. Let $(a_1, a_3)^T := a_{11}(:,1)$, $\lambda \in \lambda(a_{11})$, and let Q^H be a complex plane rotation annihilating the second element of the vector $(a_1 - \lambda, a_3)^T$. Then, simple calculations give

$$
\widehat{a}_{11} := Q^H a_{11} Q = \begin{bmatrix} \bar{\lambda} & \bar{t} \\ 0 & \lambda \end{bmatrix}. \tag{2.115}
$$

Let $u_{11}Q = Tv_{11}$, and $r_{11}Q = Pp_{11}$ be QR factorizations of $u_{11}Q$, and $r_{11}Q$, respectively. Hence, T and P are unitary matrices, and

$$v_{11} = \begin{bmatrix} v_1 & \bar{v}_2 \\ 0 & v_3 \end{bmatrix}, \qquad p_{11} = \begin{bmatrix} p_1 & \bar{p}_2 \\ 0 & p_3 \end{bmatrix}.$$

The matrices T and P may be chosen so that v_1, v_3, p_1, and p_3 are real non-negative. We prove this for P. Indeed, if Q_p^H is a (complex) plane rotation triangularizing the matrix $r_{11}Q$, hence

$$Q_p^H r_{11}Q = \begin{bmatrix} \widehat{p}_1 & \widehat{p}_2 \\ 0 & \widehat{p}_3 \end{bmatrix},$$

then, defining

$$D_p := \mathrm{diag}(d_1, d_3), \quad d_i = \begin{cases} \widehat{p}_i/|\widehat{p}_i|, & \text{if } \widehat{p}_i \neq 0, \\ 1, & \text{if } \widehat{p}_i = 0, \end{cases} \qquad i \in \{1, 3\},$$

it follows that $D_p^H Q_p^H r_{11}Q = p_{11}$ has non-negative diagonal elements, namely $p_1 = |\widehat{p}_1|$, $p_3 = |\widehat{p}_3|$. Therefore, $P := Q_p D_p$ performs the desired transformation and is unitary (since Q_p and D_p are unitary).

Now, the transformed equation (2.112) is in the form of the first equation (2.104),

$$\widehat{a}_{11}^H(v_{11}^H v_{11}) + (v_{11}^H v_{11})\widehat{a}_{11} = -p_{11}^H p_{11}, \tag{2.116}$$

so that (2.108)–(2.110) specialize as follows:

$$\begin{aligned} v_1 &= p_1/\widetilde{\alpha}, \\ v_2 &= -(p_2\widetilde{\alpha} + tv_1)/(2\lambda), \\ v_3 &= (p_3^2 + |\widetilde{y}|^2)^{1/2}/\widetilde{\alpha}, \end{aligned} \tag{2.117}$$

where

$$\widetilde{\alpha} := (-(\lambda + \bar{\lambda}))^{1/2}, \qquad \widetilde{y} := p_2 - v_2\widetilde{\alpha}. \tag{2.118}$$

Having v_{11}, the matrix u_{11} is obtained from the QR factorization of $v_{11}Q^H$, which also gives the matrix T.

Note that in (2.113),

$$u_{11}a_{11}u_{11}^{-1} = T(v_{11}\widehat{a}_{11}v_{11}^{-1})T^H, \qquad \alpha = P(p_{11}v_{11}^{-1})T^H. \tag{2.119}$$

Since

$$\begin{aligned} v_{11}\widehat{a}_{11}v_{11}^{-1} &= \begin{bmatrix} \bar{\lambda} & (\bar{v}_2(\lambda - \bar{\lambda}) + v_1\bar{t})/v_3 \\ 0 & \lambda \end{bmatrix}, \\ p_{11}v_{11}^{-1} &= \begin{bmatrix} \widetilde{\alpha} & (\bar{p}_2 - \bar{v}_2\widetilde{\alpha})/v_3 \\ 0 & p_3/v_3 \end{bmatrix}, \end{aligned} \tag{2.120}$$

simple manipulation, using (2.117) and (2.118), yields

$$v_{11}\widehat{a}_{11}v_{11}^{-1} = \begin{bmatrix} \bar{\lambda} & \bar{\gamma} \\ 0 & \lambda \end{bmatrix}, \qquad p_{11}v_{11}^{-1} = \begin{bmatrix} \widetilde{\alpha} & \bar{\delta} \\ 0 & \eta \end{bmatrix}, \tag{2.121}$$

where

$$\delta := \widetilde{y}/v_3, \quad \eta := p_3/v_3, \quad \gamma := -\widetilde{\alpha}\delta. \tag{2.122}$$

When $\widetilde{y} = 0$ and $p_3 = 0$ (hence $v_3 = 0$), we choose $\delta = 0$ and $\eta = \widetilde{\alpha}$, so that $|\delta| \le \widetilde{\alpha}$, $0 \le \eta \le \widetilde{\alpha}$, $|\gamma| \le \widetilde{\alpha}^2$, are obtained for all $\widetilde{y}$ and p_3.

Having $v_{11}\widehat{a}_{11}v_{11}^{-1}$ and $p_{11}v_{11}^{-1}$ from (2.121), we obtain $u_{11}a_{11}u_{11}^{-1}$ and α from (2.119), so that (2.113) can be solved. (There is no need for computing the inverse of the matrix u_{11}.) In the 2×2 case, (2.113) is a special Sylvester equation, where the matrices that multiply the unknown matrix u are in real Schur form. In the 1×1 case, (2.113) reduces to the system of linear algebraic equations (2.109), where the coefficient matrix is in lower real Schur form.

Finally, the Cholesky factor in (2.114) is updated by a sequence of Householder transformations. For instance, if matrix y has two columns, then $\widehat{R}$ is obtained as

$$\widehat{R} = H_{n-2} \dots H_2 H_1 \begin{bmatrix} R_1 \\ y^T \end{bmatrix}, \tag{2.123}$$

where H_i is a Householder transformation that modifies the element $(R_1)_{ii}$ and annihilates the elements y_{i1} and y_{i2}, $i = 1{:}\,n-2$. We may use modified Householder transformations of order 3 or 2, depending on if y has two columns or one column, respectively. If R_1 has the last $n - \mathtt{ord}(a_{11}) - j$ rows zero, then it suffices to apply only the first j transformations (H_1 to H_j) in (2.123) and, when y has two columns, an additional transformation that would modify $y_{j+1,1}$ and annihilate $y_{j+1,2}$; the reason is that, for the described circumstance, the non-zero part of y^T may directly be copied just below the first j rows of R_1. This situation arises when $m < n$.

Summarizing all the above, the algorithm for solving the real continuous-time Lyapunov equation is as follows.

Algorithm LyacC : $U =$ `LyacC(` A, R `)`

/* Solve for Cholesky factor U the Lyapunov equation */
/* $A^T(U^TU) + (U^TU)A = -R^TR$, */
/* with $A \in \mathbb{R}^{n\times n}$ in RSF, $\lambda(A) \subset \mathbb{C}^-$, and R upper triangular. */

/* 1. Solve (2.112), $\mathrm{diag}(A) = \mathrm{diag}(a_{11}, A_1)$, $\mathtt{ord}(a_{11}) \in \{1,2\}$. */
IF $a_{21} = 0$, THEN $\ell = 1$; ELSE $\ell = 2$;
IF $\ell = 1$, THEN

```
    α = sign(r₁₁)(−2a₁₁)^{1/2};   u₁₁ = r₁₁/α;
ELSE
    (λ, λ̄) = Schur2( a_{1:2,1:2} );
    Q = crot( a₁₁ − λ, a₂₁ );   t = conj( (Q^H a_{1:2,1:2} Q)₁₂ );
    Q_p = crot( (r_{1:2,1:2}Q)₁₁, (r_{1:2,1:2}Q)₂₁ );   [ p̂₁ p̂₂ ; 0 p̂₃ ] = Q_p^H r_{1:2,1:2} Q;
    IF p̂ᵢ = 0, THEN dᵢ = 1; ELSE dᵢ = p̂ᵢ/|p̂ᵢ|, i ∈ {1,3};
    D_p = diag(d₁, d₃);
    p₁ = |p̂₁|;   p₂ = conj( (D_p^H Q_p^H r_{1:2,1:2} Q)₁₂ );   p₃ = |p̂₃|;
    α̃ = (−(λ + λ̄))^{1/2};   v₁ = p₁/α̃;   v₂ = −(p₂α̃ + tv₁)/(2λ̄);
    ỹ = p₂ − v₂α̃;   v₃ = (p₃² + |ỹ|²)^{1/2}/α̃;
    u_{1:2,1:2} = [ v₁ v̄₂ ; 0 v₃ ] Q^H;   (u_{1:2,1:2}, T^H) = QR( u_{1:2,1:2} );
END IF
                                              /* 2. Exit when A₁ is void. */
IF n − ℓ > 0, THEN
                                              /* 3. Solve (2.113). */
    a := a^T_{1:ℓ,ℓ+1:n};   A₁ := a_{ℓ+1:n,ℓ+1:n};
    r := r^T_{1:ℓ,ℓ+1:n};   R₁ := r_{ℓ+1:n,ℓ+1:n};
    IF ℓ = 1, THEN
        u = solve( A₁^T + a₁₁I_{n−1}, −rα − au₁₁ );
    ELSE
        IF ỹ = 0 AND p₃ = 0, THEN
            δ = 0;   η = α̃;   γ = 0;
        ELSE
            δ = ỹ/v₃;   η = p₃/v₃;   γ = −α̃δ;
        END IF
        B := T [ λ̄ γ̄ ; 0 λ ] T^H;   α = P [ α̃ δ̄ ; 0 η ] T^H;   u := −rα − au^T_{1:2,1:2};
        u = Sylsc( A₁, B, u );
    END IF
    u_{1:ℓ,ℓ+1:n} = u^T;
                                              /* 4. Solve (2.114). */
    y = r − uα^T;   R̂ = [ R₁ ; y^T ];   l = ℓ + 1;
    FOR i = 1:n − ℓ,
        Hᵢ = Hous_1( (R₁)ᵢᵢ, y^T_{i,:} );   R̂ ← HᵢR̂;
    u_{ℓ+1:n,ℓ+1:n} = LyacC( A₁, R̂_{1:n−ℓ,1:n−ℓ} );
END IF
```

Note that Algorithm LyacC is a recursive one and the order of the matrix A_1

reduces to zero in $n-1$ successive calls at most. It is assumed that the Householder transformations at step 4 are imbedded into the n-dimensional space.

Consider now the discrete-time Lyapunov equation, described by the second equation in (2.103), (2.104) and (2.105). The solution procedure is similar to that for continuous-time equations, presented above. The formulas corresponding to (2.107)–(2.110) are the following:

$$\alpha := \text{sign}(r_{11})(1-|a_{11}|^2)^{1/2}, \qquad y := (r-\alpha u)/a_{11}, \tag{2.124}$$

$$\begin{aligned} u_{11} &= r_{11}/\alpha, & (2.125)\\ (a_{11}A_1^H - I_{n-1})u &= -r\alpha - a_{11}au_{11}, & (2.126)\\ A_1^H(U_1^H U_1)A_1 - U_1^H U_1 &= -\widehat{R}^H\widehat{R} := -(R_1^H R_1 + yy^H), & (2.127) \end{aligned}$$

where we assumed that $a_{11} \neq 0$. If $a_{11} = 0$, then equation (2.126) becomes $u = \text{sign}(r_{11})r$, and the definition of y from (2.124) should be replaced by $y := au_{11} + A_1^H u$.

In the real case, instead of the formulas (2.111)–(2.114), we have:

$$\begin{aligned} \alpha &:= r_{11}u_{11}^{-1}, \qquad b := u_{11}a_{11}u_{11}^{-1},\\ y &:= r - (au_{11}^T + A_1^T u + u)(I + b^T)^{-1}\alpha^T, & (2.128) \end{aligned}$$

$$\begin{aligned} a_{11}^T(u_{11}^T u_{11})a_{11} - u_{11}^T u_{11} &= -r_{11}^T r_{11}, & (2.129)\\ A_1^T ub - u &= -r\alpha - au_{11}^T b, & (2.130)\\ A_1^T(U_1^T U_1)A_1 - U_1^T U_1 &= -\widehat{R}^T\widehat{R} := -(R_1^T R_1 + yy^T). & (2.131) \end{aligned}$$

Note that the definitions of y in (2.124) and (2.128) are those in (Varga, 1990). The formulas corresponding to (2.117), (2.118) and (2.122) are:

$$\begin{aligned} v_1 &= p_1/\widetilde{\alpha},\\ v_2 &= (p_2\widetilde{\alpha} + v_1 t\bar{\lambda})/(1-\bar{\lambda}^2),\\ v_3 &= (p_3^2 + |\widetilde{y}|^2)^{1/2}/\widetilde{\alpha}, \end{aligned}$$

$$\widetilde{\alpha} := (1-|\lambda|^2)^{1/2}, \qquad \widetilde{y} := ((\lambda - \bar{\lambda})p_2 - p_1 t)/(1-\bar{\lambda}^2), \tag{2.132}$$

$$\delta := \bar{\lambda}\widetilde{y}/v_3, \quad \eta := p_3/v_3, \quad \gamma := -\widetilde{\alpha}\widetilde{y}/v_3. \tag{2.133}$$

Note that for $\lambda = 0$, the second relation in (2.132) becomes $\widetilde{y} = -p_1 t$, and for $\lambda \neq 0$, it is equivalent to $\widetilde{y} = (p_2 - v_2\widetilde{\alpha})/\bar{\lambda}$, which is similar to the corresponding definition in the continuous-time case (see (2.118)).

The algorithm for solving the real discrete-time Lyapunov equation is as follows.

Algorithm LyadC : $\boxed{U = \texttt{LyadC}(\ A, R\ )}$

```
/* Solve for Cholesky factor U the Lyapunov equation */
/* A^T(U^T U)A - U^T U = -R^T R, */
/* with A ∈ IR^{n×n} in RSF, λ(A) ⊂ C^-, and R upper triangular. */

        /* 1. Solve (2.129), diag(A) = diag(a11, A1), ord(a11) ∈ {1,2}. */
IF a21 = 0, THEN ℓ = 1; ELSE ℓ = 2;
IF ℓ = 1, THEN
   α = sign(r11)(1 - |a11|^2)^{1/2};   u11 = r11/α;
ELSE
   (λ, λ̄) = Schur2( a_{1:2,1:2} );
   Q = crot( a11 - λ, a21 );   t = conj( (Q^H a_{1:2,1:2} Q)_{12} );
   Q_p = crot( (r_{1:2,1:2}Q)_{11}, (r_{1:2,1:2}Q)_{21} );   [ p̂1 p̂2 ; 0 p̂3 ] = Q_p^H r_{1:2,1:2} Q;
   IF p̂i = 0, THEN di = 1; ELSE di = p̂i/|p̂i|, i ∈ {1,3};
   D_p = diag(d1, d3);
   p1 = |p̂1|;   p2 = conj( (D_p^H Q_p^H r_{1:2,1:2} Q)_{12} );   p3 = |p̂3|;
   α̃ = (1 - |λ|^2)^{1/2};   v1 = p1/α̃;   v2 = (p2 α̃ + v1 t λ̄)/(1 - λ̄^2);
   ỹ = ((λ - λ̄)p2 - p1 t)/(1 - λ̄^2);   v3 = (p3^2 + |ỹ|^2)^{1/2}/α̃;
   u_{1:2,1:2} = [ v1 v̄2 ; 0 v3 ] Q^H;   (u_{1:2,1:2}, T^H) = QR( u_{1:2,1:2} );
END IF
                                        /* 2. Exit when A1 is void. */
IF n - ℓ > 0, THEN
                                        /* 3. Solve (2.130). */
   a := a^T_{1:ℓ,ℓ+1:n};   A1 := a_{ℓ+1:n,ℓ+1:n};
   r := r^T_{1:ℓ,ℓ+1:n};   R1 := r_{ℓ+1:n,ℓ+1:n};
   IF ℓ = 1, THEN
      u = solve( a11 A1^T - I_{n-1}, -rα - a11 a u11 );
      IF a11 = 0, THEN y = a u11 + A1^T u; ELSE y = (r - αu)/a11;
   ELSE
      IF ỹ = 0 AND p3 = 0, THEN
         δ = 0;   η = α̃;   γ = 0;
      ELSE
         δ = λ̄ỹ/v3;   η = p3/v3;   γ = -α̃ỹ/v3;
      END IF
      B := T [ λ̄ γ̄ ; 0 λ ] T^H;   α = P [ α̃ δ̄ ; 0 η ] T^H;   u := rα + a u^T_{1:2,1:2} B;
      u = Sylsd( A1, -B, u );
      y = r - (a u^T_{1:2,1:2} + A1^T u + u)(I2 + B^T)^{-1} α^T;
   END IF
```

$u_{1:\ell,\ell+1:n} = u^T;$

/* 4. Solve (2.131). */

$\widehat{R} = \begin{bmatrix} R_1 \\ y^T \end{bmatrix}; \quad 1 = \ell + 1;$

FOR $i = 1{:}\,n - \ell$,

$H_i = \texttt{Hous\_1}(\ (R_1)_{ii}, y^T_{i,:}\ ); \quad \widehat{R} \leftarrow H_i\widehat{R};$

$u_{\ell+1:n,\ell+1:n} = \texttt{LyadC}(\ A_1, \widehat{R}_{1:n-\ell,1:n-\ell}\ );$

END IF

Note that the variables a, A_1, r, R_1, and u are used in algorithms `LyacC` and `LyadC` for convenience. In an implementation, these variables overwrite the corresponding locations of the arrays for matrices A, R, and U, respectively. The result U overwrites the given Cholesky factor R. So, the memory space needed for the data is $2n^2$, since the required working vectors are stored into the (null) lower triangular part of the arrays `A` and `R`. Only real arithmetic is used all through.

A discussion about the sensitivity of the solution of continuous-time Lyapunov equation in (2.102) to perturbations in its data is given in (Hammarling, 1982). A condition number for this problem is $||\phi^{-1}||_2||\widetilde{A}||_2$, where $\phi := I_n \otimes \widetilde{A}^H + \widetilde{A}^H \otimes I_n$. The value $||\phi^{-1}||_2$ can be estimated in about $2n^3$ *flops*, but it needs n^2 extra memory locations. When matrix A is normal, the condition number becomes $\max_i\{\,|\lambda_i|\,\}/\min_{i,j}\{\,|\bar{\lambda}_i + \lambda_j|\,\}$.

A sufficient condition for the existence of a positive semi-definite solution $\widetilde{X}$ of the *real* Lyapunov equation corresponding to (2.102) is that $\widetilde{A}$ be stable and $\widetilde{C}$ positive semi-definite ($\widetilde{C} \geq 0$). The latter requirement makes it sure that there is a factorization $\widetilde{C} = \widetilde{B}^T\widetilde{B}$, where $\widetilde{B}$ is $m \times n$, and $\operatorname{rank}(\widetilde{B}) = \operatorname{rank}(\widetilde{C})$; in particular, there is the Cholesky factorization $\widetilde{C} = R^T R$, where R is upper triangular with non-negative diagonal elements. Algorithms `LyacC` and `LyadC` will fail if the given matrix $\widetilde{A}$ is not stable (even numerically).

The algorithms discussed above need cn^3 *flops*, where c is of order of unity. The value of c depends on m and the structure of the spectrum of A. For instance, if $m = n$, R is upper triangular, and $\lambda(A) \subset \mathbb{R}$, then $c = 2/3$.

Our routines implementing Algorithms `LyacC` and `LyadC` have as input parameters the matrix $A \in \mathbb{R}^{n\times n}$ in RSF and the matrix $B \in \mathbb{R}^{m\times n}$, where $m \geq n$ or $m < n$. The subdiagonal elements of B are zeroed using Householder transformations, so that the initially required factor R is obtained. If the Lyapunov equation has a general matrix A, then A should first be reduced to RSF using the Algorithms `Hess` (§ 2.2.2) and `mSchur` or `Schur` (§ 2.2.3). Orthogonal transformations performing this reduction, which are accumulated in a matrix $\widetilde{S}$, should also be applied to B, prior to the call of `LyacC` or `LyadC`. It is worth mentioning that, for $m > 7n/6$, it is more efficient to first compute a QR factorization of $\widetilde{B}$, and then postmultiply the resulting triangular factor by the transformation matrix $\widetilde{S}$. Clearly, the matrix U, computed by `LyacC` and `LyadC`, is not the Cholesky factor of the solution $\widetilde{X}$, this factor being obtainable from the QR factorization of $U\widetilde{S}^T$. The factor U overwrites R in the array `R`.

We will give more details about our implementation of `LyacC` and `LyadC`. The first computational stage determines the Cholesky factor from the QR factorization of the given matrix B, using Algorithm `QR`. The next stage is a loop, indexed by i, for solving the reduced Lyapunov equation (2.105). The current diagonal block in A can be 1×1 or 2×2; these cases are dealt with separately. The vector or matrix y is stored in the locations $(i+1\colon n, 1)$ or $(i+1\colon n, 1\colon 2)$, respectively, of `R`. Moreover, the subarray $a_{i+1:n,1:2}$ initially contains the matrix a, and then the solution u of the system of linear algebraic equations; this solution is then copied into the corresponding elements of `R`. (To return an unmodified matrix A, the $(2,1)$ and $(3,2)$ elements of `A` are initially saved and finally restored.)

The linear system associated with (2.109) or (2.126) is solved the same way by the two algorithms. Note that, if $a_{11} = 0$, then from (2.126) it follows that $u = r\alpha$, and if $a_{11} \neq 0$, then a linear system like (2.109) is obtained by dividing both terms in (2.126) by a_{11}.

If the current submatrix A_1 has a real spectrum, then the linear system associated with (2.109) or (2.126) has a lower triangular coefficient matrix (actually, its transpose is stored), and it can simply be solved, using Level 1 BLAS operations, $z \leftarrow \alpha x + z$. The computational procedure is more complicated when A_1 has at least a pair of complex eigenvalues. The implemented variant triangularizes the coefficient matrix, using modified Householder transformations of order 2. This approach shows to be profitable, because it is conceptually simple, numerically stable, and provides a compact implementation, using the same algorithms as those needed for updating the Cholesky factor. (An alternate approach could be that of Gaussian elimination, but it would require pivoting.) The structure of A_1 matrix enables that the triangularization is done simultaneously with the computation of the solution. Indeed, if A_1 has l diagonal blocks of order 2, the linear system $(A_1^T + \lambda I)u = b$ is equivalent to

$$(A_1^T + \lambda I)H_1 \cdots H_l(H_l \cdots H_1 u) = b, \tag{2.134}$$

where H_i is a Householder transformation that annihilates the subdiagonal element of the ith 2×2 diagonal block in $(A_1 + \lambda I)$; thus, the matrix $(A_1^T + \lambda I)H_1 \cdots H_l$ is lower triangular. Due to the RSF structure, the columns in $A_1^T + \lambda I$ which H_i acts upon are distinct from those transformed by H_j, for $j \neq i$. This way, the corresponding elements of the solution can at once be determined after having found the transformation H_i. For instance, after H_1 has been obtained, one solves $(A_1^T + \lambda I)H_1 x = b$ in x, and then u_1 and u_2 follow from $H_1 x = \begin{bmatrix} u_1 & u_2 \end{bmatrix}^T$. The elements u_1 and u_2 are used for updating the right-hand side, using Level 1 BLAS operations, and the computational process goes on like this.

The Cholesky factor R_1 is updated using modified Householder transformations. If $m_r < n$, where m_r is the current number of non-zero rows of R (with $m_r = m$, initially), only $m - i$ Householder transformations are applied (when the loop index equals i); then, m_r is incremented by 1, and the last $n - m_r + 1$ elements of y (with changed sign) are copied into the m_rth row of R, beginning from the diagonal.

When the current diagonal block in A is 2×2, the computational procedure proves far more complicated than for the 1×1 case. Calculations with complex numbers are involved, but they are implemented with real arithmetic. A complex plane rotation Q is used for triangularization of matrix a_{11}. Given two numbers, c_1, and c_2 in $\mathbb{C}$, specified by their real and imaginary parts, one computes the parameters $c \in \mathbb{R}$, and $s \in \mathbb{C}$, so that $c^2 + |s|^2 = 1$, and

$$Q \begin{bmatrix} c_1 \\ c_2 \end{bmatrix} \to \begin{bmatrix} c_1 \\ 0 \end{bmatrix}, \qquad Q = \begin{bmatrix} c & \bar{s} \\ s & -c \end{bmatrix}.$$

Clearly, Q is a unitary Hermitian matrix; it is constructed as a modified complex Householder transformation. The special case when $c_2 \in \mathbb{R}$ is dealt with separately. Actually, c_2 is not modified. For constructing the matrix Q that triangularizes

$$a_{11} =: \begin{bmatrix} a_1 & a_2 \\ a_3 & a_4 \end{bmatrix},$$

one uses $c_1 := a_1 - \lambda$, $c_2 := a_3$, where $\lambda \in \lambda(a_{11})$ has $\Im(\lambda) > 0$. For determining the scalar t in (2.115), one computes

$$e_1^T Q \begin{bmatrix} a_1 - \lambda & a_2 \\ a_3 & a_4 - \lambda \end{bmatrix} = \begin{bmatrix} \tilde{a}_1 & \tilde{a}_2 \end{bmatrix},$$

and, then, $\bar{t} = \begin{bmatrix} \tilde{a}_1 & \tilde{a}_2 \end{bmatrix} Q e_2$.

For calculation of the matrices p_{11} and u_{11}, one first finds the matrix product rQ, where

$$r =: \begin{bmatrix} r_1 & r_2 \\ 0 & r_3 \end{bmatrix}$$

is upper triangular with real diagonal elements (r_2 is real or complex). Then, a QR factorization of the matrix rQ is computed as $rQ = Q_p D_p p$, where Q_p is a complex plane rotation, D_p is a unitary diagonal matrix, and p is an upper triangular matrix with real non-negative diagonal elements. Specifically,

$$Q_p = \begin{bmatrix} c_p & \bar{s}_p \\ s_p & -c_p \end{bmatrix}, \qquad D_p = \begin{bmatrix} d_1 & 0 \\ 0 & d_3 \end{bmatrix}, \qquad p = \begin{bmatrix} p_1 & \bar{p}_2 \\ 0 & p_3 \end{bmatrix},$$

where $c_p, p_1, p_3 \in \mathbb{R}$, and $s_p, d_1, d_3, p_2 \in \mathbb{C}$. The elements of p overwrite the corresponding elements of r. As D_p is unitary, the elements d_i satisfy $|d_i| = 1$, $i \in \{1, 3\}$, and, from the equality $r_1 r_3 = d_1 d_3 p_1 p_3$, obtained by taking the determinants in the relation expressing the QR factorization, it follows that $d_1 d_3 \in \mathbb{R}$, hence $\Re(d_1)\Im(d_3) + \Im(d_1)\Re(d_3) = 0$; actually, $d_1 d_3 = \pm 1$. Such relations are implicitly used in our implementation of Algorithms `LyacC` and `LyadC`, for computing the matrices $\alpha := r_{11} u_{11}^{-1}$ and $b := u_{11} a_{11} u_{11}^{-1}$. As these matrices have real elements, their imaginary parts are not evaluated. The element α_{21} is not evaluated either, because α is upper triangular. The computing formulas have been simplified as much as possible, for getting maximum efficiency. Note that the T and P matrices, defined before, are given by $P = Q_p D_p$ and $T = D_t^H Q_t$, where Q_t and D_t are obtained by a second QR factorization (of $u_{11} Q$).

Before solving the Sylvester equation (2.113) or (2.130), the matrix r is saved in the locations $(i+2{:}\,n, 1{:}\,2)$ of array `R`, and the matrix product $a u_{11}^T$ is formed in the same locations of array `A`. Then, the right-hand side of the Sylvester equation is evaluated in `R`$(i{:}\,i+1, i+2{:}\,n)$, and the solution is computed in the same memory space. The transposed Sylvester equation is actually solved. This is the only algorithmic stage when non-unitary or non-orthogonal transformations are used. But the solution of the systems of linear algebraic equations involved in the Sylvester solvers **`Sylsc`** or **`Sylsd`** is very stable in practice.

The calculation of matrix y gets more complicated in a discrete-time Lyapunov equation than in a continuous-time Lyapunov equation (compare (2.128), and (2.111)). For solving the linear system $-(I_2 + b^T)x = \alpha^T$ (which yields $-(I_2+b^T)^{-1}\alpha^T$), the matrix $-(I_2+b^T)$ is triangularized using a plane rotation, which is then applied to α^T; the solution x is found by back substitution, and then, x is premultiplied by $a u_{11}^T + A_1^T u + u$ (evaluated in `A`$(i+2{:}\,n, 1{:}\,2)$).

The Cholesky factor R_1 is updated using modified Householder transformations, as presented previously. Finally, the subdiagonal elements are effectively zeroed and the signs of the elements in the rows with negative diagonal elements, if any, are changed.

2.4 Stabilization Algorithms

Basic theoretical results for stabilization of linear systems have been presented in § 2.1.2. Kleinman's method for continuous-time systems (Kleinman, 1970; Sandell, 1974), based on Theorem 2.2, can be implemented using existing algorithms for computing matrix exponentials and their integrals (Moler and Van Loan, 1978; Van Loan, 1978), where rational diagonal Padé approximations are employed, in conjunction with a technique for matrix scaling and squaring in (Ward, 1977) to reduce the norm and estimate the accuracy. The outcome is that Δ should be small, in order to get accurate results. But a small value of Δ leads to large entries in the stabilizing matrix H, and this could in turn entail numerical and practical difficulties.

Difficulties are also encountered with Kleinman's method for discrete-time systems (Kleinman, 1974), based on Theorem 2.4, as possibly unstable matrix A should be raised to powers when forming W_N in (2.35). Even if (A, B) is controllable, W_N is often nearly singular.

The methods in (Armstrong, 1975; Armstrong and Rublein, 1976) can be taken as bases for good numerical algorithms. In the next two subsections, we shall discuss two main algorithmic variants. It is worth mentioning that the algorithms dealt with are convenient for small- and medium-size problems. Large-scale problems require special techniques, such as those based on subspace splitting via the matrix sign function method.

2.4.1 Full stabilization algorithms

For convenience, we re-write here the basic equations in § 2.1.2. We have to solve the Lyapunov equations

$$(A + \beta I_n)Z + Z(A + \beta I_n)^T = 2BB^T, \quad AZA^T = \beta^2 Z + 2BB^T, \quad (2.135)$$

and then compute the stabilizing matrix,

$$H = -B^T Z^\dagger, \qquad H = -B^T (Z + BB^T)^\dagger A. \qquad (2.136)$$

The first relation in (2.135) and in (2.136) stands for continuous-time systems, and the second relation, for discrete-time systems; β is defined either in terms of the norm of A or in terms of the eigenvalues of A (see Theorems 2.3 and 2.5, respectively).

Numerically stable algorithms for solving Lyapunov equations in (2.135), described in § 2.3.2 and § 2.3.3, and based on real Schur form decomposition, can be used. For computing H in (2.136), the symmetric matrix Z (or $Z + BB^T$) is first factored, using, for instance, its eigendecomposition, $Z = QDQ^T$, where Q is orthogonal and D diagonal (with eigenvalues of Z as diagonal elements).

The total computational effort invested in the approach just described is about $17n^3$ *flops*. More efficient algorithms can be obtained by transforming the relations (2.135) and (2.136) into their RSF correspondents (Sima, 1980b, 1981a). Let U be an orthogonal matrix so that U^TAU is in RSF, and consider the similarity transformation U^T applied to the system (A, B). Denote

$$\widetilde{A} = U^TAU, \quad \widetilde{B} = U^TB, \quad \widetilde{H} = HU, \quad \widetilde{Z} = U^TZU. \tag{2.137}$$

Using (2.137) in (2.135) and (2.136), we obtain

$$(\widetilde{A} + \beta I_n)\widetilde{Z} + \widetilde{Z}(\widetilde{A} + \beta I_n)^T = 2\widetilde{B}\widetilde{B}^T, \quad \widetilde{A}\widetilde{Z}\widetilde{A}^T = \beta^2\widetilde{Z} + 2\widetilde{B}\widetilde{B}^T, \tag{2.138}$$

and,

$$\widetilde{H} = -\widetilde{B}^T\widetilde{Z}^\dagger, \qquad \widetilde{H} = -\widetilde{B}^T(\widetilde{Z} + \widetilde{B}\widetilde{B}^T)^\dagger\widetilde{A}. \tag{2.139}$$

Hence, the required stabilizing matrix is finally

$$H = -\widetilde{B}^T\widetilde{Z}^\dagger U^T, \qquad H = -\widetilde{B}^T(\widetilde{Z} + \widetilde{B}\widetilde{B}^T)^\dagger\widetilde{A}U^T. \tag{2.140}$$

Now, the transformed Lyapunov equations in (2.138) have matrices multiplying the unknown matrix $\widetilde{Z}$ in RSF. A reduction to this form is anyhow performed by the Lyapunov solvers (see Procedure `LyaSch` in § 2.3.2). The idea of improving efficiency consisted in using the intermediate result returned by these solvers, namely the matrix $\widetilde{Z}$, for directly obtaining H via the formulas in (2.140). This way, the computational effort is by about $3n^3 - 2n^2m$ *flops* less than that in the original algorithms. If $m < 3/2n$, as usually the case is in practice, a gain in efficiency is possible.

Algorithm Stacf : $H =$ `Stacf`(A, B)

/* Stabilize continuous-time system (A, B), $A \in \mathbb{R}^{n\times n}$, $B \in \mathbb{R}^{n\times m}$. */

/* 1. Reduce A^T to RSF, $A \leftarrow U^TA^TU$, and update B. */

$\beta = 2\|A\|$; $(A, U) =$ `Hess`(A^T); $(A, U) =$ `mSchur`(A, U);
$B \leftarrow U^TB$;

/* 2. Solve transformed Lyapunov equation. */

$Z =$ `Lyac`$(A + \beta I_n, 2BB^T)$;
$H = -B^TZ^\dagger U^T$; /* 3. Compute stabilizing H. */

Algorithm Stadf : $H =$ `Stadf`(A, B)

/* Stabilize discrete-time system (A, B), $A \in \mathbb{R}^{n\times n}$, $B \in \mathbb{R}^{n\times m}$. */

/* 1. Reduce A^T to RSF, $A \leftarrow U^T A^T U$, and update B. */
$(A, U) =$ `Hess`(A^T); $(A, U, \Lambda) =$ `mSchur`(A, U); $B \leftarrow U^T B$;
$\beta = \min\{\, 1, \min\{\, |\lambda_i| : \lambda_i \in \Lambda, \lambda_i \neq 0 \,\}\}/2$;

/* 2. Solve transformed Lyapunov equation. */
$Z =$ `Lyad`($A/\beta, -2BB^T/\beta^2$);
$H = -B^T(Z + BB^T)^\dagger AU^T$; /* 3. Compute stabilizing H. */

The algorithms need about $3n^2 + 2nm$ memory locations for data and results (including workspace). Clearly, the pseudoinverse needs not be computed. One has to solve a system of linear algebraic equations with positive semi-definite coefficient matrix. This can be accomplished using either symmetric eigendecomposition, or, more efficiently, the diagonal pivoting method for indefinite matrices (the so-called UDU^T factorization, where U is a product of permutation and unit upper triangular matrices and D is a symmetric block-diagonal matrix, with diagonal blocks of order at most 2), implemented in the LINPACK package (Dongarra et al., 1979) or LAPACK package (Anderson et al., 1992).

Note that, for the continuous-time case, the matrix $A + \beta I_n$ is unstable, hence $-(A + \beta I_n)$ is stable. So, it is possible to obtain a variant of Algorithm `Stacf` that solves the Lyapunov equation by Hammarling's method (§ 2.3.3). The first equation in (2.135) is equivalent to

$$\bar{A}^T Z + Z\bar{A} = 2\bar{B}^T\bar{B}, \tag{2.141}$$

where $\bar{A} := (A + \beta I_n)^T$, $\bar{B} := B^T$. Let $\widetilde{A}$ be a RSF of $\bar{A}$, hence $\bar{A} = \bar{U}\widetilde{A}\bar{U}^T$, with $\bar{U}$ orthogonal, and let $\widetilde{B} := \bar{B}\bar{U}$. Then, (2.141) becomes

$$-\widetilde{A}^T(U^TU) - (U^TU)\widetilde{A} = -2R^TR, \tag{2.142}$$

where R is the Cholesky factor of $\widetilde{B}^T\widetilde{B}$, and $U^TU = \bar{U}^T Z\bar{U}$. Equation (2.142) can be solved by Algorithm `LyacC`, that gives the solution factor U. Then, the stabilizing matrix H follows from

$$H = -B^T Z^\dagger = -B^T(\bar{U}U^TU\bar{U}^T)^\dagger = -\widetilde{B}(U^TU)^\dagger\bar{U}^T. \tag{2.143}$$

Algorithm StacfC : $H =$ `StacfC`(A, B)

/* Stabilize continuous-time system (A, B), $A \in \mathbb{R}^{n\times n}$, $B \in \mathbb{R}^{n\times m}$, */
/* solving a stable non-negative definite Lyapunov equation. */

/* 1. Reduce A^T to RSF, $A \leftarrow U^T A^T U$, and update B. */
$\beta = 2\|A\|$; $(A, U) =$ `Hess`(A^T); $(A, U) =$ `mSchur`(A, U);
$B \leftarrow B^T U$;

/* 2. Triangularize the updated B, $B = QR$. */
$R =$ `QR`(B);
/* 3. Solve stable non-negative definite Lyapunov equation. */
$R =$ `LyacC`($-(A + \beta I_n), \sqrt{2}R$);
$H = -B(R^T R)^{\dagger} U^T$; /* 4. Compute stabilizing H. */

2.4.2 Partial stabilization algorithms

The algorithms presented in the previous subsection can move all poles of the system; this is the reason why calling them "full" stabilization algorithms. Practical systems often have much more stable poles than unstable ones. Some applications, for instance, that of initializing Newton's algorithms, require but to get a stabilizing matrix H. This suggests that stable poles could be preserved in their locations, and only the unstable poles should be moved. The algorithms presented below, based on (Varga, 1981a), perform such "partial" stabilization. Beside their low computational effort and good numerical properties, these algorithms boast about returning reduced-norm H matrices.

Given a number $\alpha \in \mathbb{R}$, consider a partition of the complex plane defined by $\mathbb{C} = \mathbb{C}_{\alpha}^{-} \oplus \mathbb{C}_{\alpha}^{+}$, $\mathbb{C}_{\alpha}^{-} \cap \mathbb{C}_{\alpha}^{+} = \emptyset$, where

$$\mathbb{C}_{\alpha}^{-} := \{\, \lambda \in \mathbb{C} : \Re(\lambda) < \alpha \,\}, \quad \text{or} \quad \mathbb{C}_{\alpha}^{-} := \{\, \lambda \in \mathbb{C} : |\lambda| < \alpha \,\}, \tag{2.144}$$

in the continuous-time, or discrete-time case, respectively. In the first relation (2.144), usually $\alpha \leq 0$, while in the second relation, $0 < \alpha \leq 1$. The parameter α represents the desired *stability degree* for the closed-loop system; taking $\alpha = 0$ or $\alpha = 1$ in (2.144), respectively, we obtain the usual stability domain, $\mathbb{C}^{-}$. Moreover, if H is a stabilizing matrix (in the usual sense) for the pair $(A + \beta I_n, B)$, with $\beta > 0$, in the continuous-time case (or for $(A/\beta, B)$, with $0 < \beta < 1$, in the discrete-time case), then $\lambda(A + BH) \subset \mathbb{C}_{-\beta}^{-}$ (or $\lambda(A + \beta BH) \subset \mathbb{C}_{\beta}^{-}$, respectively).

The stabilization algorithms below are based on the following result.

Theorem 2.9 *Let the (A, B) pair be stabilizable, with A and B partitioned as*

$$A = \begin{bmatrix} A_{11} & 0 \\ A_{21} & A_{22} \end{bmatrix}, \qquad B = \begin{bmatrix} B_1 \\ B_2 \end{bmatrix}. \tag{2.145}$$

If $\lambda(A_{11}) \subset \mathbb{C}^{+}$, then (A_{11}, B_1) is completely controllable.

The proof runs easily by contradiction. Note that when the requirements of Theorem 2.9 are satisfied, the specific results in Theorems 2.3 and 2.5

in § 2.1.2 for controllable pairs apply. Obviously, the submatrix A_{11} is nonsingular in the discrete-time case.

The partial stabilization algorithms use an ordered RSF decomposition of A^T, $A^T = U\widetilde{A}U^T$,

$$\widetilde{A} = \begin{bmatrix} A_{11} & A_{12} \\ 0 & A_{22} \end{bmatrix}, \tag{2.146}$$

so that $\lambda(A_{11}) \subset \mathbb{C}^+$, and $\lambda(A_{22}) \subset \mathbb{C}^-$. The stable and unstable parts of the system are so delimited, and only the unstable, but controllable part is stabilized. The stabilization algorithms are, therefore, applied to a reduced-order problem.

Algorithm Stac : $H = \texttt{Stac}(\ A, B, \alpha, \beta\ )$

/* Stabilize the unstable part of continuous-time system (A, B), */
/* $A \in \mathbb{R}^{n\times n}$, $B \in \mathbb{R}^{n\times m}$, with stability degree α. */

/* 1. Reduce A^T to RSF, $A \leftarrow U^TA^TU$, and order eigenvalues. */
$(A, U) = \texttt{Hess}(\ A^T\ )$; $(A, U, \Lambda) = \texttt{mSchur}(\ A, U\ )$; $k = 0$;
FOR $i = 1{:}\,n$,
  $select_i = \Re(\lambda_i) \geq \alpha$; IF $select_i =$ TRUE, THEN $k \leftarrow k + 1$;
$(A, U) = \texttt{Schord}(\ A, select, U\ )$;
$B \leftarrow U^TB$; /* 2. Update B. */
/* 3. Solve reduced Lyapunov equation of order k. */
$Z = \texttt{Lyac}(\ a_{1:k,1:k} + \beta I_k, 2b_{1:k,:}b_{1:k,:}^T\ )$;
$H = [\ -b_{1:k,:}^T Z^{-1}\ \ 0\,]U^T$; /* 4. Compute stabilizing H. */

Algorithm Stad : $H = \texttt{Stad}(\ A, B, \alpha, \beta\ )$

/* Stabilize the unstable part of discrete-time system (A, B), */
/* $A \in \mathbb{R}^{n\times n}$, $B \in \mathbb{R}^{n\times m}$, with stability degree α. */

/* 1. Reduce A^T to RSF, $A \leftarrow U^TA^TU$, and order eigenvalues. */
$(A, U) = \texttt{Hess}(\ A^T\ )$; $(A, U, \Lambda) = \texttt{mSchur}(\ A, U\ )$; $k = 0$;
FOR $i = 1{:}\,n$,
  $select_i = |\lambda_i| \geq \alpha$; IF $select_i =$ TRUE, THEN $k \leftarrow k + 1$;
$(A, U) = \texttt{Schord}(\ A, select, U\ )$;
$B \leftarrow U^TB$; /* 2. Update B. */
/* 3. Solve reduced Lyapunov equation of order k. */
$Z = \texttt{Lyad}(\ a_{1:k,1:k}/\beta, -2b_{1:k,:}b_{1:k,:}^T/\beta^2\ )$;
/* 4. Compute stabilizing H. */
$H = [\ -b_{1:k,:}^T(Z + b_{1:k,:}b_{1:k,:}^T)^{-1}a_{1:k,1:k}\ \ 0\,]U^T$;

In the implementation of `Stad`, the factor 2 at step 3 is replaced by a factor

$\gamma := \beta(1+\beta)$, where $0 < \beta \leq 1$. We may take $\beta = 1$ because A_{11} has no eigenvalues with unit modulus. The parameter β can be viewed as the desired stability degree for the reduced pair (A_{11}, B_1), and usually, $\beta \leq \alpha$.

Algorithms `Stac` and `Stad` require $12.5n^3$ *flops* at most, where $10n^3$ *flops* count for the computation of RSF, $2n^3$ *flops* for ordering the RSF, and $0.5n^3$ *flops* at most for solving the Lyapunov equation (assuming $k = n$). Usually, the dimension of the unstable part is much less than the system order, so, the effort involved in the last two steps is often negligible.

Each computational step of Algorithms `Stac` and `Stad` can be performed in a numerically stable way. Theoretically, the solution Z of the reduced-order Lyapunov equations should be positive definite, hence Cholesky factorization of Z could be used for determining H at the last step of both algorithms. In practice, the reduced pair (A_{11}, B_1) could be uncontrollable, or nearly uncontrollable, so that a matrix Z which is only positive semi-definite can be obtained. So, on implementation it is advisable to use the Cholesky factorization with column pivoting,

$$P^T Z P = R^T R, \qquad R = \begin{bmatrix} R_{11} & R_{12} \\ 0 & 0 \end{bmatrix} \begin{matrix} \} & l \\ \} & k-l \end{matrix},$$

where R_{11} has positive diagonal elements, ordered decreasingly. The pair (A_{11}, B_1) is controllable if and only if $l = k$, hence the Cholesky factorization with pivoting provides a simple mean for checking this structural property, and, implicitly, the stabilizability of (A, B). In addition, $(\text{cond}(Z))^{-1} \approx (r_{kk}/r_{11})^2$. A small value of the reciprocal condition estimate indicates an ill-conditioned stabilization problem, Z being nearly singular, or (A, B) "nearly" unstabilizable. Another usefull quantity for assessing the degree of stabilizability of (A, B) is the norm of the resulting stabilizing matrix H. For a well-scaled system, a large norm of H usually indicates that the pair (A, B) is difficult to be stabilized, or the requirement for the desired stability degree, specified by β, is too severe.

2.5 Newton-Based Riccati Solvers

The main algorithms discussed in the previous sections of this chapter will be now used to solve algebraic matrix Riccati equations. We shall first present Newton-based Riccati solvers for both continuous-time and discrete-time systems. Techniques for efficiency improvement, and computational issues (stabilization parameters, convergence and stopping criteria, etc.), are described. Finally, a general assessment of Newton algorithms, as well as their applicability and limitations, are considered.

2.5.1 Algorithmic templates

Based on the results in § 2.1.1, Procedure **Newton_k** below summarizes the computations in the kth iteration of Newton's process for continuous-time and discrete-time systems.

Procedure Newton_k

/* 1. [Compute A_k and Q_k.] */

$$A_k := A + BH_k, \qquad Q_k := H_k^T R H_k + Q. \tag{2.147}$$

/* 2. [Solve the Lyapunov equation.] Solve in X_k */

$$A_k^T X_k + X_k A_k + Q_k = 0, \tag{2.148}$$

$$X_k = A_k^T X_k A_k + Q_k. \tag{2.149}$$

/* 3. [Compute H_{k+1}.] */

$$H_{k+1} = -R^{-1} B^T X_k, \tag{2.150}$$

$$H_{k+1} = -(R + B^T X_k B)^{-1} B^T X_k A. \tag{2.151}$$

Procedure **Newton_k** can easily be implemented. Algorithms based on RSF reduction of matrices A_k (§ 2.3.2, § 2.3.3) are used for solving the Lyapunov equations (2.148) or (2.149). The Cholesky factorization (§ 1.4.3) can be employed to solve the linear systems in (2.150) or (2.151). Note that just one initial factorization of R is required in the continuous-time case. The computing effort per iteration is about $13.5n^3 + 2.5n^2m + 1.5nm^2$ *flops*, in the continuous-time case, and $14n^3 + 3.5n^2m + 2nm^2 + (1/6)m^3$ *flops*, in the discrete-time case, respectively (Sima, 1980b).

If n is sensibly larger than m, the computing effort per iteration will diminish if explicitly using the RSF of the matrix A_k, which is implicitly used in Procedure **LyaSch** for solving the Lyapunov equations. So, let U_k be an orthogonal matrix so that $\widetilde{A}_k = U_k^T A_k U_k$ is in RSF (or QTF). Multiplying (2.148) or (2.149) on the left by U_k^T and on the right by U_k, and denoting $\widetilde{X}_k := U_k^T X_k U_k$, we obtain the *transformed* Lyapunov equations (with $\widetilde{A}_k$ in RSF),

$$\widetilde{A}_k^T \widetilde{X}_k + \widetilde{X}_k \widetilde{A}_k + U_k^T Q_k U_k = 0, \tag{2.152}$$

$$\widetilde{X}_k = \widetilde{A}_k^T \widetilde{X}_k \widetilde{A}_k + U_k^T Q_k U_k, \tag{2.153}$$

and the matrix H_{k+1} in (2.150) or (2.151) can be computed as follows:

$$H_{k+1} = -R^{-1} B^T U_k \widetilde{X}_k U_k^T, \tag{2.154}$$

$$H_{k+1} = -(R + B^T U_k \widetilde{X}_k U_k^T B)^{-1} B^T U_k \widetilde{X}_k U_k^T A. \tag{2.155}$$

Avoiding the computation of $X_k = U_k \widetilde{X}_k U_k^T$ means reducing the computing effort by $1.5n^3 + 0.5n^2 - 2n^2 m$ *flops* per iteration. So, the approach just described scores better than Procedure `Newton_k` if $m \leq 0.75n + 0.25$. This condition is usually satisfied in practice, and therefore we assume it holds in the following two algorithms.

Algorithm Ntnc : $(X, H, k) = \texttt{Ntnc(}\ A, B, Q, R, H, k_{max}, \varepsilon\ \texttt{)}$

/* Solve continuous-time ARE (2.1) using Newton's method, */
/* assuming that, for the given H, $\lambda(A + BH) \subset \mathbb{C}^-$. */

$Y = RH$; $\widetilde{R} = \texttt{Chol(}\ R\ \texttt{)}$; /* 1. Factor R (Cholesky). */

/* 2. Iterate. */

FOR $k = 1\colon k_{max}$,

/* 3. Compute A_k in $\widetilde{A}$ and reduce it to RSF. */

$\widetilde{A} = A + BH$; $(\widetilde{A}, U) = \texttt{Hess(}\ \widetilde{A}\ \texttt{)}$; $(\widetilde{A}, U, \Lambda) = \texttt{mSchur(}\ \widetilde{A}, U\ \texttt{)}$;

IF $\Lambda \not\subset \mathbb{C}^-$, THEN "$H$ is not stabilizing"; STOP;

/* 4. Compute Q_k in $\widetilde{Q}$ and update it, using symmetry. */

$\widetilde{Q} = U^T(\frac{1}{2}H^T Y + \frac{1}{2}Y^T H + Q)U$;

/* 5. Compute $\widetilde{X}_k$ in $\widetilde{X}$, solving transformed Lyapunov equation. */

$\widetilde{X} = \texttt{Lyac(}\ \widetilde{A}, -\widetilde{Q}\ \texttt{)}$;

/* 6. Update Y and compute H_k in H ($\widetilde{R}$ is upper triangular). */

$Y = -B^T U \widetilde{X} U^T$; $H = \widetilde{R}^{-1}\widetilde{R}^{-T} Y$;

/* 7. Check convergence. */

IF $k > 1$, THEN
$e = \|\widetilde{X} - X\|$; $X = \widetilde{X}$;
IF $e \leq \varepsilon\|X\|$, THEN $X \leftarrow UXU^T$; STOP;
END IF
END FOR

"No convergence"; STOP; /* 8. No convergence. */

Algorithm Ntnd : $(X, H, k) = \texttt{Ntnd(}\ A, B, Q, R, H, k_{max}, \varepsilon\ \texttt{)}$

/* Solve discrete-time ARE (2.2) using Newton's method, */
/* assuming that, for the given H, $\lambda(A + BH) \subset \mathbb{C}^-$. */

$\widetilde{R} = \texttt{Chol(}\ R\ \texttt{)}$; /* 1. Factor R (Cholesky). */

/* 2. Iterate. */

FOR $k = 1\colon k_{max}$,

/* 3. Compute A_k in $\widetilde{A}$ and reduce it to RSF. */

$\widetilde{A} = A + BH$; $(\widetilde{A}, U) =$ `Hess(` $\widetilde{A}$ `)`; $(\widetilde{A}, U, \Lambda) =$ `mSchur(` $\widetilde{A}, U$ `)`;
IF $\Lambda \not\subset \mathbb{C}^-$, THEN "$H$ is not stabilizing"; STOP;

/* 4. Compute Q_k in $\widetilde{Q}$ and update it, using symmetry. */
$Y = \widetilde{R}H$; $\widetilde{Q} = U^T(Y^TY + Q)U$;

/* 5. Compute $\widetilde{X}_k$ in $\widetilde{X}$, solving transformed Lyapunov equation. */
$\widetilde{X} =$ `Lyad(` $\widetilde{A}, \widetilde{Q}$ `)`;

/* 6. Update $R + B^TX_kB$ in $\widehat{R}$ and factor it (Cholesky). */
$T = B^TU$; $Y = T\widetilde{X}$; $\widehat{R} = \frac{1}{2}YT^T + \frac{1}{2}TY^T + R$; $\widehat{R} =$ `Chol(` $\widehat{R}$ `)`;

/* 7. Compute H_k in H ($\widehat{R}$ is upper triangular). */
$Y \leftarrow -YU^TA$; $H = \widehat{R}^{-1}\widehat{R}^{-T}Y$;

/* 8. Check convergence. */
IF $k > 1$, THEN
  $e = \|\widetilde{X} - X\|$; $X = \widetilde{X}$;
  IF $e \leq \varepsilon\|X\|$, THEN $X \leftarrow UXU^T$; STOP;
END IF
END FOR
"No convergence"; STOP; /* 9. No convergence. */

A symmetric update is computed at step 4. Indeed, before multiplication by U^T and U on left and right, respectively, $\widetilde{Q}$ should be $H^TRH + Q$ for the current value of H; so, having $Y := RH$ (or $Y := \widetilde{R}H$), $\widetilde{Q}$ can be computed as shown above, using Level 3 BLAS operations (see (1.105) in § 1.4.3). A similar update is performed at step 6 of Algorithm **Ntnd**. The other calculations can also take the symmetry into account. Note that both algorithms assumed that the norm used (or its equivalent) is invariant under orthogonal transformations, i.e. $\|UXU^T\| = \|X\|$; this is true for Euclidean or Frobenius norms. It should be noticed that factorization at step 1 of Algorithm **Ntnd** is unnecessary. Doing it, however, is to reduce the computing effort by $0.5m^2n$ *flops* per iteration.

Both algorithms above must be initialized by a stabilizing matrix H. If not, the stability test at step 3 will most certainly terminate the calculations just at first iteration. This test can be optional, or only activated at first iteration.

Leaving linear and quadratic terms aside, the computing effort is about

$$(12n^3 + 4.5n^2m + nm^2)k + 1.5n^3 + (1/6)m^3 + nm^2 \textit{ flops},$$

for Algorithm **Ntnc**, and,

$$(12.5n^3 + 5.5n^2m + 2nm^2 + (1/6)m^3)k + 1.5n^3 + (1/6)m^3 \textit{ flops},$$

for Algorithm **Ntnd**, where k is the number of iterations performed (often, $k \leq 8$).

The memory requirements for data and results are of $5n^2 + m^2 + 3nm$ and $5n^2 + 2m^2 + 3nm$ locations, for Algorithm **Ntnc** and Algorithm **Ntnd**, respectively. Taking symmetry into account, these values will diminish by $0.5n^2$ and $0.5n^2 + m^2$, respectively.

Both algorithms consist in numerically stable stages, that confer reliability and accuracy to computations. The computed solutions $\widehat{X}$ are symmetric matrices. Algorithm **Ntnd** also works if matrix A is singular. As the $\widetilde{A}_k$ matrices are stable, and Q_k are positive semi-definite, it is quite possible to replace Algorithms **Lyac** or **Lyad** by **LyacC** or **LyadC** (§ 2.3.3), respectively. So, positive semi-definite solutions $\widehat{X}$ are surely obtainable.

Algorithms **Ntnc** and **Ntnd** can easily get to including the solution of the following Lyapunov equations (obtained from (1.132) or (1.133) in § 1.5.4, by neglecting the quadratic terms in E), for continuous-time or discrete-time case, respectively,

$$A_c^T E + E A_c + Y = 0, \qquad E = A_c^T E A_c + Y, \tag{2.156}$$

where $A_c := A + BH$, H being the computed optimal gain matrix, and the symmetric matrix Y being the residual corresponding to the solution $\widehat{X}$, defined by (1.130) or (1.131), and re-written here for convenience

$$Y = A^T\widehat{X} + \widehat{X}A - \widehat{X}BR^{-1}B^T\widehat{X} + Q, \tag{2.157}$$

$$Y = A^T\widehat{X}A - \widehat{X} - A^T\widehat{X}B(R + B^T\widehat{X}B)^{-1}B^T\widehat{X}A + Q. \tag{2.158}$$

Numerical experience shows that matrix E provides information on the conditioning of Riccati equations. A factor that influences the conditioning of the continuous-time Lyapunov equation in (2.156) is the proximity of the spectrum of matrix A_c to the imaginary axis (Laub, 1979). (See also § 2.3.1 and § 2.3.3.) Indeed, if $a \pm bj \in \lambda(A_c)$, with $|b/a| = 0(10^p) \gg 1$, then a condition number for the first equation in (2.156) is generally of order 10^p. Hence, the conditioning of the equation (2.1) is related to the proximity of the spectrum of closed-loop system matrix to the imaginary axis. Algorithm **Ntnc** and Algorithm **Ntnd** get results $\widehat{X}$ for which the norm $||Y||$ is usually small. For well-conditioned problems, $||E||$ is comparable with $||Y||$, but with ill-conditioned problems, $||E||$ is often much larger than $||Y||$ (Sima, 1981b, 1990).

2.5.2 Computational issues

Algorithms **Ntnc** and **Ntnd** compute the Cholesky factorization of matrix R before Newton's process start; if R is not positive definite, the computation should be terminated with a warning message. Note that the condition $R > 0$ is not necessary in the discrete-time case; $R + B^T X_k B > 0$, for $k \geq 1$, will do. Algorithm **Ntnd** can easily be modified to cope with such an instance. Most frequently one actually chooses $R > 0$, and we know that a previous factorization of R can increase efficiency. These remarks justify the approach made in the statement of algorithms.

In implementing the algorithms above, it is possible to separately deal with the cases "m is comparable to n" (more precisely, $m > 0.75n + 0.25$), and "m is small compared to n" ($m \leq 0.75n + 0.25$). This allows to obtain the maximal possible efficiency, depending on the ratio n/m.

Obviously, the implementation can be so organized as to involve as much column operations as possible. Level 1 BLAS operations $y \leftarrow x + \alpha y$ can be used. But for some machines and for large values of dimensions n and m, it is preferable to employ Level 2 and Level 3 BLAS operations. Hints are given in the presentation of the algorithms.

The matrix trace can be used to check the convergence of Newton's process, instead of norm in Algorithms **Ntnc** and **Ntnd**. The trace is efficiently evaluated and it shows all the mathematical and numerical properties asked by the context. The corresponding theoretical test, $\mathrm{tr}(X_{k-1} - X_k) \leq \varepsilon\, \mathrm{tr}(X_k)$, has the advantage of not requiring any additional memory; matrix X_k can be overwritten on X_{k-1} and only the scalar $\mathrm{tr}(X_{k-1})$ is to be transmitted from the $(k-1)$th to the kth iteration. (The property $\mathrm{tr}(A - B) = \mathrm{tr}(A) - \mathrm{tr}(B)$ is exploited.) Moreover, $\mathrm{tr}(X_{k-1} - X_k)$ is the most easily computable approximation of the distance (in the 2-norm) of matrix $X_{k-1} - X_k$ to the null matrix. It should be emphasized that $\mathrm{tr}(X_k) = \mathrm{tr}(\widetilde{X}_k)$, if $X_k = U\widetilde{X}_k U^T$ and matrix U is orthogonal. Clearly, if $\mathrm{tr}(X_k) = 0$, it follows that the computed solution of the algebraic matrix Riccati equation is $X = 0$. For $\mathrm{tr}(X_k) \neq 0$, the implemented test (mathematically equivalent to the theoretical test), is $|1 - \mathrm{tr}(X_{k-1})/\mathrm{tr}(X_k)| \leq \varepsilon$. Obviously, if at some iteration k, one obtains $\mathrm{tr}(X_k) < 0$, this indicates that the current X_k is not numerically positive semi-definite, and so, Newton's process must be stopped. This may happen, for instance, if the given H matrix is not stabilizing, but such a situation will already have been detected by the test of eigenvalues stability at step 3. Note, however, that within different context than that discussed above, let's say the use of Newton's process for improving the accuracy of an approximate solution (§ 1.5.4), it is possible to compute a correction matrix E which is not numerically positive semi-definite; therefore, the positivity test will be implemented as an option.

We have not taken the balancing option (§ 2.2.1) during reduction of a matrix A_k to QTF, because of the preliminary experiments (Sima, 1981b) that have shown this option practically missing any advantages, and the number of iterations required for achieving convergence sometimes getting larger. However, the balancing option may simply be taken when needed. From an algorithmic point of view, one should only replace U_k in relations (2.154) or (2.155) by $W_k = P_k D_k^{-1} U_k$, where P_k and D_k are the permutation and scaling matrices used for balancing. Matrix D_k is inverted by using the operations $d_i \leftarrow 1/d_i$, $i = \underline{l}: l$, where $\underline{l}$ and l are the indices in (2.43), determined by the balancing procedure ($\underline{l}$ is denoted there by k). Note that if $D_k \neq I_n$, then the code portions corresponding to the case "m is small compared to n" cannot be used, since the convergence test will then require an evaluation of the matrix X_k at each iteration, as $\mathrm{tr}(X_k)$ and $\mathrm{tr}(\widetilde{X}_k)$ may be different in this case.

Recommended values for the input parameters ε and k_{max} are $\varepsilon = \sqrt{\varepsilon_M}$, $k_{max} = 30$. Usually, five to eight iterations are sufficient for achieving convergence, and in most cases, five or six iterations suffice. (The numerical tests performed (Sima, 1981b, 1990) have shown that the number of iterations frequently increases by one, and rarely by two, if ε decreases to the value 10^{-13}.)

Generally, the stabilizing matrix H, used to initialize Newton's process, can be obtained by calling Algorithm `Stac` or `Stad`, respectively. From the numerous numerical tests performed, including (Sima, 1981b), on problems of dimensions up to $n = 30$, we found that the following values for parameters α and β of Algorithms `Stac` and `Stad` are quite suitable

`Stac` : $\alpha = -5 \times 10^{-6} \|A\|_1, \quad \beta = -0.75;$

`Stad` : $\alpha = 0.999995, \quad \beta = 0.75.$

Theoretically, one can use $\alpha = 0$, in `Stac`, or $\alpha = 1.0$, in `Stad`. However, with these latter values, systems at the stability limit can be mistaken for stable systems, due to the rounding errors made during the computations, so that `Stac` or `Stad` can erroneously provide $H = 0$ in this case. The recommended values for α prevent such undesired situations, without requiring a greater stabilizing effort than necessary.

Generally, for unstable systems, we have experimentally obtained values of parameters α and β for which Newton's process has converged in seven or eight iterations. The influence of α has proven to be small. In addition, there are ranges of values of β, as for instance, $\beta \in [-2, -0.7]$ (in the continuous-time case), for which the convergence is achieved in a small number of iterations. It is worth saying that for some hardly stabilizable continuous-time systems (for instance, Example 6 with $n = 21$, and $q = 10^4$ in (Laub, 1979)), the value $\beta = -0.5$ used in (Sima, 1981b, 1990) has determined a number of iterations twice as large as that required by the value recommended here.

2.5.3 Applicability and limitations

Algorithm **Ntnc** (**Ntnd**) can be used only if the linear dynamical system defined by the matrices A and B is stabilizable (see § 1.1.1). In general, the condition that the pair $(Q^{1/2}, A)$ is detectable should be satisfied. For instance, if $\lambda(A) \subset \mathbb{C}^-$, the matrix Q may be zero, resulting the solution $X = 0$. The detectability condition is, however, not compulsory. For example, in the optimal continuous-time problem defined by the matrices

$$A = \begin{bmatrix} -1 & 0 \\ 0 & 1 \end{bmatrix}, \quad B = \begin{bmatrix} 0 \\ 1 \end{bmatrix}, \quad Q = \begin{bmatrix} 0 & 0 \\ 0 & 0 \end{bmatrix}, \quad R = \begin{bmatrix} 1 \end{bmatrix}, \qquad (2.159)$$

the pair (A, B) is stabilizable, but the pair $(Q^{1/2}, A)$ is not detectable. However, the Riccati equation (2.1) has a positive semi-definite solution, $X = \mathrm{diag}(0, 2)$, and this is the solution delivered by **Ntnc** algorithm. Note that in this case the Hamiltonian matrix G in (1.60) (§ 1.1.2) has the spectrum $\{-1, -1, 1, 1\}$.

Algorithm **Ntnc** (**Ntnd**) provides a symmetric solution matrix X. Other algorithms may not show this property, and their result should be symmetrized.

Even if the actual implementation does not include the computation of eigenvalues Λ at step 3, information about these eigenvalues is readily available in the upper quasi-triangular part of the working array destined to $\widetilde{A}_k$; on normal return this array contains a matrix in QTF, whose eigenvalues may well approximate the eigenvalues of the matrix $A + BH$ (H being the resulted optimal stabilizing matrix), i.e. the poles of the closed-loop system.

To obtain a stabilizing matrix H, needed for initialization of Newton's process, Algorithm **Stac** (**Stad**) (§ 2.4) will precede **Ntnc** (**Ntnd**). Much more expensive, if used, will Algorithms **Schv** or **gSchv(Q)** in the next chapter be. If the optimal problem is repeatedly solved using different weighting matrices for a given pair (A, B), then matrix H needs be determined only once.

Newton's method is ideal for solving the algebraic Riccati equations in the second stage of the defect correction procedure **RicCorr** in § 1.5.4, provided an approximate solution $\widehat{X}$ is given. Indeed, the method converges quadratically to the desired stabilizing solution, if the initial matrix H is stabilizing. If the defect correction procedure is used, and matrix $\widehat{X}$ is a stabilizing solution, then Newton's method can be initialized by $H = 0$, because A_c in (1.132) or (1.133) is stable. Usually, the convergence is obtained in one or two iterations, so the corresponding computing effort is less than that of any other method for solving Riccati equations. Consider, for instance, Example 6 in (Laub, 1979), defined by

$$A = 0; \quad a_{1:n-1,2:n} = I_{n-1}; \quad B = e_n;$$

$$Q = 0; \quad q_{11} = 10^4; \quad R = 1. \tag{2.160}$$

It is known that the associated continuous-time Riccati equation is ill conditioned, and that the exact solution X has $x_{1n} = 100$. Using our implementation of Algorithm **Schv**, based on the Schur vectors of the Hamiltonian matrix of the optimal problem, the absolute error in the computed $\widehat{x}_{1n}$, for $n = 10$, was $0(10^{-8})$, but two iterations of **RicCorr** (totalizing only two Newton iterations) then gave a full accuracy result. Likewise, for $n = 21$, **Schv** computed $\widehat{x}_{1n} \approx 98.08$, but two iterations of **RicCorr** (comprising three Newton iterations) improved the absolute error in $\widehat{x}_{1n}$ to $0(10^{-13})$. This example illustrates the potential of Newton's method for improving approximate solutions.[8] On the other hand, Newton's method is an approximate defect correction method obtained by neglecting the quadratic terms in E of equations (1.132) and (1.133) in § 1.5.4. This can be acceptable if E is small, i.e. the initial approximation of solution is enough accurate.

A main limitation of Newton's algorithms is that of their ultimately quadratic convergence rate not being sometimes achieved for certain, especially ill-conditioned problems. If unproperly initialized, Newton's algorithms could waste much computational effort. Research is being intended to use recent advances in nonlinear optimization, in order to reduce the computational effort per iteration, by controlling the step size of Newton iterate, and provide a global convergence.

Note that if **Ntnc** (**Ntnd**) stops without achieving convergence, then the returned H and X contain the corresponding values of the results computed after k iterations of Newton's process. If **Ntnc** (**Ntnd**) does not converge in a reasonable number of iterations (say 50), one should use another stabilizing matrix H, or an alternate method for solving the Riccati equations, particularly, the algorithms based on (generalized) Schur vectors discussed in Chapter 3.

We conclude this section by giving some examples of Newton's algorithms use. The first example is in fact a template algorithm useful for solving a frequently encountered control problem. Other examples report numerical results obtained for specific problems.

Example 2.4 The following algorithm computes the solution of the algebraic matrix Riccati equation (2.1), if **sys** = **c** (continuous-time), or (2.2), if **sys** = **d** (discrete-time), and the corresponding optimal stabilizing matrix, using Newton's method. Algorithm **Stac** (**Stad**, respectively) is used to

---

[8] An interesting LAPACK-based Newton solver for refining approximate solutions of continuous-time Riccati equations has been developed by R. Byers and E. Barth from the University of Kansas; this code has an option for using a previous Schur factorization of the A_k matrix in Lyapunov equations, in order to reduce the total computing effort.

initialize Newton's algorithm. The system output is weighted with the matrix $\widetilde{Q}$ in the quadratic performance index, hence output regulation (Problem 1.3 in § 1.2.1) is intended. A proportional-plus-integral optimal control law (Problem 1.4) is used. It is assumed that the matrices A, B, C, $\widetilde{Q}$ and $\widehat{Q}$ in (1.66) in § 1.2.1, of dimensions $n \times n$, $n \times m$, $l \times n$, $l \times l$ and $l \times l$, respectively, as well as R, of dimension $m \times m$, are given.

Algorithm Ntn : $\boxed{(X, H, k) = \texttt{Ntn}(\ A, B, C, \widetilde{Q}, \widehat{Q}, R\ )}$

/* Solve (2.1) or (2.2) for Problem 1.4 using Newton's method. */

/* 1. Initialize parameters. */

$\varepsilon = \varepsilon_M^{1/2}$; $k_{max} = 30$; $l = \texttt{ord}(\ \widehat{Q}\ )$;
IF $\texttt{sys} = \texttt{c}$, THEN
$\alpha = -5 \times 10^{-6}$; $\beta = -.75$; $P = 0_l$;
ELSE
$\alpha = 1 - 5 \times 10^{-6}$; $\beta = .75$; $P = I_l$;
END IF

/* 2. Construct matrices for proportional-integral control. */

$$A \leftarrow \begin{bmatrix} A & 0 \\ -C & P \end{bmatrix}; \quad B \leftarrow \begin{bmatrix} B \\ 0 \end{bmatrix}; \quad C \leftarrow \begin{bmatrix} C & 0 \\ 0 & I_l \end{bmatrix}; \quad Q = \begin{bmatrix} \widetilde{Q} & 0 \\ 0 & \widehat{Q} \end{bmatrix};$$

/* 3. Save A and B. Compute state weighting matrix. Update α. */

$\widetilde{A} = A$; $\widetilde{B} = B$; $Q \leftarrow C^T Q C$; IF $\texttt{sys} = \texttt{c}$, THEN $\alpha \leftarrow \alpha \|A\|_1$;

/* 4. Compute the stabilizing matrix. */

IF $\texttt{sys} = \texttt{c}$, THEN
$H = \texttt{Stac}(\ \widetilde{A}, \widetilde{B}, \alpha, \beta\ )$;
ELSE
$H = \texttt{Stad}(\ \widetilde{A}, \widetilde{B}, \alpha, \beta\ )$;
END IF

/* 5. Solve Riccati equation and compute optimal stabilizing matrix. */

IF $\texttt{sys} = \texttt{c}$, THEN
$(X, H, k) = \texttt{Ntnc}(\ A, B, Q, R, H, k_{max}, \varepsilon\ )$;
ELSE
$(X, H, k) = \texttt{Ntnd}(\ A, B, Q, R, H, k_{max}, \varepsilon\ )$;
END IF

If matrix $\widehat{Q}$ is not given, then $l = 0$, and the construction at step 2 is not made. □

Example 2.5 An implementation of Algorithm **Ntn** in Example 2.4 was used for the continuous-time chemical reactor model in (Hung and Mac-

Farlane, 1982), defined by the matrices

$$A = \begin{bmatrix} 1.400 & -.208 & 6.715 & -5.676 \\ -.581 & -4.290 & .000 & .675 \\ 1.067 & 4.273 & -6.654 & 5.893 \\ .048 & 4.273 & 1.343 & -2.104 \end{bmatrix},$$

$$B = \begin{bmatrix} .000 & .000 \\ 5.679 & .000 \\ 1.136 & -3.146 \\ 1.136 & .000 \end{bmatrix},$$

$$C = \begin{bmatrix} 1 & 0 & 1 & -1 \\ 0 & 1 & 0 & 0 \end{bmatrix},$$

and the quadratic performance index with weighting matrices

$$\widetilde{Q} = I_2, \qquad \widehat{Q} = 0, \qquad R = 100 I_2.$$

The spectrum of A is $\Lambda(A) = \{-8.665, -5.056, .062, 2.011\}$. Algorithm `Stac` found that the unstable part of the system has order $k = 2$, and obtained the stabilizing matrix

$$H_1 = \begin{bmatrix} -.4873 & -.1707 & -.3639 & .1082 \\ 1.0447 & .1376 & .7352 & -.4151 \end{bmatrix}.$$

With $\varepsilon = 10^{-8}$, Algorithm `Ntnc` performed 6 iterations (the final value of the convergence criterion being 6.96×10^{-9}), and returned the following results (the values being rounded to four digits for the fractionar part):

$$X = \begin{bmatrix} 51.8214 & 11.0442 & 37.3455 & -17.2046 \\ 11.0442 & 3.1444 & 8.0859 & -3.1040 \\ 37.3455 & 8.0859 & 26.9415 & -12.3064 \\ -17.2046 & -3.1040 & -12.3064 & 6.1890 \end{bmatrix},$$

$$H = \begin{bmatrix} -.8560 & -.2352 & -.6255 & .2458 \\ 1.1749 & .2544 & .8476 & -.3872 \end{bmatrix}.$$

The 1-norm of the residual of computed solution was $\|Y\|_1 = 6.63 \times 10^{-13}$. For $\varepsilon = 10^{-16}$, 8 iterations were performed, the final value of the convergence criterion being zero. Note that $\|H_1\|_2 \approx 1.49$, and $\|H\|_2 \approx 1.88$, hence the norm of the initial stabilizing matrix is smaller than that of the optimal gain matrix. The eigenvalues of the optimal (closed-loop) system are: $\{-8.6686, -5.0879, -1.9924, -.3324\}$. □

Example 2.6 Consider a discrete-time system corresponding to a paper machine model, including the effect of a disturbance, defined by the following matrices ($n = 11$, $m = 2$, $l = 2$),

$$A = \begin{bmatrix} 0 & 1 & 0 & 0 & 0 & 0 & 0 & 0 & 0 & 0 & 0 \\ 0 & 0 & 1 & 0 & 0 & 0 & 0 & 0 & 0 & 0 & 0 \\ 0 & 0 & 0 & 1 & 0 & 0 & 0 & 0 & 0 & 0 & 0 \\ 0 & 0 & 0 & 0 & 1 & 0 & 0 & 0 & 0 & 0 & 0 \\ 0 & 0 & 0 & 0 & 0 & 1 & 0 & 0 & 0 & 0 & 0 \\ 0 & 0 & 0 & 0 & 0 & .7788 & 0 & 0 & 0 & 0 & 0 \\ 0 & 0 & 0 & 0 & 0 & 0 & 0 & 1 & 0 & 0 & 0 \\ 0 & 0 & 0 & 0 & 0 & 0 & -.4724 & 1.3746 & 0 & 0 & 0 \\ 0 & 0 & 0 & 0 & 0 & 0 & 0 & 0 & 0 & 1 & 0 \\ 0 & 0 & 0 & 0 & 0 & 0 & 0 & 0 & 0 & 0 & 1 \\ 0 & 0 & 0 & 0 & 0 & 0 & 0 & 0 & 0 & 0 & .8071 \end{bmatrix},$$

$$B = \begin{bmatrix} 0 & 0 & 0 & 0 & 0 & 1 & 0 & 0 & 0 & 0 & 0 \\ 0 & 0 & 0 & 0 & 0 & 0 & 0 & 1 & 0 & 0 & 0 \end{bmatrix}^T,$$

$$C = \begin{bmatrix} 3.3180 & 0 & 0 & 0 & 0 & 0 & 0 & 0 & 0 & 0 & 0 \\ 1.5484 & 0 & 0 & 0 & 0 & 0 & -.3981 & -.5113 & -5.7865 & 0 & 0 \end{bmatrix},$$

and the quadratic performance index defined by

$$\widetilde{Q} = \begin{bmatrix} 0.5 & 0 \\ 0 & 5 \end{bmatrix}, \quad \widehat{Q} = \begin{bmatrix} 0.5 & 0 \\ 0 & 5 \end{bmatrix}, \quad R = \begin{bmatrix} 400 & 0 \\ 0 & 700 \end{bmatrix}.$$

Note that A, B, and C differ from those given in § 1.2.2. Matrix A is singular, but stable; so, only the part (with order 2) corresponding to error integrators should be stabilized.

Algorithm `Stad` returned the stabilizing matrix H_1,

$$H_1 = \begin{bmatrix} -.0553 & -.0553 & -.0553 & -.0553 & -.0553 & -.2500 & .0000 \\ .0000 & .0000 & .0000 & .0000 & .0000 & .0000 & .1074 \end{bmatrix.}$$

$$\begin{matrix} .0000 & .0000 & .0000 & .0000 & .0167 & .0000 \\ -.2500 & -.1556 & -.1556 & -.8065 & .0125 & -.0269 \end{matrix}\Bigg],$$

and **`Ntnd`** delivered the following results:

$$H = \Bigg[\begin{matrix} -.1021 & -.1017 & -.1013 & -.1019 & -.1046 & -.3752 & -.0278 \\ .0973 & .1011 & .1002 & .0883 & .0700 & .0774 & .1306 \end{matrix}$$

$$\begin{matrix} .0595 & .0259 & .0245 & .1763 & .0287 & .0045 \\ -.3992 & -.3962 & -.4103 & -1.1052 & .0026 & -.0685 \end{matrix}\Bigg],$$

$$
\underline{X} = \left[\begin{array}{rrrrrrr}
.1331 & .0978 & .0800 & .0635 & .0488 & .0678 & .0138 \\
.0978 & .1149 & .0796 & .0632 & .0486 & .0675 & .0213 \\
.0800 & .0796 & .0968 & .0629 & .0483 & .0671 & .0242 \\
.0635 & .0632 & .0629 & .0813 & .0489 & .0677 & .0231 \\
.0488 & .0486 & .0483 & .0489 & .0686 & .0701 & .0193 \\
.0678 & .0675 & .0671 & .0677 & .0701 & .2092 & .0230 \\
.0138 & .0213 & .0242 & .0231 & .0193 & .0230 & .0388 \\
-.1126 & -.1116 & -.0991 & -.0799 & -.0587 & -.0572 & -.1021 \\
-.2670 & -.1758 & -.1299 & -.0889 & -.0554 & -.0384 & -.0631 \\
-.1760 & -.2191 & -.1284 & -.0876 & -.0544 & -.0373 & -.0912 \\
-.2277 & -.2234 & -.2641 & -.2202 & -.1650 & -.1652 & -.3029 \\
-.0169 & -.0153 & -.0136 & -.0120 & -.0102 & -.0173 & .0009 \\
-.0384 & -.0304 & -.0224 & -.0154 & -.0096 & -.0066 & -.0129
\end{array}\right.
$$

$$
\left.\begin{array}{rrrrrr}
-.1126 & -.2670 & -.1760 & -.2277 & -.0169 & -.0384 \\
-.1116 & -.1758 & -.2191 & -.2234 & -.0153 & -.0304 \\
-.0991 & -.1299 & -.1284 & -.2641 & -.0136 & -.0224 \\
-.0799 & -.0889 & -.0876 & -.2202 & -.0120 & -.0154 \\
-.0587 & -.0554 & -.0544 & -.1650 & -.0102 & -.0096 \\
-.0572 & -.0384 & -.0373 & -.1652 & -.0173 & -.0066 \\
-.1021 & -.0631 & -.0912 & -.3029 & .0009 & -.0129 \\
.3566 & .4349 & .4312 & .9250 & -.0011 & .0726 \\
.4349 & .9871 & .6464 & .9093 & .0009 & .1417 \\
.4312 & .6464 & .8078 & .8930 & .0009 & .1117 \\
.9250 & .9093 & .8930 & 2.6320 & -.0047 & .1571 \\
-.0011 & .0009 & .0009 & -.0047 & .0055 & .0001 \\
.0726 & .1417 & .1117 & .1571 & .0001 & .0295
\end{array}\right] .
$$

Matrix H is the optimal stabilizing matrix of the multivariable proportional-plus-integral regulator. Note that $||H_1||_2 \approx .88$, and $||H||_2 \approx 1.36$. Matrix $\underline{X}$ is the solution of the algebraic matrix Riccati equation, divided by 1000, and obtained after $k = 6$ iterations. The final value of the convergence criterion was 1.2×10^{-11}, and $||Y||_1 = 6 \times 10^{-12}$. The optimal poles are: $\{ .6362 \pm .2347j, .7614 \pm .1397j, .5837, .0006, .0002 \pm .0005j, -.0005 \pm .0003j, .0000, .0000, .8071 \}$. □

References

Ammar, G., Benner, P., and Mehrmann, V. (1993). A multishift algorithm for the numerical solution of algebraic Riccati equations, *Electronic Trans. Numer. Anal.*, *1*: 33–48.

Anderson, E., Bai, Z., Bischof, C., Demmel, J., Dongarra, J., Du Croz, J., Greenbaum, A., Hammarling, S., McKenney, A., Ostrouchov, S., and Sorensen, D. (1992). *LAPACK Users' Guide*, SIAM, Philadelphia.

Armstrong, E. S. (1975). An extension of Bass' algorithm for stabilizing linear continuous constant systems, *IEEE Trans. Automat. Control, AC-20(4)*: 153–154.

Armstrong, E. S., and Rublein, G. T. (1976). A stabilization algorithm for linear discrete constant systems, *IEEE Trans. Automat. Control, AC-21(4)*: 629–631.

Arnold, W. F., III, and Laub, A. J. (1984). Generalized eigenproblem algorithms and software for algebraic Riccati equations, *Proc. IEEE, 72(12)*: 1746–1754.

Bai, Z., and Demmel, J. W. (1989). On a block implementation of Hessenberg multishift QR iteration, *Int. J. of High Speed Comput., 1*: 97–112. Also LAPACK Working Note 8.

Barnett, S. (1975). *Introduction to Mathematical Control Theory*, Clarendon Press, Oxford.

Barraud, A. Y. (1977). A numerical algorithm to solve $A^TXA - X = Q$, *IEEE Trans. Automat. Control, AC-22(5)*: 883–885.

Bartels, R. H., and Stewart, G. W. (1972). Algorithm 432: Solution of the matrix equation $AX + XB = C$, *Comm. ACM, 15(9)*: 820–826.

Bélanger, P. R., and McGillivray, T. P. (1976). Computational experience with the solution of the matrix Lyapunov equation, *IEEE Trans. Automat. Control, AC-21*: 799–800.

Byers, R., and Nash, S. G. (1989). Approaches to robust pole assignment, *Int. J. Control, 49(1)*: 97–117.

Dahlquist, G., and Björck, A. (1974). *Numerical Methods*, Prentice-Hall, Inc., Englewood Cliffs, New Jersey.

Dongarra, J. J., Bunch, J. R., Moler, C. B., and Stewart, G. W. (1979). *LINPACK Users' Guide*, SIAM, Philadelphia.

Forsythe, G. E., Malcolm, M. A., and Moler, C. B. (1977). *Computer Methods for Mathematical Computations*, Prentice-Hall, Englewood Cliffs, New Jersey.

Golub, G. H., Nash, S., and Van Loan, C. F. (1979). A Hessenberg-Schur method for the problem $AX + XB = C$, *IEEE Trans. Automat. Control*, *AC-24(6)*: 909–913.

Golub, G. H., and Van Loan, C. F. (1983). *Matrix Computations*, The Johns Hopkins University Press, Baltimore, Maryland.

Golub, G. H., and Wilkinson, J. H. (1976). Ill-conditioned eigensystems and the computation of the Jordan canonical form, *SIAM Rev.*, *18(4)*: 578–619.

Hammarling, S. J. (1982). Numerical solution of the stable, non-negative definite Lyapunov equation, *IMA J. Numer. Anal.*, *2*: 303–323.

Hewer, G. A. (1971). An iterative technique for the computation of the steady state gains for the discrete optimal regulator, *IEEE Trans. Automat. Control*, *AC-16(4)*: 382–384.

Higham, N. J. (1988). FORTRAN codes for estimating the one-norm of a real or complex matrix, with applications to condition estimation, *ACM Trans. Math. Softw.*, *14(4)*: 381–396.

Hoskins, W. D., Meek, D. S., and Walton, D. J. (1977a). The numerical solution of $\dot{X} = A_1X + XA_2 + D$, $X(0) = C$, *IEEE Trans. Automat. Control*, *AC-22(5)*: 881–882.

Hoskins, W. D., Meek, D. S., and Walton, D. J. (1977b). The numerical solution of $A'Q + QA = -C$, *IEEE Trans. Automat. Control*, *AC-22(5)*: 882–883.

Hung, Y. S., and MacFarlane, A. G. J. (1982). *Multivariable Feedback: A Quasi-Classical Approach*, volume 40 of *Lect. Notes in Control and Information Sciences*, Springer-Verlag, Berlin.

Kautsky, J., Nichols, N. K., and Van Dooren, P. (1985). Robust pole assignment in linear state feedback, *Int. J. Control*, *41*: 1129–1155.

Kitagawa, G. (1977). An algorithm for solving the matrix equation $X = FXF^T + S$, *Int. J. Control*, *25(5)*: 745–753.

Kleinman, D. L. (1968). On an iterative technique for Riccati equation computations, *IEEE Trans. Automat. Control*, *AC-13*: 114–115.

Kleinman, D. L. (1970). An easy way to stabilize a linear constant system, *IEEE Trans. Automat. Control*, *AC-15(6)*: 692.

Kleinman, D. L. (1974). Stabilizing a discrete, constant, linear system with application to iterative methods for solving the Riccati equation, *IEEE Trans. Automat. Control*, *AC-19(3)*: 252–254.

Kwon, W. H., and Pearson, A. E. (1975). On the stabilization of a discrete constant linear system, *IEEE Trans. Automat. Control*, *AC-20(6)*: 800–801.

Laub, A. J. (1979). A Schur method for solving algebraic Riccati equations, *IEEE Trans. Automat. Control*, *AC-24(6)*: 913–921.

Mehrmann, V. (1991). *The Autonomous Linear Quadratic Control Problem. Theory and Numerical Solution*, volume 163 of *Lect. Notes in Control and Information Sciences* (M. Thoma and A. Wyner, eds.), Springer-Verlag, Berlin.

Miminis, G. S., and Paige, C. C. (1982). An algorithm for pole assignment of time-invariant linear systems, *Int. J. Control*, *35*: 341–354.

Miminis, G. S., and Paige, C. C. (1988). A direct algorithm for pole assignment of time-invariant multi-input linear systems using state feedback, *Automatica*, *24*: 343–356.

Moler, C. B., and Stewart, G. W. (1973). An algorithm for generalized matrix eigenvalue problems, *SIAM J. Numer. Anal.*, *10(2)*: 241–256.

Moler, C. B., and Van Loan, C. F. (1978). Nineteen dubious ways to compute the exponential of a matrix, *SIAM Rev.*, *20(4)*: 801–836.

Nash, J. C. (1979). *Compact Numerical Methods for Computers. Linear Algebra and Function Minimization*, Adam Hilger Ltd., New York.

Parlett, B. N., and Reinsch, C. (1969). Balancing a matrix for calculating of eigenvalues and eigenvectors, *Numer. Math.*, *13*: 293–304.

Petkov, P. H., Christov, N. D., and Konstantinov, M. M. (1984). A computational algorithm for pole assignment of linear single-input systems, *IEEE Trans. Automat. Control*, *29(11)*: 1045–1048.

Petkov, P. H., Konstantinov, M. M., and Christov, N. D. (1985). Computational algorithms for linear control systems: A brief survey, *Int. J. Systems Sci.*, *16(4)*: 465–477.

Sandell, N. R., Jr. (1974). On Newton's method for Riccati equation solution, *IEEE Trans. Automat. Control*, *AC-19(3)*: 254–255.

Schreiber, R., and Van Loan, C. F. (1989). A storage efficient *WY* representation for products of Householder transformations, *SIAM J. Sci. Stat. Comput.*, *10*: 53–57.

Sima, V. (1980a). Comparison of some algorithms for solving the Lyapunov-type equations, *Rev. Roum. Sci. Techn.—Electrotechn. et Energ.*, *25*(*2*): 235–246.

Sima, V. (1980b). On the real Schur form in linear control system design, *Rev. Roum. Sci. Techn.—Electrotechn. et Energ.*, *25*(*4*): 625–632.

Sima, V. (1981a). An efficient Schur method to solve the stabilizing problem, *IEEE Trans. Automat. Control*, *AC-26*(*3*): 724–725.

Sima, V. (1981d). Computational experience in solving the algebraic matrix Riccati equations, Report ICI–TR–07.81, I.C.I., Bucharest. 27 pag.

Sima, V. (1990a). Comparison of some algorithms to solve the algebraic Riccati equations, *Studies and Researches in Computers and Informatics*, *1*(*2*): 77–96.

Smith, B. T., Boyle, J. M., Dongarra, J. J., Garbow, B. S., Ikebe, Y., Klema, V. C., and Moler, C. B. (1976). *Matrix Eigensystem Routines—EISPACK Guide*, volume 6 of *Lect. Notes in Computer Science*, Springer-Verlag, New York, second edition.

Stewart, G. W. (1973). *Introduction to Matrix Computations*, Academic Press, New York.

Stewart, G. W. (1976). Algorithm 506: HQR3 and EXCHNG, Fortran subroutines for calculating and ordering the eigenvalues of a real upper Hessenberg matrix, *ACM Trans. Math. Softw.*, *2*(*3*): 275–280.

Stoer, J., and Bulirsch, R. (1980). *Introduction to Numerical Analysis*, Springer-Verlag, New York.

Van Loan, C. F. (1978). Computing integrals involving the matrix exponential, *IEEE Trans. Automat. Control*, *AC-23*(*3*): 395–404.

Varah, J. M. (1979). On the separation of two matrices, *SIAM J. Numer. Anal.*, *16*(*2*): 216–222.

Varga, A. (1981a). On stabilization algorithms for linear time-invariant systems, *Rev. Roum. Sci. Techn.—Electrotechn. et Energ.*, *26*(*1*): 115–124.

Varga, A. (1981b). A Schur method for pole assignment, *IEEE Trans. Automat. Control*, *AC-26*(*2*): 517–519.

Varga, A. (1990). A note on Hammarling's algorithm for the discrete Lyapunov equation, *Systems & Control Letters*, *15*: 273–275.

Ward, R. C. (1977). Numerical computation of the matrix exponential with accuracy estimate, *SIAM J. Numer. Anal.*, *14*(*4*): 600–610.

Wilkinson, J. H. (1965). *The Algebraic Eigenvalue Problem*, Oxford University Press (Clarendon), London.

Wonham, W. M. (1979). *Linear Multivariable Control: A Geometric Approach*, Springer-Verlag, New York, second edition.

3

Schur and Generalized Schur Algorithms

This chapter deals with *Schur and generalized Schur algorithms* for solving algebraic matrix Riccati equations. These algorithms essentially determine a basis for the stable *invariant* or *deflating subspace* of an extended matrix, or matrix pencil, associated with the optimal problem concerned; the solution of Riccati equations is then obtained by solving a system of linear algebraic equations. The required basis is determined by the Schur vectors or "generalized" Schur vectors to be found using QR or QZ algorithms, respectively. These calculations involve the ordering of real Schur form (RSF) of a matrix, or real Schur-triangular form (generalized RSF) of a pair of matrices, so that the stable eigenvalues appear in the left-top corner. The ordering can be performed in a numerically stable way (using unitary transformations). The computation and ordering of RSF of a matrix have already been described in Section 2.2 of Chapter 2. The computation and ordering of generalized RSF of a pair of matrices are dealt with here in an ample section. The *ordering algorithms* in the generalized problem, hardly discussed in the literature, are not included in usual implementations, such as EISPACK package (Garbow et al., 1977). Moreover, the first release of LAPACK package (Anderson et al., 1992) had not contained any routines for solving nonsymmetric generalized eigenvalue problems, but such routines were added in the second release (1995). Two other sections describe algorithmic templates for solving Riccati equations for continuous-time and discrete-time systems. The new stability-enhancing scaling procedure is also discussed. Techniques for efficiency improvement, and computational issues are presented.

3.1 Basic Theory

This chapter investigates Schur and generalized Schur algorithms for the numerical solution of continuous-time and discrete-time algebraic matrix Riccati equations, defined by (1.55) and (1.59) in § 1.1.2, here re-written

$$A^T X + XA - XBR^{-1}B^T X + Q = 0, \tag{3.1}$$

$$X = A^T XA - A^T XB(R + B^T XB)^{-1}B^T XA + Q, \tag{3.2}$$

where $A \in \mathbb{R}^{n\times n}$, $B \in \mathbb{R}^{n\times m}$, $Q \in \mathbb{R}^{n\times n}$, and $R \in \mathbb{R}^{m\times m}$ are given matrices. Under certain conditions (see, for instance, Theorems 1.16, 1.17, and 1.18 in § 1.1.2), there is a unique symmetric positive semi-definite solution, $X \in \mathbb{R}^{n\times n}$, for each of these Riccati equations. As already mentioned, Schur and generalized Schur algorithms compute this solution using a basis for the stable invariant or deflating subspace of an extended matrix, or matrix pencil, respectively, associated with the optimal problem.

Using the solution X, the optimal gain matrix H can be computed by the formulas below

$$H \;:=\; -R^{-1}B^T X, \tag{3.3}$$

$$H \;:=\; -(R + B^T XB)^{-1}B^T XA, \tag{3.4}$$

for continuous-time and discrete-time case, respectively.

Schur and generalized Schur algorithms for solving algebraic matrix Riccati equations are variants of the classical eigenvector approach (Fath, 1969; Gaalman, 1980; MacFarlane, 1963; Potter, 1966; Vaughan, 1969, 1970), and have been investigated in several works, for instance, (Arnold and Laub, 1984; Bender and Laub, 1987a, 1987b; Gohberg, Lancaster and Rodman, 1986; Laub, 1979; Mehrmann, 1991; Michelsen, 1979; Pappas, Laub and Sandell, 1980; Van Dooren, 1980, 1981). See also the survey papers (Laub, 1985; Van Dooren, 1989). The basic theoretical results are stated in the next two subsections.

3.1.1 Schur vectors method

Consider the following extended matrices (for continuous-time and discrete-time systems, respectively), as defined in § 1.1.2 in terms of the system and performance index matrices:

$$G = \begin{bmatrix} A & -F \\ -Q & -A^T \end{bmatrix}, \qquad F := BR^{-1}B^T, \tag{3.5}$$

$$G = \begin{bmatrix} A + FA^{-T}Q & -FA^{-T} \\ -A^{-T}Q & A^{-T} \end{bmatrix}. \tag{3.6}$$

Also define

$$J := \begin{bmatrix} 0 & I_n \\ -I_n & 0 \end{bmatrix}. \tag{3.7}$$

Note that $J^T = J^{-1} = -J$.

A matrix G is *Hamiltonian* if $G^T J = -JG$ (or $J^T G^T J = -G$, or $JGJ = G^T$, or $JG = (JG)^T$). A matrix G is *symplectic* if $G^T J = JG^{-1}$ (or $J^T G^T J = G^{-1}$, or $G^T JG = J$).

It can easily be verified that G in (3.5) is Hamiltonian, and G in (3.6) is symplectic and has the inverse

$$G^{-1} = \begin{bmatrix} A^{-1} & A^{-1}F \\ QA^{-1} & A^T + QA^{-1}F \end{bmatrix}. \tag{3.8}$$

Note that we have assumed that matrix A is nonsingular in the discrete-time. The singular case will be dealt with in § 3.1.2.

Sums and inverses of Hamiltonian matrices are Hamiltonian matrices. Symplectic matrices are nonsingular; products and inverses of symplectic matrices are symplectic. These properties follow easily from the above definitions. Hamiltonian matrices and symplectic matrices have other useful properties as well, among which there will be mentioned but two.

Theorem 3.1 *Let $G \in \mathbb{R}^{2n \times 2n}$. If G is a Hamiltonian matrix, then $\lambda \in \lambda(G)$ implies $-\lambda \in \lambda(G)$, with the same (algebraic) multiplicity. If G is a symplectic matrix, then $\lambda \in \lambda(G)$ implies $1/\lambda \in \lambda(G)$, with the same multiplicity.*

Theorem 3.2 *Let $G \in \mathbb{R}^{2n \times 2n}$ be a Hamiltonian (or symplectic) matrix, and let $V \in \mathbb{R}^{2n \times 2n}$ be a symplectic matrix. Then $V^{-1}GV$ is Hamiltonian (or symplectic).*

Theorem 3.1 is proven by noting that if G is Hamiltonian, and $Gx = \lambda x$, then $Gx = -\lambda x$ too, because $-G = J^T G^T J$ is similar to G^T and G. If G is symplectic, then $G^{-1} = J^T G^T J$ is similar to G; so, $Gx = \lambda x$ implies

$1/\lambda x = G^{-1}x$, hence $1/\lambda \in \lambda(G)$. The proof of Theorem 3.2 is an easy exercise.

Let us make the basic assumption needed in the sequel, which involves just the conditions (i) and (iii) in § 1.1.2.

Assumption 3.1 The pair (A, B) is stabilizable and the pair (D, A) is detectable, where $D^T D = Q$, and $\mathrm{rank}(D) = \mathrm{rank}(Q)$. □

With Assumption 3.1, it is known that each Riccati equation (3.1) and (3.2) has a unique symmetric positive semi-definite solution X (see Theorem 1.18, § 1.1.2, and (Wonham, 1968), for a proof).

Remark 3.1 According to Theorem 1.11 in § 1.1.1, (A, B) is stabilizable if its uncontrollable part is stable, or formally (Anderson and Moore, 1979), if $w^H B = 0$ and $w^H A = \lambda w^H$ for a constant λ imply $\lambda \in \mathbb{C}^-$ or $w = 0$. A similar formal definition is obtained for detectability, using Remark 1.9 in § 1.1.1. □

With Assumption 3.1, the matrices G in (3.5) and (3.6) are dichotomic (see the condition (ii) in § 1.1.2). From Theorem 3.1 it follows that these matrices have exactly n eigenvalues in $\mathbb{C}^-$. Using Theorem 2.7 in § 2.2.1, one deduces that there is an orthogonal matrix $U \in \mathbb{R}^{2n\times 2n}$, reducing G to a RSF so that

$$U^T G U =: S = \begin{bmatrix} S_{11} & S_{12} \\ 0 & S_{22} \end{bmatrix}, \qquad S_{ij} \in \mathbb{R}^{n\times n}, \tag{3.9}$$

where $\lambda(S_{11}) \subset \mathbb{C}^-$, $\lambda(S_{22}) \subset \mathbb{C}^+$. By conformably partitioning the matrix U into blocks $U_{ij} \in \mathbb{R}^{n\times n}$, $i, j = 1{:}2$, the basic outcome, to be used in the sequel, is yielded by the following theorem, adapted from (Laub, 1979).

Theorem 3.3 *With Assumption 3.1, for $R > 0$, and with the notations above, it follows that:*

1. *Matrix U_{11} is nonsingular and $X = U_{21}U_{11}^{-1}$ is a solution of Riccati equations (3.1) or (3.2) with the property $X = X^T \geq 0$.*

2. *$\lambda(S_{11}) = \lambda(A + BH)$, with H in (3.3) or (3.4), is the spectrum of the closed-loop system.*

3. *In addition, for the discrete-time case,*
   $\lambda(S_{11}) = \lambda(A - F(X^{-1} + F)^{-1}A)$, if X is nonsingular, or
   $\lambda(S_{11}) = \lambda(A - FA^{-T}(X - Q))$, if A is nonsingular,
   where F is defined in (3.5).

We summarize the proof suggested in (Laub, 1979) for the continuous-time case, based on the classical eigenvector approach (MacFarlane, 1963; Potter, 1966; Vaughan, 1969), noting that the Schur vectors are related to eigenvectors by a nonsingular transformation. Let $V \in \mathbb{R}^{2n\times 2n}$ be a nonsingular matrix which takes G to the real Jordan form

$$V^{-1}GV = \mathrm{diag}(-\Lambda, \Lambda), \tag{3.10}$$

where Λ is the real Jordan form of the eigenvalues of G with positive real parts. Then, with V partitioned like U, the submatrix V_{11} is nonsingular, and $V_{21}V_{11}^{-1}$ solves (3.1). Let $T \in \mathbb{R}^{n\times n}$ be a nonsingular matrix which renders S_{11} in (3.9) to the real Jordan form $-\Lambda$, hence $T^{-1}S_{11}T = -\Lambda$. Obviously, (3.10) and (3.9) imply

$$\begin{aligned} G\begin{bmatrix} V_{11} \\ V_{21} \end{bmatrix} &= \begin{bmatrix} V_{11} \\ V_{21} \end{bmatrix}(-\Lambda), \\ G\begin{bmatrix} U_{11} \\ U_{21} \end{bmatrix} &= \begin{bmatrix} U_{11} \\ U_{21} \end{bmatrix} S_{11}, \end{aligned} \tag{3.11}$$

or

$$G\begin{bmatrix} U_{11} \\ U_{21} \end{bmatrix} T = \begin{bmatrix} U_{11} \\ U_{21} \end{bmatrix} T(-\Lambda). \tag{3.12}$$

Comparing (3.11) and (3.12), and noting that eigenvectors are unique up to a non-zero scalar multiple, we obtain

$$\begin{bmatrix} U_{11} \\ U_{21} \end{bmatrix} T = \begin{bmatrix} V_{11} \\ V_{21} \end{bmatrix} \Delta,$$

with Δ a diagonal nonsingular matrix. This implies that $U_{11} = V_{11}\Delta T^{-1}$ is nonsingular, and

$$U_{21}U_{11}^{-1} = V_{21}\Delta T^{-1}(V_{11}\Delta T^{-1})^{-1} = V_{21}V_{11}^{-1},$$

and so, $U_{21}U_{11}^{-1}$ must be a solution of Riccati equation (3.1).

Now, equating the (1,1) blocks of the relation $GV = V\mathrm{diag}(-\Lambda, \Lambda)$, obtained from (3.10), we get

$$AV_{11} - FV_{21} = -V_{11}\Lambda,$$

hence, using (3.3), $X = V_{21}V_{11}^{-1}$, and definition of F in (3.5),

$$A + BH = A - FX = -V_{11}\Lambda V_{11}^{-1},$$

which proves part 2 for continuous-time case, because S_{11} is similar to $-\Lambda$.

The discrete-time case is analogously dealt with. The proof of part 3 will be given in § 3.1.2 (see Theorem 3.6).

Note that for the solution of the algebraic matrix Riccati equations (3.1) and (3.2) it is sufficient to determine the submatrices U_{11} and U_{21} of the matrix U; but, from (3.9) it follows that the columns of $\begin{bmatrix} U_{11}^T & U_{21}^T \end{bmatrix}^T$, representing the first n Schur vectors from U, form an orthogonal basis for the stable G-invariant subspace, associated with the stable (that is, included in $\mathbb{C}^-$) eigenvalues of G. The techniques presented in § 2.2.4 are used for ordering RSF as required by (3.9). Algorithms for solving (3.1) and (3.2) will be given in Section 3.3. The special structure of G will be exploited in Chapter 4.

3.1.2 Generalized Schur vectors method

The Schur vectors method presented in § 3.1.1 is conceptually appealing and efficient in practice, but it has some limitations concerning the class of problems which can be solved. The most important limitation is the requirement for matrix A to be nonsingular in the discrete-time case. (Singular state transition matrices may appear, for instance, by discretizing continuous-time systems with dead times. An example has been given in § 1.2.2.) Moreover, even if this requirement is fulfilled, but matrix A is ill conditioned with respect to inversion, the possibility of obtaining numerically an acceptable solution is drastically reduced, just from the beginning, i.e. from the construction of G in (3.6) or (3.8). The numerical difficulties encountered when some eigenvalues are close to zero are discussed, for instance, in (Michelsen, 1979). It is therefore necessary to avoid inverting matrix A, or solving systems of linear algebraic equations having A as coefficient matrix. To get such a variant, it is useful to present the relationship between the discrete-time algebraic matrix Riccati equation and the two-point boundary value problem associated with the optimal problem to be solved. Denoting by $p(t)$ the adjoint (or the co-state) vector at the time t, the discrete maximum principle, applied to the linear-quadratic optimization problem, leads to the following equation (Gaalman, 1980; Pappas, Laub and Sandell, 1980), also valid when A is singular,

$$\begin{bmatrix} I & F \\ 0 & A^T \end{bmatrix} \begin{bmatrix} x(t+1) \\ p(t+1) \end{bmatrix} = \begin{bmatrix} A & 0 \\ -Q & I \end{bmatrix} \begin{bmatrix} x(t) \\ p(t) \end{bmatrix}, \tag{3.13}$$

where $F := BR^{-1}B^T$. Note that if the matrix A is nonsingular, (3.13) can be re-written as

$$\begin{bmatrix} x(t+1) \\ p(t+1) \end{bmatrix} = G \begin{bmatrix} x(t) \\ p(t) \end{bmatrix}, \qquad G := \begin{bmatrix} I & F \\ 0 & A^T \end{bmatrix}^{-1} \begin{bmatrix} A & 0 \\ -Q & I \end{bmatrix}, \tag{3.14}$$

where G is just the matrix in (3.6). Numerical difficulties raised by the inversion of A can be shun by considering, instead of the standard eigenvalue problem for G, the generalized eigenvalue problem[1]

$$Kz = \lambda Lz, \qquad K := \begin{bmatrix} A & 0 \\ -Q & I \end{bmatrix}, \qquad L := \begin{bmatrix} I & F \\ 0 & A^T \end{bmatrix}. \tag{3.15}$$

Note that $K - \lambda L$ in (3.15) is a *symplectic* pencil, i.e. $KJK^T = LJL^T$, with J defined in (3.7), § 3.1.1. Specifically,

$$KJK^T = LJL^T = \begin{bmatrix} 0 & A \\ -A^T & 0 \end{bmatrix}.$$

A basis for the stable deflating subspace (see § 3.2.2) of the problem (3.15) can be used to construct the solution of the Riccati equation (3.2), instead of a basis for the stable G-invariant subspace. The following results, which show the symplectic nature of the eigenvalues of the problem $Kz = \lambda Lz$, are proven in (Pappas, Laub and Sandell, 1980); the proofs are adapted and summarized below.

Theorem 3.4 *With Assumption 3.1, and for $R > 0$, the generalized eigenvalue problem $Kz = \lambda Lz$, with K and L defined in (3.15), has no eigenvalues on the unit circle in $\mathbb{C}$.*

The proof runs by contradiction. Assume that λ with $|\lambda| = 1$ is an eigenvalue. Then, there is a vector z, $z \neq 0$, so that $Kz = \lambda Lz$. By conformably partitioning $z = \begin{bmatrix} z_1^T & z_2^T \end{bmatrix}^T$, from (3.15) it follows that

$$Az_1 = \lambda z_1 + \lambda F z_2, \qquad -Qz_1 + z_2 = \lambda A^T z_2. \tag{3.16}$$

Premultiplying the first relation (3.16) by $\bar{\lambda} z_2^H$, and postmultiplying the complex conjugate of the second relation by z_1, then adding the results, and using $|\lambda| = 1$, we obtain

$$z_2^H F z_2 + z_1^H Q z_1 = 0.$$

But $F = BR^{-1}B^T$, and $Q = D^T D$, and so,

$$B^T z_2 = 0, \qquad Dz_1 = 0, \tag{3.17}$$

and combining (3.17) with (3.16), we have

$$Az_1 = \lambda z_1, \qquad A^T z_2 = 1/\lambda z_2. \tag{3.18}$$

---

[1] The terminology and properties of generalized eigenvalue problems will be discussed in Section 3.2, but some main concepts and results are used here in advance.

But (3.17) and (3.18) can be intermixed and re-arranged as

$$z_2^H B = 0, \qquad z_2^H A = 1/\bar{\lambda} z_2^H,$$

and

$$z_1^H D^T = 0, \qquad z_1^H A^T = \bar{\lambda} z_1^H.$$

So, Assumption 3.1 and $|\lambda| = 1$ imply that $z_2 = 0$, and $z_1 = 0$, hence $z = 0$, which is a contradiction.

Theorem 3.5 *With Assumption 3.1, and for $R > 0$, if $\lambda \neq 0$ is an eigenvalue with the (algebraic) multiplicity r of the generalized eigenvalue problem $Kz = \lambda Lz$, then $1/\lambda$ is also an eigenvalue with the same multiplicity. If $\lambda = 0$ is an eigenvalue with multiplicity r, then the problem $Kz = \lambda Lz$ has only $2n-r$ finite eigenvalues, and the missing r eigenvalues are infinite.*

For proof, remember that $KJK^T = LJL^T$ and $\det(K - \lambda L) \not\equiv 0$. (Otherwise, the determinant would be zero for $|\lambda| = 1$, which would contradict Theorem 3.4.)

Consider also the generalized eigenvalue problem $L^T x = \mu K^T x$. As $\det(L^T - \mu K^T) \equiv \det(L - \mu K)$, if λ_1 is an eigenvalue of $Kz = \lambda Lz$ with multiplicity r, then, clearly, $\mu_1 := 1/\lambda_1$ is an eigenvalue of $L^T x = \mu K^T x$ with multiplicity r.

If $\{ x_1, \dots, x_l \}$, $l \leq r$, is any chain of principal vectors corresponding to μ_1 for problem $L^T x = \mu K^T x$, then

$$L^T x_1 = \mu_1 K^T x_1, \qquad (L^T - \mu_1 K^T) x_k = K^T x_{k-1}, \quad k = 2{:}\, l.$$

Premultiplying these relations by KJ, setting $z_k := JL^T x_k$, and using the property of K and L ($KJK^T = LJL^T$), we get a chain relation for z_k,

$$Kz_1 = \mu_1 L z_1, \qquad (K - \mu_1 L) z_k = L z_{k-1}, \quad k = 2{:}\, l.$$

So, μ_1 is also an eigenvalue of $Kz = \lambda Lz$ with multiplicity r. The independence of the vectors z_k for all chains corresponding to μ_1 derives from $\det(K - \lambda L) \not\equiv 0$.

When A is nonsingular, $Kz = \lambda Lz$ has non-zero eigenvalues only, because if $\lambda = 0$ were an eigenvalue, then Kz would be zero, for $z \neq 0$, or, equivalently,

$$\begin{bmatrix} A & 0 \\ -Q & I \end{bmatrix} \begin{bmatrix} z_1 \\ z_2 \end{bmatrix} = 0,$$

which would imply $z_1 = z_2 = 0$.

When A is singular, there must be at least one zero eigenvalue, because $Kz = 0$ has a nontrivial solution z. So, let $\lambda_1 = 0$ be an eigenvalue with multiplicity r. With generality preserved, we may assume that K and L are both upper triangular (see Theorem 3.8 and the equivalence property in § 3.2.1), with diagonal elements α_i and β_i, respectively. Then,

$$\det(K - \lambda L) = \prod_{i=1}^{2n} (\alpha_i - \lambda\beta_i).$$

We may consider $\alpha_1 = \cdots = \alpha_r = 0$. As above, it follows that $\mu_1 = 0$ is an eigenvalue with multiplicity r of the problem $L^T x = \mu K^T x$, and hence

$$\det(L^T - \mu K^T) = \det(L - \mu K) = \prod_{i=1}^{2n} (\beta_i - \mu\alpha_i).$$

Note that $\alpha_i = \beta_i = 0$ is not possible, because $\det(K - \lambda L) \not\equiv 0$. Thus,

$$\beta_{2n-r+1} = \cdots = \beta_{2n} = 0,$$

and

$$\det(K - \lambda L) = \prod_{i=1}^{r} (-\lambda\beta_i) \prod_{i=r+1}^{2n-r} (\alpha_i - \lambda\beta_i) \prod_{i=2n-r+1}^{2n} \alpha_i.$$

With convention that the reciprocal of zero is infinity, the missing r eigenvalues are infinite.

Note that Theorem 3.5 generalizes Theorem 3.1 from the nonsingular case. If conditions like (i) and (iii) in § 1.1.2 are fulfilled, then, according to Theorem 3.4, the generalized problem $Kz = \lambda Lz$ has no eigenvalues on the unit circle in $\mathbb{C}$. (This proves that the matrix pencil in (3.15) is regular, but it could have infinite eigenvalues). According to Theorem 3.5, the eigenvalues may be arranged as follows:

$$0, \cdots, 0, \lambda_{r+1}, \cdots, \lambda_n, 1/\lambda_n, \cdots, 1/\lambda_{r+1}, \infty, \cdots, \infty,$$

where $0 < |\lambda_i| < 1$, $i = r + 1\colon n$. One can show that if the columns of the matrix $\widetilde{U} := \begin{bmatrix} U_{11}^T & U_{21}^T \end{bmatrix}^T \in \mathbb{R}^{2n\times n}$ form a basis for the stable deflating subspace, corresponding to the eigenvalues λ_i, with $|\lambda_i| < 1$, $i = 1\colon n$, then results as in Theorem 3.3 hold. This will be proven in the sequel.

From the relation $K\widetilde{U} = L\widetilde{U}S_{11}$ (where $\lambda(S_{11}) \subset \mathbb{C}^-$), expressing that $\widetilde{U}$ is such a basis, we obtain

$$AU_{11} = U_{11}S_{11} + FU_{21}S_{11}, \qquad -QU_{11} + U_{21} = A^T U_{21}S_{11}. \tag{3.19}$$

If U_{11} is nonsingular, then we get

$$A = U_{11}S_{11}U_{11}^{-1} + FU_{21}S_{11}U_{11}^{-1}, \tag{3.20}$$

$$U_{21}U_{11}^{-1} = A^T U_{21}S_{11}U_{11}^{-1} + Q. \tag{3.21}$$

Remark 3.2 Note that matrix S_{11} has stable eigenvalues only, i.e. $|\lambda| < 1$. Relations like (3.19)–(3.21) apply to any set of n generalized eigenvectors and generalized principal eigenvectors $\begin{bmatrix} V^T & W^T \end{bmatrix}^T$, and a corresponding matrix $\widehat{S}$ in "JCF". This fact reflects in the lemmas below, adapted from (Pappas, Laub and Sandell, 1980), which we provide alternate, shorter proofs on. □

Lemma 3.1 *All solutions of the Riccati equation (3.2) are of the form* $X = WV^{-1}$, *where* $\begin{bmatrix} V^T & W^T \end{bmatrix}^T$ *is a set of n generalized eigenvectors and generalized principal eigenvectors of the problem* $Kz = \lambda Lz$.

Indeed, let X be a solution of the Riccati equation (3.2), hence

$$A^T X A_c = X - Q, \tag{3.22}$$

where $A_c := A + BH$, H being defined in (3.4). Premultiplying (3.4) by $R + B^T XB$ and re-arranging, we get

$$RH = -(B^T XA + B^T XBH) = -B^T XA_c, \tag{3.23}$$

or, equivalently, $H = -R^{-1}B^T XA_c$, so that, using the definition of F ($F := BR^{-1}B^T$), here results

$$A - A_c = FXA_c. \tag{3.24}$$

Now, let V be nonsingular so that $V^{-1}A_cV = \widehat{S}$ (in JCF), and define $W = XV$. Then, (3.24) implies

$$AV = V\widehat{S} + FW\widehat{S}. \tag{3.25}$$

Replacing $A_c = V\widehat{S}V^{-1}$ in (3.22), we get $A^T W\widehat{S} = W - QV$, and combining it with (3.25),

$$\begin{bmatrix} A & 0 \\ -Q & I \end{bmatrix} \begin{bmatrix} V \\ W \end{bmatrix} = \begin{bmatrix} I & F \\ 0 & A^T \end{bmatrix} \begin{bmatrix} V \\ W \end{bmatrix} \widehat{S},$$

the proof is through.

Lemma 3.2 *Consider any set of n generalized eigenvectors and generalized principal eigenvectors* $\begin{bmatrix} V^T & W^T \end{bmatrix}^T$, *and a corresponding matrix* $\widehat{S}$ *in "JCF". If V nonsingular, and with Assumption 3.1, then*

(i) $X = WV^{-1}$ *solves (3.2), assuming* $R + B^T XB$ *as nonsingular;*

(ii) $X = X^T \geq 0$ *if and only if* $\widehat{S}$ *is stable.*

Indeed, using analogs of (3.20) and (3.21), we get

$$A^T W V^{-1} A - W V^{-1} + Q = A^T W V^{-1} F W \widehat{S} V^{-1}.$$

So, assertion (i) is proven if demonstrated that

$$A^T W V^{-1} F W \widehat{S} V^{-1} = A^T W V^{-1} B (R + B^T W V^{-1} B)^{-1} B^T W V^{-1} A,$$

or, equivalently, that

$$R^{-1} B^T W \widehat{S} V^{-1} = (R + B^T W V^{-1} B)^{-1} B^T W V^{-1} A. \tag{3.26}$$

But, re-using an analog of (3.20), we have

$$\begin{aligned}
&(R + B^T W V^{-1} B)^{-1} B^T W V^{-1} A - R^{-1} B^T W \widehat{S} V^{-1} = \\
&(R + B^T W V^{-1} B)^{-1} (B^T W \widehat{S} V^{-1} + B^T W V^{-1} F W \widehat{S} V^{-1} \\
&- B^T W \widehat{S} V^{-1} - B^T W V^{-1} B R^{-1} B^T W \widehat{S} V^{-1}) = 0,
\end{aligned}$$

hence (i) holds.

Now, from the analogs of (3.21) and the complex conjugate of (3.20), we obtain

$$W V^{-1} - (V \widehat{S} V^{-1})^H W V^{-1} (V \widehat{S} V^{-1}) = (W \widehat{S} V^{-1})^H F (W \widehat{S} V^{-1}) + Q =: \widetilde{Q}.$$

Denoting $\widetilde{A} := V \widehat{S} V^{-1}$, $X = W V^{-1}$, the above equality gets equivalent to the following Lyapunov equation

$$X - \widetilde{A}^H X \widetilde{A} = \widetilde{Q}, \tag{3.27}$$

where $\widetilde{Q} = \widetilde{Q}^H \geq 0$, hence there is a matrix $\widetilde{D}$ so that $\widetilde{Q} = \widetilde{D}\widetilde{D}^H$. For any such $\widetilde{D}$, the pair $(\widetilde{A}^H, \widetilde{D})$ is stabilizable, i.e. $w^H \widetilde{D} = 0$ and $w^H \widetilde{A}^H = \lambda w^H$ for some λ imply $|\lambda| < 1$ or $w = 0$ (see Remark 3.1). Indeed, $w^H \widetilde{D} = 0$ implies $w^H \widetilde{Q} w = 0$, and consequently, given the definition of $\widetilde{Q}$,

$$w^H (W \widehat{S} V^{-1})^H B R^{-1/2} = 0, \qquad w^H D^T = 0,$$

where $Q = D^T D$. From the analog of the complex conjugate of (3.20), we have

$$w^H A^T = w^H \widetilde{A}^H + w^H (W \widehat{S} V^{-1})^H B R^{-1} B^T = w^H \widetilde{A}^H = \lambda w^H.$$

But (D, A) is detectable, hence (A^T, D^T) is stabilizable (see Remark 1.9 in § 1.1.1); so, $w^H D^T = 0$ and $w^H A^T = \lambda w^H$ imply $|\lambda| < 1$ or $w = 0$. Thus, $(\widetilde{A}^H, \widetilde{D})$ is stabilizable. Under these circumstances, the Lyapunov equation (3.27) has a unique symmetric positive semi-definite solution, and conversely, if there is a symmetric positive semi-definite solution of (3.27), then $\widetilde{A}$ is stable (Anderson and Moore, 1979; see also Theorem 1.6, § 1.1.1). But $\lambda(\widetilde{A}) = \lambda(\widehat{S})$, and so, if $\widehat{S}$ is stable, then $X = WV^{-1}$, by definition solving (3.27), is the unique symmetric positive semi-definite solution of the Lyapunov equation (3.27). Conversely, if X (symmetric positive semi-definite) solves (3.27), then $\widehat{S}$ must be stable. The proof is through.

Remark 3.3 Clearly, $X \in \mathbb{R}^{n\times n}$, because one can always choose real vectors in $\begin{bmatrix} V^T & W^T \end{bmatrix}^T$ to span the subspace corresponding to complex conjugate eigenvalues. □

Remind of the matrix $\widetilde{U} := \begin{bmatrix} U_{11}^T & U_{21}^T \end{bmatrix}^T$, whose columns form a basis for the stable deflating subspace of $K - \lambda L$; so, U_{11} and U_{21} satisfy (3.19). The main outcome is the following theorem, which is a slight reformulation of Theorem 3.3.

Theorem 3.6 *With Assumption 3.1, and for $R > 0$, the following results hold*

1. *Matrix U_{11} is nonsingular and $X = U_{21}U_{11}^{-1}$ is a solution of the Riccati equation (3.2) with the property $X = X^T \geq 0$ ($X = X^T > 0$ if (D, A) is completely reconstructible).*

2. *With H in (3.4), and F defined in (3.5), $\lambda(S_{11}) = \lambda(A + BH)$, or $\lambda(S_{11}) = \lambda(A - F(X^{-1} + F)^{-1}A)$, if X is nonsingular, or $\lambda(S_{11}) = \lambda(A - FA^{-T}(X - Q))$, if A is nonsingular, is the spectrum of the closed-loop system.*

First assume that U_{11} is nonsingular. Then $X = X^T = U_{21}U_{11}^{-1} \geq 0$, from Lemma 3.2, (ii). But $X \geq 0$ implies $R + B^T XB > 0$, and so $X = U_{21}U_{11}^{-1}$ solves (3.2), according to Lemma 3.2, (i). Assume now that U_{11} is singular. Our assumptions and Theorem 1.18 make sure that the Riccati equation (3.2) has a unique symmetric positive semi-definite solution that, by Lemma 3.1, must be of the form $X = WV^{-1} \geq 0$, where $\begin{bmatrix} V^T & W^T \end{bmatrix}^T$ is a set of eigenvectors of the problem $Kz = \lambda Lz$, which is different from $\begin{bmatrix} U_{11}^T & U_{21}^T \end{bmatrix}^T$. But this implies that the corresponding matrix $\widehat{S}$ in "JCF" is unstable, and so, using Lemma 3.2, (ii), $X = WV^{-1}$ would not be positive semi-definite, which contradicts the above statement.

If (D, A) is completely reconstructible, then it is known that (3.2) has a unique symmetric positive definite solution, which should also be the unique positive semi-definite solution X.

Using (3.20), we have

$$\begin{aligned} B^T XA &= B^T X(U_{11}S_{11}U_{11}^{-1} + FU_{21}S_{11}U_{11}^{-1}) \\ &= (R + B^T XB)R^{-1}B^T U_{21}S_{11}U_{11}^{-1}; \end{aligned}$$

so, with (3.4), $BH = -FU_{21}S_{11}U_{11}^{-1}$, hence, from the first relation (3.19), we get

$$U_{11}S_{11} = AU_{11} - FU_{21}S_{11} = AU_{11} + BHU_{11},$$

or, equivalently, $U_{11}S_{11}U_{11}^{-1} = A+BH$. This certifies that $\lambda(S_{11})$ coincides with the closed-loop spectrum.

When X is nonsingular, by the matrix inversion lemma (see, for instance, (Goodwin and Payne, 1977)), or by direct calculations, it follows that

$$\begin{aligned}(X^{-1} + BR^{-1}B^T)^{-1} &= X - XB(I_m + R^{-1}B^TXB)^{-1}R^{-1}B^TX \\ &= X - XB(R + B^TXB)^{-1}B^TX.\end{aligned}$$

Premultiplying this result by B^T, postmultiplying by A, and using (3.23), we obtain $B^T(X^{-1}+F)^{-1}A = -RH$, or $F(X^{-1}+F)^{-1}A = -BH$, so that

$$A - F(X^{-1} + F)^{-1}A = A + BH.$$

Moreover, when A is nonsingular,

$$A - FA^{-T}(X - Q) = A + BH$$

derives from (3.22) and (3.24). So, part 2 follows.

The generalized eigenvalue problem approach can be extended to *descriptor systems* described by (1.71) in § 1.2.1 (see Remark 2.1 in § 2.1.1), to solve general algebraic matrix Riccati equation (2.17)—for continuous-time systems—or (2.18)—for discrete-time systems. In this case, the matrix pencil of interest can be written, in our notation, as $K - \lambda L$, where (Bender and Laub, 1987a, 1987b; Mehrmann, 1991)

$$K := \begin{bmatrix} A & 0 & B \\ Q & A^H & S \\ S^H & B^H & R \end{bmatrix}, \qquad L := \begin{bmatrix} E & 0 & 0 \\ 0 & -E^H & 0 \\ 0 & 0 & 0 \end{bmatrix}, \tag{3.28}$$

in the continuous-time case, and

$$K := \begin{bmatrix} A & 0 & B \\ Q & -E^H & S \\ S^H & 0 & R \end{bmatrix}, \qquad L := \begin{bmatrix} E & 0 & 0 \\ 0 & -A^H & 0 \\ 0 & -B^H & 0 \end{bmatrix}, \tag{3.29}$$

in the discrete-time case. These pencils are obtainable just from the linear boundary value problems derived when applying the maximum principle (Pontryagin et al., 1962) to the optimal problems concerned. When matrix Q weights the output vector (Problem 1.3 in § 1.2.1), then Q and S are replaced by C^HQC and C^HS, respectively. Assuming for the moment that

R is nonsingular, the pencils $K - \lambda L$, with K and L in (3.28) and (3.29) are equivalent to

$$\begin{bmatrix} A - BR^{-1}S^H & -BR^{-1}B^H & 0 \\ Q - SR^{-1}S^H & A^H - SR^{-1}B^H & 0 \\ S^H & B^H & R \end{bmatrix} - \lambda \begin{bmatrix} E & 0 & 0 \\ 0 & -E^H & 0 \\ 0 & 0 & 0 \end{bmatrix}, \quad (3.30)$$

and

$$\begin{bmatrix} A - BR^{-1}S^H & 0 & 0 \\ Q - SR^{-1}S^H & -E^H & 0 \\ S^H & 0 & R \end{bmatrix} - \lambda \begin{bmatrix} E & BR^{-1}B^H & 0 \\ 0 & -A^H + SR^{-1}B^H & 0 \\ 0 & -B^H & 0 \end{bmatrix}, (3.31)$$

respectively. These two pencils are regular if and only if their upper-left $2n \times 2n$ pencils are regular. If E is nonsingular, the reduced $2n \times 2n$ pencils generalize the usual case, which is obtained for $E = I_n$, and $S = 0$. When $E = I_n$, the reduced pencils are Hamiltonian or symplectic pencils, respectively (Paige and Van Loan, 1981; Pappas, Laub and Sandell, 1980). Results are given in (Mehrmann, 1991) for $E \neq I_n$, and E and/or R possibly singular. It is worth mentioning that the general pencils $K - \lambda L$, with K and L in (3.28) and (3.29) inherit most of the spectral properties or the corresponding reduced Hamiltonian or symplectic pencils. There is a pairing of Jordan blocks corresponding to *finite* eigenvalues, which are not on the imaginary axis or on the unit circle, respectively. If the triple (E, A, B) is *strongly stabilizable*[2] and if

$$\mathcal{R} := \begin{bmatrix} Q & S \\ S^H & R \end{bmatrix} > 0,$$

then the general pencils, defined by (3.28) and (3.29), are dichotomic. The conditions above are sufficient, but not necessary, for regularity. A necessary condition is that $\operatorname{rank}(\begin{bmatrix} B^T & S^T & R^T \end{bmatrix}^T) = m$. If the rank is less than m, then there are components of u which do not influence the state trajectory of the dynamical system, and so, these components can be done away with. If $\operatorname{rank}(\begin{bmatrix} S^T & R^T \end{bmatrix}^T) < m$, then there are cost-free control trajectories, i.e. those for which the performance index may be zero. An

---

[2]The triple (E, A, B) and the corresponding descriptor system is *strongly stabilizable* if

$$\begin{aligned} \operatorname{rank}\begin{bmatrix} \lambda E - A & B \end{bmatrix} &= n, \qquad \forall \lambda \in \mathbb{C}^+, \\ \operatorname{rank}\begin{bmatrix} E & A\mathcal{S}_{E,B} & B \end{bmatrix} &= n, \end{aligned}$$

where the columns of $\mathcal{S}_{E,B}$ span $\mathcal{N}(E) \cap \mathcal{N}(B^H)$ and $\mathcal{S}_{E,B}^H \mathcal{S}_{E,B} = I$ (see (Mehrmann, 1991)).

important class of control problems which can fit into this setting is that of the deadbeat control problems (Emami-Naeini and Franklin, 1980, 1982).

There are certain relationships between bases of certain deflating subspaces of the general pencils and Hermitian solutions of the associated Riccati equations. For instance, the following results are proven in (Mehrmann, 1991).

Theorem 3.7 *Let $K - \lambda L$, with K and L in (3.28), be a regular pencil and let X be an Hermitian solution of (2.17). Then, the columns of*

$$V = \begin{bmatrix} I_n \\ XE \\ -R^{-1}(B^H XE + S^H) \end{bmatrix},$$

span an n-dimensional deflating subspace of $K - \lambda L$. Conversely, if E is nonsingular, $\mathcal{R}$ positive definite, and the columns of

$$V = \begin{bmatrix} I_n \\ XE \\ Z \end{bmatrix}$$

span an n-dimensional deflating subspace of $K - \lambda L$, then X is a solution to (2.17).

There is a similar theorem in the discrete-time case.

The presentation in this subsection points out to the need for powerful algorithms capable of solving generalized eigenvalue problems, and motivates the investigation in the next section.

3.2 Computation of Generalized Real Schur Form and Deflating Subspaces

This section deals with efficient methods and algorithms for finding out the eigenvalues and generalized Schur vectors of the generalized eigenvalue problem $Ax = \lambda Bx$, where A and B are $n \times n$ real matrices. In essence, the techniques for solving generalized eigenvalue problems are extensions of those used for standard eigenvalue problems, discussed in Section 2.2. The basic algorithm, the so-called *QZ algorithm* (Moler and Stewart, 1973), is actually an interesting adaptation of the QR algorithm. The transformation matrices involved in the calculation and the ordering of generalized real Schur form give bases for certain *deflating subspaces* of special interest in many applications, including the solution of algebraic matrix Riccati equations, and particularly, singular Riccati equations.

3.2.1 Basic definitions and properties

Let us first summarize some definitions and results, which are essential to developing efficient algorithms for solving Riccati equations. Details and full proofs can be found, for instance, in (Golub and Van Loan, 1983; Moler and Stewart, 1973; Stewart, 1973a—for an introduction to nonsymmetric eigenproblems; Wilkinson, 1965—for symmetric eigenproblems).

The *generalized eigenvalue problem* for two real or complex $n \times n$ matrices A and B asks for finding the scalars $\lambda \in \mathbb{C}$ for which the equation $Ax = \lambda Bx$ has solutions $x \neq 0$. Such scalars λ are called *generalized eigenvalues.* A solution $x \neq 0$, corresponding to a generalized eigenvalue λ, is a *generalized eigenvector.* Clearly, a generalized eigenvalue satisfies the relation

$$\lambda \in \lambda(A, B) := \{ \mu \in \mathbb{C} : \det(A - \mu B) = 0 \}. \tag{3.32}$$

The set $\{ A - \mu B : \mu \in \mathbb{C} \}$ is called a *matrix pencil,* and the elements of $\lambda(A, B)$ and the corresponding solution vectors x are the eigenvalues and (right) eigenvectors of the pencil, respectively.

If $B = I_n$, the generalized eigenvalue problem reduces to the standard problem dealt with at Section 2.2. So, $\lambda(A, I_n) = \lambda(A)$. Moreover, if B is nonsingular, using the *generalized characteristic equation,* we obtain

$$0 = \det(A - \lambda B) = \det(B)\det(B^{-1}A - \lambda I_n), \tag{3.33}$$

hence $\lambda(A, B) = \lambda(B^{-1}A) = \lambda(AB^{-1})$, where was made use of similarity $(AB^{-1} = B(B^{-1}A)B^{-1})$.

The algebraic and analytic theory of the generalized eigenvalue problem is considerably more complicate than that of the standard problem. We shall mention several outstanding characteristics of the generalized eigenvalue problem. First, it is possible that $Ax = \lambda Bx = 0$, $x \neq 0$, for any $\lambda \in \mathbb{C}$. This happens, for instance, if $x \neq 0$ satisfies $x \in \mathcal{N}(A) \cap \mathcal{N}(B)$, i.e. x is a non-zero null vector for both A and B. Second, the matrix pencil $A - \lambda B$ has n eigenvalues if and only if $\operatorname{rank}(B) = n$. If B is singular, then the set $\lambda(A, B)$ may be empty, finite, or infinite, as shown by the following example.

Example 3.1 Let

$$A = \begin{bmatrix} 3 & 2 \\ 0 & 1 \end{bmatrix}, \quad A' = \begin{bmatrix} 3 & 2 \\ 0 & 0 \end{bmatrix}, \quad B = \begin{bmatrix} 1 & 0 \\ 0 & 0 \end{bmatrix}, \quad B' = \begin{bmatrix} 0 & 1 \\ 0 & 0 \end{bmatrix}.$$

Then, $\lambda(A, B) = \{3\}$, $\lambda(A, B') = \emptyset$, $\lambda(A', B) = \lambda(A', B') = \mathbb{C}$. Note that $\mathcal{N}(A') \cap \mathcal{N}(B) = \mathcal{N}(A') \cap \mathcal{N}(B') = 0$. □

The case when $\lambda(A, B) = \mathbb{C}$ can be regarded as a pathological instance. In this case, $\det(A - \lambda B) \equiv 0$, and one says that the pencil $A - \lambda B$ is *singular*. Otherwise, the pencil is called *regular*. The case when the generalized characteristic polynomial has a degree less than n should not be seen as an abnormal one. If A is nonsingular, then the equation $Ax = \lambda Bx$ is equivalent to $A^{-1}Bx = \lambda^{-1}x =: \mu x$, and the missing eigenvalues correspond to the null eigenvalues of the reciprocal problem $B - \mu A$. For instance, for the matrices in Example 3.1, it follows that $\lambda(B, A) = \{1/3, 0\}$. If we adopt the convention that the reciprocal of 0 is ∞, then the missing eigenvalues can be referred to as *infinite* eigenvalues. So, for Example 3.1, we have $\lambda(A, B) = \{3, \infty\}$, and $\lambda(A, B') = \{\infty, \infty\}$. This convention will be reinforced by noting that if B is slightly perturbed, to become nonsingular, then very large eigenvalues will appear, which grow unboundedly as the perturbation is reduced to zero. Other eigenvalues will not be much affected by such perturbations, if well conditioned.

When A and B matrices are symmetric (or Hermitian), and $B > 0$, then the generalized eigenvalues of the problem $Ax = \lambda Bx$ are real. Indeed, $B > 0$ implies that there is an orthogonal (or unitary) matrix U, so that[3]

$$U^T BU = \operatorname{diag}(\beta_i^2) =: D^2,$$

where β_i^2, $i = 1\colon n$, are the (positive) eigenvalues of B. It follows that

$$A - \lambda B = UD(P - \lambda I_n)DU^T, \qquad P := D^{-1}U^T AUD^{-1}, \tag{3.34}$$

hence $\det(A - \lambda B) = (\det(U))^2(\det(D))^2 \det(P - \lambda I_n)$. As

$$(\det(U))^2 = 1, \qquad (\det(D))^2 = \prod_{i=1}^{n} \beta_i^2 > 0,$$

the roots of $\det(A - \lambda B)$ coincide with those of $\det(P - \lambda I_n)$, i.e. they are the eigenvalues of P, and so, they are real. In addition, P has a full system of eigenvectors, which can be chosen as orthonormal. Setting $Pz_i = \lambda_i z_i$, $i = 1\colon n$, from the definition of P we get

$$A(UD^{-1}z_i) = \lambda_i UDz_i = \lambda_i UD(DU^T UD^{-1})z_i = \lambda_i B(UD^{-1}z_i), \tag{3.35}$$

hence $x_i = UD^{-1}z_i$ is an eigenvector of the problem $Ax = \lambda Bx$, corresponding to λ_i. As the matrix UD^{-1} is real and nonsingular, if z_i, $i = 1\colon n$, are orthonormal, then x_i, $i = 1\colon n$, form a full system of real eigenvectors of the generalized eigenvalue problem. Moreover, the x_i vectors are orthogonal with respect to B, because

$$0 = z_i^T z_j = (DU^T x_i)^T DU^T x_j = x_i^T UD^2U^T x_j = x_i^T Bx_j. \tag{3.36}$$

[3] For the Hermitian case, the superscript T in the formulas below must be replaced by H.

If, in addition, $A > 0$, then $Ax_i = \lambda_i Bx_i$ implies $x_i^T Ax_i = \lambda_i x_i^T Bx_i$, hence $\lambda_i > 0$.

If matrix B is nonsingular, a natural approach in the numerical solution of the generalized eigenvalue problem is by solving the standard problem for the matrix $B^{-1}A$. (Obviously, one should actually not compute B^{-1} and then form $X := B^{-1}A$, but solve the matrix equation $BX = A$, using a factorization of B.) This approach is only acceptable when B is not too ill conditioned with respect to inversion. When B is nearly singular, the matrix $C := B^{-1}A$ cannot be accurately computed, and, even if it could, its elements would be large if compared with the elements of A and B. So, $\|C\|$ will be large. As even numerically stable algorithms cannot themselves guarantee more than computing eigenvalues with small relative errors with respect to $\|C\|_2$, it follows that the small eigenvalues of $B^{-1}A$, which may be well conditioned in the original problem, cannot be accurately determined. Obviously, when A is well conditioned, one can work with the matrix $A^{-1}B$.[4] But if A is in turn nearly singular, this is not advisable. Some of the above issues are illustrated in the following examples.

Example 3.2 Let

$$A = \begin{bmatrix} 6 & 6.917 \\ 1 & 1.152 \end{bmatrix}, \quad B = \begin{bmatrix} 1 & 3.853 \\ 0 & 1 \end{bmatrix}.$$

Using *single precision* calculations on a computer with $\varepsilon_M \approx 1.19 \times 10^{-7}$, we obtained

$$\begin{aligned} \lambda(I_n, A^{-1}B) &\approx \{3.300234, -1.515022 \times 10^{-3}\}, \\ \lambda(B^{-1}A) &\approx \{3.300515, -1.514912 \times 10^{-3}\}. \end{aligned}$$

The differences between the results produced based on two equivalent mathematical formulations, which are about three orders of magnitude greater than ε_M, can be explained by the different conditions of A and B. Matrix B is well conditioned ($\mathrm{cond}_2(B) \approx 16.79$), but A is relatively ill conditioned ($\mathrm{cond}_2(A) \approx 1.72 \times 10^4$). The values returned for $\lambda(B^{-1}A)$ have six exact significant digits. □

The difference between the conditions of A and B is more clear-cut in the next example, with A and B matrices taken from (Golub and Van Loan, 1983).

---

[4] Another approach for the case when B is nearly singular is described in (Peters and Wilkinson, 1970), where an approximate null space of B is computed and then removed from the problem. This approach must, however, determine the rank of B, a delicate numerical problem.

Example 3.3 Let

$$A = \begin{bmatrix} 1.746 & 0.940 \\ 1.246 & 1.898 \end{bmatrix}, \qquad B = \begin{bmatrix} 0.780 & 0.563 \\ 0.913 & 0.659 \end{bmatrix},$$

for which $\mathrm{cond}_2(A) \approx 3.99$, $\mathrm{cond}_2(B) \approx 2.19 \times 10^6$. As for the previous example, we obtained

$$\begin{aligned} \lambda(AB^{-1}) &\approx \{1.9375, 1.097314 \times 10^6\}, \\ \lambda(I_n, A^{-1}B) &\approx \{2, 1.1001453 \times 10^6\}. \end{aligned}$$

The smallest eigenvalue in the latter case is exact. □

The results inaccuracy is sometimes incurred by ill conditioning of the problem. But loss of accuracy is not always inherent, as the following example shows.

Example 3.4 Consider the symmetric matrices

$$A = \begin{bmatrix} 2 & 2 \\ 2 & 1 \end{bmatrix}, \qquad B = \begin{bmatrix} 1 & 2 \\ 2 & 4+\varepsilon \end{bmatrix}.$$

The generalized characteristic equation is $\varepsilon\lambda^2 - (1+2\varepsilon)\lambda - 2 = 0$, hence $\lambda_{1,2} = (1 + 2\varepsilon \pm (1 + 12\varepsilon + 4\varepsilon^2)^{1/2})/(2\varepsilon)$. When $\varepsilon \to 0$, it follows that $\lambda_1 \to \infty$, and $\lambda_2 \to -2$. But it can be verified that the eigenvalue λ_2 is insensitive to small perturbations in A and B. □

In practice, it is important that well-conditioned eigenvalues (like λ_2 in Example 3.4) be accurately determined. An alternative to forming $B^{-1}A$ or $A^{-1}B$ and solving a standard eigenproblem, is to reduce A and B to simpler forms, which the solution of the problem can readily be computed from. For instance, choosing some nonsingular matrices Q and Z, such that

$$\widetilde{A} = Q^{-1}AZ, \qquad \widetilde{B} = Q^{-1}BZ \tag{3.37}$$

be in certain canonical forms, then $\lambda(A, B) = \lambda(\widetilde{A}, \widetilde{B})$, because the system $Ax = \lambda Bx$ is equivalent to $\widetilde{A}y = \lambda\widetilde{B}y$, for $x = Zy$. One says that the matrix pencils $A - \lambda B$ and $\widetilde{A} - \lambda\widetilde{B}$, satisfying (3.37), are *equivalent*, and that the transformation (3.37) is an *equivalence transformation*. The analogue of the Jordan canonical form (JCF) for the standard eigenproblem (see (2.40), § 2.2.1) is here the *Kronecker canonical form*, where $\widetilde{A}$ and $\widetilde{B}$ are both block-diagonal matrices, the blocks being analogous to the Jordan blocks.[5] There are algorithms for the computation of the Kronecker structure of a matrix pencil (Demmel and Kågström, 1987; Kågström, 1986;

[5] For singular pencils, the Kronecker decomposition also contains other kinds of blocks (see, for instance, (Gantmacher, 1959; Mehrmann, 1991—for applications in control theory; Wilkinson, 1978)). It is advisable to extract the singular part before applying the QZ algorithm (Van Dooren, 1979).

Van Dooren, 1979). Unfortunately, the Kronecker canonical form produces the same difficulties as JCF does, because it is very sensitive to perturbations. More attractive, from a numerical point of view, is the *generalized Schur decomposition* described by the following result (Moler and Stewart, 1973):

Theorem 3.8 (Generalized Schur decomposition) *If* $A, B \in \mathbb{C}^{n \times n}$, *then there are unitary matrices* Q *and* Z, *so that*

$$Q^H AZ = \widetilde{A}, \qquad Q^H BZ = \widetilde{B}$$

are upper triangular; if there is an index k *so that* $\widetilde{a}_{kk} = \widetilde{b}_{kk} = 0$, *then the matrix pencil* $A - \lambda B$ *is singular, hence* $\lambda(A, B) = \mathbb{C}$, *and conversely; otherwise,* $\lambda(A, B) = \{ \widetilde{a}_{ii}/\widetilde{b}_{ii} \}$[6].

Therefore, for a regular pencil, the generalized eigenvalues can be infinite, but cannot be undetermined (i.e. of the form 0/0). The generalized Schur decomposition also clarifies the problem of *sensitivity* of generalized eigenvalues. The backward error analysis shows that $\widetilde{A}$ and $\widetilde{B}$ are the exact matrices that would be obtained by applying unitary matrices Q and Z on the pair $(A + \delta A, B + \delta B)$, where $||\delta A||_F = 0(\varepsilon_M)||A||_F$, and $||\delta B||_F = 0(\varepsilon_M)||B||_F$. So, any eigenvalues insensitive to small perturbations in A and B will be well-determined by the computed matrices $\widetilde{A}$ and $\widetilde{B}$. Clearly, small perturbations in A and B may induce large perturbations in $\lambda_i = \widetilde{a}_{ii}/\widetilde{b}_{ii}$, if $\widetilde{b}_{ii}$ is small. But such an eigenvalue should not be regarded as ill conditioned, because its reciprocal, $\mu_i = \widetilde{b}_{ii}/\widetilde{a}_{ii}$, could be a well-conditioned eigenvalue of the pencil $B - \mu A$. If $\widetilde{a}_{ii}$ and $\widetilde{b}_{ii}$ are both small (with regard to $||A||$ and $||B||$, respectively), the ratios λ_i can be of normal size, and hence they do not offer any information about the unstable manner in which they have been determined (Stewart, 1978). Therefore, it is preferable to conceive the generalized eigenvalues as *ordered pairs* $(\widetilde{a}_{ii}, \widetilde{b}_{ii})$, rather than ratios λ_i. From this viewpoint, the perturbations in generalized eigenvalues are well measured by the *harmonic metric*,

$$\text{chord}(\lambda_i, \lambda_j) := |\lambda_i - \lambda_j| / \sqrt{(1 + |\lambda_i|^2)(1 + |\lambda_j|^2)}.$$

It can be shown that if λ is a simple eigenvalue of the pencil $A - \lambda B$ and λ_ε is the corresponding eigenvalue of the ε-perturbed pencil $\widehat{A} - \lambda \widehat{B}$ (i.e. $||A - \widehat{A}||_2 = 0(\varepsilon)$, $||B - \widehat{B}||_2 = 0(\varepsilon)$), then

$$\text{chord}(\lambda, \lambda_\varepsilon) \leq \varepsilon / ((y^H Ax)^2 + (y^H Bx)^2) + 0(\varepsilon^2),$$

[6] with the convention that the reciprocal of 0 is ∞.

where $Ax = \lambda Bx$, $y^H A = \lambda y^H B$, and $\|x\|_2 = \|y\|_2 = 1$. Note that the denominator is symmetric in A and B. The really ill-conditioned eigenvalues are those of which such denominator is small. The extreme case is when $\widetilde{a}_{ii} = \widetilde{b}_{ii} = 0$, for at least an i, the pencil $A - \lambda B$ being singular; see (Wilkinson, 1979), for details.

When $A, B \in \mathbb{R}^{n\times n}$, the matrix $\widetilde{A}$ in the generalized Schur decomposition is in real Schur form (RSF), and the complex eigenvalues are finite (Stewart, 1976).

Theorem 3.9 (Generalized real Schur decomposition) *In the real case, $A, B \in \mathbb{R}^{n\times n}$, there are orthogonal matrices Q and Z, so that $Q^T AZ$ is in RSF, and $Q^T BZ$ is upper triangular.*

We say that the matrix pair $(Q^T AZ, Q^T BZ)$ in Theorem 3.9 is the *generalized real Schur decomposition*, or *Schur-triangular decomposition*. It can be obtained using the numerically stable algorithm QZ, described in the next subsection. Only real arithmetic is employed all through.

3.2.2 Computation of generalized real Schur form

The QZ algorithm has been proposed by Moler and Stewart (1973). The first stage of this algorithm reduces A and B to the matrices $\widetilde{A}$ and $\widetilde{B}$, where $\widetilde{A}$ is in upper Hessenberg form (UHF) and $\widetilde{B}$ is upper triangular, using orthogonal equivalence transformations. The next stage essentially applies the implicit QR algorithm (§ 2.2.3) to the matrix $\widetilde{A}\widetilde{B}^{-1}$, without explicitly forming it.

Let us now describe the reduction to the *Hessenberg-triangular form*. An orthogonal matrix U is first determined as a product of $n-1$ Householder transformations (§ 1.4.3), so that $U^T B$ is upper triangular. These transformations are also applied to A, as soon as they are computed. (Note that the operations $B \leftarrow U^T B$, and $A \leftarrow U^T A$ are equivalence transformations.) Then A is reduced to UHF, maintaining the triangular structure of B. To do this, elements below the subdiagonal of A are annihilated by successive premultiplications by plane rotations $Q_{i-1,i}^{(j)}$, in the plane $(i-1, i)$, for $i = n{:}-1{:}j+2$, $j = 1{:}n-2$. These rotations are similarly applied to B. Each rotation $Q_{i-1,i}^{(j)}$ zeroes an element of A, a_{ij}, but introduces a nonzero element, $b_{i,i-1}$, on the subdiagonal of B. This element is immediately annihilated by postmultiplication by another plane rotation, $Z_{i,i-1}^{(j)}$, in the plane $(i, i-1)$; when applied to A, this rotation does not perturb any of the zeroed elements. The transformations performed are partly illustrated in the diagram below (for $n = 5$), which should be explored in parallel for A and B. The asynchronism in applying the transformations to A and B

is meant to indicate the right sequence of the operations.

$$\begin{array}{l} A \leftarrow Q_{45}^{(1)} A \quad \Rightarrow \quad A \leftarrow A Z_{54}^{(1)} \Rightarrow A \leftarrow Q_{34}^{(1)} A \quad \Rightarrow \quad A \leftarrow A Z_{43}^{(1)} \cdots \\ B \leftarrow Q_{45}^{(1)} B \Rightarrow B \leftarrow B Z_{54}^{(1)} \quad \Rightarrow \quad B \leftarrow Q_{34}^{(1)} B \Rightarrow B \leftarrow B Z_{43}^{(1)} \cdots \end{array}$$

The computational process is summarized in the algorithm below.

Algorithm QZHess : $(A, B, Q, Z) = \mathtt{QZHess}(\ A, B\ )$

/* Reduce A to UHF and B to upper triangular form, */
/* $A \leftarrow Q^T A Z$, $B \leftarrow Q^T B Z$ $(A, B \in \mathbb{R}^{n \times n})$, */
/* using orthogonal equivalence transformations; */
/* accumulate the transformations in Q and Z. */

/* 1. Triangularize B and update A. Set Z to I_n. */
$(B, Q) = \mathtt{QR}(\ B\ )$; $A \leftarrow Q^T A$; $Z = I_n$;
FOR $j = 1{:}\, n-1$,
  $b_{j+1:n,j} = 0$;

/* 2. Reduce A to UHF, maintaining B upper triangular. */
FOR $j = 1{:}\, n-2$,

  FOR $i = n{:} -1{:}\, j+2$,

    $(c, s) = \mathtt{rot}(\ a_{i-1,j}, a_{ij}\ )$;

/* 3. Premultiply by $Q_{i-1,i}^{(j)}$. */

$$a_{i-1:i,j+1:n} \leftarrow \begin{bmatrix} c & s \\ -s & c \end{bmatrix} a_{i-1:i,j+1:n}; \quad a_{ij} = 0;$$

$$b_{i-1:i,i-1:n} \leftarrow \begin{bmatrix} c & s \\ -s & c \end{bmatrix} b_{i-1:i,i-1:n};$$

$$q_{:,i-1:i} \leftarrow q_{:,i-1:i} \begin{bmatrix} c & -s \\ s & c \end{bmatrix};$$ /* Update Q. */

    $(c, s) = \mathtt{rot}(\ b_{ii}, b_{i,i-1}\ )$;

/* 4. Postmultiply by $Z_{i,i-1}^{(j)}$. */

$$a_{:,i-1:i} \leftarrow a_{:,i-1:i} \begin{bmatrix} c & s \\ -s & c \end{bmatrix};$$

$$b_{1:i-1,i-1:i} \leftarrow b_{1:i-1,i-1:i} \begin{bmatrix} c & s \\ -s & c \end{bmatrix}; \quad b_{i,i-1} = 0;$$

$$z_{:,i-1:i} \leftarrow z_{:,i-1:i} \begin{bmatrix} c & s \\ -s & c \end{bmatrix};$$ /* Update Z. */

  END FOR
END FOR

If the orthogonal transformations performed need not be accumulated, then the output parameters Q and Z should be omitted. For instance, the

EISPACK routines do not accumulate the transformations giving Q. We sometimes used modifications of these routines which optionally apply the row transformations performed to another matrix, M. For instance, if on input to the corresponding `QZHess` routine, M is set to I_n, then on output, M contains the matrix Q^T. The implementation details could differ from the calculations suggested by Algorithm `QZHess`. Clearly, B and A can be transformed in the same loop at step 1. Modified Householder transformations of order 2 (see § 1.4.3) can be used instead of plane rotations. For instance, if $\bar{Q} = I_2 + vw^T$, where $v, w \in \mathbb{R}^2$, and $w_1 = 1$, is such a transformation, then the operation $(b_{i-1,k}, b_{ik})^T \leftarrow \bar{Q}(b_{i-1,k}, b_{ik})^T$ is performed as follows

$$t = b_{i-1,k} + w_2 b_{ik}; \quad b_{i-1,k} \leftarrow b_{i-1,k} + tv_1; \quad b_{ik} \leftarrow b_{ik} + tv_2;$$

Note that the corresponding transformation $\bar{Z}$ can be obtained with the statement

$(v, w) =$ `Hous_2`$(b_{ii}, b_{i,i-1})$;

hence the indices 1 and 2 in v and w correspond to the columns i and $i-1$, respectively, in A and B, when that transformation is applied. This is somewhat more difficult to formalize (see Algorithm `idQZs` below), so plane rotations have been used in the template algorithm.

Algorithm `QZHess` requires about $17/3n^3$ *flops*. The accumulations of Q (including the initial triangularization of B), and Z need about $7/3n^3$ and $3/2n^3$ *flops*, respectively. Matrix Q is not generally useful, but Z is necessary for finding the generalized eigenvectors or generalized Schur vectors of the original generalized eigenvalue problem from those of the reduced problem.

Remark 3.4 The reduction to Hessenberg-triangular form can alternately be accomplished by stabilized elementary transformations, but the numerical stability will not be guaranteed theoretically in this case. This approach is preferred for complex matrices. □

Let us now describe the second stage of the QZ algorithm, the reduction to the *quasi-triangular–triangular form.* Starting with A in UHF, and B upper triangular, this stage performs an iterative reduction of A to QTF, while maintaining the upper triangular structure of B. The *QZ iteration*, also called the *QZ step*, or *generalized QR iteration*, is used. To describe it, we can assume, without reducing the generality, that matrix A is unreduced (see § 2.2.3), and B is nonsingular. The first assumption is clearly not restrictive, since if $a_{k+1,k} = 0$, one may continue by solving two smaller subproblems. To motivate the second assumption, we show that if there is an index k such that $b_{kk} = 0$, then it is possible to transform the problem so that $b_{kk} \neq 0$, but $b_{nn} = a_{n,n-1} = 0$ in the equivalent problem so obtained; therefore, one can use deflation to reduce the problem order. A "chasing" technique, illustrated below for $n = 5$, and $k = 3$, is used to annihilate the undesired non-zero elements (note the asynchronism in applying the transformations to A and B):

$$\begin{array}{ccccccc} A \leftarrow Q_{34}A & \Rightarrow & A \leftarrow AZ_{32} & \Rightarrow & A \leftarrow Q_{45}A & \Rightarrow & A \leftarrow AZ_{43}Z_{54} \\ B \leftarrow Q_{34}B & \Rightarrow & B \leftarrow BZ_{32} & \Rightarrow & B \leftarrow Q_{45}B & \Rightarrow & B \leftarrow BZ_{43}Z_{54} \end{array}$$

As we shall see later, a simpler technique consists in annihilating the element $a_{k+1,k}$, using a suitable transformation $A \leftarrow Q_{k,k+1}A$, $B \leftarrow Q_{k,k+1}B$.

The basic idea of the QZ step is to update A and B by orthogonal equivalence transformations, so that $\widetilde{A} - \lambda\widetilde{B} = \widetilde{Q}^T(A - \lambda B)\widetilde{Z}$, where $\widetilde{A}$ is in UHF, $\widetilde{B}$ is upper triangular, $\widetilde{Q}$ and $\widetilde{Z}$ are orthogonal, and $\widetilde{A}\widetilde{B}^{-1}$ is *essentially* the same matrix that would result if the double QR step (§ 2.2.3) were applied to AB^{-1}. This can be accomplished using a certain chasing technique and the known property on unreduced Hessenberg matrices (see Theorem 2.8 in § 2.2.3). Let $C := AB^{-1}$, which is an unreduced matrix in UHF, and let v be the first column of the matrix $(C - \delta_s I_n)(C - \delta_{s+1} I_n)$, where δ_s and δ_{s+1} are the eigenvalues of the 2×2 trailing submatrix of C. Let Q_1 be a Householder transformation so that $Q_1 v = \alpha e_1$, $\alpha \in \mathbb{R}$. From § 2.2.3 we know that a double QR step applied to C actually accomplishes the reduction of the matrix $Q_1 C Q_1$ to UHF, giving $Q^T C Q$ in UHF, where the first column of Q coincides with the first column of Q_1. As only the first three elements of v are non-zero, noting the special structure of Q_1, A, and B, it follows that $\widetilde{A} := Q_1 A$ and $\widetilde{B} := Q_1 B$ have extra (possibly) non-zero elements in positions (3,1), and (2,1), (3,1), (3,2), respectively. These elements can be annihilated, to restore the Hessenberg-triangular form. To do this, two Householder transformations, Z_1 and Z_2, are first used to zero the elements $\widetilde{b}_{31}$ and $\widetilde{b}_{32}$, and $\widetilde{b}_{21}$, respectively, so that $\widetilde{B} \leftarrow \widetilde{B}Z_1Z_2$ is upper triangular. Therefore, $\widetilde{A} \leftarrow \widetilde{A}Z_1Z_2$ has three extra non-zero elements (if compared to A) in positions (3,1), (4,1), and (4,2). Using a Householder transformation Q_2, which annihilates $\widetilde{a}_{31}$ and $\widetilde{a}_{41}$, the resulting matrices, $\widetilde{A} \leftarrow Q_2\widetilde{A}$ and $\widetilde{B} \leftarrow Q_2\widetilde{B}$, will have the undesired non-zero elements moved down with a position along the diagonal, if compared to their initial structure. Hence, we can again apply the procedure just described to the submatrices obtained by removing the first row and column in $\widetilde{A}$ and $\widetilde{B}$, and then continue the same manner. Let $\widetilde{Q} := Q_1 Q_2 \cdots Q_{n-1}$ and $\widetilde{Z} := Z_1 Z_2 \cdots Z_{2n-3}$ (the number of transformations being motivated below) be so that $\widetilde{A} := \widetilde{Q}^T A \widetilde{Z}$ is in UHF, and $\widetilde{B} := \widetilde{Q}^T B \widetilde{Z}$ is upper triangular. (Note that these definitions coincide with the preceding ones.) Then,

$\widetilde{A}\widetilde{B}^{-1}$ is in UHF, $\widetilde{A}\widetilde{B}^{-1} = \widetilde{Q}^T AB^{-1}\widetilde{Q}$, and the first column of $\widetilde{Q}$ coincides with the first column of Q_1. Consequently, the property of unreduced Hessenberg matrices implies that the matrix $\widetilde{A}\widetilde{B}^{-1}$ is essentially the same with that obtained by applying a double QR step directly to the matrix $C = AB^{-1}$.

The above described procedure is partially illustrated in the diagram below ($n = 5$):

$$\begin{array}{llll} \widetilde{A} = Q_1 A \;\Rightarrow & \widetilde{A} \leftarrow \widetilde{A}Z_1 \;\Rightarrow & \widetilde{A} \leftarrow \widetilde{A}Z_2 \;\Rightarrow & \widetilde{A} \leftarrow Q_2\widetilde{A} \;\Rightarrow \\ \widetilde{B} = Q_1 B \;\Rightarrow & \widetilde{B} \leftarrow \widetilde{B}Z_1 \;\Rightarrow & \widetilde{B} \leftarrow \widetilde{B}Z_2 \;\Rightarrow & \widetilde{B} \leftarrow Q_2\widetilde{B} \;\Rightarrow \end{array}$$

$$\begin{array}{ll} \Rightarrow \widetilde{A} \leftarrow \widetilde{A}Z_3Z_4 \;\Rightarrow & \widetilde{A} \leftarrow Q_3\widetilde{A} \\ \Rightarrow \widetilde{B} \leftarrow \widetilde{B}Z_3Z_4 \;\Rightarrow & \widetilde{B} \leftarrow Q_3\widetilde{B} \end{array}$$

Applying then Z_5 and Z_6, matrix $\widetilde{B}$ becomes upper triangular, and $\widetilde{A}$ preserves its structure. The process terminates by using the transformations Q_4 and Z_7, which annihilate the elements $\widetilde{a}_{53}$ and $\widetilde{b}_{54}$ (generated by Q_4), respectively.

Note that matrix $C = AB^{-1}$ is not explicitly formed. Only the first two columns in C should be computed in order to obtain the vector v, hence Q_1. These columns follow from the relation

$$(c_1, c_2) = (a_1, a_2) \begin{bmatrix} b_{11} & b_{12} \\ 0 & b_{22} \end{bmatrix}^{-1},$$

where c_k and a_k denote the kth column in C and A, respectively. Therefore, only the 2×2 leading principal submatrix of B should be inverted to compute Q_1. Denoting

$$\widehat{a}_{k1} = a_{k1}/b_{11},\ k = 1\colon 2, \quad \widehat{a}_{k2} = a_{k2}/b_{22},\ k = 1\colon 3, \quad \widehat{b}_{12} = b_{12}/b_{22},$$ [7]

it is easy to see that the possibly non-zero elements in c_1 and c_2 are given by

$$\begin{array}{lll} c_{11} = \widehat{a}_{11}; & c_{12} = \widehat{a}_{12} - \widehat{a}_{11}\widehat{b}_{12}; & \\ c_{21} = \widehat{a}_{21}; & c_{22} = \widehat{a}_{22} - \widehat{a}_{21}\widehat{b}_{12}; & c_{32} = \widehat{a}_{32}. \end{array} \tag{3.38}$$

Moreover, the shifts δ_s and δ_{s+1} are the eigenvalues of the 2×2 current trailing submatrix of C, hence of the matrix

$$\begin{bmatrix} c_{i-1,i-1} & c_{i-1,i} \\ c_{i,i-1} & c_{ii} \end{bmatrix} = \begin{bmatrix} a_{i-1,i-1} & a_{i-1,i} \\ a_{i,i-1} & a_{ii} \end{bmatrix} \begin{bmatrix} b_{i-1,i-1} & b_{i-1,i} \\ 0 & b_{ii} \end{bmatrix}^{-1}$$

[7] If $b_{jj} = 0$, for $j > 1$, then one sets $b_{jj} = \varepsilon\|B\|_\infty$. If $b_{11} = 0$, a suitable transformation is applied, as mentioned above.

$$= \begin{bmatrix} \widehat{a}_{i-1,i-1} & \widehat{a}_{i-1,i} - \widehat{a}_{i-1,i-1}\widehat{b}_{i-1,i} \\ \widehat{a}_{i,i-1} & \widehat{a}_{ii} - \widehat{a}_{i,i-1}\widehat{b}_{i-1,i} \end{bmatrix},$$

where $i = n$ initially, $\widehat{a}_{kj} := a_{kj}/b_{jj}$ (see the footnote), for $k = i-1\colon i$, $j = i-1\colon i$, and $\widehat{b}_{i-1,i} = b_{i-1,i}/b_{ii}$. The relations defining the shifts δ_s and δ_{s+1}, and the first column of the matrix $(C - \delta_s I_n)(C - \delta_{s+1} I_n)$, hence the vector v, are readily obtained from (2.54) and (2.55) in § 2.2.3, by replacing the elements a_{kj} by c_{kj} and operating possible simplifications. This way, neither δ_s and δ_{s+1}, nor the coefficients of the associated characteristic equation, are explicitly computed, which is an important aspect from a numerical point of view.

Summarizing and generalizing the foregoing facts, we obtain the following algorithm, which can be used on a pair of diagonal principal submatrices of A and B. (For the moment, assume that, on input, $l = \underline{i} = 1$, $i = \bar{\imath} = n$, and $s = 2$.)

Algorithm idQZs : $(A, B, Q, Z) = \texttt{idQZs}(\ A, B, a_{1:3,0}, Q, Z, l, i, \underline{i}, \bar{\imath}, s\ )$

/* Apply an implicit single (if $s = 1$) or double (if $s = 2$) QZ step */
/* with parameters a_{10}, a_{20}, a_{30}, to submatrices $(l\colon i, l\colon i)$ of */
/* A and B, $A, B \in \mathbb{R}^{n\times n}$ ($a_{l:i,l:i}$ unreduced Hessenberg and */
/* $b_{l:i,l:i}$ nonsingular), preserving their UHF and upper */
/* triangular form, respectively; apply the transformations to the */
/* rows $\underline{i}\colon i$ and columns $l\colon \bar{\imath}$ ($\underline{i} \le l$, $\bar{\imath} \ge i$); update Q and Z. */

$P_3 := \begin{bmatrix} e_3 & e_2 & e_1 \end{bmatrix} \in \mathbb{R}^{3\times 3}$; $P_2 := \begin{bmatrix} e_2 & e_1 \end{bmatrix} \in \mathbb{R}^{2\times 2}$;

FOR $k = l\colon i-1$,

- $m = \min\{k+s+1, i\}$; $\underline{k} = \max\{k-1, l\}$;
  - /* 1. Find and apply a transformation Q. */
- IF $k = l$, THEN $x = a_{1:s+1,0}$; ELSE $x = a_{k:k+s,k-1}$;
- IF $k = i-1$ OR $s = 1$, THEN
  - $(v, w) = \texttt{Hous\_2}(\ x\ )$;
  - $a_{k:k+1,k:\bar{\imath}} \leftarrow (I_2 + vw^T)a_{k:k+1,k:\bar{\imath}}$;
  - $b_{k:k+1,\underline{k}:\bar{\imath}} \leftarrow (I_2 + vw^T)b_{k:k+1,\underline{k}:\bar{\imath}}$;
  - $q_{:,k:k+1} \leftarrow q_{:,k:k+1}(I_2 + vw^T)$; /* Update Q. */
  - /* $a_{k+1,k-1} = 0$. */
  - IF $k > l$, THEN $a_{k,k-1} = x_1$; $a_{k+1,k-1} = 0$;
- ELSE
  - $(v, w) = \texttt{Hous\_3}(\ x\ )$;
  - $a_{k:k+2,k:\bar{\imath}} \leftarrow (I_3 + vw^T)a_{k:k+2,k:\bar{\imath}}$;
  - $b_{k:k+2,\underline{k}:\bar{\imath}} \leftarrow (I_3 + vw^T)b_{k:k+2,\underline{k}:\bar{\imath}}$;
  - $q_{:,k:k+2} \leftarrow q_{:,k:k+2}(I_3 + vw^T)$; /* Update Q. */

```
                                          /* a_{k+1:k+2,k-1} = 0. */
    IF  k > l,  THEN  a_{k,k-1} = x_1;   a_{k+1:k+2,k-1} = 0;
                               /* 2. Find and apply a transformation Z_1. */
    (v, w) = Hous_3( b_{k+2,k+2}, b_{k+2,k+1}, b_{k+2,k} );
    a_{i:m,k:k+2} <- a_{i:m,k:k+2} P_3 (I_3 + v w^T) P_3;
    b_{i:k+1,k:k+2} <- b_{i:k+1,k:k+2} P_3 (I_3 + v w^T) P_3;
    z_{:,k:k+2} <- z_{:,k:k+2} P_3 (I_3 + v w^T) P_3;          /* Update Z. */
    b_{k+2,k:k+1} = 0;                                  /* b_{k+2,k:k+1} = 0. */
  END IF
                               /* 3. Find and apply a transformation Z_2. */
  (v, w) = Hous_2( b_{k+1,k+1}, b_{k+1,k} );
  a_{i:m,k:k+1} <- a_{i:m,k:k+1} P_2 (I_2 + v w^T) P_2;
  b_{i:k,k:k+1} <- b_{i:k,k:k+1} P_2 (I_2 + v w^T) P_2;
  z_{:,k:k+1} <- z_{:,k:k+1} P_2 (I_2 + v w^T) P_2;             /* Update Z. */
  b_{k+1,k} = 0;                                           /* b_{k+1,k} = 0. */
END FOR
```

The (almost) upper triangular structure of A and B is taken into account. If $s = 2$ and $\widetilde{A}$ and $\widetilde{B}$ are the matrices returned by Algorithm **idQZs**, then $\widetilde{A}\widetilde{B}^{-1}$ is essentially the matrix that would be obtained by applying Algorithm **idQRs** with parameters a_{10}, a_{20}, a_{30}, to AB^{-1}. If the orthogonal transformations performed need not be accumulated, then the input and output parameters Q and Z should be overlooked.

For $l = \underline{i} = 1$, $i = \bar{\imath} = n$, and $s = 2$, Algorithm **idQZs** requires about $13n^2$ *flops*, if the orthogonal transformations are not accumulated. If Q (or Z) are accumulated, then additional $5n^2$ *flops* (or $8n^2$ *flops*) are required.

Remark 3.5 The usual procedure for calculating the generalized eigenvectors, or generalized Schur vectors, is based on the accumulation of transformations Z. The operation counts above suggest that it would be more economical to solve the transposed problem, $x^T A^T = \lambda x^T B^T$, finding the left eigenvectors or Schur vectors of the pair (A^T, B^T), which need the cheaper accumulation of Q. □

By applying a sequence of QZ steps (using Algorithm **idQZs**) to the Hessenberg-triangular pencil $A - \lambda B$, the matrix A can be reduced to QTF, while maintaining the upper triangular structure of B. Indeed, this way we obtain a sequence of matrices $\{A_s\}$, and $\{B_s\}$, for $s = 1, 2, \ldots$, where $A_1 = A$, $B_1 = B$, and A_s and B_s preserve their initial forms, $s = 1, 2, \ldots$. Setting $C_s = A_s B_s^{-1}$, then $\{C_s\}$ is just the sequence that would result if using the implicit QR algorithm for the matrix AB^{-1}. As the shifts involved determine $\{C_s\}$ to approach a matrix in QTF, one deduces that $\lim_{s\to\infty}\{A_s\}$ is in QTF too, because all matrices B_s are upper triangular. In addition, the strategy for choosing the shifts enables an exclusive use of real arithmetic. To make sure that the requirements for Algorithm **idQZs** are satisfied, it is necessary to inspect the subdiagonal elements of A_s, and the diagonal elements of B_s before starting each new iteration, and to split into subproblems whenever possible. The subdiagonal elements of A_s, $a_{l,l-1}^{(s)}$, which satisfy the second negligibility criterion in (2.51) (§ 2.2.3), can be considered as zero, and be effectively zeroed. The implementation included in the EISPACK package uses the criterion $|a_{l,l-1}^{(s)}| \leq \varepsilon \|A\|_\infty$, $l \leq i$, where i is defined below, and ε is either given or set by default to $\varepsilon_u/2$. The indices l and i are so defined that if we partition

$$A_s = \begin{bmatrix} A_{11} & A_{12} & A_{13} \\ 0 & A_{22} & A_{23} \\ 0 & 0 & A_{33} \end{bmatrix} \begin{matrix} \} & l-1 \\ \} & i-l+1 \\ \} & n-i \end{matrix} ,$$

where A_{kk} are square matrices, A_{33} is in QTF, and A_{22} is an unreduced Hessenberg matrix. Therefore, the submatrix A_{22} lies on the intersection of the rows and columns from l to i. Initially, $i = n$, and l is the largest index that satisfies the above mentioned negligibility criterion, if possible, or $l = 1$, otherwise. Let us partition B according to A. If $|b_{ll}| \leq \varepsilon \|B\|_\infty$, then one sets $b_{ll} = 0$, annihilates $a_{l+1,l}$ by premultiplying A (and B) by a transformation $Q :=$ block-diag$(I_{l-1}, I_2 + vw^T, I_{n-l-1})$, sets $l \leftarrow l+1$, and resumes this procedure until $b_{ll} > \varepsilon \|B\|_\infty$. So, A_{22} is unreduced and B_{22} nonsingular. If A_{22} has the order 1 or 2, then it is included into A_{33}, one sets $i \leftarrow l-1$, and, if $i > 2$, the process continues with the submatrix A_{11}. Otherwise, Algorithm **idQZs** is applied to A_{22}. Note that if one plans to determine generalized eigenvalues only, the computed transformations Q and Z should only be applied altogether to the matrix pair (A_{22}, B_{22}). The computational process is terminated when matrix A is split into 1×1 and 2×2 diagonal blocks.

In the EISPACK implementation, the double shift strategy (for $s = 2$ in Algorithm **idQZs**) is only used if the eigenvalues δ_s and δ_{s+1} are complex conjugate. If they are real, the eigenvalue $\delta_s' \in \{\delta_s, \delta_{s+1}\}$, the nearest to $a_{ii}^{(s)}$, is used as single shift; in this case, the initial transformation Q_1 is defined by a_{10} and a_{20}, where $(a_{10}, a_{20}, 0, \ldots, 0)^T := (C - \delta_s' I_n)e_1$. Consequently, when $\delta_s, \delta_{s+1} \in \mathbb{R}$, an iteration requires $6n^2$ *flops*, if the orthogonal transformations are not accumulated; each of the accumulations of Q and Z further involves $9n^2$ *flops*. Moreover, in this case, a procedure similar to that in § 2.2.3 for searching two consecutive small (but non-negligible) subdiagonal elements in A_s, is applied. (These refinements have been proposed by Ward (1975).) The criterion used looks like

$$|a_{m,m-1}a_{m+1,m}| \leq |a_{mm} - \delta_s' b_{mm}|\, \varepsilon\, \|A\|_\infty, \qquad m = i-2\colon -1\colon l+1. \quad (3.39)$$

If this condition holds for an m, the QZ iteration may start from the row and column m, instead of l. Indeed, one can show that if b_{mm} is non-negligible, the initial transformation Q_1, defined by $a_{10} = a_{mm}/b_{mm} - \delta'_s$ and $a_{20} = a_{m+1,m}/b_{mm}$, generates a negligible element in position $(m+1, m)$. Note that the element $(m, m-1)$ in Q_1A has an opposite value to that of the respective element in A, because the Householder transformations preserve the 2-norm, and the element $(m+1, m-1)$ in Q_1A is negligible. If b_{mm} is negligible, then it is considered as zero in (3.39), and $a_{m+1,m}$ is annihilated in the usual manner, by premultiplying A and B by a suitable transformation Q. The element $(m, m-1)$ in A also changes its sign in this case. In Algorithm `QZSch` below, l is reset to the first m that satisfies (3.39).

Actually, Algorithm `QZSch` does somewhat more than performing QZ iterations; for simplifying the later use, it also calls Algorithm `QZSch2`, to be described thereafter, for obtaining generalized eigenvalues of a pair of 2×2 diagonal submatrices.

Algorithm QZSch : $(A, B, Q, Z, \alpha, \beta) =$ `QZSch(` A, B, Q, Z `)`

/* Reduce $A \in \mathbb{R}^{n\times n}$ in upper Hessenberg form to QTF and */
/* then to RSF, $\widetilde{A} = Q^TAZ$, preserving $B \in \mathbb{R}^{n\times n}$ upper triangular, */
/* $\widetilde{B} = Q^TBZ$, using the implicit QZ algorithm; */
/* compute generalized eigenvalues (α, β); update Q and Z. */

$\varepsilon_A = \varepsilon_M\|A\|_\infty$; IF $\|A\|_\infty = 0$, THEN $\varepsilon_A = \varepsilon_M$;
$\varepsilon_B = \varepsilon_M\|B\|_\infty$; IF $\|B\|_\infty = 0$, THEN $\varepsilon_B = \varepsilon_M$;

/* 1. Start main loop: i decreases from n to 1 in steps of 1 or 2. */
$\underline{i} = 1$; $\bar{\imath} = n$; $itn = 50$; $i = n$;

WHILE $i > 2$, DO
/* 2. Perform QZ iterations on $a_{1:i,1:i}$ and $b_{1:i,1:i}$ until */
/* trailing submatrix of order 1 or 2 in A splits off. */
FOR $itr = 0{:}\,itn$,
$s = 2$; /* Shift strategy: double shift. */
/* 3. Look for a single small subdiagonal element. */
FOR $k = i{:}-1{:}2$,
IF $|a_{k,k-1}| \le \varepsilon_A$, THEN EXIT LOOP;
$l = k$; $\underline{l} = l$;
IF $l > 1$, THEN $a_{l,l-1} = 0$; /* $a_{l,l-1}$ negligible. */
/* 4. Use chasing if b_{ll} is negligible. */
WHILE $|b_{ll}| \le \varepsilon_B$ AND $l < i - 1$, DO
$b_{ll} = 0$; $(v, w) =$ `Hous_2(` $a_{ll}, a_{l+1,l}$ `)`;
$a_{l:l+1,l+1:\bar{\imath}} \leftarrow (I_2 + vw^T)a_{l:l+1,l+1:\bar{\imath}}$;

```
    b_{l:l+1,l:ī} ← (I_2 + vw^T) b_{l:l+1,l:ī};
    q_{:,l:l+1} ← q_{:,l:l+1} (I_2 + vw^T);
    IF  l > 1,  THEN  a_{l,l-1} = -a_{l,l-1};
    l ← l + 1;   l_ = l;   a_{l,l-1} = 0;
END WHILE
IF  l ≥ i - 1,  THEN  EXIT LOOP;                  /* a_{l:i,l:i} splits off. */
  /* Active submatrices (l: i, l: i), with l = 1, or a_{l,l-1} negligible. */
b̂_11 = b_ll;   â_11 = a_ll / b̂_11;   â_21 = a_{l+1,l} / b̂_11;
IF  s = 1,  THEN
  a_10 = â_11 - δ;   a_20 = â_21;
  IF  l ≠ l_,  THEN  a_{l,l-1} ← -a_{l,l-1};
ELSE
  IF  itr = 10,  THEN
                                    /* 5. Compute exceptional shifts. */
    a_10 = 0;   a_20 = 1;   a_30 = 1.1605;
  ELSE
                                    /* 6. Compute usual shifts. */
    b̂_22 = b_{l+1,l+1};   IF  |b̂_22| < ε_B,  THEN  b̂_22 = ε_B;
    b̂_33 = b_{i-1,i-1};   IF  |b̂_33| < ε_B,  THEN  b̂_33 = ε_B;
    b̂_44 = b_ii;   IF  |b̂_44| < ε_B,  THEN  b̂_44 = ε_B;
    â_33 = a_{i-1,i-1} / b̂_33;   â_34 = a_{i-1,i} / b̂_44;
    â_43 = a_{i,i-1} / b̂_33;   â_44 = a_ii / b̂_44;   b̂_34 = b_{i-1,i} / b̂_44;
    t = (â_43 b̂_34 - â_33 - â_44)/2;   r = t^2 + â_34 â_43 - â_33 â_44;
    IF  r ≥ 0,  THEN
                                    /* Shift strategy: single shift. */
      s = 1;   r ← r^{1/2};   δ = -t + r;
      IF  |t + r + â_44| < |-t + r - â_44|,  THEN  δ = -t - r;
      /* 7. Look for two consecutive small subdiagonal elements. */
      FOR l = i - 2: -1: l_ + 1,
        t = a_ll;   IF  |b_ll| > ε_B,  THEN  t ← t - δ b_ll;
        IF  |a_{l,l-1}| ≤ |t / a_{l+1,l}| ε_A,  THEN  GO TO step 4;
      END FOR
      a_10 = â_11 - δ;   a_20 = â_21;
      IF  l ≠ l_,  THEN  a_{l,l-1} ← -a_{l,l-1};
    ELSE
                                    /* Shift strategy: double shift. */
      â_12 = a_{l,l+1} / b̂_22;   â_22 = a_{l+1,l+1} / b̂_22;   b̂_12 = b_{l,l+1} / b̂_22;
      a_00 = ((â_33 - â_11)(â_44 - â_11) - â_34 â_43 + â_43 b̂_34 â_11) / â_21;
      a_10 = a_00 + â_12 - â_11 b̂_12;
```

```
        a_20 = (â_22 − â_11) − â_21 b̂_12 − (â_33 − â_11) − (â_44 − â_11) + â_43 b̂_34;
        a_30 = a_{l+2,l+1} / b̂_22;
      END IF
    END IF
  END IF
          /* 8. Apply a single or double QZ step on a_{l:i,l:i} and b_{l:i,l:i}. */
                          /* If only eigenvalues are desired, set i_ = l, ī = i. */
  (A, B, Q, Z) = idQZs( A, B, a_{1:3,0}, Q, Z, l, i, i_, ī, s );
END FOR
                                                   /* 9. No convergence. */
IF itr > itn, THEN  "No convergence at eigenvalue i";   STOP;
                                          /* Here, l is either i or i − 1. */
IF l = i, THEN
  α_i = sign(b_ii) a_ii;   β_i = |b_ii|;
ELSE
  (A, B, Q, Z, α, β) = QZSch2( A, B, Q, Z, l, i_, ī, ε_B );
END IF
i = l − 1;
END WHILE
```

The convergence properties of the QZ algorithm presented above, are the same as those of the QR algorithm. The convergence rate is not affected by the rank defficiency of B, i.e. B may be singular. Assuming that an average of two QZ iterations is necessary for an eigenvalue, the algorithm requires about $15n^3$ *flops*. The accumulation of the transformations Q and Z adds about $8n^3$ *flops* and $10n^3$ *flops*, respectively.

The QZ algorithm is numerically stable. The computed matrices $\widehat{A}$ and $\widehat{B}$ satisfy

$$Q_0^T(A+E)Z_0 = \widehat{A}, \quad ||E|| \approx \varepsilon_M ||A||,$$
$$Q_0^T(B+F)Z_0 = \widehat{B}, \quad ||F|| \approx \varepsilon_M ||B||,$$

where Q_0 and Z_0 are exactly orthogonal.

Remark 3.6 In the complex case, $A,\ B \in \mathbb{C}^{n\times n}$, it is reasonable to use complex arithmetic, and so, one may apply the single QZ step with a complex shift. Some implementations employ the LZ algorithm of (Kaufmann, 1975), based on elementary transformations, stabilized by row and column interchanges. □

Consider now the third stage of the QZ algorithm, the reduction to the *Schur-triangular form.* Starting with A in QTF, and B upper triangular, this stage performs the reduction of A to RSF, while maintaining the upper triangular structure of B, and determines the numerators and denominators whose ratios give the generalized eigenvalues. It is therefore necessary to triangularize the 2×2 diagonal blocks in A with real eigenvalues. In essence, A is fully triangularized, but without actually applying all the due transformations. The diagonal elements which would result if A and B were triangularized (some elements in A possibly being complex), are instead computed. This way, maximum information about generalized eigenvalues is obtained.

Let $A = \text{block-diag}(A_{11}, \ldots, A_{qq}) + N$ be given, where N is strict upper triangular, and A_{kk}, $k = 1\colon q$, are square matrices of orders 1 or 2. (The structure of A can be determined by examining its subdiagonal elements.) Matrix B is partitioned according to the partitioning of A. Then,

$$\lambda(A, B) = \bigcup_{k=1}^{q} \lambda(A_{kk}, B_{kk}).$$

For preserving the pencil structure, the eigenvalues of the subproblem (A_{kk}, B_{kk}) are determined in backward order (for $k = q\colon -1\colon 1$). Let $\alpha/\beta \in \lambda(A_{kk}, B_{kk})$ be a generalized eigenvalue of the kth subproblem, expressed as a ratio. (In practice, the numbers α and β are returned, and one requires that $\beta \geq 0$.) If $\alpha \in \mathbb{C}$, then $\bar{\alpha}/\beta \in \lambda(A_{kk}, B_{kk})$. For convenience, we shall denote

$$A_{kk} = \begin{bmatrix} a_{11} & a_{12} \\ a_{21} & a_{22} \end{bmatrix}, \qquad B_{kk} = \begin{bmatrix} b_{11} & b_{12} \\ 0 & b_{22} \end{bmatrix},$$

where the last row and column of A_{kk} and B_{kk} may be missing.

If $\alpha \in \mathbb{R}$, i.e. A_{kk} and B_{kk} have order 1, then $\alpha = \text{sign}(b_{11})a_{11}$, and $\beta = |b_{11}|$.

If A_{kk} and B_{kk} have order 2, it is possible that $\lambda(A_{kk}, B_{kk}) \subset \mathbb{R}$. If there is an infinite eigenvalue, it is possible to employ a special, efficient and reliable procedure. So, one first checks if one of the diagonal elements of B_{kk} is negligible. The tolerance of the test is $\varepsilon_B := \varepsilon\|B\|_\infty$. If $|b_{11}| \leq \varepsilon_B$, then a row transformation, $\bar{Q} = \text{block-diag}(I_*, I_2 + vw^T, I_*)$, can be applied to triangularize A_{kk}, so that the problem is reduced to two subproblems of order 1. The unspecified orders of the identity matrices in $\bar{Q}$ depend on k and on the structure of A. The calculation is stably accomplished as follows:

$(v, w) =$ `Hous_2`$(\ a_{11}, a_{21}\ );\quad \bar{Q} = \text{block-diag}(I_*, I_2 + vw^T, I_*);$
$A \leftarrow \bar{Q}A;\quad B \leftarrow \bar{Q}B;\quad Q \leftarrow Q\bar{Q};\quad a_{21} = b_{21} = 0;$
$\alpha_1 = \text{sign}(b_{11})a_{11};\quad \alpha_2 = \text{sign}(b_{22})a_{22};\quad \beta_1 = |b_{11}|;\quad \beta_2 = |b_{22}|;$

This approach is motivated by the argument that b_{21} remains negligible after having applied the Householder transformation $\bar{Q}$, because $b_{21} = v_2 b_{11}$, hence $|b_{21}| \leq |b_{11}| \leq \varepsilon_B$.

If $|b_{22}| \le \varepsilon_B$, then one may proceed similarly, using a column transformation, $\bar{Z} = \text{block-diag}(I_*, P_2(I_2 + vw^T)P_2, I_*)$, where $P_2 = [\, e_2 \;\; e_1 \,] \in \mathbb{R}^{2\times 2}$, and v and w are defined by $(v, w) = \texttt{Hous\_2}(\; a_{22}, a_{21} \;)$; the transformation is applied as $A \leftarrow A\bar{Z}$, $B \leftarrow B\bar{Z}$, $Z \leftarrow Z\bar{Z}$.

If B_{kk} is nonsingular, the generalized eigenvalues are finite, real or complex. The theoretical procedure for reducing A_{kk} to triangular form may be described as follows. Let δ be an eigenvalue of the pencil $A_{kk} - \lambda B_{kk}$. Form the matrix $E = A_{kk} - \delta B_{kk}$. Clearly, one can choose a matrix $\widetilde{Z}$ so that it zeroes either e_{11} or e_{21} (by postmultiplying E). As the rows of E are linearly dependent, because $\det(E) = 0$, it follows that by annihilating any of e_{11} or e_{21}, the other element is also annihilated. Then a matrix $\widetilde{Q}$ is chosen so that either $\widetilde{Q}A_{kk}\widetilde{Z}$ or $\widetilde{Q}B_{kk}\widetilde{Z}$ is upper triangular. Since the first column of $\widetilde{Q}E\widetilde{Z}$ is zero, and $\widetilde{Q}E\widetilde{Z} = \widetilde{Q}A_{kk}\widetilde{Z} - \delta\widetilde{Q}B_{kk}\widetilde{Z}$, we deduce that both $\widetilde{Q}A_{kk}\widetilde{Z}$ and $\widetilde{Q}B_{kk}\widetilde{Z}$ will be upper triangular, no matter how $\widetilde{Q}$ is chosen. Obviously, the transformations $\widetilde{Q}$ and $\widetilde{Z}$ should be optionally accumulated.

In the presence of rounding-errors, the computing method for δ, and the choice of $\widetilde{Z}$ and $\widetilde{Q}$ are critical for the stability of the process. It can be shown (Moler and Stewart, 1973) that, under some reasonable conditions, the *practical* procedure described below is stable. First, one finds an approximate eigenvalue of $C := A_{kk}B_{kk}^{-1}$, using the characteristic equation. For enhancing robustness, in implementations one uses $C := \widetilde{A}_{kk}\widetilde{B}_{kk}^{-1}$, where $\widetilde{A}_{kk} = A_{kk}/\|A_{kk}\|$ and $\widetilde{B}_{kk} = B_{kk}/\|B_{kk}\|$, and $\|X_{kk}\| := |x_{11}| + |x_{12}| + |x_{21}| + |x_{22}|$, hence the elements $\widetilde{a}_{ij}$ and $\widetilde{b}_{ij}$ are used instead of a_{ij} and b_{ij}, respectively, for computing δ. (The effect of this modification is finally compensated for. This is only needed for complex eigenvalues.) If $\lambda(C) = \{\, \lambda_1, \lambda_2 \,\}$, then from (3.38), with notations as defined there (i.e. $\widehat{a}_{kj} = \widetilde{a}_{kj}/\widetilde{b}_{jj}$, $\widehat{b}_{kj} = \widetilde{b}_{kj}/\widetilde{b}_{jj}$), it follows that

$$\lambda_1 + \lambda_2 = \widehat{a}_{11} + \widehat{a}_{22} - \widehat{a}_{21}\widehat{b}_{12}, \quad \lambda_1\lambda_2 = \det(C) = \widehat{a}_{11}\widehat{a}_{22} - \widehat{a}_{12}\widehat{a}_{21}.$$

The discriminant of the characteristic equation is $d := p^2 - \det(C)$, where $p := (\lambda_1 + \lambda_2)/2$. The following equivalent relations (originating from (Moler and Stewart, 1973)) are used in the EISPACK package:

$s = \widetilde{a}_{21}/(\widetilde{b}_{11}\widetilde{b}_{22})$; $\quad e = \widetilde{a}_{11}/\widetilde{b}_{11}$; $\quad t = (\widetilde{a}_{22} - e\widetilde{b}_{22})/\widetilde{b}_{22}$;
IF $|e| > |\widetilde{a}_{22}/\widetilde{b}_{22}|$, THEN $e = \widetilde{a}_{22}/\widetilde{b}_{22}$; $\quad t = (\widetilde{a}_{11} - e\widetilde{b}_{11})/\widetilde{b}_{11}$;
$c = (t - s\widetilde{b}_{12})/2$; $\quad d = c^2 + s(\widetilde{a}_{12} - e\widetilde{b}_{12})$;

Note that $t + 2e = \widehat{a}_{11} + \widehat{a}_{22}$, hence $2p = t + 2e - sb_{12}$, i.e. $p = e + c$; in addition, $et + e^2 = \widehat{a}_{11}\widehat{a}_{22}$, so that the concordance of the two above definitions can easily be verified.

If $d \geq 0$, then $\lambda_1, \lambda_2 \in \mathbb{R}$. Among these two values, the eigenvalue $\delta = e + (c + \text{sign}(c) d^{1/2})$ is used. The transformations $\widetilde{Q}$ and $\widetilde{Z}$, as applied by the following algorithm, have the aforementioned form. Note that each $\widetilde{Q}$ and $\widetilde{Z}$ uses the vectors v and w obtained just before having applied the corresponding transformation. The parameters $\underline{i}$ and $\bar{\imath}$ are defined at Algorithm QZSch.

Algorithm reig2 : $(A, B, Q, Z, \alpha, \beta) = \texttt{reig2}(\ A, B, Q, Z, \delta, l, \underline{i}, \bar{\imath}\ )$

/* Triangularize $A_{kk} := a_{l:l+1,l:l+1}$, maintaining $B_{kk} := b_{l:l+1,l:l+1}$ */
/* upper triangular, $A, B \in \mathbb{R}^{n \times n}$, and */
/* find $\lambda(A_{kk}, B_{kk}) \subset \mathbb{R}$ (returned as $\{ (\alpha_l, \beta_l), (\alpha_{l+1}, \beta_{l+1}) \}$), */
/* where δ is an approximate eigenvalue of $C := \widetilde{A}_{kk} \widetilde{B}_{kk}^{-1}$, */
/* and $\widetilde{A}_{kk} = A_{kk}/\|A_{kk}\|$, $\widetilde{B}_{kk} = B_{kk}/\|B_{kk}\|$; */
/* update Q and Z. */

/* 1. Find and apply a column transformation Z. */
$P_2 = \begin{bmatrix} e_2 & e_1 \end{bmatrix} \in \mathbb{R}^{2 \times 2}$;
$\widetilde{A} := a_{l:l+1,l:l+1}/\|a_{l:l+1,l:l+1}\|$; $\widetilde{B} := b_{l:l+1,l:l+1}/\|b_{l:l+1,l:l+1}\|$;
$e_{11} := \widetilde{a}_{11} - \delta \widetilde{b}_{11}$; $e_{12} := \widetilde{a}_{12} - \delta \widetilde{b}_{12}$; $e_{22} := \widetilde{a}_{22} - \delta \widetilde{b}_{22}$;
$a_1 = e_{12}$; $a_2 = e_{11}$;
IF $|e_{11}| + |e_{12}| < |\widetilde{a}_{21}| + |e_{22}|$, THEN $a_1 = e_{22}$; $a_2 = \widetilde{a}_{21}$;
$(v, w) = \texttt{Hous\_2}(\ a_1, a_2\ )$; $a_{\underline{i}:l+1,l:l+1} \leftarrow a_{\underline{i}:l+1,l:l+1} P_2 (I_2 + v w^T) P_2$;
$b_{\underline{i}:l+1,l:l+1} \leftarrow b_{\underline{i}:l+1,l:l+1} P_2 (I_2 + v w^T) P_2$;
$z_{:,l:l+1} \leftarrow z_{:,l:l+1} P_2 (I_2 + v w^T) P_2$; /* Update Z. */

/* 2. Find and apply a row transformation Q. */
$a_1 = b_{ll}$; $a_2 = b_{l+1,l}$;
IF $\|a_{l:l+1,l:l+1}\| < |\delta|\, \|b_{l:l+1,l:l+1}\|$, THEN $a_1 = a_{ll}$; $a_2 = a_{l+1,l}$;
$(v, w) = \texttt{Hous\_2}(\ a_1, a_2\ )$; $a_{l:l+1,l:\bar{\imath}} \leftarrow (I_2 + v w^T) a_{l:l+1,l:\bar{\imath}}$;
$b_{l:l+1,l:\bar{\imath}} \leftarrow (I_2 + v w^T) b_{l:l+1,l:\bar{\imath}}$;
$a_{l+1,l} = 0$; $b_{l+1,l} = 0$;
$q_{:,l:l+1} \leftarrow q_{:,l:l+1} (I_2 + v w^T)$; /* Update Q. */
$\alpha_l = \text{sign}(b_{ll}) a_{ll}$; $\beta_l = |b_{ll}|$;
$\alpha_{l+1} = \text{sign}(b_{l+1,l+1}) a_{l+1,l+1}$; $\beta_{l+1} = |b_{l+1,l+1}|$;

Algorithm **reig2** computes $\lambda(A_{kk}, B_{kk})$, but applies the transformations to A and B, as needed. Note that the column transformation Z annihilates the (1,1) or (2,1) element of the matrix $E := \widetilde{A}_{kk} - \delta \widetilde{B}_{kk}$, depending on the 1-norm of the first row of E being greater than (or equal to), or less than that of the second row, respectively. Moreover, the row transformation Q annihilates the (2,1) element of the matrix $\widetilde{B}_{kk}$ or $\widetilde{A}_{kk}$, depending on $\|A_{kk}\| \geq |\delta|\, \|B_{kk}\|$, or $\|A_{kk}\| < |\delta|\, \|B_{kk}\|$, respectively.

Finally, if $d < 0$, then $\lambda_1, \lambda_2 \in \mathbb{C}$; one considers that $\lambda(A_{kk}, B_{kk}) \subset \mathbb{C}$.

In this case, the procedure is similar, but complex rotations are required; transformations are used to compute the would be diagonal elements in A_{kk} and B_{kk}, but they are not actually applied, and the final matrix A is in RSF. The shift $\delta = \Re(\delta) - j\Im(\delta) = e + c - j(-d)^{1/2}$ is used to obtain the initial transformation. Consequently, the matrix $E := \widetilde{A}_{kk} - \delta\widetilde{B}_{kk}$ is complex (only $e_{21} \in \mathbb{R}$). For efficient calculations, the modulus is redefined here as $|e_{11}| = |\Re(e_{11})| + |\Im(e_{11})|$ (hence $|e_{11}| = |\widetilde{a}_{11} - \Re(\delta)\widetilde{b}_{11}| + |\Im(\delta)\widetilde{b}_{11}|$). The corresponding values a_1 and a_2 (see Algorithm **reig2**) are also complex. Complex rotations can be obtained with the statement

$$(c, s) = \texttt{rotc}(\ a_1, a_2\ );$$

where $c \in \mathbb{R}$, and $s \in \mathbb{C}$. So, the first two operations at step 2 of **reig2** are replaced by

$$\Re(a_1) = \widetilde{b}_{11}c + \widetilde{b}_{12}\Re(s); \quad \Im(a_1) = \widetilde{b}_{12}\Im(s);$$
$$\Re(a_2) = \widetilde{b}_{22}\Re(s); \quad \Im(a_2) = \widetilde{b}_{22}\Im(s);$$

corresponding to the rotation computed at step 1,

$$\begin{bmatrix} \widetilde{b}_{11} & \widetilde{b}_{12} \\ \widetilde{b}_{21} & \widetilde{b}_{22} \end{bmatrix} \leftarrow \begin{bmatrix} \widetilde{b}_{11} & \widetilde{b}_{12} \\ 0 & \widetilde{b}_{22} \end{bmatrix} \begin{bmatrix} c & \bar{s} \\ s & -c \end{bmatrix}.$$

The diagonal elements of A_{kk} and B_{kk} would result by evaluating

$$\begin{bmatrix} \gamma & \bar{\sigma} \\ \sigma & -\gamma \end{bmatrix} \begin{bmatrix} \widetilde{a}_{11} & \widetilde{a}_{12} \\ \widetilde{a}_{21} & \widetilde{a}_{22} \end{bmatrix} \begin{bmatrix} c & \bar{s} \\ s & -c \end{bmatrix},$$
$$\begin{bmatrix} \gamma & \bar{\sigma} \\ \sigma & -\gamma \end{bmatrix} \begin{bmatrix} \widetilde{b}_{11} & \widetilde{b}_{12} \\ 0 & \widetilde{b}_{22} \end{bmatrix} \begin{bmatrix} c & \bar{s} \\ s & -c \end{bmatrix}, \qquad (3.40)$$

where $(\gamma, \sigma) = \texttt{rotc}(\ a_1, a_2\ )$ is the complex rotation corresponding to the second Householder transformation in Algorithm **reig2**. But in order to obtain real diagonal elements in B_{kk}, the matrices A_{kk} and B_{kk}, which would result, are premultiplied by a suitably chosen diagonal matrix. So, if (a, b) is a pair of diagonal elements computed with the relations (3.40), where $a, b \in \mathbb{C}$, then the pair of diagonal elements finally returned is (α, β), where $\alpha \in \mathbb{C}$, $\beta \in \mathbb{R}$ $(\beta > 0)$, given by

$$\begin{aligned} \Re(\alpha) &= \|A_{kk}\|(\Re(a)\Re(b) + \Im(a)\Im(b))/r, \quad r := (\Re(b)^2 + \Im(b)^2)^{1/2}, \\ \Im(\alpha) &= \|A_{kk}\|(\Im(a)\Re(b) - \Re(a)\Im(b))/r, \quad \beta = \|B_{kk}\|r. \end{aligned}$$

It is a simple exercise to write down the corresponding algorithm for triangularizing 2×2 diagonal blocks with complex conjugate eigenvalues. So, the calling statement only will be given:

$(\alpha, \beta) =$ `ceig2`$(A, B, \delta, l);$

Note that the matrices Q and Z are not parameters because the transformations performed are not accumulated. Moreover, A and B are only input parameters.

Now, Algorithm `QZSch2` is as follows:

Algorithm QZSch2 : $(A, B, Q, Z, \alpha, \beta) =$ `QZSch2`$(A, B, Q, Z, l, \underline{i}, \bar{\imath}, \varepsilon_B)$

/* Triangularize $A_{kk} := a_{l:l+1,l:l+1}$, maintaining $B_{kk} := b_{l:l+1,l:l+1}$ */
/* upper triangular, $A, B \in \mathbb{R}^{n \times n}$, and */
/* find $\lambda(A_{kk}, B_{kk})$ (returned as $\{(\alpha_l, \beta_l), (\alpha_{l+1}, \beta_{l+1})\}$), */
/* using an approximate eigenvalue of $C := \widetilde{A}_{kk} \widetilde{B}_{kk}^{-1}$, */
/* where $\widetilde{A}_{kk} = A_{kk}/\|A_{kk}\|$, $\widetilde{B}_{kk} = B_{kk}/\|B_{kk}\|$; */
/* update Q and Z. */

$P_2 = \begin{bmatrix} e_2 & e_1 \end{bmatrix} \in \mathbb{R}^{2\times 2};$

/* 1. Check for negligible diagonal elements in B_{kk}. */
IF $|b_{ll}| \le \varepsilon_B$, THEN
/* Find and apply a row transformation Q (b_{ll} negligible). */
  $(v, w) =$ **Hous_2**$(a_{ll}, a_{l+1,l});$
  $a_{l:l+1,l+1:\bar{\imath}} \leftarrow (I_2 + vw^T) a_{l:l+1,l+1:\bar{\imath}}; \quad a_{l+1,l} = 0;$
  $b_{l:l+1,l:\bar{\imath}} \leftarrow (I_2 + vw^T) b_{l:l+1,l:\bar{\imath}}; \quad b_{l+1,l} = 0;$
  $q_{:,l:l+1} \leftarrow q_{:,l:l+1}(I_2 + vw^T);$ /* Update Q. */
  $\alpha_l = \text{sign}(b_{ll}) a_{ll}; \quad \beta_l = |b_{ll}|;$
  $\alpha_{l+1} = \text{sign}(b_{l+1,l+1}) a_{l+1,l+1}; \quad \beta_{l+1} = |b_{l+1,l+1}|;$
ELSE IF $|b_{l+1,l+1}| \le \varepsilon_B$, THEN
/* Find and apply a transformation Z ($b_{l+1,l+1}$ negligible). */
  $(v, w) =$ **Hous_2**$(a_{l+1,l+1}, a_{l+1,l});$
  $a_{\underline{i}:l,l:l+1} \leftarrow a_{\underline{i}:l,l:l+1} P_2 (I_2 + vw^T) P_2; \quad a_{l+1,l} = 0;$
  $b_{\underline{i}:l+1,l:l+1} \leftarrow b_{\underline{i}:l+1,l:l+1} P_2 (I_2 + vw^T) P_2; \quad b_{l+1,l} = 0;$
  $z_{:,l:l+1} \leftarrow z_{:,l:l+1} P_2 (I_2 + vw^T) P_2;$ /* Update Z. */
  $\alpha_l = \text{sign}(b_{ll}) a_{ll}; \quad \beta_l = |b_{ll}|;$
  $\alpha_{l+1} = \text{sign}(b_{l+1,l+1}) a_{l+1,l+1}; \quad \beta_{l+1} = |b_{l+1,l+1}|;$
ELSE
/* 2. Triangularize A_{kk} (B_{kk} nonsingular). */
  $\widetilde{A} := a_{l:l+1,l:l+1}/\|a_{l:l+1,l:l+1}\|; \quad \widetilde{B} := b_{l:l+1,l:l+1}/\|b_{l:l+1,l:l+1}\|;$
  $s = \widetilde{a}_{21}/(\widetilde{b}_{11}\widetilde{b}_{22}); \quad e = \widetilde{a}_{11}/\widetilde{b}_{11}; \quad f = \widetilde{a}_{22}/\widetilde{b}_{22};$
  IF $|e| \le |f|$, THEN
    $t = (\widetilde{a}_{22} - e\widetilde{b}_{22})/\widetilde{b}_{22};$
  ELSE
    $e = f; \quad t = (\widetilde{a}_{11} - e\widetilde{b}_{11})/\widetilde{b}_{11};$

```
  END IF
  c = (t − s b̃12)/2;   d = c² + s(ã12 − e b̃12);
  IF d ≥ 0, THEN
     δ = e + (c + sign(c)d^{1/2});                     /* Real eigenvalues. */
     (A, B, Q, Z, α, β) = reig2( A, B, Q, Z, δ, l, i̲, ī );
  ELSE
     δ = e + c − j(−d)^{1/2});          /* Complex conjugate eigenvalues. */
     (α, β) = ceig2( A, B, δ, l );
  END IF
END IF
```

Remark 3.7 It is advisable to use the computed values α_j and β_j, instead of their ratios λ_j. As unitary transformations are exclusively used, the errors in α_j and β_j are of size $0(\varepsilon)$, if λ_j is a simple generalized eigenvalue, and of size $0(\varepsilon^{1/m})$, if λ_j has algebraic multiplicity m (or it belongs to a cluster of m nearly equal generalized eigenvalues), where ε is the size of the errors in the input data, or of the rounding-errors in the calculations. Therefore, an α_j or β_j which has about the same size as the corresponding given matrix elements, can be regarded as being accurately determined by data and computations; otherwise, it should be considered as having a larger relative error. In the extreme case when both α_j and β_j are $0(\varepsilon)$, then any value λ makes $A - \lambda B$ singular to working precision. If β_j is very small, but α_j is not, then λ can be considered as an infinite eigenvalue. □

Remark 3.8 A more general algorithm, called VZ, for solving the problem $ACx = \lambda BDx$, where $A, B \in \mathbb{R}^{n\times m}$, $C, D \in \mathbb{R}^{m\times n}$ $(m \geq n)$, has been proposed in (Van Loan, 1975); it does not explicitly form AC, BD, or $(BD)^{-1}$, and it generalizes the QR and QZ algorithms. □

3.2.3 Ordering generalized real Schur form

A *deflating subspace* (also called *generalized invariant subspace*), $\mathcal{S}$, of a regular pencil $A - \lambda B$ has the property that $\dim(A\mathcal{S} + B\mathcal{S}) = \dim(\mathcal{S})$. (One also says that $A\mathcal{S} + B\mathcal{S}$ and $\mathcal{S}$ form a pair of left and right deflating subspaces of $A - \lambda B$.) When $B = I_n$, this definition reduces to that of an invariant subspace of A, because the relation $\dim(A\mathcal{S} + \mathcal{S}) = \dim(\mathcal{S})$ is equivalent to $A\mathcal{S} \subset \mathcal{S}$.

Let $A, B \in \mathbb{C}^{n\times n}$. Assume that the deflating subspace $\mathcal{S}$ has dimension s, and the first s columns of the unitary matrices Q and Z, partitioned into the form $Z = \begin{bmatrix} Z_1 & Z_2 \end{bmatrix}$, $Q = \begin{bmatrix} Q_1 & Q_2 \end{bmatrix}$, where $Z_1, Q_1 \in \mathbb{C}^{n\times s}$, generate the subspaces $\mathcal{S}$ and $A\mathcal{S} + B\mathcal{S}$, respectively. Then, it follows that

$Q_2^H A Z_1 = Q_2^H B Z_1 = 0$, or

$$Q^H(A - \lambda B)Z = \begin{bmatrix} \widetilde{A}_{11} & \widetilde{A}_{12} \\ 0 & \widetilde{A}_{22} \end{bmatrix} - \lambda \begin{bmatrix} \widetilde{B}_{11} & \widetilde{B}_{12} \\ 0 & \widetilde{B}_{22} \end{bmatrix}, \tag{3.41}$$

where $\widetilde{A}_{11}, \widetilde{B}_{11} \in \mathbb{C}^{s\times s}$. Conversely, if (3.41) holds, the columns of Z_1 span a deflating subspace $\mathcal{S}$ for $A - \lambda B$. For $s = 1$, $\mathcal{S}$ is spanned by an eigenvector of $A - \lambda B$, corresponding to the single eigenvalue in $\lambda(\widetilde{A}_{11}, \widetilde{B}_{11})$. For arbitrary s, $\lambda(\widetilde{A}_{11}, \widetilde{B}_{11})$ is a subset of the spectrum $\lambda(A, B)$, and $\mathcal{S}$ is uniquely determined when this subset is disjoint from the rest of the spectrum. (In this case, $\mathcal{S}$ is generated by the eigenvectors and principal generalized vectors corresponding to $\lambda(\widetilde{A}_{11}, \widetilde{B}_{11})$.) It follows that the matrix Z in the generalized Schur decomposition provides orthogonal bases for the deflating subspaces of dimensions $1{:}\,n-1$, because the transformed A and B matrices, being upper triangular, satisfy (3.41) for $s = 1{:}\,n-1$. This also holds in the case $A, B \in \mathbb{R}^{n\times n}$, for the generalized real Schur decomposition, if

$$s = \sum_{j=1}^{k} d_j, \qquad k = 1{:}\,q,$$

where $d_j \in \{1, 2\}$ are the orders of the resulting diagonal blocks.

As for the standard eigenvalue problem (see § 2.2.4), one can obtain an orthogonal basis for a deflating subspace $\mathcal{S}$, corresponding to a given subset of s generalized eigenvalues, by "ordering" these eigenvalues so that they appear into the first s positions. (In the real case, this subset will always include both elements of a complex pair of eigenvalues, to satisfy the above mentioned constraint.)

It would suffice to consider the exchange of two consecutive diagonal blocks of A and B, where A and B are in a Schur-triangular form. To do this, some unitary matrices Q and Z should be constructed so that

$$\begin{aligned} Q^H A Z &= Q^H \begin{bmatrix} A_{11} & A_{12} \\ 0 & A_{22} \end{bmatrix} Z =: \begin{bmatrix} \widetilde{A}_{11} & \widetilde{A}_{12} \\ 0 & \widetilde{A}_{22} \end{bmatrix} =: \widetilde{A}, \\ Q^H B Z &= Q^H \begin{bmatrix} B_{11} & B_{12} \\ 0 & B_{22} \end{bmatrix} Z =: \begin{bmatrix} \widetilde{B}_{11} & \widetilde{B}_{12} \\ 0 & \widetilde{B}_{22} \end{bmatrix} =: \widetilde{B}, \end{aligned} \tag{3.42}$$

where $\lambda(A_{11}, B_{11}) = \lambda(\widetilde{A}_{22}, \widetilde{B}_{22})$ and $\lambda(A_{22}, B_{22}) = \lambda(\widetilde{A}_{11}, \widetilde{B}_{11})$, and where the dimensions d_1 and d_2 are either 1 or 2.

In case $d_1 = d_2 = 1$, in (3.42) we have $A_{ij} = a_{ij}$, $j = 1{:}\,2$, $i = 1{:}\,j$, and similarly for $\widetilde{A}_{ij}$, B_{ij}, and $\widetilde{B}_{ij}$. We may assume that $|b_{22}| \geq |a_{22}|$, since otherwise the roles of A and B can be interchanged. In order to exchange the eigenvalues so that $\lambda(a_{22}, b_{22}) = \lambda(\widetilde{a}_{11}, \widetilde{b}_{11})$, it is necessary that the first column, z_1, of Z in (3.41) be an eigenvector of the pencil $A - \lambda B$ corresponding to $\lambda(a_{22}, b_{22})$, i.e.

$$(b_{22}A - a_{22}B)Z = \begin{bmatrix} 0 & x \\ 0 & x \end{bmatrix}. \tag{3.43}$$

But the second row of the matrix $H := b_{22}A - a_{22}B$ is zero, so, choosing Z to annihilate h_{11} (for instance, Z can be a plane rotation, possibly complex, in the plane (2,1)), then HZ will have the second row and first column zero. As from (3.43) it follows that the vectors Az_1 and Bz_1 are parallel, the desired form is obtained by applying a transformation Q so that $(Q^H Bz_1)_2 = 0$. (The assumption that $|\widetilde{b}_{11}|/|\widetilde{a}_{11}| = |b_{22}|/|a_{22}| \geq 1$ implies $\widetilde{b}_{11} \neq 0$, hence $Q^H Az_1$ can be parallel with $Q^H Bz_1$ only if $Q^H AZ$ is upper triangular.) It is possible to prove that the algorithm just described is numerically stable (Van Dooren, 1980, 1981).

In case $d_1 = 2$, $d_2 = 1$, the configuration (3.42) is as follows[8]

$$\begin{aligned} Q^T AZ &= Q^T \begin{bmatrix} a_{11} & a_{12} & a_{13} \\ a_{21} & a_{22} & a_{23} \\ 0 & 0 & a_{33} \end{bmatrix} Z =: \begin{bmatrix} \widetilde{a}_{11} & \widetilde{a}_{12} & \widetilde{a}_{13} \\ 0 & \widetilde{a}_{22} & \widetilde{a}_{23} \\ 0 & \widetilde{a}_{32} & \widetilde{a}_{33} \end{bmatrix} =: \widetilde{A}, \\ Q^T BZ &= Q^T \begin{bmatrix} b_{11} & b_{12} & b_{13} \\ 0 & b_{22} & b_{23} \\ 0 & 0 & b_{33} \end{bmatrix} Z =: \begin{bmatrix} \widetilde{b}_{11} & \widetilde{b}_{12} & \widetilde{b}_{13} \\ 0 & \widetilde{b}_{22} & \widetilde{b}_{23} \\ 0 & 0 & \widetilde{b}_{33} \end{bmatrix} =: \widetilde{B}. \end{aligned}$$

We may assume that $|b_{33}| \geq |a_{33}|$, since otherwise the roles of A and B can be interchanged by initially transforming the first two rows of A and, finally, the last two rows of $\widetilde{A}$, to annihilate a_{21} and $\widetilde{a}_{32}$, and create b_{21} and $\widetilde{b}_{32}$. As in the previous case, in order to get $\lambda(a_{33}, b_{33}) = \lambda(\widetilde{a}_{11}, \widetilde{b}_{11})$, the matrix $\widetilde{Q}^T(b_{33}A - a_{33}B)Z$ should have the first column zero, where $\widetilde{Q}$ is any nonsingular matrix. $\widetilde{Q}^T$ is chosen so that to annihilate the element h_{21} of the matrix $H := b_{33}A - a_{33}B$, still preserving the last row of H zero. (For instance, $\widetilde{Q}^T$ can be a rotation in the plane (1,2).) Hence, $\widetilde{Q}^T H$ has the structure

$$\widetilde{Q}^T H = \begin{bmatrix} x & x & x \\ 0 & x & x \\ 0 & 0 & 0 \end{bmatrix},$$

and its first column can be zeroed by taking $Z = Z_1 Z_2$, where Z_1 and Z_2 are, for instance, rotations in the planes (3,2) and (2,1), respectively, which

[8]Note that cases $d_1 = 2$, $d_2 = 1$, or $d_1 = 1$, $d_2 = 2$, or $d_1 = d_2 = 2$, only appear if $A, B \in \mathbb{R}^{n \times n}$.

annihilate the elements (2,2) and (1,1) in $\widetilde{Q}^T H$, respectively. Consequently, $\widetilde{Q}^T(b_{33}A - a_{33}B)Z$ has only the elements (1,2), (1,3), and (2,3) non-zero, hence $(b_{33}A - a_{33}B)Z$ has the first column zero. Finally, a matrix Q is constructed so that $Q^T BZ$ is upper triangular. One takes $Q = Q_1Q_2$, where Q_1 and Q_2 are rotations in the planes (2,3) and (1,2), chosen to annihilate the element (3,2), created by Z_1 (i.e. $Q_1^T BZ_1$ is upper triangular), and the element (2,1), created by Z_2 (i.e. $Q_2^T Q_1^T BZ_1Z_2$ is upper triangular), respectively. The new non-zero element introduced in B by a transformation Z is immediately annihilated by a transformation Q. Clearly, the transformations are applied in the same order to the A matrix too. As for case $d_1 = d_2 = 1$, it can be shown that $\widetilde{A}$ and $\widetilde{B}$ have the required properties.

For efficiency reasons, modified Householder transformations are used in the algorithm below, instead of standard plane rotations.

Algorithm swap21 : $(A, B, Q, Z) = \texttt{swap21}(\ A, B, Q, Z, l\ )$

/* Swap two pairs of adjacent diagonal blocks of $A, B \in \mathbb{R}^{n\times n}$, */
/* starting from the diagonal position l, where the first pair is 2×2 */
/* and the second pair 1×1; update Q and Z. */

$P_2 = \begin{bmatrix} e_2 & e_1 \end{bmatrix} \in \mathbb{R}^{2\times 2}$;

/* 1. Find and apply a row transformation $\widetilde{Q}$. */
$g = \max\{\ |a_{l+2,l+2}|, |b_{l+2,l+2}|\ \}$; $ich = g \neq |b_{l+2,l+2}|$;
IF ich, THEN
/* Interchange the roles of A and B. */
  $(v, w) = \texttt{Hous\_2}(\ a_{ll}, a_{l+1,l}\ )$;
  $a_{l:l+1,l+1:n} \leftarrow (I_2 + vw^T)a_{l:l+1,l+1:n}$; $a_{l+1,l} = 0$;
  $b_{l:l+1,l:n} \leftarrow (I_2 + vw^T)b_{l:l+1,l:n}$;
  $q_{:,l:l+1} \leftarrow q_{:,l:l+1}(I_2 + vw^T)$; /* Update Q. */
END IF

$\alpha = a_{l+2,l+2}/g$; $\beta = b_{l+2,l+2}/g$;
$H := \beta a_{l:l+1,l:l+2} - \alpha b_{l:l+1,l:l+2}$; $(v, w) = \texttt{Hous\_2}(\ h_{11}, h_{21}\ )$;
$H \leftarrow (I_2 + vw^T)H$;

/* 2. Find and apply a column transformation Z_1. */
$(v, w) = \texttt{Hous\_2}(\ h_{23}, h_{22}\ )$; $h_{12} \leftarrow h_{12} + (h_{13} + v_2h_{12})w_2$;
$a_{1:l+2,l+1:l+2} \leftarrow a_{1:l+2,l+1:l+2}P_2(I_2 + vw^T)P_2$;
$b_{1:l+2,l+1:l+2} \leftarrow b_{1:l+2,l+1:l+2}P_2(I_2 + vw^T)P_2$;
$z_{:,l+1:l+2} \leftarrow z_{:,l+1:l+2}P_2(I_2 + vw^T)P_2$; /* Update Z. */

/* 3. Find and apply a row transformation Q_1. */
$a_1 = b_{l+1,l+1}$; $a_2 = b_{l+2,l+1}$;
IF ich, THEN $a_1 = a_{l+1,l+1}$; $a_2 = a_{l+2,l+1}$;
$(v, w) = \texttt{Hous\_2}(\ a_1, a_2\ )$;

$a_{l+1:l+2,l:n} \leftarrow (I_2 + vw^T)a_{l+1:l+2,l:n}$;
$b_{l+1:l+2,l:n} \leftarrow (I_2 + vw^T)b_{l+1:l+2,l:n}$;
$q_{:,l+1:l+2} \leftarrow q_{:,l+1:l+2}(I_2 + vw^T)$; /* Update Q. */

/* 4. Find and apply a column transformation Z_2. */
$(v, w) =$ `Hous_2`(h_{12}, h_{11});
$a_{1:l+2,l:l+1} \leftarrow a_{1:l+2,l:l+1}P_2(I_2 + vw^T)P_2$;
$b_{1:l+2,l:l+1} \leftarrow b_{1:l+2,l:l+1}P_2(I_2 + vw^T)P_2$;
$z_{:,l:l+1} \leftarrow z_{:,l:l+1}P_2(I_2 + vw^T)P_2$; /* Update Z. */

/* 5. Find and apply a row transformation Q_2. */
$a_1 = b_{ll}$; $a_2 = b_{l+1,l}$;
IF *ich*, THEN $a_1 = a_{ll}$; $a_2 = a_{l+1,l}$;
$(v, w) =$ `Hous_2`(a_1, a_2);
$a_{l:l+1,l:n} \leftarrow (I_2 + vw^T)a_{l:l+1,l:n}$;
$b_{l:l+1,l:n} \leftarrow (I_2 + vw^T)b_{l:l+1,l:n}$;
$q_{:,l:l+1} \leftarrow q_{:,l:l+1}(I_2 + vw^T)$; /* Update Q. */
IF *ich*, THEN
/* Interchange the roles of A and B. */
  $(v, w) =$ `Hous_2`($b_{l+1,l+1}, b_{l+2,l+1}$);
  $a_{l+1:l+2,l+1:n} \leftarrow (I_2 + vw^T)a_{l+1:l+2,l+1:n}$;
  $b_{l+1:l+2,l+2:n} \leftarrow (I_2 + vw^T)b_{l+1:l+2,l+2:n}$;
  $q_{:,l+1:l+2} \leftarrow q_{:,l+1:l+2}(I_2 + vw^T)$; /* Update Q. */
END IF
$a_{l+1:l+2,l} = 0$; $b_{l+1:l+2,l} = 0$; $b_{l+2,l+1} = 0$;

Note that, for later use, Algorithm `swap21` works on the full $n \times n$ matrices.

Case $d_1 = 1$, $d_2 = 2$ is dual to case $d_1 = 2$, $d_2 = 1$, and can be solved by *pertransposition* (a reflection about the secondary diagonal). The corresponding algorithm is called `swap12` and has the same parameters as Algorithm `swap21`.

For case $d_1 = d_2 = 2$, all blocks in (3.42) have order 2. In order to get $\lambda(A_{22}, B_{22}) = \lambda(\widetilde{A}_{11}, \widetilde{B}_{11})$, the first two columns of Z must span the deflating subspace of $A - \lambda B$ corresponding to $\lambda(A_{22}, B_{22}) = \{ \lambda_2, \bar{\lambda}_2 \}$. Equivalently, the first two columns of the matrix $(B^{-1}A - \lambda_2 I_4)(B^{-1}A - \bar{\lambda}_2 I_4)Z$ must be zero. This can be accomplished by using the QZ algorithm, but adapting it to a 4×4 pencil. According to the properties of the QZ algorithm with implicit double shift, one will obtain $\widetilde{a}_{32} = 0$, and $\lambda(\widetilde{A}_{22}, \widetilde{B}_{22}) = \lambda(A_{11}, B_{11})$ if the eigenvalues λ_1 and $\bar{\lambda}_1$, where $\{ \lambda_1, \bar{\lambda}_1 \} := \lambda(A_{11}, B_{11})$, are used as shifts, and if, in addition, $a_{32} \neq 0$. As the second condition does not hold, a QZ step with an "arbitrary" shift, constructed using the parameters $a_{10} = a_{20} = a_{30} = 1$, is first performed, so that $a_{32} \neq 0$. In general, a second QZ step with the double shift mentioned above will do for obtaining the desired result. Algorithm **idQZs**, also used for obtaining the reduction to Schur-triangular form, can be called.

Clearly, QZ steps can also be used for exchanging 1×1 by 2×2, or 2×2 by 1×1 blocks, but this approach would be more inefficient than that discussed above, because two QZ steps at least are needed, and each QZ step for a 3×3 problem involves five transformations.

Summarizing and generalizing the above discussion, the following algorithm is obtained.

Algorithm gswap2 : $(A, B, Q, Z) = \texttt{gswap2}(\ A, B, Q, Z, l, n_1, n_2\ )$

/* Swap two pairs of adjacent diagonal blocks of $A, B \in \mathbb{R}^{n\times n}$, */
/* of orders n_1 and n_2 ($n_1 \le 2$, $n_2 \le 2$), starting from the diagonal */
/* position l; update Q and Z. */

IF $n_1 = 1$ AND $n_2 = 1$, THEN
/* 1. Swap two pairs of 1×1 blocks. */
/* Find and apply a column transformation Z. */
$g = \max\{\,|a_{l+1,l+1}|, |b_{l+1,l+1}|\,\}$; $\alpha = a_{l+1,l+1}/g$; $\beta = b_{l+1,l+1}/g$;
$(v, w) = \texttt{Hous\_2}(\ \beta a_{l,l+1} - \alpha b_{l,l+1}, \beta a_{ll} - \alpha b_{ll}\ )$;
$a_{1:l+1,l:l+1} \leftarrow a_{1:l+1,l:l+1} P_2 (I_2 + vw^T) P_2$;
$b_{1:l+1,l:l+1} \leftarrow b_{1:l+1,l:l+1} P_2 (I_2 + vw^T) P_2$;
$z_{:,l:l+1} \leftarrow z_{:,l:l+1} P_2 (I_2 + vw^T) P_2$; /* Update Z. */
/* Find and apply a row transformation Q. */
$a_1 = b_{ll}$; $a_2 = b_{l+1,l}$;
IF $g \neq |b_{l+1,l+1}|$, THEN $a_1 = a_{ll}$; $a_2 = a_{l+1,l}$;
$(v, w) = \texttt{Hous\_2}(\ a_1, a_2\ )$;
$a_{l:l+1,l:n} \leftarrow (I_2 + vw^T) a_{l:l+1,l:n}$; $a_{l+1,l} = 0$;
$b_{l:l+1,l:n} \leftarrow (I_2 + vw^T) b_{l:l+1,l:n}$; $b_{l+1,l} = 0$;
$q_{:,l:l+1} \leftarrow q_{:,l:l+1} (I_2 + vw^T)$; /* Update Q. */
ELSE IF $n_1 = 2$ AND $n_2 = 1$, THEN
/* 2. Swap a pair of 2×2 blocks by a pair of 1×1 blocks. */
$(A, B, Q, Z) = \texttt{swap21}(\ A, B, Q, Z, l\ )$;
ELSE IF $n_1 = 1$ AND $n_2 = 2$, THEN
/* 3. Swap a pair of 1×1 blocks by a pair of 2×2 blocks. */
$(A, B, Q, Z) = \texttt{swap12}(\ A, B, Q, Z, l\ )$;
ELSE
/* 4. Swap two pairs of 2×2 blocks. */
$i = l + 3$; $\underline{i} = 1$; $\bar{i} = n$; $s = 2$; $\varepsilon_A = \varepsilon_M \|a_{l:i,l:i}\|_\infty$;
$\widehat{a}_{33} = a_{ll}/b_{ll}$; $\widehat{a}_{34} = a_{l,l+1}/b_{l+1,l+1}$;
$\widehat{a}_{43} = a_{l+1,l}/b_{ll}$; $\widehat{a}_{44} = a_{l+1,l+1}/b_{l+1,l+1}$; $\widehat{b}_{34} = b_{l,l+1}/b_{l+1,l+1}$;
FOR $k_1 = 1\!:\!5$,

$a_{10} = 1;\quad a_{20} = 1;\quad a_{30} = 1;$
FOR $k_2 = 1{:}\,10$,
$(A, B, Q, Z) =$ `idQZs`$(\ A, B, a_{1:3,0}, Q, Z, l, i, \underline{i}, \overline{\imath}, s\ );$
IF $|a_{l+2,l+1}| \leq \varepsilon_A$, THEN EXIT LOOPS;
$\widehat{a}_{11} = a_{ll}/b_{ll};\quad \widehat{a}_{12} = a_{l,l+1}/b_{l+1,l+1};$
$\widehat{a}_{21} = a_{l+1,l}/b_{ll};\quad \widehat{a}_{22} = a_{l+1,l+1}/b_{l+1,l+1};\quad \widehat{b}_{12} = b_{l,l+1}/b_{l+1,l+1};$
$a_{00} = ((\widehat{a}_{33} - \widehat{a}_{11})(\widehat{a}_{44} - \widehat{a}_{11}) - \widehat{a}_{34}\widehat{a}_{43} + \widehat{a}_{43}\widehat{b}_{34}\widehat{a}_{11})/\widehat{a}_{21};$
$a_{10} = a_{00} + \widehat{a}_{12} - \widehat{a}_{11}\widehat{b}_{12};$
$a_{20} = (\widehat{a}_{22} - \widehat{a}_{11}) - \widehat{a}_{21}\widehat{b}_{12} - (\widehat{a}_{33} - \widehat{a}_{11}) - (\widehat{a}_{44} - \widehat{a}_{11}) + \widehat{a}_{43}\widehat{b}_{34};$
$a_{30} = a_{l+2,l+1}/b_{l+1,l+1};$
END FOR
END FOR
$a_{l+2,l:l+1} = 0;\quad a_{l+3,l:l+1} = 0;$
$b_{l+1,l} = 0;\quad b_{l+2,l:l+1} = 0;\quad b_{l+3,l:l+2} = 0;$
END IF

Note that Algorithm `idQZs` is called with $\underline{i} = 1$ and $\overline{\imath} = n$, because it seems that other values serve no purpose when eigenvalues are to be ordered.

More generally, let Algorithm `gswap` be defined by the statement:

$(A, B, Q, Z) =$ `gswap`$(\ A, B, i, j, Q, Z\ );$

which orders a pair of real matrices (A, B) in generalized real Schur form so that the diagonal blocks of A and B with row index i are moved to row j $(1 \leq i \leq n; 1 \leq j \leq n)$. Algorithm `gswap` consists of a sequence of interchanges of pairs of adjacent blocks using calls to Algorithm `gswap2`. If $i < j$ each pair of blocks is swapped with the next one below; if $i > j$ each pair is swapped with the next one above.

Algorithm `gswap` is used in Algorithm `QZSchord` below, that orders a generalized real Schur decomposition of a pair of real matrices, (A, B), so that a selected cluster of generalized eigenvalues appears in the leading diagonal blocks, and the leading columns of Z form an orthonormal basis of the corresponding deflating subspace. If the cluster includes a pair of complex conjugate generalized eigenvalues, then both are automatically selected.

Algorithm QZSchord : $(A, B, Q, Z) =$ `QZSchord`$(\ A, B, \mathit{select}, Q, Z\ )$

/* Order the generalized real Schur decomposition of $A, B \in \mathbb{R}^{n\times n}$, */
/* with A in RSF and B upper triangular, so that the generalized */
/* eigenvalues λ_i, for which select_i is TRUE, appear in the leading */
/* diagonal blocks; update Q and Z. */

$\widetilde{k} = 0;\quad \mathit{pair} =$ FALSE;

```
/* 1. Collect the selected blocks at the upper-left corner of A and B. */
FOR k = 1:n,
    IF pair = TRUE, THEN
        pair = FALSE;
    ELSE
        s = select_k;
        IF k < n, THEN
            IF a_{k+1,k} ≠ 0, THEN pair = TRUE;    s = s OR select_{k+1};
        IF s = TRUE, THEN
            k̃ ← k̃ + 1;   k̂ = k;   /* 2. Swap the kth blocks to position k̃. */
            IF k ≠ k̃, THEN (A, B, Q, Z) = gswap( A, B, k̂, k̃, Q, Z );
            IF pair = TRUE, THEN k̃ ← k̃ + 1;
        END IF
    END IF
END FOR
```

Ordering the Schur-triangular form requires a small multiple of n^3 *flops*, a reasonable estimate being $5n^3$ *flops*.

The algorithms for ordering the diagonal blocks of a matrix pair in generalized RSF are stable in practice. Perturbation analysis for deflating subspaces in dealt with, for instance, in (Stewart, 1972).

Several remarks about implementations will follow. Our implementation of the ordering algorithms, used in Algorithm `gSchv(Q)` in § 3.4.1, is based on the codes in (Van Dooren, 1980), also discussed in (Van Dooren, 1982), but largely modified. The ordering routine analyzes the diagonal blocks in A and B, starting from the bottom-right corner. After the size of the current diagonal block in A (which could be one or two) has been determined, the code checks if an exchange is to be made or not. The blocks can be ordered so that the generalized eigenvalues with either real parts or moduli larger than or equal to, or smaller than a given non-negative number will be moved to the first positions. All exchanges are performed by an auxiliary routine, `EXCQZS`, implementing Algorithm `gswap2`. This routine has distinct code portions for each different type of block exchange. The row and column transformations are optionally accumulated in Q and Z, respectively. Only modified Householder transformations of order two are used all through, contrary to the implementation in (Van Dooren, 1980, 1982), based on plane rotations.

The `DSUBSP` routine in (Van Dooren, 1982) uses an integer function subprogram to specify the eigenvalues which should be moved to the left-top corner. Like the EISPACK subroutines, it does not accumulate the transformations Q. Modifications of the `SROTG` and `SROT` subprograms from the BLAS collection (Lawson et al., 1979) are used for the computation and application of plane rotations. The eigenvalues are sorted by a "bubble sort" technique. For exchanging pairs of 2×2 adjacent blocks, an adaptation of the QZ algorithm is called.

Ordering the Schur-triangular form has important applications in the solution of some linear-quadratic optimization problems, spectral factorization, etc. We shall consider here the use of this process for solving some categories of algebraic matrix Riccati equations.

Consider a discrete-time algebraic Riccati equation with a singular matrix A. In this case, the standard Schur vectors method cannot be used, because of not being possible to construct the symplectic matrix G in (3.6) or (3.8). A feasible approach is based on the solution of the generalized eigenvalue problem $\widehat{A}x = \lambda\widehat{B}x$ of order $2n$ (see (3.15)), where

$$\widehat{A} = \begin{bmatrix} A & 0 \\ -Q & I_n \end{bmatrix}, \qquad \widehat{B} = \begin{bmatrix} I_n & F \\ 0 & A^T \end{bmatrix}, \qquad F := BR^{-1}B^T. \tag{3.44}$$

According to Theorems 3.4 and 3.5 in § 3.1.2, with usual assumptions, the pencil $\widehat{A} - \lambda\widehat{B}$ has no eigenvalues on the unit circle in $\mathbb{C}$, and if λ is an eigenvalue, then $1/\lambda$ is also an eigenvalue with the same multiplicity. (For $\lambda = 0$, infinite eigenvalues correspond.) In addition, by Theorem 3.6, the solution of the Riccati equation (3.2) can be reached by solving the linear system of order n, $U_{11}^T X = U_{21}^T$, where the matrix $\begin{bmatrix} U_{11}^T & U_{21}^T \end{bmatrix}^T$ forms an orthogonal basis for the stable deflating subspace, corresponding to the eigenvalues λ_i of the pencil with the property $|\lambda_i| < 1$.

Another application is the solution of the continuous-time and discrete-time Riccati equations without any matrix inversions, hence with possibly singular matrix R. To this aim, one considers the pencils below, of order $2n + m$, associated with the equations (3.1) and (3.2), respectively,

$$\begin{bmatrix} A & 0 & B \\ -Q & -A^T & 0 \\ 0 & B^T & R \end{bmatrix} - \lambda \begin{bmatrix} I_n & 0 & 0 \\ 0 & I_n & 0 \\ 0 & 0 & 0 \end{bmatrix}, \tag{3.45}$$

$$\begin{bmatrix} A & 0 & B \\ -Q & I_n & 0 \\ 0 & 0 & R \end{bmatrix} - \lambda \begin{bmatrix} I_n & 0 & 0 \\ 0 & A^T & 0 \\ 0 & B^T & 0 \end{bmatrix}. \tag{3.46}$$

Obviously, these pencils are specializations of the pencils implied by (3.28) and (3.29) in § 3.1.2. Note that no matrix inversions appear. For both cases, if $\operatorname{rank}(\begin{bmatrix} B^T & 0 & R^T \end{bmatrix}^T) = m$, a regular generalized eigenvalue problem of order $2n$ can be obtained by applying Householder transformations on rows, to annihilate the submatrix B in the first matrix of each pencil, while maintaining the zeros in the next n rows. The solution of each Riccati equation is then determined using a basis for the stable deflating subspace associated with the stable generalized eigenvalues of the corresponding reduced problem, as shown above. The technique is also useful for the deadbeat control problem (control in a minimum number of steps) of a discrete-time dynamical system (Emami-Naeini and Franklin, 1980, 1982), where a Riccati equation with $R = 0$ is solved.

Example 3.5 Consider the discrete-time algebraic Riccati equation, where

$$A = \begin{bmatrix} 0.5 & 1 & 0 & 0 \\ 0 & 0 & 1 & 0 \\ 0 & 0 & 0 & 1 \\ 0 & 0 & 0 & 0 \end{bmatrix}, \quad B = \begin{bmatrix} 0 \\ 0 \\ 0 \\ 1 \end{bmatrix}, \quad Q = \begin{bmatrix} 1 & 0 & 0 & 0 \\ 0 & 0 & 0 & 0 \\ 0 & 0 & 0 & 0 \\ 0 & 0 & 0 & 0 \end{bmatrix},$$

and $R = 0.25$. Clearly, matrix A is singular. The exact generalized eigenvalues of the pencil $\widehat{A} - \lambda \widehat{B}$ in (3.44) are

$$\lambda_1 = \lambda_2 = \lambda_3 = 0, \ \lambda_{4,5} = (21 \pm 5 \times (17)^{1/2})/4, \ \lambda_6 = \lambda_7 = \lambda_8 = \infty,$$

and an exact orthogonal basis for the deflating subspace associated with the stable eigenvalues λ_1 to λ_4 is known (Pappas, Laub and Sandell, 1980). The numerical solution computed in double precision based on the *inverse-free* approach described above, is given below (rounded to 13 decimal digits),

$$X = \begin{bmatrix} 1.3289138320938 & .6578276641876 & .3156553283752 & .1313106567503 \\ .6578276641876 & 1.3156553283752 & .6313106567503 & .2626213135007 \\ .3156553283752 & .6313106567503 & 1.2626213135007 & .5252426270014 \\ .1313106567503 & .2626213135007 & .5252426270014 & 1.0504852540028 \end{bmatrix}.$$

The estimate of the reciprocal condition number of U_{11} (using the 1-norm) is 0.344579. The computed solution has a residual 1-norm of about 1.86×10^{-15}. The stable generalized eigenvalues are given by the ratios α_i/β_i, $i = 1{:}4$, where

$$\alpha = \{ .101770532788603, 0, 0, 0 \},$$
$$\beta = \{ 1.058808617468660, 1.006757213621159, .966352881862065, .851109746843553 \},$$

so that, their ratios are $\lambda = \{ .096117967977924, 0, 0, 0 \}$. The gain matrix of the optimal regulator is

$$H = \begin{bmatrix} -.0504852540028 & -.1009705080055 & -.2019410160110 & -.4038820320221 \end{bmatrix}.$$

The solution X coincides with that computed using the pencil $\widehat{A} - \lambda \widehat{B}$ in (3.44), reported in (Pappas, Laub and Sandell, 1980), where ten decimal digits are given. □

The approaches presented above are advisable not only when the matrices A and/or R are singular, but also when they are nearly singular, because the numerical difficulties caused by ill conditioning can thus be avoided. This topic will be further addressed in Section 3.4, where specific algorithms for solving Riccati equations will be described. An example with a singular R matrix will also be given.

3.3 Schur-Based Riccati Solvers

The Schur vectors approach presented in § 3.1.1 will be now translated into algorithms to solve both continuous-time and discrete-time algebraic matrix Riccati equations. Efficiency improvement for discrete-time case, and the recently proposed scaling strategy for continuous-time case are considered. Computational issues, as well as the scope of the Schur-based Riccati solvers are also discussed.

3.3.1 Algorithmic templates

Based on the results in § 3.1.1, the following algorithm is obtained.

Algorithm Schv : $\boxed{X = \texttt{Schv}(\ A, B, Q, R, \texttt{sys}\ )}$

/* Solve Riccati equations (3.1) (if $\texttt{sys} = \texttt{c}$) or (3.2) (if $\texttt{sys} = \texttt{d}$), */
/* using Schur vectors method. */

/* 1. Construct the matrix G from (3.5) or (3.6), respectively. */
$F = BR^{-1}B^T$;
IF $\texttt{sys} = \texttt{c}$, THEN

$$G = \begin{bmatrix} A & -F \\ -Q & -A^T \end{bmatrix};$$

ELSE

$$G = \begin{bmatrix} A + FA^{-T}Q & -FA^{-T} \\ -A^{-T}Q & A^{-T} \end{bmatrix};$$

END IF

/* 2. Reduce G to RSF, accumulating the transformations in U. */
$(G, U) = \texttt{Hess}(\ G\ )$; $(G, U, \Lambda) = \texttt{mSchur}(\ G, U\ )$;

/* 3. Order RSF so that $\lambda(S_{11}) \subset \mathbb{C}^-$ (see (3.9)), */
/* accumulating the transformations in U. */

FOR $i = 1{:}\,2n$,
  IF $\texttt{sys} = \texttt{c}$, THEN
    $select_i = \Re(\lambda_i) < 0$;
  ELSE

$select_i = |\lambda_i| < 1;$
END IF
END FOR
$(G, U) = $ `Schord`$(G, select, U);$

/* 4. Solve matrix equation $U_{11}^T X = U_{21}^T$ using the LU factorization. */
$X = $ `solve`$(u_{1:n,1:n}^T, u_{n+1:2n,1:n}^T);$

Note that the LU factorization at stage 4 can be replaced, for instance, by QR factorization. As a preprocessing option, Algorithm `balanc` (§ 2.2.2) can also be called before calling `Hess`, to detect isolated eigenvalues and equilibrate the row and column norms of G. Algorithm `Schv` may be used for solving both equation (3.1) and equation (3.2). The treatment of the continuous-time and discrete-time cases differs only at stages 1 and 3. It is necessary to order RSF so that $\lambda(S_{11})$ contains the eigenvalues with negative real parts, in the continuous-time case, and those with moduli less than 1, in the discrete-time case.

Since the QR algorithm has the natural tendency to delivering eigenvalues in the decreasing order of their moduli along the main diagonal, the situation when $\lambda(S_{11})$ contains the eigenvalues with moduli greater than 1 would rather be in the discrete-time case. In this way, the effort required to the RSF ordering would generally be smaller than otherwise. For this reason, the following algorithm is implemented in our routines for the discrete-time case with a nonsingular matrix A.

Algorithm Schvd : $X = $ `Schvd`(A, B, Q, R)

/* Solve Riccati equation (3.2) using Schur vectors method, */
/* selecting the unstable part. */

/* 1. Construct the matrix $\bar{G} = G^{-1}$ from (3.8). */

$$F = BR^{-1}B^T; \quad G = \begin{bmatrix} A^{-1} & A^{-1}F \\ QA^{-1} & A^T + QA^{-1}F \end{bmatrix};$$

/* 2. Reduce $\bar{G}$ to RSF, accumulating the transformations in U. */
$(G, U) = $ `Hess`$(G); \quad (G, U, \Lambda) = $ `mSchur`$(G, U);$

/* 3. Order RSF so that $\lambda(U_{11}) \subset \mathbb{C}^+$, */
/* accumulating the transformations in U. */

FOR $i = 1{:}\,2n$,
$select_i = |\lambda_i| \geq 1;$
$(G, U) = $ `Schord`$(G, select, U);$

/* 4. Solve matrix equation $U_{11}^T X = U_{21}^T$ using the LU factorization. */
$X = $ `solve`$(u_{1:n,1:n}^T, u_{n+1:2n,1:n}^T);$

Algorithm `Schvd` is justified by noting that a transformation matrix U that realizes $\lambda((U^T G^{-1} U)_{11}) \subset \mathbb{C}^+$, also accomplishes $\lambda((U^T G U)_{11}) \subset \mathbb{C}^-$.

Algorithms `Schv` and `Schvd` are conceptually very simple and easily implementable. Numerical difficulties can appear when matrix U_{11} at stage 4 is ill conditioned. In such a case, the computed result X can be inaccurate. Ill-scaled Riccati equations can also cause inaccuracy.

The following example, proposed in (Petkov, Christov and Konstantinov, 1987), also studied in (Kenney, Laub and Wette, 1989), illustrates the influence of poor scaling.

Example 3.6 Let $v = (1,1,1)^T$, $V = I_3 - \frac{2}{3}vv^T$, $A_0 = \lambda\,\mathrm{diag}(1,2,3)$, $Q_0 = \mathrm{diag}(1/\lambda, 1, \lambda)$, $A = VA_0V$, $Q = VQ_0V$, $B = I_3$, $R = \lambda I_3$. The exact solution to the Riccati equation (3.1) is given by $X = VX_0V$, where $X_0 = \mathrm{diag}(x_1, x_2, x_3)$, and

$$x_1 = \lambda^2 + \sqrt{\lambda^4 + 1}, \quad x_2 = 2\lambda^2 + \sqrt{4\lambda^4 + \lambda}, \quad x_3 = 3\lambda^2 + \sqrt{9\lambda^4 + \lambda^2}.$$

The underlying Riccati equation is well conditioned (its condition number reported in the second cited reference being about 2). For $\lambda = 10^6$ the problem is ill-scaled. The closed-loop poles are: -3×10^6, -2×10^6, -10^6. The solution computed in MATLAB by the analytic formulas above has an absolute residual of size $0(10^4)$; the relative residual is of the order of 10^{-9}. (Note that $\|A^T X + XA\|$ and $\|XFX\|$ are $0(10^{19})$, while $\|Q\|$ is $0(10^6)$.) If Algorithm `Schv` is used in double precision on a computer with $\varepsilon_M \approx 2.22 \times 10^{-16}$, the computed solution has only 3 digits of accuracy in its largest entry. Indeed, the computed result was

$$\widehat{X} = \begin{bmatrix} 4.6678595 \times 10^{12} & 1.3336996 \times 10^{12} & -5.4400180 \times 10^{08} \\ 1.3336996 \times 10^{12} & 3.9996718 \times 10^{12} & -1.3350550 \times 10^{12} \\ -5.4400180 \times 10^{08} & -1.3350550 \times 10^{12} & 3.3343031 \times 10^{12} \end{bmatrix}.$$

The relative error, $\|X - \widehat{X}\|_2 / \|X\|_2$, is about 4×10^{-4}. Note that matrix U_{11} is well conditioned, because $1/\mathrm{cond}_1(U_{11}) \approx 0.26$. Balancing improved the relative error to about 2×10^{-4}. The generalized Schur vectors algorithm (Algorithm `gSchv` in § 3.4.1) gave a slightly better result

$$\widehat{X} = \begin{bmatrix} 4.6666370 \times 10^{12} & 1.3330036 \times 10^{12} & 6.3160624 \times 10^{08} \\ 1.3330036 \times 10^{12} & 3.9998759 \times 10^{12} & -1.3332587 \times 10^{12} \\ 6.3160624 \times 10^{08} & -1.3332587 \times 10^{12} & 3.3333723 \times 10^{12} \end{bmatrix},$$

with a relative error of about 10^{-4}. Three iterations of the defect correction procedure `RicCorr` in § 1.5.4 (totalizing six Newton iterations, for $\varepsilon = 10^{-6}$), initialized by the approximate solution computed by Algorithm `Schv`, produced a fully accurate result (rounded below to 8 significant digits):

$$\bar{X} = \begin{bmatrix} 4.6666667 \times 10^{12} & 1.3333333 \times 10^{12} & -3.6951358 \times 10^{-2} \\ 1.3333333 \times 10^{12} & 4.0000000 \times 10^{12} & -1.3333333 \times 10^{12} \\ -3.6951358 \times 10^{-2} & -1.3333333 \times 10^{12} & 3.3333333 \times 10^{12} \end{bmatrix}.$$

The relative error of $\bar{X}$ is of size $0(10^{-16})$. The problem can also be solved using the stability-enhancing scaling procedure below. The effect of varying the scaling factor is described in (Kenney, Laub and Wette, 1989). □

Let us discuss the *stability-enhancing scaling procedure* proposed in (Kenney, Laub and Wette, 1989) for continuous-time Riccati equations.[9] The presentation will also highlight several related results of independent interest.

Let $U = U_{ij} \in \mathbb{R}^{n \times n}$, $i, j = 1{:}2$, be a matrix that reduces G in (3.5) to its RSF S in (3.9). The solution is given by $X = U_{21}U_{11}^{-1} = -U_{12}U_{22}^{-1}$ (if U_{22} is nonsingular). Indeed, if equating the $(1,1)$ and $(2,1)$ blocks in $U^TGU = S$, it follows that

$$\begin{aligned} U_{11}^TAU_{11} - U_{21}^TQU_{11} - U_{11}^TFU_{21} - U_{21}^TA^TU_{21} &= S_{11}, \qquad (3.47) \\ U_{12}^TAU_{11} - U_{22}^TQU_{11} - U_{12}^TFU_{21} - U_{22}^TA^TU_{21} &= 0; \end{aligned}$$

while premultiplying the second equality by U_{22}^{-T} and postmultiplying by U_{11}^{-1}, with $Y := U_{22}^{-T}U_{12}^T$, we obtain

$$YA - Q - YFX - A^TX = 0,$$

hence, comparing with (3.1), we get $Y = -X$, or $X = -U_{22}^{-T}U_{12}^T = -U_{12}U_{22}^{-1}$, because $X = X^T$. As $U_{21}U_{11}^{-1} = U_{11}^{-T}U_{21}^T$, it follows that $U_{11}^TU_{21} = U_{21}^TU_{11}$. Thus, simple calculations prove that the matrix

$$\widetilde{U} := \begin{bmatrix} U_{11} & -U_{21} \\ U_{21} & U_{11} \end{bmatrix}, \qquad (3.48)$$

is orthogonal and symplectic (as $\widetilde{U}^TJ\widetilde{U} = J$). Consequently (Paige and Van Loan, 1981), there exist orthogonal matrices, W and V, at once performing singular value decomposition (SVD) of U_{11} and U_{21},

$$U_{11} = W\Sigma V^T, \qquad U_{21} = W\Delta V^T, \qquad (3.49)$$

where $\Sigma = \mathrm{diag}(\sigma_1, \ldots, \sigma_n)$, $\Delta = \mathrm{diag}(\delta_1, \ldots, \delta_n)$, $0 < \sigma_1 \le \cdots \le \sigma_n$, and $\sigma_i^2 + \delta_i^2 = 1$. The latter equality derives from

$$I_n = U_{11}^TU_{11} + U_{21}^TU_{21} = V(\Sigma^2 + \Delta^2)V^T.$$

[9] For discrete-time Riccati equations, it is not yet clear how the problem could be scaled.

Moreover, $\widetilde{U}$ transforms G into the so-called "Hamiltonian-Schur" form,

$$\widetilde{U}^T G \widetilde{U} = \widetilde{S} = \begin{bmatrix} S_{11} & \widetilde{S}_{12} \\ 0 & S_{22} \end{bmatrix}, \tag{3.50}$$

where $\lambda(S_{11}) \subset \mathbb{C}^-$, S_{11} and S_{22} are the same as the matrices in (3.9), and $\widetilde{S}_{12}$ is symmetric. Indeed, (3.50) results by direct calculations, using (3.47); for instance, if the computed (2,1) block of $\widetilde{U}^T G \widetilde{U}$ is premultiplied by U_{11}^{-T} and postmultiplied by U_{11}^{-1}, with $X = U_{21}U_{11}^{-1}$, we get $-XA - Q + XFX - A^T X$, which is zero because X is a solution of Riccati equation. The symmetry of $\widetilde{S}_{12}$ is proven by noting that

$$(-U_{11}^T A U_{21} - U_{21}^T A^T U_{11})^T = -U_{21}^T A^T U_{11} - U_{11}^T A U_{21},$$

and using the symmetry of Q and F. Note that (3.50), with $\widetilde{U}$ in (3.48), can be deduced from (3.9), when G is dichotomic, without resorting to the solution X of algebraic Riccati equation (3.1). Proofs are given in (Mehrmann, 1991; Paige and Van Loan, 1981). An analogous result exists for discrete-time case.

From (3.49), it follows that $X = W\text{diag}(\delta_1/\sigma_1, \ldots, \delta_n/\sigma_n)W^T$. Therefore, if σ_1 is small, then large inaccuracies can occur. The idea of the scaling procedure lies in re-writing the Riccati equation (3.1) so that the corresponding solution should have a norm less than or equal to 1. This makes sure that matrix U_{11} has a small condition number, according to the following lemma, adapted from (Kenney, Laub and Wette, 1989):

Lemma 3.3 *For an orthogonal matrix U transforming G as in (3.9), where $\lambda(S_{11}) \subset \mathbb{C}^-$, and for $X = U_{21}U_{11}^{-1}$, it follows that $\|U_{11}^{-1}\|_2 \leq 1 + \|X\|_2^2$, and therefore,*

$$\text{cond}(U_{11}) := \|U_{11}\|_2 \|U_{11}^{-1}\|_2 \leq 1 + \|X\|_2^2.$$

Indeed, $U_{11}^T U_{11} + U_{21}^T U_{21} = I_n$ implies $U_{11}^T U_{11} \leq I_n$, hence $\|U_{11}\|_2 \leq 1$. In addition,

$$\begin{aligned} I_n &= U_{11}^T U_{11} + U_{11}^T (U_{21}U_{11}^{-1})^T U_{21} U_{11}^{-1} U_{11} \\ &= U_{11}^T U_{11} + U_{11}^T X^2 U_{11} = U_{11}^T (I_n + X^2) U_{11}. \end{aligned}$$

This yields $\|U_{11}^{-1}\|_2 \leq \|U_{11}\|_2 (1 + \|X\|_2^2)$, which implies the stated result.

Note that the upper bound on $\text{cond}(U_{11})$ can be very pessimistic (see Example 3.6). The proof indicates that if $\|X\|_2$ is small, then U_{11} is nearly orthogonal. So, it would be better to transform the problem so that the corresponding solution X has a small norm. This idea is exploited by the procedure below.

Procedure SclSch

/* 1. [Scale the matrices A and B.] */

$$\widetilde{A} = \rho A, \qquad \widetilde{B} = \rho B, \qquad \text{with } \rho \geq \|X\|_2. \tag{3.51}$$

/* 2. [Solve the transformed Riccati equation.] Solve in $\widetilde{X}$ */

$$\widetilde{A}^T \widetilde{X} + \widetilde{X}\widetilde{A} - \widetilde{X}\widetilde{B}R^{-1}\widetilde{B}^T \widetilde{X} + Q = 0. \tag{3.52}$$

/* 3. [Unscale the solution X.] */

$$X = \rho \widetilde{X}. \tag{3.53}$$

The equation (3.52) can be solved by Algorithm `Schv`. Let $\widetilde{G}$ be the transformed Hamiltonian matrix, associated with the optimal problem defined by $\widetilde{A}$, $\widetilde{B}$, Q, and R, and let $\widetilde{U}$ be an orthogonal matrix transforming $\widetilde{G}$ into an ordered RSF, as in (3.9). Then, by Lema 3.3, $\|\widetilde{U}_{11}^{-1}\|_2 \leq 1 + \|\widetilde{X}\|_2^2 \leq 2$, because $\rho \geq \|X\|_2$, and $X = \rho\widetilde{X}$. Therefore, $\text{cond}(\widetilde{U}_{11}) \leq 2$, and the inversion step for $\widetilde{X}$ (step 4 in Algorithm `Schv`) is well conditioned.

It can be proven that the condition number of the Riccati equation is invariant with this type of scaling (see Lemma 2 in (Kenney, Laub and Wette, 1989)). Thus, Procedure `SclSch` does not alter conditioning. If we apply this procedure to Example 3.6, with $\rho = \|X\|_2 = 6 \times 10^{12}$, then the computed result will be fully accurate, the relative error being of size $0(10^{-15})$.

Procedure `SclSch` has a main drawback: it requires a good estimate of the 2-norm of the unknown solution X. Indeed, if ρ is chosen too large compared to $\|X\|_2$, then the scaled Schur vectors method will lose accuracy, due to truncation errors during calculations. The optimal scaling factor is $\rho = \|X\|_2$, as proven in the cited reference. Overscaling can as much cause loss of accuracy as underscaling. It is advisable to apply a two-step approach: first solve the Riccati equation using Algorithm `Schv`, thus obtaining a tight estimate of $\|X\|_2$; provided the estimate is large, then use it in the scaling procedure in order to increase accuracy, or apply the defect correction procedure in § 1.5.4.

Let us mark by the "hat" sign the computed matrices. Based on the error and perturbation bounds for invariant/deflating subspaces, developed in (Stewart, 1973b), a relative error bound for Riccati equation solution has been obtained by Kenney, Laub and Wette (1989). They proved that if X (or scaled X) has unit 2-norm, i.e., using (3.49), $\sigma_1 = \delta_1 = 1/\sqrt{2}$, and $k \leq 1/4$, with $k := \|\widehat{\Omega}^{-1}(\widehat{S}_{21})\|_2 \, \|\widehat{\Omega}^{-1}\|_2 \, \|\widehat{S}_{12}\|_2$, and $\widehat{\Omega}(Z) := Z\widehat{S}_{11} - \widehat{S}_{22}Z$, then

$$\|\widehat{X} - X\|_2 / \|X\|_2 \leq \eta_1 + \|\widehat{R}\|_2 \, \|\widehat{U}_{11}^{-1}\|_2, \tag{3.54}$$

where

$$\eta_1 := \frac{2\sqrt{2}\eta_0}{1-\sqrt{2}\eta_0}, \quad \eta_0 := \sqrt{2}\left(\frac{1+\sqrt{1-4k}}{1-2k+\sqrt{1-4k}}\right) \|\widehat{\Omega}^{-1}(\widehat{S}_{21})\|_2,$$

and $\widehat{R} := \widehat{X}\widehat{U}_{11} - \widehat{U}_{21}$ is the residual error of the computed solution $\widehat{X}$.

The results allow us to determine if the first step of Newton's process, initialized by the approximate solution $\widehat{X}$, can improve the accuracy of the computed solution. Specifically, this happens if

$$\|X - \widehat{X}\|_2 \le r := 1/(3\|F\|_2\|\Omega^{-1}\|_2),$$

where $F := BR^{-1}B^T$, and Ω is the Lyapunov operator used in § 1.4.2, $\Omega(Z) = (A - FX)^T Z + Z(A - FX)$. If $\|X - \widehat{X}\|_2 \le \frac{3}{2}r$, then $A - F\widehat{X}$ is stable, and the conditions $\|X - \widehat{X}\|_2 < r$ and $4\|F\|_2\,\|\Omega^{-1}\|_2^2\,\|Y\|_2 < 1$ guarantee that the residual can bound the initial error,

$$\|X - \widehat{X}\|_2 \le 2\|\Omega^{-1}\|_2\,\|Y\|_2, \tag{3.55}$$

with $Y := A^T\widehat{X} + \widehat{X}A - \widehat{X}F\widehat{X} + Q$.

The cited reference proposes an efficient technique for estimating the condition number of the Riccati equation (3.1), expressed by

$$\kappa(A, Q, F) = \|D\|_2/\|X\|_F, \tag{3.56}$$

where $D = \begin{bmatrix} D_1^T & D_2^T & D_3^T \end{bmatrix}^T$, and, using Kronecker's product,

$$\begin{aligned} D_1 &= \|Q\|_F(A_c^T \otimes I_n + I_n \otimes A_c^T)^{-1}, \\ D_2 &= \|A\|_F(A_c^T \otimes I_n + I_n \otimes A_c^T)^{-1}(I_n \otimes X + (X \otimes I_n)S), \\ D_3 &= \|F\|_F(A_c^T \otimes I_n + I_n \otimes A_c^T)^{-1}(X \otimes X), \end{aligned}$$

with $A_c := A - FX$, and S a permutation matrix relating $\mathrm{vec}(M^T)$ and $\mathrm{vec}(M)$ (for any matrix M), $\mathrm{vec}(M^T) = S\,\mathrm{vec}(M)$ (see § 2.3.1); matrix S depends on the dimension of M. The norm of D can be estimated without forming D, an $n^2 \times 3n^2$ matrix, using an inverse power method for such problems, developed in (Byers, 1984a; Cline et al., 1979; Kenney and Laub, 1989). An initial matrix triple $(\Delta A_0, \Delta Q_0, \Delta F_0)$ of unit Frobenius norm is formed. Denote

$$Q' = \|Q\|_F\Delta Q_0 + \|A\|_F(\Delta A_0^T X + X\Delta A_0) - \|F\|_F X\Delta F_0 X,$$

and solve for W_0, and then for W_1, the Lyapunov equations

$$A_c^T W_0 + W_0 A_c = -Q', \qquad A_c W_1 + W_1 A_c^T = -W_0. \tag{3.57}$$

Then, $||D||_2 \approx ||(\Delta A_1, \Delta Q_1, \Delta F_1)||_F^{1/2}$, where

$$\Delta A_1 := 2||A||_F X W_1, \quad \Delta Q_1 := ||Q||_F W_1, \quad \Delta F_1 := 2||F||_F X W_1 X.$$

Better estimates for $||D||_2$ can be computed by replacing ΔA_0, ΔQ_0, and ΔF_0 by normalized ΔA_1, ΔQ_1, and ΔF_1, respectively, and by restarting the calculations. If $\Delta A_0 = \Delta Q_0 = \Delta F_0 = I_n/\sqrt{3n}$, then one or two steps of the procedure just described will give tight estimates for $||D||_2$ and $\kappa(A, Q, F)$.

It is worth mentioning that the Lyapunov equations in (3.57) can directly be transformed into reduced Lyapunov equations, having their coefficient matrices in RSF. Indeed, from (3.50), (3.47), and (3.1), we have

$$\begin{aligned} S_{11} &= U_{11}^T(A - FX)U_{11} - U_{21}^T(Q + A^T X)U_{11} \\ &= U_{11}^T(A - FX)U_{11} + U_{21}^T X(A - FX)U_{11} \\ &= (U_{11}^T + U_{21}^T X)A_c U_{11} = U_{11}^{-1} A_c U_{11}. \end{aligned}$$

Thus, the solution to the first equation (3.57) is $W_0 = U_{11}^{-T} V U_{11}^{-1}$, where V solves the reduced Lyapunov equation $S_{11}^T V + V S_{11} = -U_{11}^T Q' U_{11}$, with S_{11} in RSF. Similarly, for the second equation (3.57), one solves $S_{11} Z + Z S_{11}^T = -U_{11}^{-1} W_0 U_{11}^{-T}$, and then $W_1 = U_{11} Z U_{11}^T$. So, Algorithm **Lyac** (§ 2.3.2) or Algorithm **LyacC** (§ 2.3.3) can be used.

Defining $\Omega_M(Z) := M^T Z + ZM$, it can be proven that

$$||\Omega_{S_{11}}^{-1}||_2/\mathrm{cond}^2(U_{11}) \le ||\Omega_{A_c}^{-1}||_2 \le \mathrm{cond}^2(U_{11})||\Omega_{S_{11}}^{-1}||_2.$$

This shows the relation between the condition numbers of the first Lyapunov equation in (3.57) and the transformed equation.

Other results concerning perturbation analysis for Riccati equations and related topics can be found, for instance, in (Arnold and Laub, 1984; Byers, 1984b; Kenney and Laub, 1989; Laub, 1980; Laub et al., 1987). In general, computational problems which are *ill-possed*, i.e., for which small perturbations destroy the essential structural properties (like stabilizability or detectability), are often ill conditioned (Demmel, 1987).

3.3.2 Computational issues

The bulk of computation in Algorithms **Schv** and **Schvd** takes place at stages 2 and 3; the QR-based tools described in §§ 2.2.2–2.2.4 are used. The construction of the matrices G and G^{-1} does not involve algorithmic subtleties. Level 3 BLAS operations (see (1.105) in § 1.4.1) can be implemented. The computation of matrix F in (3.5) and (3.8), $F = BR^{-1}B^T$, is performed using the Cholesky factorization of the matrix R, $R = \bar{R}^T \bar{R}$, with $\bar{R}$ upper triangular. The solution of the system $X\bar{R} = B$ is obtained and then one determines $F = XX^T$, using a symmetric rank m update. The LU factorization is used in discrete-time case to obtain the inverse of the matrix A in (3.6) or (3.8). (Note that A^T should *a priori* be saved in $g_{n+1:2n,n+1:2n}$, because A will no longer be available.) The submatrix $g_{1:n,n+1:2n} = A^{-1}F$ can be calculated either by solving systems of linear algebraic equations, based on the factorization of A, or by a matrix product, after A^{-1} has been found and coppied into $g_{1:n,1:n}$. The first approach is to be preferred from a numerical point of view, because it guarantees a greater accuracy. After the submatrix $g_{1:n,1:2n}$ has been constructed, the rest of the matrix $\bar{G} := G^{-1}$ is obtained using two matrix-matrix operations, namely:

$$g_{n+1:2n,1:n} = Qg_{1:n,1:n}; \quad g_{n+1:2n,n+1:2n} \leftarrow Qg_{1:n,n+1:2n} + g_{n+1:2n,n+1:2n}.$$

Algorithms `Schv` and `Schvd` may be completed with the determination of the optimal gain matrix H, defined by the formulas (3.3) or (3.4), respectively. The Cholesky factorization is also used. Some particular results obtained at stage 1 of the Algorithm `Schv` may be exploited in the continuous-time case. A similar technique applies to the discrete-time case. Note that, in order to reduce the memory requirements, usual implementations construct G (or G^{-1}) over some other matrix, for instance, A; similarly, matrix U, computed at stages 2 and 3, can overwrite Q. Therefore, in the discrete-time case matrix A should be saved into an auxiliary array, prior to call `Schvd`. Note also that `Schv` (or `Schvd`) can modify the upper triangle of the matrix R.

At the same time, Algorithms `Schv` and `Schvd` enable the eigenvalues of the closed-loop system be obtained, based on Theorem 3.3.

In practice, it is often useful to apply the balancing procedure (§ 2.2.2) to the matrix G or $\bar{G}$. Note that if balancing is used, the required back transformation of the Schur vectors should precede the RSF ordering. An indirect scaling of the optimal control problem can be accomplished just by a suitable selection of the matrices Q and R in the performance index (1.53) or (1.57) in § 1.1.2 (see (Laub, 1979)).

The continuous-time and discrete-time cases can be implemented in a single routine, by replacing the portions for discrete-time case in Algorithms `Schv` by the corresponding portions in Algorithm `Schvd`. The (near) singularity of matrix U_{11} should be controlled and the computations should be stopped if U_{11} is (numerically) singular.

The total cost is of about $97.3n^3 + 0.5n^2m + 0.5nm^2 + (1/6)m^3$ *flops*, in the continuous-time case, and with $4.3n^3$ *flops* in addition in the discrete-time case, provided that the reduction of an $n \times n$ matrix to RSF involves $10n^3$ *flops* (§ 2.2.4), and the RSF ordering and balancing require an extra effort of 15 % and 5 %, respectively. This effort is equivalent to seven or eight iterations of Newton's method (§ 2.5.1).

Algorithms **Schv** and **Schvd** require $8n^2 + m^2 + nm$ memory locations for data and results, in the continuous-time case, and with n additional locations in the discrete-time case. Therefore, in the usual situation with m much less than n, the memory requirements are almost twice as large as those corresponding to Newton's method. If matrix H is also computed, then the memory needs and the computing effort are somewhat larger.

Algorithms **Schv** and **Schvd** contain, in essence, only practically stable steps: reduction to and ordering of RSF, and the solution of the linear algebraic equations $U_{11}^T X = U_{21}^T$. One cannot state that the computational process, as a whole, is stable. It is obviously useful to have some means for assessing the condition of Riccati equations. The condition estimators briefly described in § 3.3.1 are relatively expensive, so cheaper, possibly less accurate estimators are desirable. Experimentally, it was observed (Laub, 1979; Sima, 1981, 1990) that an ill conditioning of the Riccati equations manifested by an ill conditioning of U_{11}. When cond(U_{11}) is about 10^p, then a loss of accuracy of about p decimal digits usually follows in X. Therefore, it is advisable to evaluate cond(U_{11}), using a condition estimator like that described in (Cline et al., 1979) for systems of linear algebraic equations. When cond(U_{11}) is very large, one could resort to specialized, sophisticated condition estimators for Riccati equations.

A drawback of the Schur vectors approach is that it does not guarantee the symmetry of the computed solution X. The deviation from symmetry is more radical with ill-conditioned Riccati equations. In addition, the residual of solution can be large. As Example 3.6 shows, large residuals and inaccuracies can also appear for well-conditioned, but ill-scaled Riccati equations. The scaling procedure in § 3.3.1, or the defect correction procedure in § 1.5.4 may be used to improve accuracy.

Another difficulty can be met with in the discrete-time case. If matrix A is almost singular, then matrix $\bar{G}$ (or G) will not be accurately computed and hence all computations will be compromised even from the first stage of the algorithm. Therefore, it is important to estimate the condition number of matrix A with respect to inversion, and use alternate algorithms if cond(A) is too large.

3.3.3 Applicability and limitations

To use Algorithms **Schv** and **Schvd**, it is necessary that $R > 0$. In the discrete-time case, matrix A must be nonsingular. In addition, if the estimate $1/\text{cond}(A)$ is small (that is if A is ill conditioned), then the solution of the Riccati equation (3.2), computed by **Schvd**, can be largely inaccurate; in this case, it is advisable to call **Ntnd** (§ 2.5.1) or **gSchvQ** (§ 3.4.1), instead of **Schvd**.

Algorithm `Schv(d)` can be used if the pair (A, B) is stabilizable and if matrix G (or $\bar{G}$) is dichotomic. (A sufficient, but not necessary condition for dichotomy (see § 1.1.2 and § 2.5.3) is that the pair $(Q^{1/2}, A)$ is detectable.) If such conditions do not hold, the execution of the corresponding routine will usually terminate with an error return, due to failing to separate n stable eigenvalues.

Generally, the Schur vectors method, based on `Schv(d)` is more efficient than Newton's method, or than the generalized Schur vectors method. In our previous experiments on the Romanian-made **FELIX C–256** computer (Sima, 1981; see also Sima, 1990) the computing times for the first two methods were comparable for most examples; sometimes (for 7 examples out of 33), the first method was two to four times faster. In fact, great values of the timing ratios were obtained for examples seldom occurring in practice, for which $m = n$. Moreover, on an IBM PC compatible computer, the values of the timing ratios for the first two methods were less (sometimes even twice less) than those obtained on **FELIX** computers. Using the values in § 2.5.2 of the stabilizing parameters α and β, the superiority of the first method is even less obvious, since the number of iterations of Newton's process diminishes.

Algorithm `Schv(d)` does not guarantee that the computed matrix is a symmetric positive semi-definite solution, although such properties are theoretically provided. The deviation from symmetry is more pronounced when the estimate of the number $1/\text{cond}(U_{11})$ is small. Although matrix U is orthogonal, the submatrix U_{11} may be ill conditioned. Experimentally, this fact was observed for ill-conditioned problems, such as almost unstabilizable problems, or for discrete-time problems having $\text{cond}(A)$ large (Sima, 1981). In the tests performed, we computed the norms of both the residual Y, defined by (1.130) or (1.131) in § 1.5.4, and the deviation from symmetry, $N := \widehat{X} - \widehat{X}^T$, where $\widehat{X}$ is the computed solution. It was noted that $\|N\|_\infty$ was large for the examples for which $\text{cond}(U_{11})$ was large, and, for such examples, $\|Y\|_\infty$ was 10^2 to 10^5 times greater than for Newton's method. (In addition, $\|Y\|_\infty$ reduces in the latter case if a smaller tolerance ε in the termination criterion for Newton's process is used.) Clearly, a smaller value for $\|Y\|_\infty$ does not generally indicate better accuracy, except for well-conditioned problems. The deviation N can be zeroed by replacing $\widehat{X}$ by its Frobenius projection onto the space of symmetric matrices, $(\widehat{X} + \widehat{X}^T)/2$. This is called *symmetrization* of the computed solution, and it should always be done.

The fact that Newton's method may ensure symmetry, positivity (if Hammarling's algorithm is employed, see § 2.3.3) and a small residual, and on the other hand, the Schur vectors method is usually more efficient, suggests that it would be profitable to use a hybrid method. The Schur vectors method can make a suitable initialization of Newton's method, that may, in two or three iterations, improve the numerical properties of the solution. The hybrid method was tested in (Sima, 1981) for three examples of practical significance. The norm of residual was improved by two to five significant decimal digits after two iterations of Newton's process.

Experimentally, it was observed (Sima, 1981, 1990) that better results could be obtained, especially for ill-conditioned examples, by using the option of balancing the matrix G (or $\bar{G}$) before its reduction to RSF. The first reference gives the solutions computed in both simple precision without and with balancing, and double precision for Example 6 in (Laub, 1979), with $n = 10$, $q_{11} = 10^4$. The solution computed in the first case was totally erroneous. Balancing sensibly improved the result; however, there were differences even in the first significant digit, compared to the solution obtained in double precision.

It should be remembered that the Schur vectors method can also help in obtaining other solutions of the algebraic matrix Riccati equations. For instance, equation (3.1) has a finite number of solutions if the characteristic polynomial and the minimal polynomial of $A-FX$ coincide, where X is the unique symmetric positive semi-definite solution. In this case, the number of solutions equals the number of combinations of n elements taken from $2n$ elements, corresponding to the possible selections of n eigenvalues from the $2n$ eigenvalues of G. All these solutions can be generated by a suitable ordering in RSF of G.

Besides, the method can be used for solving non-symmetric quadratic equations, as for example,

$$BX + XA - XFX + Q = 0, \tag{3.58}$$

where $B \in \mathbb{R}^{m\times m}$, $A \in \mathbb{R}^{n\times n}$, $F \in \mathbb{R}^{n\times m}$, and $Q, X \in \mathbb{R}^{m\times n}$. In this case, one should replace the matrix G at stage 1 of Algorithm `Schv` by

$$G = \begin{bmatrix} A & -F \\ -Q & -B \end{bmatrix}.$$

The various solutions are obtained by generating the combinations corresponding to n eigenvalues from the diagonal of a RSF of G. The associated Schur vectors deliver the solution $X = U_{21}U_{11}^{-1}$, where $U_{11} \in \mathbb{R}^{n\times n}$, $U_{21} \in \mathbb{R}^{m\times n}$. The "discrete-time" case can be approached similarly.

By definition, Algorithm `Schv(d)` does not work for optimal problems with a non-dichotomic matrix G (or $\bar{G}$). So, numerical difficulties are to be expected when the separation between the stable and unstable subspaces is small.

Example 3.7 Consider the matrix G in Example 2.1 (§ 2.2.1), with A, F, and Q defined in (Ammar, Benner and Mehrmann, 1993; Arnold and Laub, 1984). For $\varepsilon = 10^{-10}$, $\varepsilon = 10^{-11}$, and $\varepsilon = 10^{-13}$, Algorithm `Schv` could not separate n stable eigenvalues, and for $\varepsilon = 10^{-8}$, $\varepsilon = 10^{-14}$, and $\varepsilon = 10^{-15}$, it returned two unstable eigenvalues as closed-loop poles. For $\varepsilon = 10^{-12}$, the algorithm apparently succeeded to compute an accurate result:

$$\widehat{X} = \begin{bmatrix} .4379 & .0000 & .5621 & .0000 \\ .0000 & .4379 & .0000 & .5621 \\ .5621 & .0000 & .4379 & .0000 \\ .0000 & .5621 & .0000 & .4379 \end{bmatrix}.$$

The absolute and relative residual 1-norm were very close to the machine precision. The reciprocal condition number of U_{11} was `RCOND` $\approx .47$. The optimal gain matrix was $H = \begin{bmatrix} -1 & -1 & -1 & -1 \end{bmatrix}$, and the closed-loop poles were $\{-3.7321, -.2680, -2 \times 10^{-16} \pm \mathrm{j}\}$. These results seem to be good, but $\widehat{X}$ is not positive semi-definite, its eigenvalues being $\{-.1242, -.1242, 1, 1\}$. No indication about the numerical difficulties are given by the algorithm. A good solution is obtained for $\varepsilon = 10^{-9}$:

$$\widehat{X} = \begin{bmatrix} .5390 & .0000 & .4610 & .0000 \\ .0000 & .5390 & .0000 & .4610 \\ .4610 & .0000 & .5390 & .0000 \\ .0000 & .4610 & .0000 & .5390 \end{bmatrix}.$$

Good solutions have also been obtained for $\varepsilon \leq 10^{-7}$. □

Let us now present some examples of Schur vectors algorithms use. The first example is in fact a template algorithm useful for solving a frequently encountered control problem. Other examples report numerical results obtained for specific problems.

Example 3.8 The following algorithm computes the solution of the algebraic matrix Riccati equation (3.1), if `sys` = `c` (continuous-time), or (3.2), if `sys` = `d` (discrete-time), and the corresponding optimal gain matrix, using Schur vectors method. The system output is weighted with the matrix $\widetilde{Q}$ in the quadratic performance index, hence output regulation (Problem 1.3 in § 1.2.1) is accomplished. A proportional-plus-integral optimal control law (Problem 1.4) is used. It is assumed that the matrices A, B, C, $\widetilde{Q}$ and $\widehat{Q}$ in (1.66) in § 1.2.1, of dimensions $n \times n$, $n \times m$, $l \times n$, $l \times l$ and $l \times l$, respectively, as well as R, of dimension $m \times m$, are given.

Algorithm Sch : $(X, H) = \mathtt{Sch}(\ A, B, C, \widetilde{Q}, \widehat{Q}, R, \mathtt{sys}\ )$

/* Solve (3.1) or (3.2) for Problem 1.4 using Schur vectors method. */

/* 1. Construct matrices for proportional-plus-integral control. */
IF **sys** = **c**, THEN $P = 0_l$; ELSE $P = I_l$;

$$A \leftarrow \begin{bmatrix} A & 0 \\ -C & P \end{bmatrix}; \quad B \leftarrow \begin{bmatrix} B \\ 0 \end{bmatrix}; \quad C \leftarrow \begin{bmatrix} C & 0 \\ 0 & I_l \end{bmatrix}; \quad Q = \begin{bmatrix} \widetilde{Q} & 0 \\ 0 & \widehat{Q} \end{bmatrix};$$

/* 2. Save R. Save A in discrete-time case. */
$\widetilde{R} = R$; IF **sys** = **d**, THEN $\widetilde{A} = A$;
$Q \leftarrow C^T Q C$; /* Compute state weighting matrix. */

/* 3. Solve Riccati equation and compute optimal gain matrix. */
IF **sys** = **c**, THEN
  $X =$ **Schv**($A, B, Q, R,$ **sys**); $H = -\widetilde{R}^{-1} B^T X$;
ELSE
  $X =$ **Schvd**(A, B, Q, R); $H = -(\widetilde{R} + B^T X B)^{-1} B^T X \widetilde{A}$;
END IF

If matrix $\widehat{Q}$ is not given, then $l = 0$, and the construction at step 1 is not made. □

Example 3.9 An implementation of Algorithm **Sch** in Example 3.8 was used for the continuous-time turbogenerator model in (Hung and MacFarlane, 1982), with $n = 10$, $m = 2$, $l = 2$, defined by the matrices

$$A = \left[\begin{array}{rrrrr} 0 & 1 & 0 & 0 & 0 \\ 0 & -.11323 & -.98109 & -11.847 & -11.847 \\ 324.121 & -1.17550 & -29.101 & .12722 & 2.83448 \\ -127.300 & .46167 & 11.4294 & -1.0379 & 13.1237 \\ -186.050 & .67475 & 16.7045 & .86092 & -17.068 \\ 341.917 & 1.09173 & 1052.750 & 756.465 & 756.465 \\ -30.748 & -.09817 & -94.674 & -68.029 & -68.029 \\ -302.360 & -.96543 & -930.960 & -668.950 & -668.950 \\ 0 & 0 & 0 & 0 & 0 \\ 0 & 0 & 0 & 0 & 0 \end{array}\right.$$

$$\left.\begin{array}{rrrrr} 0 & 0 & 0 & 0 & 0 \\ -63.080 & -34.339 & -34.339 & -27.645 & 0 \\ -967.730 & -678.140 & -678.140 & 0 & -129.29 \\ 380.079 & 266.341 & 266.341 & 0 & 1054.85 \\ 555.502 & 389.268 & 389.268 & 0 & -874.92 \\ -29.774 & .16507 & 3.27626 & 0 & 0 \\ 2.67753 & -2.65580 & 4.88497 & 0 & 0 \\ 26.32920 & 2.42028 & -9.56030 & 0 & 0 \\ 0 & 0 & 0 & -1.66667 & 0 \\ 0 & 0 & 0 & 0 & -10 \end{array}\right],$$

$$B = \begin{bmatrix} 0 & 0 & 0 & 0 & 0 & 0 & 0 & 0 & 1.66667 & 0 \\ 0 & 0 & 0 & 0 & 0 & 0 & 0 & 0 & 0 & 10 \end{bmatrix}^T,$$

$$C = \begin{bmatrix} 1 & 0 & 0 & 0 & 0 & 0 & 0 & 0 & 0 & 0 \\ -.49134 & 0 & -.63203 & 0 & 0 & -.20743 & 0 & 0 & 0 & 0 \end{bmatrix},$$

and the quadratic performance index with $\widetilde{Q} = 10I_2$, $\widehat{Q} = I_2$, $R = 1000I_2$. The following results were obtained:

$$X = \left[\begin{matrix} 2.6296 & .0904 & .2806 & .2712 & .2466 & -1.2229 \\ .0904 & .1050 & -.3729 & -.3858 & -.3709 & .8462 \\ .2806 & -.3729 & 4.6919 & 4.8146 & 4.6138 & -.5114 \\ .2712 & -.3858 & 4.8146 & 4.9861 & 4.7735 & -.4782 \\ .2466 & -.3709 & 4.6138 & 4.7735 & 4.5716 & -.4530 \\ -1.2229 & .8462 & -.5114 & -.4782 & -.4530 & 20.6283 \\ -1.3594 & .9884 & -.5926 & -.5564 & -.5286 & 24.3092 \\ -1.1897 & .8591 & -.5426 & -.5126 & -.4864 & 20.7168 \\ .8797 & -.6459 & 1.4596 & 1.5016 & 1.4348 & -13.3976 \\ 3.4895 & -1.3065 & 20.2484 & 21.5118 & 20.3576 & -3.0294 \\ -.3129 & -.9313 & .7068 & .6745 & .6530 & -21.9435 \\ .2906 & -.6986 & -.3911 & -.4717 & -.4470 & -19.3333 \end{matrix}\right.$$

$$\left.\begin{matrix} -1.3594 & -1.1897 & .8797 & 3.4895 & -.3129 & .2906 \\ .9884 & .8591 & -.6459 & -1.3065 & -.9313 & -.6986 \\ -.5926 & -.5426 & 1.4596 & 20.2484 & .7068 & -.3911 \\ -.5564 & -.5126 & 1.5016 & 21.5118 & .6745 & -.4717 \\ -.5286 & -.4864 & 1.4348 & 20.3576 & .6530 & -.4470 \\ 24.3092 & 20.7168 & -13.3976 & -3.0294 & -21.9435 & -19.3333 \\ 28.7387 & 24.4626 & -15.8955 & -3.3634 & -26.1167 & -23.0057 \\ 24.4626 & 20.8374 & -13.5330 & -3.1312 & -22.1486 & -19.5029 \\ -15.8955 & -13.5330 & 9.1782 & 6.9887 & 14.3604 & 12.4007 \\ -3.3634 & -3.1312 & 6.9887 & 135.0512 & 2.0668 & -2.3934 \\ -26.1167 & -22.1486 & 14.3604 & 2.0668 & 29.0778 & 22.5452 \\ -23.0057 & -19.5029 & 12.4007 & -2.3934 & 22.5452 & 23.1671 \end{matrix}\right],$$

$$H = \left[\begin{matrix} -.0015 & .0011 & -.0024 & -.0025 & -.0024 & .0223 \\ -.0349 & .0131 & -.2025 & -.2151 & -.2036 & .0303 \end{matrix}\right.$$

$$\left.\begin{matrix} .0265 & .0226 & -.0153 & -.0116 & -.0239 & -.0207 \\ .0336 & .0313 & -.0699 & -1.3505 & -.0207 & .0239 \end{matrix}\right].$$

The value of the estimate of $1/\text{cond}(U_{11})$ was **RCOND** $\approx .004$. The 1-norm of the residual was about 8.4×10^{-10}, and the relative residual, 3.8×10^{-12}. Two iterations of the defect correction procedure `RicCorr` in § 1.5.4 (totalizing two Newton iterations, for $\varepsilon = 10^{-6}$), then reduced the residual and relative residual 1-norms to about 9.4×10^{-12}, and 4.2×10^{-14}, respectively.

The eigenvalues of the optimal (closed-loop) system are:

$$\begin{aligned}\lambda &= \{-29.4625 \pm 313.9367\mathrm{j}, -19.2672, -9.3718 \pm 7.8653\mathrm{j}, -10.7164, \\ &\quad -1.9160 \pm 7.0035\mathrm{j}, -1.6681, -1.0181, -.0208, -.3162\}.\end{aligned}$$

The first two eigenvalues, with large moduli and significant imaginary parts, practically coincide with two poles of the open-loop system. □

We ought to mention that the same problem was also solved using the Algorithm **Ntn** in Example 2.4 (§ 2.5.3). Results identical to those listed here were obtained. (In fact, the matrices X and H computed with the two methods have at least 10 identical significant decimal digits, and most elements have even 12 identical digits. There are more pronounced, but explainable differences only to some complex conjugate eigenvalues.) Ten iterations were performed, and the final value of the convergence criterion was about 8.99×10^{-7}. Note that Algorithm `Sch` in Example 3.8 does not work for the data in Example 2.6 (§ 2.5.3), because matrix A is singular.

Example 3.10 Consider the discrete-time tubular ammonia reactor model in (Patnaik, Viswanadham and Sarma, 1980), defined by the matrices

$$A = \left[\begin{array}{ccccc} .8701 & .1350 & .01159 & .0005014 & -.03722 \\ .07665 & .8974 & .01272 & .0005504 & -.04016 \\ -.1272 & .3575 & .81700 & .001455 & -.10280 \\ -.3635 & .6339 & .07491 & .796600 & -.27350 \\ -.9600 & 1.6459 & -.12890 & -.005597 & .07142 \\ -.6644 & 1.1296 & -.08889 & -.003854 & .08447 \\ -.4102 & .6930 & -.05471 & -.002371 & .06649 \\ -.1799 & .3017 & -.02393 & -.001035 & .06059 \\ -.3451 & .5804 & -.04596 & -.001989 & .10560 \end{array}\right.$$

$$\left.\begin{array}{cccc} .0003484 & 0 & .004242 & .007249 \\ .0003743 & 0 & .004530 & .007499 \\ .0009870 & 0 & .01185 & .01872 \\ .002653 & 0 & .03172 & .04882 \\ .007108 & 0 & .08452 & .1259 \\ .01360 & 0 & .1443 & .1016 \\ .01249 & .0001063 & .09997 & .06967 \\ .02216 & 0 & .2139 & .03554 \\ .01986 & 0 & .2191 & .2152 \end{array}\right],$$

$$B = \begin{bmatrix} .0004760 & -.00005701 & -.0083680 \\ .0000879 & -.0004773 & -.0002730 \\ .0001482 & -.001312 & .0008876 \\ .0003892 & -.003513 & .0024800 \\ .0010340 & -.009275 & .0066800 \\ .0007203 & -.006159 & .0038340 \\ .0004454 & -.003683 & .0020290 \\ .0001971 & -.001554 & .0006937 \\ .0003773 & -.003028 & .0014690 \end{bmatrix},$$

$$C = \begin{bmatrix} 1 & 0 & 0 & 0 & 0 & 0 & 0 & 0 & 0 \\ 0 & 0 & 0 & 0 & 1 & 0 & 0 & 0 & 0 \end{bmatrix},$$

and the quadratic performance index defined by $Q = 50I_9$, $R = I_3$ (without error integrators). Rounded to four decimal digits, the results obtained are:

$$X = 10^3 \left[\begin{matrix} .8698 & .0411 & .1381 & -.0192 & -.0593 \\ .0411 & 1.8658 & .2075 & .1095 & -.1125 \\ .1381 & .2075 & .2281 & .0401 & -.0518 \\ -.0192 & .1095 & .0401 & .1380 & -.0385 \\ -.0593 & -.1125 & -.0518 & -.0385 & .0783 \\ -.0024 & .0055 & -.0001 & .0002 & .0002 \\ .0000 & .0000 & .0000 & .0000 & .0000 \\ -.0245 & .0583 & -.0009 & .0024 & .0017 \\ -.0182 & .0493 & .0005 & .0031 & .0005 \end{matrix} \right.$$

$$\left. \begin{matrix} -.0024 & .0000 & -.0245 & -.0182 \\ .0055 & .0000 & .0583 & .0493 \\ -.0001 & .0000 & -.0009 & .0005 \\ .0002 & .0000 & .0024 & .0031 \\ .0002 & .0000 & .0017 & .0005 \\ .0501 & .0000 & .0008 & .0005 \\ .0000 & .0500 & .0000 & .0000 \\ .0008 & .0000 & .0581 & .0054 \\ .0005 & .0000 & .0054 & .0547 \end{matrix} \right],$$

$$H = \left[\begin{matrix} -.2257 & -.3495 & -.0619 & -.0179 & .0246 \\ -1.2104 & 2.2299 & -.0240 & .1877 & .0126 \\ 7.3284 & .2113 & 1.1825 & -.1783 & -.4802 \end{matrix} \right.$$

$$\left. \begin{matrix} -.0019 & .0000 & -.0198 & -.0190 \\ .0183 & .0000 & .1921 & .1692 \\ -.0161 & .0000 & -.1670 & -.1192 \end{matrix} \right].$$

The elements $h_{:,7}$ are of order of 10^{-5}; specifically, $h_{:,7} \approx (-2.4 \times 10^{-6}, 1.9 \times 10^{-5}, -9.9 \times 10^{-6})^T$. Matrix A is well conditioned, the estimate of its reciprocal condition number being about 0.95. The value of the estimate of $1/\mathrm{cond}(U_{11})$ was `RCOND` $\approx 1.19 \times 10^{-2}$. The 1-norm of the residual was about 1.13×10^{-5}, and the relative residual, 4.62×10^{-9}. Two iterations of the defect correction procedure `RicCorr`, totalizing two Newton iterations (for $\varepsilon = 10^{-6}$), then reduced the residual and relative residual 1-norms to about 2.91×10^{-13}, and 1.19×10^{-16}, respectively. The eigenvalues of the optimal (closed-loop) system are: $\{ .9466, .8182, .8061, .7236, .3945, .1054, .0340, -.0002, .0001 \}$. □

3.4 Generalized Schur-Based Riccati Solvers

This section describes specific algorithms for solving both continuous-time and discrete-time algebraic matrix Riccati equations. A special eigenvalue ordering strategy is shown to improve efficiency for discrete-time case. Extensions for solving equations with possibly singular matrix R are considered. Computational issues, as well as applicability and limitations of the generalized Schur-based Riccati solvers are also discussed.

3.4.1 Algorithmic templates

Let us start with the discrete-time case, A possibly singular, but R nonsingular. (This case was the original motivation for the generalized Schur vectors approach for solving Riccati equations.) It was proven in § 3.1.2 that, with the usual assumptions of stabilizability and detectability, the matrix pencil $Kz - \lambda Lz$ in (3.15) has no eigenvalues on the unit circle in $\mathbb{C}$ (see Theorems 3.4 and 3.5). Moreover, if the columns of the matrix $\widetilde{U} := [\, U_{11}^T \;\; U_{21}^T \,]^T \in \mathbb{R}^{2n \times n}$ form a basis for the stable deflating subspace of $K - \lambda L$, corresponding to the eigenvalues λ_i, with $|\lambda_i| < 1$, $i = 1\!:\!n$, then (see Theorem 3.6) U_{11} is nonsingular and $X = U_{21}U_{11}^{-1}$ is a positive semi-definite solution of the Riccati equation (3.2). Consequently, one may proceed similarly on to the case when matrix A is nonsingular.

As for the standard problem, it is necessary to avoid determination of the generalized eigenvectors (including the principal vectors), due to possible numerical difficulties associated with the multiple or nearly equal generalized eigenvalues. It is better to use the generalized Schur vectors as a basis for the stable deflating subspace. According to Theorem 3.9 in § 3.2.1 (see (Golub and Van Loan, 1983; Moler and Stewart, 1973)), there are some orthogonal matrices V and Z, so that $\widetilde{K} := VKZ$ is in a RSF, $\widetilde{L} := VLZ$ is upper triangular, and $\lambda(\widetilde{K}_{11}, \widetilde{L}_{11}) \subset \mathbb{C}^-$, where $\widetilde{K}$ and $\widetilde{L}$ are partitioned into $n \times n$ blocks, $\widetilde{K}_{ij}$ and $\widetilde{L}_{ij}$, respectively, $i, j = 1\!:\!2$ ($\widetilde{K}_{21} = \widetilde{L}_{21} = 0$).

The submatrix $\widetilde{L}_{11}$ has nonzero diagonal elements, as otherwise infinite, hence unstable eigenvalues would correspond to it. So, $\widetilde{L}_{11}$ is nonsingular. Denoting $S_{11} = \widetilde{L}_{11}^{-1}\widetilde{K}_{11}$, where S_{11} is a matrix in RSF, the identity

$$\begin{bmatrix} \widetilde{K}_{11} \\ 0 \end{bmatrix} = \begin{bmatrix} \widetilde{L}_{11} \\ 0 \end{bmatrix} \widetilde{L}_{11}^{-1}\widetilde{K}_{11}$$

is equivalent to $VK\widetilde{U} = VL\widetilde{U}S_{11}$, where $\widetilde{U} := [\,U_{11}^T \;\; U_{21}^T\,]^T := Z[\,I_n \;\; 0\,]^T$. Since V is a nonsingular matrix, we obtain $K\widetilde{U} = L\widetilde{U}S_{11}$. Theorem 3.6 is clearly valid for this interpretation of $\widetilde{U}$, having as columns the generalized Schur vectors corresponding to the stable eigenvalues.

The following algorithm is obtained.

Algorithm gSchv : $X =$ `gSchv`$(\,A, B, Q, R\,)$

/* Solve Riccati equation (3.2) using generalized Schur */
/* vectors method (R nonsingular). */

/* 1. Construct the matrices K and L in (3.15). */

$F = BR^{-1}B^T; \quad K = \begin{bmatrix} A & 0 \\ -Q & I \end{bmatrix}; \quad L = \begin{bmatrix} I & F \\ 0 & A^T \end{bmatrix};$

/* 2. Reduce K to RSF and L to upper triangular form; */
/* accumulate in Z the column transformations performed. */

$(K, L, Z) =$ `QZHess`$(\,K, L\,)$; $\quad (K, L, Z, \alpha, \beta) =$ `QZSch`$(\,K, L, Z\,)$;

/* 3. Order the Schur-triangular form of K and L so that stable */
/* generalized eigenvalues correspond to submatrices $(1\!:\!n, 1\!:\!n)$; */
/* accumulate in Z the column transformations performed. */

FOR $i = 1\!:\!2n$,
  IF $\beta_i \neq 0$, THEN $select_i = |\alpha_i| < \beta_i$;
$(K, L, Z) =$ `QZSchord`$(\,K, L, select, Z\,)$;

/* 4. Solve matrix equation $U_{11}^T X = U_{21}^T$, where */
/* $[\,U_{11}^T \;\; U_{21}^T\,]^T := Z[\,I_n \;\; 0\,]^T$. */

$X =$ `solve`$(\,z_{1:n,1:n}^T, z_{n+1:2n,1:n}^T\,)$;

The computation of the optimal gain matrix H in (3.4) can be performed based on its definition, and taking the symmetry into account.

The QZ algorithm (§ 3.2.2) and its extension for eigenvalues ordering (§ 3.2.3) are used to accomplish stages 2 and 3. According to § 3.2.2, stage 2 of Algorithm **gSchv** requires about $(\frac{17}{3} + \frac{3}{2} + 25)(2n)^3 \approx 250n^3$ *flops*. As for RSF ordering, if we consider that stage 3 takes about 15 % of the effort needed for reduction (or using the estimate $5(2n)^3$ *flops*), we obtain a total effort of about $290n^3$ *flops*, that is almost three times greater than that required for Algorithm `Schvd`. However, in our experiments performed on an IBM PC compatible microcomputer, Algorithm **gSchv** never required more than twice the time spent on Algorithm **Schvd**.

The efficiency of Algorithm **gSchv** can still be improved. Indeed, it is useful to take into account the natural tendency of the QZ algorithm to determining the generalized eigenvalues in a decreasing order of magnitude along the diagonals of K and L. (Note that this is simply feasible by dealing with the problem $Lx = \lambda Kx$.) In addition, as derived from § 3.2.2, the accumulation of the transformations V would be somewhat cheaper than the accumulation of the transformations Z. To obtain these improvements, Algorithm **gSchv** will be modified as follows (Sima, 1991).

Algorithm gSchvQ : $\boxed{X = \texttt{gSchvQ}(\ A, B, Q, R\ )}$

/* Solve Riccati equation (3.2) using improved generalized Schur */
/* vectors method (R nonsingular). */

/* 1. Construct the matrices $\bar{K} := K^T$ and $\bar{L} := L^T$, K, L in (3.15). */

$$F = BR^{-1}B^T; \quad K = \begin{bmatrix} A & 0 \\ -Q & I \end{bmatrix}^T; \quad L = \begin{bmatrix} I & F \\ 0 & A^T \end{bmatrix}^T;$$

/* 2. Reduce $\bar{K}$ to RSF and $\bar{L}$ to upper triangular form; */
/* accumulate in V the *row* transformations performed. */

$(K, L, V) = \texttt{QZHess}(\ K, L\ ); \quad (K, L, V, \alpha, \beta) = \texttt{QZSch}(\ K, L, V\ );$

/* 3. Order the Schur-triangular form of $\bar{K}$ and $\bar{L}$ so that *unstable* */
/* generalized eigenvalues correspond to submatrices $(1{:}n, 1{:}n)$; */
/* accumulate in V the row transformations performed. */

```
FOR i = 1:2n,
   IF β_i = 0, THEN
      select_i = TRUE;
   ELSE
      select_i = |α_i| ≥ β_i;
   END IF
END FOR
(K, L, V) = QZSchord( K, L, select, V );
```

/* 4. Solve matrix equation $V_{12}X = -V_{11}$, where */
/* $\begin{bmatrix} V_{11} & V_{12} \end{bmatrix} := \begin{bmatrix} I_n & 0 \end{bmatrix} V$. */

$X = \texttt{solve}(\ v_{1:n,n+1:2n}, -v_{1:n,1:n}\ );$

To justify this algorithm, let $\widetilde{K} := V\bar{K}Z$ and $\widetilde{L} := V\bar{L}Z$ be the matrices obtained after the first three stages. It follows that $\widetilde{K}^T = Z^T K V^T$, $\widetilde{L}^T = Z^T L V^T$, and the first n columns in V^T form a basis for the deflating subspace $\mathcal{S}$, corresponding to the generalized eigenvalues in $\mathbb{C}^+ := \mathbb{C} \setminus \mathbb{C}^-$. Note that if $x := \begin{bmatrix} x_1^T & x_2^T \end{bmatrix}^T \in \mathcal{S} \subset \mathbb{R}^{2n}$ $(x \neq 0)$, where

Table 3.1 Sample of execution times of Algorithm **gSchv** and Algorithm **gSchvQ** for some discrete-time problems.

n	Algorithm **gSchv**	Algorithm **gSchvQ**
9	.55	.50
11	1.26	.72
13	1.26	1.05
26	13.35	12.03

$x_1, x_2 \in \mathbb{R}^n$, then $Jx \in \mathcal{S}^\perp$, since $x^T J x = x_1^T x_2 - x_2^T x_1 = 0$. Therefore, $JV^T [\, I_n \;\; 0 \,]^T = [\, V_{12} \;\; -V_{11} \,]^T$ form a basis for $\mathcal{S}^\perp$, i.e. the deflating subspace corresponding to stable generalized eigenvalues. Taking this basis, one determines the solution of the Riccati equation as in Algorithm **gSchv**.

An implementation of Algorithm **gSchv** had been included in the BIMAS library (Varga and Sima, 1985), but Algorithm **gSchvQ** was implemented much later. In our preliminary experiments performed using these two implementations, the computing times were often practically equal, but sometimes the time required by Algorithm **gSchvQ** was by 25 % less than that required for Algorithm **gSchv**.

Table 3.1 compares the execution times for Algorithm **gSchv** and Algorithm **gSchvQ** for some discrete-time problems taken from the literature. Algorithm **gSchvQ** is more efficient than Algorithm **gSchv** for all these problems. The increase in speed varies between about 10 % to 40 %.

Table 3.2 presents an additional timing comparison of the two algorithms. The examples used were in fact continuous-time examples, but taken as discrete-time ones, just for enabling more comparisons. (Note that most of examples in the literature are continuous-time systems.) The results also illustrate that Algorithm **gSchvQ** is by 10 % to 40 % more efficient than Algorithm **gSchv**. If times in Table 3.2 are cummulated, then the increase in efficiency is about 30 %. The conclusion is that one must always use Algorithm **gSchvQ** (for solving discrete-time problems).

Algorithm **gSchv** and Algorithm **gSchvQ** do not need the inverse of matrix A. Consequently, these algorithms can be used in both nonsingular and singular cases. Clearly, a cheaper alternative in the nonsingular case, using the Schur vectors, is to employ Algorithm **Schvd** in § 3.3.1, but the requirement for computing A^{-1} yields a significant loss of accuracy in X when A is ill conditioned with respect to inversion.

An important application of the generalized Schur vectors approach is the solution of the Riccati equations (3.1) or (3.2) without any matrix inversion. In § 3.3.1, but also above, the matrix R had to be inverted

Table 3.2 Sample of execution times of Algorithm `gSchv` and Algorithm `gSchvQ` for other "discrete-time" problems.

n	Algorithm `gSchv`	Algorithm `gSchvQ`
9	.39	.27
19	2.91	2.09
29	9.56	6.70
15	2.64	2.36
10	.28	.22
21	2.25	1.37
10	.83	.66
20	5.05	4.23
30	23.90	16.75

(since $F := BR^{-1}B^T$) for constructing the matrices G, G^{-1}, K, and L. It was therefore assumed that R is nonsingular and well conditioned. This is a limitation of the discussed methods and, in the sequel, how it can be circumvented will be shown. The Riccati equations (3.1) and (3.2) where R is (almost) singular, are sometimes called *(almost) singular*. The essential idea (Van Dooren, 1980; see also § 3.2.3) is to define a generalized eigenvalue problem $\bar{K} - \lambda \bar{L}$ of dimension $2n + m$, whose spectrum should include $\lambda(K, L)$, and then reduce it to a generalized problem of order $2n$, with spectrum $\lambda(K, L)$. In the continuous-time case, the matrices $\bar{K}$ and $\bar{L}$ are defined by (see (3.45))

$$\bar{K} = \begin{bmatrix} A & 0 & B \\ -Q & -A^T & 0 \\ 0 & B^T & R \end{bmatrix}, \qquad \bar{L} = \begin{bmatrix} I & 0 & 0 \\ 0 & I & 0 \\ 0 & 0 & 0 \end{bmatrix}, \tag{3.59}$$

and in the discrete-time case, by (see (3.46))

$$\bar{K} = \begin{bmatrix} A & 0 & B \\ -Q & I & 0 \\ 0 & 0 & R \end{bmatrix}, \qquad \bar{L} = \begin{bmatrix} I & 0 & 0 \\ 0 & A^T & 0 \\ 0 & B^T & 0 \end{bmatrix}. \tag{3.60}$$

To obtain the problem $K - \lambda L$ (of dimension $2n$), from the problem $\bar{K} - \lambda \bar{L}$, an orthogonal transformation $\bar{V} \in \mathbb{R}^{(2n+m)\times(2n+m)}$ is resorted to, so that to annihilate the elements (i, j), $j = 2n + m\colon -1\colon 2n + 1$, $i = 1\colon j - 1$ in $\bar{V}\bar{K}$. Note that the last block-column in $\bar{V}\bar{L}$ remains zero. It follows that

$$\bar{V}(\bar{K} - \lambda \bar{L}) = \begin{bmatrix} K - \lambda L & 0 \\ * & P - \lambda 0 \end{bmatrix}, \tag{3.61}$$

where the matrix P, of order m, is lower triangular. According to the definition of equivalence (§ 3.2.1), the problems $\bar{K} - \lambda\bar{L}$ and $\bar{V}\bar{K} - \lambda\bar{V}\bar{L}$ are equivalent, but from (3.61) it follows that $\lambda(\bar{K}, \bar{L}) = \lambda(K, L) \cup \lambda(P, 0)$. Assuming that the problem $\bar{K} - \lambda\bar{L}$ is regular, then $\lambda(P, 0)$ contains only infinite (not undetermined) eigenvalues, and these can be removed from the problem. It can be said that infinite eigenvalues due to possible "singularities" of R are so deflated. (Conditions for regularity are mentioned in § 3.1.2. A necessary condition is that $\text{rank}([\,B^T \quad R^T\,]^T) = m$. If $\bar{K} - \lambda\bar{L}$ is singular, then a "regularization" algorithm, like the one detailed in (Mehrmann, 1991), based on singular value decomposition, should be used as a preprocessing step.) The problem $K - \lambda L$, obtained in this manner, may then be approached as in Algorithms **gSchv** or **gSchvQ**, for both the continuous-time and discrete-time cases.[10] See Example 3.11 for a complete solution technique.

Algorithm gext : $(K, L) = \texttt{gext}(\ A, B, Q, R, \texttt{sys}\ )$

/* Construct the problem $\bar{K} - \lambda\bar{L}$ in (3.59) or (3.60) */
/* and reduce it to the problem $K - \lambda L$ of order $2n$. */

/* 1. Construct the problem. */

IF **sys** = c, THEN

$$K = \begin{bmatrix} A & 0 & B \\ -Q & -A^T & 0 \\ 0 & B^T & R \end{bmatrix}; \quad L = \begin{bmatrix} I & 0 & 0 \\ 0 & I & 0 \\ 0 & 0 & 0 \end{bmatrix};$$

ELSE

$$K = \begin{bmatrix} A & 0 & B \\ -Q & I & 0 \\ 0 & 0 & R \end{bmatrix}; \quad L = \begin{bmatrix} I & 0 & 0 \\ 0 & A^T & 0 \\ 0 & B^T & 0 \end{bmatrix};$$

END IF

FOR $j = 2n + m: -1: 2n + 1,$

/* 2. Find Householder transformation U_j */
/* annihilating $\bar{k}_{1:j-1,j}$, and modifying $\bar{k}_{jj}$. */

$(U, r) = \texttt{Hous}(\ k_{jj}, k_{1:j-1,j}\ );$

/* 3. Apply U_j to $\bar{K}$ and $\bar{L}$. */

$$\begin{bmatrix} k_{j,1:j-1} \\ k_{1:j-1,1:j-1} \end{bmatrix} \leftarrow U \begin{bmatrix} k_{j,1:j-1} \\ k_{1:j-1,1:j-1} \end{bmatrix}; \quad k_{jj} = -r;$$

$$\begin{bmatrix} l_{j,1:2n} \\ l_{1:j-1,1:2n} \end{bmatrix} \leftarrow U \begin{bmatrix} l_{j,1:2n} \\ l_{1:j-1,1:2n} \end{bmatrix};$$

[10] For the continuous-time case, Algorithm **gSchv(Q)** must be modified so that $select_i$ be TRUE if the eigenvalue corresponding to the pair (α_i, β_i) belongs to the required (in)stability domain.

The matrix $\bar{V}$ (the product of Householder transformations U_j) must not be accumulated. This is obvious in the context of using Algorithm `gSchv`. For Algorithm `gSchvQ`, one would transform the problem $\bar{K}^T - \lambda \bar{L}^T$ using $(\bar{K}^T - \lambda \bar{L}^T)\bar{Z}$ (without accumulating $\bar{Z}$), where $\bar{Z}$ annihilates the undesired nonzero elements in the last block-row. But it is easy to verify that the matrices $\bar{K}$ and $\bar{L}$, obtained in this way, coincide with those resulted when using Algorithm `gext`.

Algorithm `gext` requires $8n^2 + 5nm + m^2$ memory locations, assuming that the last block-column in $\bar{K}$ and $\bar{L}$ is not stored. Algorithms `gSchv` and `gSchvQ` require $12n^2 + 6n$ locations.

It is easy to modify `gext` for solving more general optimization problems, for descriptor systems (see (3.28) and (3.29) in § 3.1.2).

The method presented above looks more advantageous from a numerical point of view than the method in § 3.3.1 since there are no matrix inversions. This is important especially in the discrete-time case.

3.4.2 Computational issues

The calculations in Algorithm `gext` can be organized to take advantage of modern computer architectures. The continuous-time and discrete-time cases are distinguished only at stage 1, which constructs the matrices K and L. Householder transformations (see § 1.4.3) are used to annihilate the elements in the last block-column of matrix K in (3.59) or (3.60). Our implementation exploits the presence of the zero submatrix $k_{n+1:2n,2n+1:2n+m}$. Moreover, the last block-column in K and L is not actually constructed. The information stored in arrays `B` and `R`, corresponding to the matrices B and R, respectively, is used for both constructing and applying the Householder transformations. Consequently, the vector that defines a Householder transformation is stored just in the corresponding column of the arrays `B` and `R`. This approach requires to double the number of routine calls. By applying the respective transformations, the initial contents of the arrays `B` and `R` is entirely changed.

The implementation of Algorithm `gSchv(Q)` does not construct the matrices K and L, so that the output of Algorithm `gext` can directly be used. Moreover, each variant can solve both continuous-time and discrete-time problems. The continuous-time and discrete-time cases are distinguished only in setting the parameters which determine the strategy for ordering the Schur-triangular form. The corresponding routines control the (near) singularity of the matrix U_{11} (V_{12}), and stop the computation if this matrix is (numerically) singular.

The variant of Algorithm `gSchv` is recommended for continuous-time systems, while the variant of Algorithm `gSchvQ` should be used for discrete-time systems. Clearly, Algorithm `gSchvQ` can also be called for solving continuous-time problems. There is usually a slight increase in speed as to Algorithm `gSchv`, due to accumulation of transformations Q, but occasionally the ordering strategy in Algorithm `gSchvQ` can produce sensibly larger execution times.

3.4.3 Applicability and limitations

Algorithm `gext` should be followed by an adaptation of `gSchv(Q)`, to compute the solution of the continuous-time or discrete-time algebraic matrix Riccati equations (3.1) or (3.2), respectively. No matrix inversions or linear systems solving are performed by `gext`. Consequently, it may be used even if the matrices R and/or A in (3.1) and (3.2) are singular, contrary to Algorithm `Schv(d)` in § 3.3.1. Moreover, R and/or A can be nearly singular, without affecting the numerical stability of the computations.

Algorithms `gSchv` and `gSchvQ` can be used only if the linear dynamical system defined by the matrices A and B is stabilizable and the associated matrix pencil is dichotomic. However, these algorithms can be modified to solve more general problem categories.

Algorithm `gSchv(Q)` operates upon the generalized eigenvalue problem $K - \lambda L$, of dimension $2n$, by ordering the Schur-triangular form corresponding to the pair (K, L) and determining the stable generalized eigenvalues (the closed-loop poles), as well as the solution of the Riccati equation (3.1) or (3.2).

As for Algorithm `Schv` (§ 3.3.1), `gSchv(Q)` does not ensure that the computed solution is symmetric positive semi-definite (properties theoretically guaranteed). In our experiments, the deviation from symmetry was more pronounced when the estimate of the reciprocal condition number of the linear system to be solved was small. This estimate was usually greater, hence better, than the corresponding value returned by implementations of `Schv` or `Schvd`. It is advisable to use symmetrization of the computed solution (see § 3.3.3).

Algorithm `gSchv(Q)` is less efficient than `Schv(d)`, but it can be used even if the matrices R and/or A are singular, contrary to `Schv(d)`. In addition, although `Schv(d)` can compute solutions when cond(R) and/or cond(A) are large (but not so large as the respective matrices could be considered numerically singular), the results obtained in such situations are usually disturbed by significant rounding-errors. In our experiments carried out on an IBM-PC compatible computer, the ratio between the computing times required by `Schv(d)` and `gSchv` was less than 2, and often less than 1.5. As already mentioned, `gSchvQ` is comparable with `gSchv`, but typically by 10 %–40 % faster.

The generalized Schur vectors method can also be used for solving the *deadbeat control* problem for discrete-time systems (Emami-Naeini and Franklin, 1980, 1982). An optimal problem with $R = 0$ and $Q = D^T D$, $D \in \mathbb{R}^{m \times n}$, is to be solved.

Another important application of the generalized problem $K - \lambda L$ is the spectral factorization of a positive real transfer-function matrix. Algorithms `gSchv` and `gSchvQ` can be used with appropriate definitions for K and L matrices (Van Dooren, 1980).

Examples of generalized Schur vectors algorithms use are now given.

Example 3.11 The following algorithm computes the solution of the algebraic matrix Riccati equation (3.1), if `sys` = `c` (continuous-time), or (3.2), if `sys` = `d` (discrete-time), and the corresponding optimal gain matrix, using the generalized Schur vectors method. The system output is weighted with the matrix $\widetilde{Q}$ in the quadratic performance index, hence output regulation (Problem 1.3 in § 1.2.1) is accomplished. A proportional-plus-integral optimal control law (Problem 1.4) is used. It is assumed that the matrices A, B, C, $\widetilde{Q}$ and $\widehat{Q}$ in (1.66) in § 1.2.1, of dimensions $n \times n$, $n \times m$, $l \times n$, $l \times l$ and $l \times l$, respectively, as well as R, of dimension $m \times m$, are given.

Algorithm gSch : $(X, H) = \texttt{gSch}(\ A, B, C, \widetilde{Q}, \widehat{Q}, R, \texttt{sys}\ )$

/* Solve (3.1) or (3.2) for Problem 1.4 using */
/* generalized Schur vectors method. */

/* 1. Construct matrices for proportional-plus-integral control. */
IF `sys` = `c`, THEN $P = 0_l$; ELSE $P = I_l$;

$$A \leftarrow \begin{bmatrix} A & 0 \\ -C & P \end{bmatrix};\quad B \leftarrow \begin{bmatrix} B \\ 0 \end{bmatrix};\quad C \leftarrow \begin{bmatrix} C & 0 \\ 0 & I_l \end{bmatrix};\quad Q = \begin{bmatrix} \widetilde{Q} & 0 \\ 0 & \widehat{Q} \end{bmatrix};$$

/* 2. Save A, B and R. Compute state weighting matrix. */
$\widetilde{A} = A$; $\widetilde{B} = B$; $\widetilde{R} = R$;
$Q \leftarrow C^T Q C$;

/* 3. Construct the generalized problem of order $2n$. */
$(K, L) = \texttt{gext}(\ A, B, Q, R, \texttt{sys}\ )$;

/* 4. Solve Riccati equation and compute optimal gain matrix. */
IF `sys` = `c`, THEN
/* Reduce K to RSF and L to upper triangular form. */
$(K, L, Z) = \texttt{QZHess}(\ K, L\ )$; $(K, L, Z, \alpha, \beta) = \texttt{QZSch}(\ K, L, Z\ )$;
/* Order the Schur-triangular form of K and L so that stable */
/* generalized eigenvalues correspond to submatrices $(1{:}\,n, 1{:}\,n)$. */
FOR $i = 1{:}\,2n$,
$select_i = \Re(\alpha_i) < 0$;

$(K, L, Z) = \texttt{QZSchord}(\ K, L, select, Z\ );$

/* Solve matrix equation $U_{11}^T X = U_{21}^T$. */

$X = \texttt{solve}(\ z_{1:n,1:n}^T, z_{n+1:2n,1:n}^T\ );\quad H = -\widetilde{R}^{-1}\widetilde{B}^T X;$

ELSE

/* Reduce K to RSF and L to upper triangular form. */

$(K, L, V) = \texttt{QZHess}(\ K, L\ );\quad (K, L, V, \alpha, \beta) = \texttt{QZSch}(\ K, L, V\ );$

/* Order the Schur-triangular form of K and L so that unstable */
/* generalized eigenvalues correspond to submatrices $(1\!:n, 1\!:n)$. */

FOR $i = 1\!:2n$,
  IF $\beta_i = 0$, THEN
    $select_i$ = TRUE
  ELSE
    $select_i = |\alpha_i| \geq \beta_i;$
  END IF
END FOR

$(K, L, V) = \texttt{QZSchord}(\ K, L, select, V\ );$

/* Solve matrix equation $V_{12}X = -V_{11}$. */

$X = \texttt{solve}(\ v_{1:n,n+1:2n}, -v_{1:n,1:n}\ );\quad H = -(\widetilde{R} + \widetilde{B}^T X \widetilde{B})^{-1}\widetilde{B}^T X \widetilde{A};$

END IF

If matrix $\widehat{Q}$ is not given, then $l = 0$, and the construction at step 1 is not made. No matrix inversions occur. □

Example 3.12 Consider the continuous-time unstable and non-minimum phase helicopter model in (Yue, Postlethwaite and Padfield, 1989), defined by the following matrices ($n = 8$, $m = 4$, $l = 6$)

$$A_{:,1:4} = \begin{bmatrix} 0 & 0 & 0 & .9985791 \\ 0 & 0 & 0 & -.5343404 \\ 0 & 0 & 1 & -.003935174 \\ 0 & 0 & -10.98004 & -2.6432820 \\ 0 & 0 & .4558556 & -1.8968060 \\ 0 & 0 & -1.9231580 & -.4810042 \\ -37.07327 & 0 & -.5238533 & 2.1821260 \\ .1262140 & 32.02769 & -2.2615190 & -.5253553 \end{bmatrix},$$

$$A_{:,5:8} = \begin{bmatrix} .05328893 & 0 & 0 & 0 \\ 1.00129800 & 0 & 0 & 0 \\ .07374105 & 0 & 0 & 0 \\ -.06251947 & .10350120 & -.08205526 & .007183709 \\ .007408097 & .01426707 & .01792734 & .001814035 \\ -.3766521 & .01859729 & .000561662 & .002725934 \\ -.0112772 & -.01983032 & -.02005057 & .021491190 \\ .4419744 & .02055577 & -.04019620 & .0002727032 \end{bmatrix},$$

$$B = \begin{bmatrix} 0 & 0 & 0 & 0 \\ 0 & 0 & 0 & 0 \\ .12279450 & .1312919 & -2.741718 & -.01958942 \\ .01663534 & .4732456 & .02266005 & 0 \\ .30181100 & .02372016 & -.49457870 & -.2638639 \\ .39724110 & -.5457401 & -.02614260 & 0 \\ -.02058744 & .02613068 & -.54584740 & .3211498 \\ -5.3717610 & -.001144409 & 0 & 0 \end{bmatrix},$$

$$C_{:,1:4} = \begin{bmatrix} 0 & 0 & 0 & 0 \\ 0 & 0 & 2 & 0 \\ 0 & 0 & 0 & 5 \\ .0000716085 & .00030701775 & 0 & -2.671702 \\ 5 & 0 & 0 & 0 \\ 0 & 2 & 0 & 0 \end{bmatrix},$$

$$C_{:,5:8} = \begin{bmatrix} 0 & .01472909 & .010628844 & -.19917352 \\ 0 & 0 & 0 & 0 \\ 0 & 0 & 0 & 0 \\ 5.00649 & 0 & 0 & 0 \\ 0 & 0 & 0 & 0 \\ 0 & 0 & 0 & 0 \end{bmatrix},$$

and the performance index defined by the matrices $Q = 10I_8$, $R = I_4$. (No error integrators are used.) Rounded to four digits after the decimal point, the results obtained using an implementation of Algorithm **gSch** in Example 3.11 are:

$$X = 10^3 \begin{bmatrix} 584.5279 & 113.0409 & -20.7957 & 30.8656 \\ 113.0409 & 99.6027 & -7.4413 & -3.0235 \\ -20.7957 & -7.4413 & 3.5745 & -1.6999 \\ 30.8656 & -3.0235 & -1.6999 & 4.3321 \\ 74.4446 & 53.1256 & -5.6590 & -1.5977 \\ 5.6753 & 5.6644 & -.5037 & -.5027 \\ -35.6356 & -.5886 & .1208 & -2.1372 \\ 4.0052 & 4.0088 & -.3017 & -.1744 \end{bmatrix}$$

$$\begin{bmatrix} 74.4446 & 5.6753 & -35.6356 & 4.0052 \\ 53.1256 & 5.6644 & -.5886 & 4.0088 \\ -5.6590 & -.5037 & .1208 & -.3017 \\ -1.5977 & -.5027 & -2.1372 & -.1744 \\ 39.5323 & 1.2104 & .0897 & 2.0433 \\ 1.2104 & 5.0906 & -.2050 & .3789 \\ .0897 & -.2050 & 4.3270 & -.0240 \\ 2.0433 & .3789 & -.0240 & .7123 \end{bmatrix},$$

$$H = \left[\begin{array}{rrrr} -1.9013 & 4.2024 & -.1205 & -.1624 \\ -9.6095 & 4.2589 & .1910 & -2.0078 \\ -40.1999 & 5.7682 & 7.0927 & -6.7287 \\ 30.6802 & 14.0612 & -1.4620 & .2315 \end{array}\right.$$

$$\left.\begin{array}{rrrr} -.7125 & -.2859 & .0351 & 3.0984 \\ 1.2219 & 3.0592 & .7685 & .2819 \\ 4.1534 & -.7497 & 2.7805 & .1843 \\ 10.2915 & .3754 & -1.3636 & .5410 \end{array}\right].$$

The value of the estimate of $1/\mathrm{cond}(U_{11})$ was `RCOND` $\approx 1.45 \times 10^{-3}$. The 1-norm of the residual was about 9.07×10^{-11}, and the relative residual, 1.04×10^{-13}. Two iterations of the defect correction procedure `RicCorr` (§ 1.5.4), totalizing two Newton iterations (for $\varepsilon = 10^{-6}$), then reduced the residual and relative residual 1-norms to about 2.27×10^{-13}, and 2.62×10^{-16}, respectively. The eigenvalues of the optimal (closed-loop) system are: $\{-16.9396,\ -8.3841 \pm 5.0396\mathrm{j},\ -4.3241 \pm 4.2084\mathrm{j},\ -1.4884 \pm .6243\mathrm{j},\ -2.1167\}$. □

Example 3.13 Using Algorithm `gSch` in Example 3.11 for the continuous-time system in Example 3.9, results which coincide with those listed there were obtained. (In fact, 10 or 11 significant decimal digits coincide.) The value of the estimate of $1/\mathrm{cond}(U_{11})$ was `RCOND` ≈ 0.0032. □

Example 3.14 Using Algorithm `gSch` in Example 3.11 for the discrete-time system in Example 2.6, results which coincide with those listed there were obtained. (In fact, 9 or 10 significant decimal digits coincide in the matrices X and H computed by the two methods.) Here, `RCOND` estimates $1/\mathrm{cond}(V_{12})$ and has the value 1.34×10^{-4}. □

Example 3.15 Algorithm `gSch` in Example 3.11 was used for a discrete-time unstable system with the matrices (Van Dooren, 1980):

$$A = \begin{bmatrix} 2 & -1 \\ 1 & 0 \end{bmatrix}, \qquad B = \begin{bmatrix} 1 \\ 0 \end{bmatrix},$$

and the performance index defined by

$$Q = \begin{bmatrix} 0 & 0 \\ 0 & 1 \end{bmatrix}, \qquad R = 0.$$

Therefore, the matrix Q weights the state-vector and no error integrators are used. We obtained $X = I_2$ and $H = \begin{bmatrix} -2 & 1 \end{bmatrix}$. Both poles of the optimal closed-loop system are practically in the origin, since they are $\{.0000, 10^{-16} \times .2534\}$. The value of the estimate `RCOND` is exactly 1. □

Note that the latter problem cannot be solved using algorithms in Examples 2.4 and 3.8, since matrix R is singular. However, the optimal gain matrix can be computed, because $R + B^T XB = B^T B = 1 > 0$.

Example 3.16 An implementation of Algorithm `gSch` in Example 3.11 was used for the discrete-time power plant model in (Katayama et al., 1985), defined by the matrices ($n = 20$, $m = 6$, $l = 6$)

$$A_{:,1:10} = \begin{bmatrix}
0 & -.4607 & 0 & .0045 & 0 & .1304 & 0 & .0731 & 0 & .0608 \\
1 & 1.4269 & 0 & .0034 & 0 & -.1702 & 0 & -.0728 & 0 & -.0527 \\
0 & .0179 & 0 & -.1242 & 0 & -.1065 & 0 & -.0351 & 0 & -.0603 \\
0 & -.0090 & 1 & 1.0126 & 0 & .2113 & 0 & .0411 & 0 & .0543 \\
0 & .0406 & 0 & -.2096 & 0 & -.0777 & 0 & .0492 & 0 & -.0334 \\
0 & -.0431 & 0 & .1458 & 1 & 1.1320 & 0 & -.0392 & 0 & .0343 \\
0 & .1253 & 0 & .1610 & 0 & .0953 & 0 & -.6278 & 0 & -.0066 \\
0 & -.1222 & 0 & -.2340 & 0 & -.0159 & 1 & 1.5797 & 0 & .0551 \\
0 & .0086 & 0 & -.1020 & 0 & -.1038 & 0 & .0057 & 0 & .1812 \\
0 & -.0094 & 0 & .1284 & 0 & .0851 & 0 & -.0079 & 1 & .7771 \\
0 & .0095 & 0 & -.0669 & 0 & -.0342 & 0 & .0658 & 0 & -.0341 \\
0 & .0237 & 0 & .0813 & 0 & .0268 & 0 & -.0848 & 0 & .0164 \\
0 & .0189 & 0 & -.3275 & 0 & -.1496 & 0 & .0689 & 0 & -.1201 \\
0 & -.0059 & 0 & .2417 & 0 & .1944 & 0 & -.0739 & 0 & .3157 \\
0 & -.0373 & 0 & .0986 & 0 & -.1332 & 0 & .0657 & 0 & -.0546 \\
0 & .0610 & 0 & -.0931 & 0 & .1297 & 0 & -.0714 & 0 & .0254 \\
0 & .0000 & 0 & .0000 & 0 & .0000 & 0 & .0000 & 0 & .0000 \\
0 & .0000 & 0 & .0000 & 0 & .0000 & 0 & .0000 & 0 & .0000 \\
0 & .0000 & 0 & .0000 & 0 & .0000 & 0 & .0000 & 0 & .0000 \\
0 & .0000 & 0 & .0000 & 0 & .0000 & 0 & .0000 & 0 & .0000
\end{bmatrix},$$

$$A_{:,11:20} = \begin{bmatrix}
0 & .0178 & 0 & .0067 & 0 & -.0090 & 0 & .0000 & 0 & .0000 \\
0 & -.0595 & 0 & .0011 & 0 & .0065 & 0 & .0000 & 0 & .0000 \\
0 & -.0939 & 0 & .0004 & 0 & .0693 & 0 & .0000 & 0 & .0000 \\
0 & .0521 & 0 & .0013 & 0 & -.0728 & 0 & .0000 & 0 & .0000 \\
0 & -.0421 & 0 & -.0082 & 0 & -.1448 & 0 & .0000 & 0 & .0000 \\
0 & .0290 & 0 & .0002 & 0 & .1535 & 0 & .0000 & 0 & .0000 \\
0 & .0144 & 0 & .0047 & 0 & .1116 & 0 & .0000 & 0 & .0000 \\
0 & -.0192 & 0 & -.0004 & 0 & -.1173 & 0 & .0000 & 0 & .0000 \\
0 & -.0301 & 0 & .0048 & 0 & -.0517 & 0 & .0000 & 0 & .0000 \\
0 & .0253 & 0 & .0081 & 0 & .0529 & 0 & .0000 & 0 & .0000 \\
0 & .2095 & 0 & .0107 & 0 & .0923 & 0 & .0000 & 0 & .0000 \\
1 & .6173 & 0 & -.0138 & 0 & -.0945 & 0 & .0000 & 0 & .0000 \\
0 & .1359 & 0 & .3119 & 0 & -.0085 & 0 & .0000 & 0 & .0000 \\
0 & .1776 & 1 & .3995 & 0 & .0516 & 0 & .0000 & 0 & .0000 \\
0 & .0440 & 0 & -.0111 & 0 & -.0391 & 0 & .0000 & 0 & .0000 \\
0 & -.0848 & 0 & .0055 & 1 & 1.0233 & 0 & .0000 & 0 & .0000 \\
0 & .0000 & 0 & .0000 & 0 & .0000 & 0 & -.1228 & 0 & 2.0351 \\
0 & .0000 & 0 & .0000 & 0 & .0000 & 1 & .8965 & 0 & 3.0747 \\
0 & .0000 & 0 & .0000 & 0 & .0000 & 0 & -.0041 & 0 & .2600 \\
0 & .0000 & 0 & .0000 & 0 & .0000 & 0 & .0042 & 1 & .1704
\end{bmatrix},$$

$$B = \begin{bmatrix} -.0026 & .1205 & -.0022 & -.0603 & .0091 & -.0362 \\ .0010 & -.0096 & .0038 & -.0378 & .0004 & .0019 \\ -.0025 & .0800 & -.0014 & -.0647 & .0188 & .0001 \\ -.0011 & .0399 & .0034 & -.0165 & .0057 & .0087 \\ -.0014 & .0547 & -.0205 & -.0001 & .0070 & -.0005 \\ -.0004 & .0105 & .0056 & -.0131 & -.0014 & .0023 \\ -.0086 & .0656 & .0191 & -.0307 & -.0617 & .0095 \\ -.0101 & .2088 & -.0095 & -.0300 & .0038 & .0086 \\ -.0033 & -.0021 & -.0294 & -.0086 & .0035 & -.0007 \\ .0013 & .0050 & .0187 & -.0022 & -.0013 & .0002 \\ .0081 & .0446 & -.0092 & .0964 & .0258 & -.0081 \\ .0244 & .1400 & .0039 & -.5574 & -.0419 & .0152 \\ .0147 & -.1752 & -.0013 & -1.2201 & -.0057 & -.0014 \\ .0103 & .2547 & .0006 & 1.9353 & -.0402 & .0117 \\ .0008 & .1287 & .0077 & -.1102 & .0106 & .0075 \\ .0006 & .0210 & .0002 & .0126 & .0066 & .0004 \\ .0000 & .0901 & -.0152 & .0000 & -.2026 & .0000 \\ .0000 & .0810 & .0026 & .0000 & -.0255 & .0000 \\ .0000 & .0544 & .0011 & .0000 & -.0037 & .0000 \\ .0000 & -.0668 & -.0035 & .0000 & .0039 & .0000 \end{bmatrix}$$

$$C = \begin{bmatrix} 0&1&0&0&0&0&0&0&0&0&0&0&0&0&0&0&0&0&0&0 \\ 0&0&0&1&0&0&0&0&0&0&0&0&0&0&0&0&0&0&0&0 \\ 0&0&0&0&0&0&0&0&0&0&0&1&0&0&0&0&0&0&0&0 \\ 0&0&0&0&0&0&0&0&0&0&0&0&0&1&0&0&0&0&0&0 \\ 0&0&0&0&0&0&0&0&0&0&0&0&0&0&0&1&0&0&0&0 \\ 0&0&0&0&0&0&0&0&0&0&0&0&0&0&0&0&0&1&0&0 \end{bmatrix},$$

and the performance index defined by

$$\widetilde{Q} = 0, \qquad \widehat{Q} = \mathrm{diag}(2,1,3,3,4,1), \qquad R = \mathrm{diag}(.1,.1,10,.1,10,1).$$

We do not list the solution matrix X, but only the optimal gain matrix, including the proportional (H) and integral parts (H_i):

$$H = \left[\begin{array}{ccccccc} 2.3056 & 3.0277 & 2.6554 & 2.6256 & 3.8078 & 4.0192 & .5064 \\ -2.1616 & -3.4193 & -.7843 & -1.3334 & -1.2145 & -1.5129 & .2908 \\ -.3147 & -.3791 & .4699 & .3174 & 5.8848 & 6.3278 & .7876 \\ .1541 & .2852 & .0589 & .0228 & .0562 & -.0028 & -.0118 \\ -.0897 & -.1307 & .3917 & .3964 & .0106 & .0654 & .1061 \\ 9.3825 & 10.7888 & -1.3226 & -1.2971 & -3.8211 & -4.8834 & -.6654 \end{array}\right.$$

$$\begin{array}{ccccccc} .8685 & -.7932 & -1.5419 & -10.0459 & -12.3655 & -3.4818 & -4.1008 \\ .7166 & -.2369 & -.3062 & -.6462 & -1.1528 & -.4062 & -.5934 \\ .8263 & 1.9352 & 2.0500 & 1.0170 & 1.0568 & .1739 & .1708 \\ -.0355 & .0080 & -.1188 & .3742 & .4850 & -.3862 & -.5416 \\ .1181 & .0707 & .1010 & .2074 & .2586 & .0459 & .0540 \\ -.7398 & -.3442 & -.3255 & -.6827 & -.8403 & .1886 & .2258 \end{array}$$

$$\left.\begin{array}{cccccc} 10.2728 & 12.1412 & 1.6255 & 1.7049 & 11.5659 & 12.7308 \\ -4.5725 & -6.3124 & -.1801 & -.2219 & -1.3064 & -1.3908 \\ -5.4487 & -5.5841 & -.3646 & -.3625 & -3.3071 & -3.2937 \\ .2320 & .3050 & .0002 & .0020 & .0217 & .0110 \\ -1.0023 & -1.1187 & .9616 & 1.0289 & 6.2009 & 7.1377 \\ -1.9065 & -2.3337 & .0776 & .0810 & .5424 & .6018 \end{array}\right],$$

$$H_i = \begin{bmatrix} -.4752 & -1.0223 & 2.8090 & 1.9118 & -1.8735 & -.4278 \\ .6243 & .5229 & 1.1088 & .4559 & 1.7616 & .0779 \\ -.0813 & -.2201 & -.0287 & -.0381 & .3386 & .0804 \\ -.0652 & -.0425 & -.3182 & .3894 & -.1152 & -.0018 \\ .0040 & -.0317 & -.0491 & -.0204 & .1284 & -.2691 \\ -1.0725 & .2865 & .1141 & -.0112 & .3077 & -.0201 \end{bmatrix}.$$

The estimate of the reciprocal condition number of matrix V_{12} was **RCOND** $\approx 1.67 \times 10^{-5}$. The residual 1-norm was about 1.74×10^{-8}, and that of the relative residual, 6.62×10^{-13}. Two iterations of the defect correction procedure **RicCorr**, comprising two Newton iterations (for $\varepsilon = 10^{-6}$), then reduced the residual and relative residual 1-norms to about 6.05×10^{-12}, and 2.30×10^{-16}, respectively. The closed-loop poles are:

$$\begin{aligned} \lambda = \{ & -.4318, -.3090, -.2144, -.1768, -.0581, .0678, .3003 \pm .2584j, .1213, \\ & .1589, .2326, .4828, .5954, .7241 \pm .1929j, .6551, .8170 \pm .2208j, \\ & .7352 \pm .1232j, .8052 \pm .1274j, .9456 \pm .0703j, .9712, .9307 \}. \end{aligned}$$

Note that we have changed the (14,13) element of A from 0 to 1 in order to get a nonsingular matrix A, because the authors of the cited reference state that the problem has been solved by the Schur vectors approach. We have also solved the problem with that element set to zero, and obtained the optimal gain matrices:

$$H = \left[\begin{array}{rrrrrrr} 3.2536 & 4.0965 & 3.8185 & 3.3863 & 4.0183 & 4.2404 & .8700 \\ -2.1101 & -3.3522 & -.6447 & -1.1614 & -1.3234 & -1.6062 & .3216 \\ -.3613 & -.4323 & .7300 & .5688 & 6.1060 & 6.5965 & .7104 \\ .1299 & .2622 & .0408 & -.0026 & .1527 & .0987 & -.0190 \\ -.1118 & -.1540 & .3192 & .3251 & .0091 & .0563 & .0884 \\ 9.3333 & 10.7361 & -1.4210 & -1.3613 & -3.9282 & -4.9994 & -.6955 \end{array} \right.$$

$$\begin{array}{rrrrrrr} 1.3327 & -.5742 & -1.1033 & -10.1346 & -12.1923 & .0000 & -2.0243 \\ .7405 & -.1976 & -.2601 & -.5716 & -1.0901 & .0000 & -.5724 \\ .7510 & 1.8126 & 1.9270 & 1.1240 & 1.1653 & .0000 & .0771 \\ -.0421 & .0294 & -.1060 & .2926 & .3654 & .0000 & -.5874 \\ .0994 & .0496 & .0739 & .1867 & .2339 & .0000 & .0238 \\ -.7790 & -.3796 & -.3740 & -.6567 & -.8216 & .0000 & .1109 \end{array}$$

$$\left. \begin{array}{rrrrrr} 8.1681 & 9.9105 & 1.4799 & 1.5486 & 10.6784 & 11.6812 \\ -4.8304 & -6.5741 & -.2090 & -.2496 & -1.5549 & -1.6441 \\ -5.7702 & -5.9177 & -.3269 & -.3247 & -2.9709 & -2.9565 \\ .3902 & .5121 & .0129 & .0163 & .0972 & .1013 \\ -.8668 & -.9748 & .9716 & 1.0391 & 6.2825 & 7.2225 \\ -1.7427 & -2.1718 & .0834 & .0876 & .5677 & .6377 \end{array} \right],$$

$$H_i = \begin{bmatrix} -.5943 & -1.2385 & 3.1560 & 1.3028 & -1.6459 & -.3868 \\ .6111 & .4997 & 1.1137 & .4186 & 1.7686 & .0833 \\ -.0835 & -.2077 & -.0854 & -.0389 & .3597 & .0718 \\ -.0624 & -.0476 & -.2772 & .4175 & -.1608 & -.0059 \\ .0075 & -.0217 & -.0523 & -.0118 & .1165 & -.2715 \\ -1.0664 & .3043 & .0803 & .0104 & .2993 & -.0219 \end{bmatrix}.$$

The residual and relative residual 1-norms were of about 7.48×10^{-9}, and 2.79×10^{-13}, respectively (without defect correction). □

References

Ammar, G., Benner, P., and Mehrmann, V. (1993). A multishift algorithm for the numerical solution of algebraic Riccati equations, *Electronic Trans. Numer. Anal.*, *1*: 33–48.

Anderson, E., Bai, Z., Bischof, C., Demmel, J., Dongarra, J., Du Croz, J., Greenbaum, A., Hammarling, S., McKenney, A., Ostrouchov, S., and Sorensen, D. (1992). *LAPACK Users' Guide*, SIAM, Philadelphia.

Anderson, B. D. O., and Moore, J. B. (1979). *Optimal Filtering*, Prentice-Hall, Englewood Cliffs, New Jersey.

Arnold, W. F., III, and Laub, A. J. (1984). Generalized eigenproblem algorithms and software for algebraic Riccati equations, *Proc. IEEE*, *72(12)*: 1746–1754.

Bender, D. J., and Laub, A. J. (1987a). The linear-quadratic optimal regulator for descriptor systems, *IEEE Trans. Automat. Control*, *AC-32(8)*: 672–688.

Bender, D. J., and Laub, A. J. (1987b). The linear-quadratic optimal regulator for descriptor systems: Discrete-time case, *Automatica*, *23(1)*: 71–85.

Byers, R. (1984a). A LINPACK-style condition estimator for the equation $AX - XB^T = C$, *IEEE Trans. Automat. Control*, *AC-29(10)*: 926–928.

Byers, R. (1984b). Numerical condition of the algebraic Riccati equation, *Linear Algebra and Its Role in Systems Theory*, volume 47 of *AMS Contemporary Math.*, pp. 35–49.

Cline, A. K., Moler, C. B., Stewart, G. W., and Wilkinson, J. H. (1979). An estimate for the condition number of a matrix, *SIAM J. Numer. Anal.*, *16(2)*: 368–375.

Demmel, J. W. (1987). On condition numbers and the distance to the nearest ill-posed problem, *Numer. Math.*, *51*: 251–289.

Demmel, J. W., and Kågström, B. (1987). Computing stable eigendecompositions of matrix pencils, *Lin. Alg. Appl.*, *88/89*: 139–186.

Emami-Naeini, A., and Franklin, G. F. (1980). Comments on "On the numerical solution of the discrete-time algebraic Riccati equation", *IEEE Trans. Automat. Control*, *AC–25(5)*: 1015–1016.

Emami-Naeini, A., and Franklin, G. F. (1982). Deadbeat control and tracking of discrete-time systems, *IEEE Trans. Automat. Control*, *AC–27(1)*: 176–181.

Fath, A. F. (1969). Computational aspects of the linear optimal regulator problem, *IEEE Trans. Automat. Control*, *AC–14*: 547–550.

Gaalman, G. J. (1980). Comments on "A nonrecursive algebraic solution for the discrete Riccati equation", *IEEE Trans. Automat. Control*, *AC–25(3)*: 610–612.

Gantmacher, F. R. (1959). *The Theory of Matrices*, volumes 1 and 2, Chelsea, New York.

Garbow, B. S., Boyle, J. M., Dongarra, J. J., and Moler, C. B. (1977). *Matrix Eigensystem Routines—EISPACK Guide Extension*, volume 51 of *Lect. Notes in Computer Science*, Springer-Verlag, Berlin.

Gohberg, I., Lancaster, P., and Rodman, L. (1986). On Hermitian solutions of the symmetric algebraic Riccati equations, *SIAM J. Control & Optimiz.*, *24*: 1323–1334.

Golub, G. H., and Van Loan, C. F. (1983). *Matrix Computations*, The Johns Hopkins University Press, Baltimore, Maryland.

Goodwin, G. C., and Payne, R. L. (1977). *Dynamic System Identification: Experiment Design and Data Analysis*, volume 136 of *Math. in Sci. & Engineering*, Academic Press, New York.

Hung, Y. S., and MacFarlane, A. G. J. (1982). *Multivariable Feedback: A Quasi-Classical Approach*, volume 40 of *Lect. Notes in Control and Information Sciences*, Springer-Verlag, Berlin.

Kågström, B. (1986). RGSVD—An algorithm for computing the Kronecker structure and reducing subspaces of singular $A - \lambda B$ pencils, *SIAM J. Sci. Stat. Comput.*, *7*: 185–211.

Katayama, T., Ohki, T., Inoue, T., and Kato, T. (1985). Design of an optimal controller for a discrete-time system subject to previewable demand, *Int. J. Control*, *41(3)*: 677–699.

Kaufman, L. C. (1975). The LZ algorithm to solve the generalized eigenvalue problem for complex matrices, *ACM Trans. Math. Softw.*, *1(3)*: 271–281.

Kenney, C., and Laub, A. J. (1989). Condition estimates for matrix functions, *SIAM J. Matrix Anal. Appl.*, *10*: 191–209.

Kenney, C., Laub, A. J., and Wette, M. (1989). A stability-enhancing scaling procedure for Schur-Riccati solvers, *Systems Control Lett.*, *12*: 241–250.

Laub, A. J. (1979). A Schur method for solving algebraic Riccati equations, *IEEE Trans. Automat. Control*, *AC–24(6)*: 913–921.

Laub, A. J. (1980). "On computing "balancing" transformations", Proceedings of 1980 Joint Automat. Contr. Conf., Sess. FA8-E, San Francisco, California.

Laub, A. J. (1985). Numerical linear algebra aspects of control design computations, *IEEE Trans. Automat. Control*, *AC–30(2)*: 97–108.

Laub, A. J., Heath, M. T., Paige, C. C., and Ward, R. C. (1987). Computation of system balancing transformations and other applications of simultaneous diagonalization algorithms, *IEEE Trans. Automat. Control*, *AC–32(2)*: 115–122.

Lawson, C. L., Hanson, R. J., Kincaid, D. R., and Krogh, F. T. (1979). Basic Linear Algebra Subprograms for Fortran usage, *ACM Trans. Math. Softw.*, *5(3)*: 308–323.

MacFarlane, A. G. J. (1963). An eigenvector solution of the optimal linear regulator problem, *J. Electron. Contr.*, *14*: 643–654.

Mehrmann, V. (1991). *The Autonomous Linear Quadratic Control Problem. Theory and Numerical Solution*, volume 163 of *Lect. Notes in Control and Information Sciences* (M. Thoma and A. Wyner, eds.), Springer-Verlag, Berlin.

Michelsen, M. L. (1979). On the eigenvalue-eigenvector method for solution of the stationary discrete matrix Riccati equation, *IEEE Trans. Automat. Control*, *AC–24(3)*: 480–481.

Moler, C. B., and Stewart, G. W. (1973). An algorithm for generalized matrix eigenvalue problems, *SIAM J. Numer. Anal.*, *10(2)*: 241–256.

Paige, C., and Van Loan, C. F. (1981). A Schur decomposition for Hamiltonian matrices, *Lin. Alg. Appl.*, *41*: 11–32.

Pappas, T., Laub, A. J., and Sandell, N. R. (1980). On the numerical solution of the discrete-time algebraic Riccati equation, *IEEE Trans. Automat. Control, AC-25(4)*: 631–641.

Patnaik, L. M., Viswanadham, N., and Sarma, I. G. (1980). Computer control algorithms for a tubular ammonia reactor, *IEEE Trans. Automat. Control, AC-25(4)*: 642–651.

Peters, G., and Wilkinson, J. H. (1970). $Ax = \lambda Bx$ and the generalized eigenproblem, *SIAM J. Numer. Anal., 7*: 479–492.

Petkov, P. H., Christov, N. D., and Konstantinov, M. M. (1987). On the numerical properties of the Schur approach for solving the matrix Riccati equation, *Systems Control Lett., 9*: 197–201.

Pontryagin, L. S., Boltyanskii, V., Gamkrelidze, R., and Mishenko, E. (1962). *The Mathematical Theory of Optimal Processes*, Interscience, New York.

Potter, J. E. (1966). Matrix quadratic solutions, *SIAM J. Appl. Math., 14*: 496–501.

Sima, V. (1981). Computational experience in solving the algebraic matrix Riccati equations, Report ICI–TR–07.81, I.C.I., Bucharest.

Sima, V. (1990). Comparison of some algorithms to solve the algebraic Riccati equations, *Studies and Researches in Computers and Informatics, 1(2)*: 77–96.

Sima, V. (1991). "On Hamiltonian-symplectic algorithms for solving algebraic Riccati equations", Proc. of the First European Control Conference ECC'91, July 2–5 1991, Grenoble, France (I. D. Landau et al., eds.), volume 3, Edition Hermes, Paris, pp. 2202–2207.

Stewart, G. W. (1972). On the sensitivity of the eigenvalue problem $Ax = \lambda Bx$, *SIAM J. Numer. Anal., 9*: 669–686.

Stewart, G. W. (1973a). *Introduction to Matrix Computations*, Academic Press, New York.

Stewart, G. W. (1973b). Error and perturbation bounds for subspaces associated with certain eigenvalue problems, *SIAM Rev., 15*: 727–764.

Stewart, G. W. (1976). Simultaneous iteration for computing invariant subspaces of non Hermitian matrices, *Numer. Math., 25*: 123–136.

Stewart, G. W. (1978). Perturbation theory for the generalized eigenvalue problem, *Recent Advances in Numerical Analysis* (C. De Boor and G. H. Golub, eds.), Academic Press, New York, pp. 193–206.

Van Dooren, P. (1979). The computation of Kronecker's canonical form of a singular pencil, *Lin. Alg. Appl.*, *27*: 103–141.

Van Dooren, P. (1980). A generalized eigenvalue approach for solving Riccati equations, Technical Report NA-80-02, Computer Science Department, Stanford University, California.

Van Dooren, P. (1981). A generalized eigenvalue approach for solving Riccati equations, *SIAM J. Sci. Stat. Comput.*, *2*(*2*): 121–135.

Van Dooren, P. (1982). Algorithm 590. DSUBSP and EXCHQZ: FORTRAN subroutines for computing deflating subspaces with specified spectrum, *ACM Trans. Math. Softw.*, *8*(*4*): 376–382.

Van Dooren, P. (1989). Numerical aspects of system and control algorithms, *Journal A*, *30*(*1*): 25–32.

Van Loan, C. F. (1975). A general matrix eigenvalue algorithm, *SIAM J. Numer. Anal.*, *12*: 819–834.

Varga, A., and Sima, V. (1985). "BIMAS — A basic mathematical package for computer-aided systems analysis and design", A Bridge between Control Science and Technology (J. Gertler and L. Keviczky, eds.), volume 1, Pergamon Press, Oxford.

Vaughan, D. R. (1969). A negative exponential solution for the matrix Riccati equation, *IEEE Trans. Automat. Control*, *AC-14*: 72–75.

Vaughan, D. R. (1970). A nonrecursive algebraic solution for the discrete Riccati equation, *IEEE Trans. Automat. Control*, *AC-15*(*5*): 597–599.

Ward, R. C. (1975). The combination shift QZ-algorithm, *SIAM J. Numer. Anal.*, *12*: 835–853.

Wilkinson, J. H. (1965). *The Algebraic Eigenvalue Problem*, Oxford University Press (Clarendon), London.

Wilkinson, J. H. (1978). Linear differential equations and Kronecker's canonical form, *Recent Advances in Numerical Analysis* (C. De Boor and G. H. Golub, eds.), Academic Press, New York, pp. 231–265.

Wilkinson, J. H. (1979). Kronecker's canonical form and the QZ algorithm, *Lin. Alg. Appl.*, *28*: 285–303.

Wonham, W. M. (1968). On a matrix Riccati equation of stochastic control, *SIAM J. Control*, *6*: 681–697.

Yue, A., Postlethwaite, I., and Padfield, G. (1989). H^∞ design and the improvement of helicopter handling qualities, *Vertica*, *13*(*2*): 119–132.

4

Structure-Preserving Algorithms

This chapter deals with the newly proposed *structure-preserving algorithms* for solving algebraic matrix Riccati equations. These algorithms exploit and/or preserve the Hamiltonian or symplectic structure of the optimal problem. Special emphasis is put here on continuous-time systems. One class of algorithms is based on the matrix sign function approach. Essentially, a variant of Newton process for finding a square root of identity matrix is used; on exploiting the structure, the general matrix inverses in the iterative process are replaced by symmetric indefinite matrix inverses and, finally, a linear system of equations is solved in a least squares sense. Special devices are incorporated to accelerate the convergence and enhance the numerical properties. Another class of algorithms determine a basis for the stable *invariant* or *deflating subspace* of an extended matrix, or matrix pencil, associated with the optimal problem, using variants of the QR or QZ algorithms based on symplectic transformations only. An algorithm mostly uses orthogonal symplectic transformations, but also needs some non-orthogonal transformations. The presentation focuses on another algorithm based on condition-controlled Gaussian symplectic transformations. Advanced numerical techniques, like symmetric updating scheme for computing the solution, the use of accurate approximations of eigenvalues as shifts in a QR-like process, and defect correction, are employed. The new multishift algorithm, that uses orthogonal transformations only, is also considered. Techniques for improving convergence speed and accuracy are discussed, and the numerical behaviour of algorithms is analysed.

4.1 Basic Theory

This chapter mainly investigates structure-exploiting/preserving algorithms for the numerical solution of continuous-time algebraic matrix Riccati equations, defined by (1.55) in § 1.1.2, re-written here for convenience

$$A^T X + XA - XBR^{-1}B^T X + Q = 0, \tag{4.1}$$

where $A \in \mathbb{R}^{n\times n}$, $B \in \mathbb{R}^{n\times m}$, $Q \in \mathbb{R}^{n\times n}$, and $R \in \mathbb{R}^{m\times m}$ are given matrices. Under certain conditions, like those in Theorems 1.16, 1.17, and 1.18 in § 1.1.2, the equation (4.1) has a unique symmetric positive semi-definite solution, $X \in \mathbb{R}^{n\times n}$. As shown in Chapter 3, this solution can be computed using a basis for the stable (or unstable) invariant subspace of an extended matrix G (of order $2n$), associated with the optimal problem,

$$G = \begin{bmatrix} A & -F \\ -Q & -A^T \end{bmatrix}, \qquad F := BR^{-1}B^T. \tag{4.2}$$

This basis is provided by the first n Schur vectors of G, $\begin{bmatrix} U_{11}^T & U_{21}^T \end{bmatrix}^T \in \mathbb{R}^{2n\times n}$; then $X = U_{21}U_{11}^{-1}$ (Laub, 1979). The QR algorithm has been used for determining and ordering the RSF of matrix G, to obtain such a basis. However, this approach has the disadvantage of not exploiting the particular structure of G, by involving special properties of its spectrum. (Matrix G is Hamiltonian.) Moreover, the symmetric matrix X is essentially obtained as a product of two matrices, U_{21} and U_{11}^{-1}, that could lead to non-symmetry and inaccuracy.

This section presents three main methods enabling the Hamiltonian structure to be taken into account: the matrix sign function approach, structure-preserving QR-type approach, and multishift approach. Specific algorithms will be described in the next sections. Each algorithm has both appealing features and limitations.

Having the solution X, the optimal gain matrix H can be computed by the formula

$$H := -R^{-1}B^T X. \tag{4.3}$$

For later reference, also define

$$J := \begin{bmatrix} 0 & I_n \\ -I_n & 0 \end{bmatrix}. \tag{4.4}$$

Clearly, JG is a symmetric matrix; for this reason, G is sometimes called *J-symmetric*. The basic theoretical results are stated in the next three subsections.

4.1.1 Matrix sign function method

Let $G \in \mathbb{R}^{p\times p}$ be a matrix having no purely imaginary eigenvalues. Let $K = V^{-1}GV$ be the Jordan canonical form of G; hence K is a block-diagonal matrix with Jordan blocks on its diagonal. The *matrix sign function* of G is defined by $\text{sign}(G) = VSV^{-1}$, where $S = \text{diag}(s_1, \ldots, s_p)$ and

$$s_i = \begin{cases} 1, & \text{if } \Re(k_{ii}) > 0, \\ -1, & \text{if } \Re(k_{ii}) < 0, \end{cases} \qquad i = 1{:}p.$$

Note that $\text{sign}(G)$ is not defined if G has purely imaginary eigenvalues. Moreover, if $\text{sign}(G)$ is defined, and U is a nonsingular matrix, then

$$\text{sign}(U^{-1}GU) = U^{-1}\text{sign}(G)U.$$

Since $\text{sign}(G)\text{sign}(G) = VS^2V^{-1} = I_p$, it follows that $\text{sign}(G)$ is a square root of I_p. Then, Newton's method for computing a square root of I_p, described by the relations

$$W_1 = G, \qquad W_{k+1} = W_k - (W_k - W_k^{-1})/2, \qquad k = 1, 2, \ldots, \tag{4.5}$$

converges to $\text{sign}(G)$. Note that such formulation of the iteration is numerically preferable to the mathematically equivalent formulation,

$$W_{k+1} = (W_k + W_k^{-1})/2. \tag{4.6}$$

If $\text{sign}(G)$ is not defined, then either the iterative process (4.5) does not converge, or one of the matrices W_k is singular.

The process (4.5) (or (4.6)) is the basis for the matrix sign function method. This method has been investigated in many papers, including (Balzer, 1980; Barraud, 1979, 1980; Beavers and Denman, 1974a, 1974b, 1974c; Byers, 1987; Denman and Beavers, 1976; Gardiner and Laub, 1986; Howland, 1983; Mattheys, 1978; Mehrmann, 1991; Roberts, 1980). One of the most elegant and efficient matrix sign function algorithms for solving continuous-time algebraic Riccati equations is that proposed by Byers

(1987). Our numerical experience ranked this Riccati solver on a prominent position among various solvers tested, in terms of efficiency and accuracy (Sima, 1994, 1995b).

Although the process (4.5) finally has quadratic convergence, the initial progress may be slow. The convergence rate can be improved by scaling W_k so that its eigenvalues are as close to ± 1 as possible. Since for $c > 0$, $\text{sign}(cG) = \text{sign}(G)$, each iterate W_k may be suitably scaled by a factor c_k. Scaling strategies for accelerating convergence have been proposed, for instance, in (Balzer, 1980; Barraud, 1979; Byers, 1987). According to Byers (1987), by choosing $c_k = |\det(W_k)|^{-1/p}$, i.e. the reciprocal of the geometric mean of the eigenvalues of W_k, the function $d(cW_k)$ gets minimized, where

$$d(W) := \sum_{\lambda \in \lambda(W)} (\ln |\lambda|)^2$$

is a measure of the distance between the spectrum $\lambda(W)$ and the unit circle in $\mathbb{C}$. (Moreover, $d(W) = 0$ if and only if $\lambda(W) \subset \{z \in \mathbb{C} : |z| = 1\}$, $d(W)$ increases with $|1 - |\lambda||$ for any $\lambda \in \lambda(W)$, and $d(W_k) \to 0$ when (4.5) does converge.) The iterative process (4.5) can therefore be replaced by

$$W_{k+1} = Z_k - (Z_k - Z_k^{-1})/2, \qquad Z_k := |\det(W_k)|^{-1/p} W_k. \tag{4.7}$$

The process (4.7) has the advantage of being invariant under scaling, i.e. the multiplication of W_1 by a positive constant does not incur the modification of the iterates W_k, $k = 2, 3, \ldots$. The determinant in (4.7) can be conveniently computed from the triangular factors of the factorization used to invert Z_k. Moreover, with this scaling strategy, for $n = 1$ the iteration reduces to the square root formula, and for $n = 2$ and $\lambda(G) \subset \mathbb{R}$, the iterative process converges in just two iterations.

An important argument for analyzing the matrix sign function method is its potential for exploiting the capabilities of modern computer architectures. An evolving toolbox for solving dense nonsymmetric eigenproblems on massively parallel computers, is based on matrix sign function algorithms (Bai and Demmel, 1994; Bai, Demmel and Gu, 1994). The techniques can be used for determining bases for invariant or deflating subspaces (for regular pencils). The basic operations involved are matrix multiplication, QR decomposition and matrix inversion. Moreover, it is possible to avoid all explicit matrix inversions, at the expense of a significantly increased computational effort (Bai, Demmel and Gu, 1994). The cited references give details on the perturbation theory of the matrix sign function, on the conditioning of its computation, and on the numerical stability of the associated algorithms. We shall present some of these new results, having in mind their applicability in solving Riccati equations.

The discussion below particularly considers the matrix G in (4.2), but the results are general, and can be applied to any invariant subspace corresponding to the eigenvalues of G in a specified domain. So, assume that $p = 2n$, as needed for solving Riccati equations. By appropriate permutations, sign(G) can be written as

$$\mathrm{sign}(G) = \widetilde{V}\mathrm{diag}(-I_n, I_n)\widetilde{V}^{-1}.$$

Then, the spectral projection corresponding to the stable eigenvalues of G is defined by $P = (I_{2n} - \mathrm{sign}(G))/2$. If the first n columns of an orthogonal matrix U span the range of P, then U yields the spectral decomposition

$$U^T G U = \begin{bmatrix} \widetilde{G}_{11} & \widetilde{G}_{12} \\ 0 & \widetilde{G}_{22} \end{bmatrix} =: \widetilde{G}, \tag{4.8}$$

where $\lambda(\widetilde{G}_{11}) \subset \mathbb{C}^-$, $\lambda(\widetilde{G}_{22}) \subset \mathbb{C}^+$. Matrix U can be computed, for example, from the QR factorization of P. Let R be the solution of the Sylvester equation

$$\widetilde{G}_{11}R - R\widetilde{G}_{22} = -\widetilde{G}_{12}. \tag{4.9}$$

Clearly, R exists and is unique, because $\widetilde{G}_{11}$ and $\widetilde{G}_{22}$ have no common eigenvalues (Section 2.3). Then, the spectral projector P is given by (§ 2.2.1)

$$P = \begin{bmatrix} I_n & R \\ 0 & 0 \end{bmatrix},$$

and $\|P\| = \sqrt{1 + \|R\|^2}$. (Here, $\|\cdot\|$ denotes the Euclidean norm.) It follows that

$$\mathrm{sign}(\widetilde{G}) = \begin{bmatrix} -I_n & 2R \\ 0 & I_n \end{bmatrix}.$$

Indeed, let

$$V := \begin{bmatrix} I_n & R \\ 0 & I_n \end{bmatrix},$$

where R satisfies (4.9). Then, $V^{-1}\widetilde{G}V = \mathrm{diag}(\widetilde{G}_{11}, \widetilde{G}_{22})$, and so,

$$\begin{aligned} \mathrm{sign}(\widetilde{G}) &= \mathrm{sign}(V\mathrm{diag}(\widetilde{G}_{11}, \widetilde{G}_{22})V^{-1}) \\ &= V\mathrm{diag}(-I_n, I_n)V^{-1} = \begin{bmatrix} -I_n & 2R \\ 0 & I_n \end{bmatrix}. \end{aligned}$$

Using the singular value decomposition (SVD) of R (see, for instance, (Golub and Van Loan, 1983)), $YRZ^H = \Sigma =: \mathrm{diag}(\sigma_i)$, the calculation of $\mathrm{sign}(\widetilde{G})$ (and $\mathrm{sign}(G)$) can be reduced to computing the SVD of the matrix

$$\begin{bmatrix} Y & 0 \\ 0 & Z \end{bmatrix} \mathrm{sign}(\widetilde{G}) \begin{bmatrix} Y^H & 0 \\ 0 & Z^H \end{bmatrix} = \begin{bmatrix} -I_n & 2\Sigma \\ 0 & I_n \end{bmatrix},$$

which, by permutations, is equivalent to computing the SVD of the 2×2 matrices

$$\begin{bmatrix} -1 & 2\sigma_i \\ 0 & 1 \end{bmatrix}.$$

It can be shown (Bai and Demmel, 1994) that

$$\|\mathrm{sign}(G)\| = \|\mathrm{sign}(G^{-1})\| = \rho := \|R\| + \sqrt{1 + \|R\|^2}.$$

So, $\mathrm{cond}(\mathrm{sign}(G)) = \rho^2$. On the other hand,

$$\|R\| \le \|\widetilde{G}_{12}\| / \mathrm{sep}(\widetilde{G}_{11}, \widetilde{G}_{22}),$$

where sep has been defined at § 2.2.1. Therefore, the conditioning of the matrix sign function computations is closely related to the quantity $\mathrm{sep}(\widetilde{G}_{11}, \widetilde{G}_{22})$. If $\widetilde{G}_{12}$ is of moderate size, an ill-conditioned matrix sign function means large $\|R\|$, which in turn means small $\mathrm{sep}(\widetilde{G}_{11}, \widetilde{G}_{22})$.

The analysis in (Bai and Demmel, 1994) reveals that if G has eigenvalues close to the imaginary axis (say, with distance to this axis less than $\varepsilon_M^{1/2}$), the intermediate matrices in Newton's process (4.6) can become numerically singular, and the process can fail to converge or can misconverge. Another important result is that for G close "enough" to $\mathrm{sign}(G)$, and $\mathrm{cond}(\mathrm{sign}(G)) < \varepsilon_M^{-1/2}$, then Newton's process in floating point arithmetic will compute $\mathrm{sign}(G)$ with an absolute error bounded by $\varepsilon_M^{1/2} \|\mathrm{sign}(G)\|$.

It is also shown in (Bai and Demmel, 1994) that a condition number for matrix sign function is $1/d_G^2$, where d_G is the distance of the eigenvalues of G to the imaginary axis,

$$d_G := \min_{\tau \in \mathbb{R}} \sigma_{\min}(\mathrm{j}\tau I_{2n} - G),$$

$\sigma_{\min}(\cdot)$ denoting the minimal singular value. In practice, d_G can be replaced by the easy to compute quantity

$$\Delta(G) := \min_i |\Re(\lambda_i(G))|.$$

If $\Delta(G) < \varepsilon_M^{1/2}$, then Newton's process may fail to converge or to compute the matrix sign function within the absolute error $\varepsilon_M^{1/2} \|\mathrm{sign}(G)\|$, even if $\mathrm{sign}(G)$ is well conditioned. If $\mathrm{cond}(\mathrm{sign}(G)) > \varepsilon_M^{-1/2}$, then Newton's process may similarly fail, even if $\Delta(G)$ is not small.

The matrix sign function algorithm can also be used in the generalized eigenvalue problem $K - \lambda L$, by applying, for instance, the process (4.6) to KL^{-1}. An implicit scheme is used, for avoiding inversion of L,

$$K_{k+1} = (K_k + LK_k^{-1}L)/2, \qquad k = 1, 2, \ldots,$$

where $K_1 = K$. The sequence $\{ K_k \}$ converges quadratically if L is nonsingular and $K - \lambda L$ has no purely imaginary eigenvalues. There are extensions in case L is singular. This technique suggests how discrete-time optimal problems can be approached. Another technique is considered in (Gardiner and Laub, 1986).

An efficient 3-step algorithm is proposed in (Bai, Demmel and Gu, 1994) for computing bases for invariant or deflating subspaces, which tries the fastest but least stable method first, and resorts to slower but more stable methods if necessary. The matrix sign function algorithm is used first to split the spectrum. If it succeeds, the calculations are terminated. If it fails, an inverse-free algorithm is similarly tried. If it also fails, the QR (or QZ) algorithm is used. The inverse-free algorithms are based on rank-revealing QR factorizations, and so, in the general (non-Hamiltonian, or non-symplectic) case, they are by 6–7 times more expensive than matrix sign function algorithms.

The results presented above stand for general eigenproblems. A specialization of the matrix sign function algorithm for solving continuous-time algebraic Riccati equations, which exploits the particular Hamiltonian structure of G, will be discussed at § 4.2.1.

4.1.2 Structure-preserving QR-type methods

In the following (sub)sections, we shall use the matrix

$$M := \begin{bmatrix} A & F \\ Q & -A^T \end{bmatrix},$$

($F := BR^{-1}B^T$), that slightly differs from G in (4.2). We shall generally refer to the $n \times n$ blocks of such $2n \times 2n$ matrices by $(\cdot)_{ij}$, $i, j = 1\!:\!2$. Assume that M has no purely imaginary eigenvalues. If $[\, S_{11}^T \;\; S_{21}^T \,]^T$ generates the stable M-invariant subspace and S_{11} is nonsingular, then $X = -S_{21}S_{11}^{-1}$ is the symmetric positive semi-definite solution of the Riccati equation (4.1).

In this subsection, we shall point to how to efficiently use the structure of the matrix M for obtaining an algorithm for solving the continuous-time Riccati equations (4.1). The key idea is to compute a basis for the stable invariant subspace by a structure-preserving QR-like algorithm, exclusively using similarity transformations with symplectic matrices. Remember that a matrix S is *symplectic* if $S^T JS = J$, where J is defined in (4.4). Moreover, an interesting property of symplectic matrices $S = (S_{ij}), S_{ij} \in \mathbb{R}^{n\times n}$, $i, j = 1{:}2$, is that $S_{11} = S_{22}$ and $S_{21} = -S_{12}$ if and only if S is orthogonal (see (3.48), § 3.3.1).

Let us denote by $\mathcal{S}_{2n}$ the class of symplectic matrices of order $2n$. By definition, if $S, S' \in \mathcal{S}_{2n}$, then $S^{-1} \in \mathcal{S}_{2n}$ and $SS' \in \mathcal{S}_{2n}$. (In the latter case, we have $(SS')^T J(SS') = S'^T(S^T JS)S' = S'^T JS' = J$.) Consider first symplectic matrices which are also orthogonal. Two main classes of such matrices can be identified. The first one is the class of *Householder symplectic* matrices, which have the form

$$H(k,u) = \begin{bmatrix} U & 0 \\ 0 & U \end{bmatrix}, \qquad U = I_n - 2uu^T/u^T u, \tag{4.10}$$

where $u = (0, \ldots, 0, u_k, \ldots, u_n)^T$. The second class is formed by the *Jacobi* or *Givens symplectic* matrices, which have the form

$$J(k,c,s) = \begin{bmatrix} \Gamma & \Sigma \\ -\Sigma & \Gamma \end{bmatrix}, \qquad \Gamma, \Sigma \in \mathbb{R}^{n\times n}, \tag{4.11}$$

where

$$\Gamma = \operatorname{diag}(\underbrace{1, \ldots, 1}_{k-1}, c, 1, \ldots, 1), \qquad \Sigma = \operatorname{diag}(\underbrace{0, \ldots, 0}_{k-1}, s, 0, \ldots, 0),$$

and $c^2 + s^2 = 1$. Paige and Van Loan (1981) proved that any orthogonal symplectic matrix is a product of Householder symplectic and Jacobi symplectic matrices.

The basic theoretical result is the following theorem, adapted from (Paige and Van Loan, 1981).

Theorem 4.1 *If $M \in \mathbb{R}^{2n\times 2n}$ is a Hamiltonian matrix without eigenvalues on the imaginary axis in $\mathbb{C}$, then there is an* orthogonal *symplectic matrix S, so that*

$$S^T MS =: \bar{T} = \begin{bmatrix} T & R \\ 0 & -T^T \end{bmatrix}, \qquad T, R \in \mathbb{R}^{n\times n}, \tag{4.12}$$

where T is in QTF or RSF, $R = R^T$, and $\lambda(T) \subset \mathbb{C}^-$.

One says that $\bar{T}$ in Theorem 4.1 is in the *Hamiltonian (quasi-)triangular form*, or *Hamiltonian-Schur form*.

The proof of Theorem 4.1 is done by induction. For the $n = 1$ case, M has the form

$$M = \begin{bmatrix} a & f \\ q & -a \end{bmatrix},$$

hence its eigenvalues are $\lambda = (a^2 + fq)^{1/2}$ and $-\lambda$. By assumption, M has two real eigenvalues of opposite signs. From § 2.2.4, we know that a plane rotation (which is a symplectic matrix), U, can be chosen so that $U^T MU$ be upper triangular, with diagonal elements $-\lambda$ and λ.

Assume now that a Hamiltonian quasi-triangular form exists for Hamiltonian matrices of order less than or equal to $2(n-1)$ that have no purely imaginary eigenvalues. Let $\lambda \in \lambda(M)$, hence

$$M \begin{bmatrix} y \\ z \end{bmatrix} = \lambda \begin{bmatrix} y \\ z \end{bmatrix}, \qquad y, z \in \mathbb{C}^n, \tag{4.13}$$

where $y \neq 0$, $z \neq 0$, and suppose $\Re(\lambda) < 0$. By writing down (4.13) by blocks, premultiplying the first equality so obtained by z^H, postmultiplying the conjugate transpose of the second equality by y, and adding the results, we have

$$y^H Qy + z^H Fz = (\lambda + \bar{\lambda}) y^H z, \tag{4.14}$$

which implies that $y^H z \in \mathbb{R}$. Note that this is a property of the eigensystem of any Hamiltonian matrix. Assume first that λ is real, hence $y, z \in \mathbb{R}^n$. We can choose a Householder transformation, $U_1 = I_n - 2u_1 u_1^T / u_1^T u_1$, so that $z \leftarrow U_1 z = \alpha_1 e_1$ ($\alpha_1 \in \mathbb{R}$), next, a plane rotation, G, defined by c and s, so that $\begin{bmatrix} y_1 & z_1 \end{bmatrix}^T \leftarrow G \begin{bmatrix} y_1 & z_1 \end{bmatrix}^T$ has the second element zero, and, finally, another Householder transformation, $U_2 = I_n - 2u_2 u_2^T / u_2^T u_2$, so that $y \leftarrow U_2 y = \alpha_2 e_1$ ($\alpha_2 \in \mathbb{R}$). Clearly, the matrix $Z := H(1, u_2) J(1, c, s) H(1, u_1)$ is orthogonal and symplectic, and, by construction, we have for y and z in (4.13)

$$Z \begin{bmatrix} y \\ z \end{bmatrix} = \alpha \begin{bmatrix} e_1 \\ 0 \end{bmatrix}, \qquad \alpha \neq 0, \quad e_1 \in \mathbb{R}^n.$$

This implies, by similarity, that $(ZMZ^T)e_1 = \lambda e_1$ ($e_1 \in \mathbb{R}^{2n}$), and, using the Hamiltonian structure,

$$M \leftarrow ZMZ^T = \begin{bmatrix} \lambda & w^T & \beta & s^T \\ 0 & A_1 & s & F_1 \\ 0 & 0 & -\lambda & 0 \\ 0 & Q_1 & -w & -A_1^T \end{bmatrix},$$

where $\beta \in \mathbb{R}$, and A_1, F_1, and Q_1 are $(n-1) \times (n-1)$ matrices, with F_1 and Q_1 symmetric. Thus, the eigenvalues of M are λ and $-\lambda$, together with the eigenvalues of the Hamiltonian matrix

$$M_1 := \begin{bmatrix} A_1 & F_1 \\ Q_1 & -A_1^T \end{bmatrix}. \tag{4.15}$$

Consider now that $\lambda = \gamma + \mathrm{j}\mu$, with $\gamma,\ \mu \in \mathbb{R}$, $\gamma \neq 0$, $\mu \neq 0$. (Clearly, $\bar{\lambda}$ is also an eigenvalue.) Then (4.13) is equivalent to

$$M\begin{bmatrix} u & v \\ r & s \end{bmatrix} = \begin{bmatrix} u & v \\ r & s \end{bmatrix}\begin{bmatrix} \gamma & \mu \\ -\mu & \gamma \end{bmatrix}, \tag{4.16}$$

where

$$y = u + \mathrm{j}v, \qquad z = r + \mathrm{j}s, \qquad u, v, r, s \in \mathbb{R}^n.$$

As above, we can find an orthogonal symplectic matrix Z_1, as a product of three transformations, so that

$$Z_1\begin{bmatrix} u & v \\ r & s \end{bmatrix} = \begin{bmatrix} \alpha e_1 & g \\ 0 & h \end{bmatrix}, \qquad g, h \in \mathbb{R}^n, \quad 0 \neq \alpha \in \mathbb{R}.$$

But $\left[(\alpha e_1 + \mathrm{j}g)^T \quad \mathrm{j}h^T\right]^T$ is an eigenvector of the Hamiltonian matrix $Z_1 M Z_1^T$ corresponding to the eigenvalue λ with $\Re(\lambda) \neq 0$. From the property implied by (4.14), it follows that $(\alpha e_1 + \mathrm{j}g)^H(\mathrm{j}h) \in \mathbb{R}$, and so, $h_1 = 0$. Similarly, we can determine an orthogonal symplectic matrix Z_2, so that

$$Z_2\begin{bmatrix} \alpha e_1 & g \\ 0 & h \end{bmatrix} = \begin{bmatrix} V \\ 0 \end{bmatrix}, \qquad V = \begin{bmatrix} \alpha & \eta \\ 0 & \kappa \end{bmatrix} \in \mathbb{R}^{2\times 2}.$$

Matrix V must be nonsingular, because if not, we had $\eta u = \alpha v$ and $\eta r = \alpha s$, which would imply that $\mu = 0$ (since $y = (1 + \mathrm{j}\eta/\alpha)u$, $z = (1 + \mathrm{j}\eta/\alpha)r$). Defining $Z = Z_2 Z_1$ and using (4.16), we obtain

$$MZ^T\begin{bmatrix} I_2 \\ 0 \end{bmatrix}V = MZ^T Z\begin{bmatrix} u & v \\ r & s \end{bmatrix} = \begin{bmatrix} u & v \\ r & s \end{bmatrix}\begin{bmatrix} \gamma & \mu \\ -\mu & \gamma \end{bmatrix},$$

which is equivalent to

$$ZMZ^T\begin{bmatrix} I_2 \\ 0 \end{bmatrix} = \begin{bmatrix} I_2 \\ 0 \end{bmatrix}V\begin{bmatrix} \gamma & \mu \\ -\mu & \gamma \end{bmatrix}V^{-1} =: \begin{bmatrix} L \\ 0 \end{bmatrix},$$

where $L \in \mathbb{R}^{2\times 2}$. Consequently,

$$M \leftarrow ZMZ^T = \begin{bmatrix} L & W^T & N & S^T \\ 0 & A_1 & S & F_1 \\ 0 & 0 & -L^T & 0 \\ 0 & Q_1 & -W & -A_1^T \end{bmatrix},$$

where A_1, F_1, $Q_1 \in \mathbb{R}^{(n-2)\times(n-2)}$. Now, for both $\lambda \in \mathbb{R}$ or $\lambda \in \mathbb{C}$, let

$$S = Z\begin{bmatrix} I & 0 & 0 & 0 \\ 0 & \widetilde{Z}_{11} & 0 & \widetilde{Z}_{12} \\ 0 & 0 & I & 0 \\ 0 & -\widetilde{Z}_{12} & 0 & \widetilde{Z}_{11} \end{bmatrix}, \qquad \widetilde{Z} = \begin{bmatrix} \widetilde{Z}_{11} & \widetilde{Z}_{12} \\ -\widetilde{Z}_{12} & \widetilde{Z}_{11} \end{bmatrix},$$

where $\widetilde{Z}$ is an orthogonal symplectic matrix reducing M_1 in (4.15) to the form (4.12), and I has order one or two. The induction argument proves the theorem.

An algorithm that uses transformations like those applied above will be described in § 4.3.1. The result stated by Theorem 4.1 has been extended by W.-W. Lin to Hamiltonian matrices having eigenvalues with even multiplicity on the imaginary axis (see (Ammar, Benner and Mehrmann, 1993) and the references therein). Similar results are reported for symplectic matrices in the discrete-time case (Mehrmann, 1991).

The Hamiltonian-triangular form is very appealing in both theory and practice. The original idea in (Paige and Van Loan, 1981) has been that of using it in a QR-like process for computing invariant subspaces. Unfortunately, the Hamiltonian-triangular form is not invariant under the QR process, and restoring it would involve an $0(n^3)$ computational effort. For this reason, other "canonical" triangular-like forms for Hamiltonian and symplectic matrices have also been investigated, for instance, in (Byers, 1986; Laub and Meyer, 1974; Mehrmann, 1988). A numerically stable QR-type algorithm, called the *Hamiltonian QR-algorithm*, has been proposed by Byers (1986) for the special case when either F or Q is of rank one. (This corresponds to an optimal problem for a single-input or single-output system.) A similar algorithm for discrete-time systems is studied in (Mehrmann, 1988). Byers' algorithm iterates on a matrix M in *Hamiltonian-Hessenberg form*, i.e. for which M_{11} is in upper Hessenberg form (UHF), $M_{12} = M_{12}^T$, $M_{22} = -M_{11}^T$, and $M_{21} = \alpha e_n e_n^T$. The Hamiltonian-Hessenberg form is *unreduced* if $m_{i+1,i} \neq 0$, for $i = 1{:}n-1$, and $\alpha \neq 0$. This special form is invariant under the Hamiltonian-QR algorithm, because it is in fact a permuted version of a matrix in UHF. Indeed, using a similarity transformation $\bar{P} = \operatorname{diag}(I_n, \bar{I}_n)$, where $\bar{I}_n = \begin{bmatrix} e_n & \dots & e_2 & e_1 \end{bmatrix}$ is the identity matrix with columns taken in the reverse order, it is easy-to-see that a matrix in Hamiltonian-Hessenberg form is similar to the usual UHF. Thus, the property of unreduced Hessenberg matrices (see Theorem 2.8 in § 2.2.3) implies that an orthogonal matrix that transforms M into unreduced Hamiltonian-Hessenberg form is essentially determined by its first column.

Unfortunately, no numerically stable algorithm of complexity $0(n^3)$, that operates with orthogonal symplectic similarities, has so far been known for reducing a general Hamiltonian (or symplectic) matrix (rank$(F) > 1$ and rank$(Q) > 1$) to Hamiltonian-Hessenberg form. An explanation why this is a difficult task is given in (Ammar and Mehrmann, 1991). Specifically, the following theorem is proven.

Theorem 4.2 *Let $M \in \mathbb{R}^{2n\times 2n}$ be Hamiltonian. Then there is an orthogonal symplectic matrix S so that $S^T MS$ is an unreduced Hamiltonian-*

Hessenberg matrix if and only if the set of nonlinear equations

$$x^T JM^{2i-1}x = 0, \qquad i = 1\colon n-1; \qquad x^T x = 1, \tag{4.17}$$

has a solution vector that is not contained in an M-invariant subspace of dimension less than or equal to n.

We prove the "only if" part of the theorem. Let $S \in \mathcal{S}_{2n}$, $S^T S = I_{2n}$, be so that $\widetilde{M} := S^T MS$ is in unreduced Hamiltonian-Hessenberg form, and $x := Se_1$. Clearly, $x^T x = 1$. As $\widetilde{M}$ is unreduced, it follows that x does not belong to an M-invariant subspace of dimension less than or equal to n. Now, using $JS = SJ$ and $S^T x = e_1$, we obtain

$$\begin{aligned} x^T JM^{2i-1}x &= x^T JS(S^T MS)^{2i-1}S^T x \\ &= e_{n+1}^T (S^T MS)^{2i-1} e_1 = e_{2n}^T Z^{2i-1} e_1, \end{aligned}$$

where $Z := \bar{P}S^T MS\bar{P}$ is in UHF. Thus, $z_{2n,1}^j = 0$, for $j = 1\colon 2n-2$, and so (4.17) holds.

So, the first column x of the orthogonal symplectic matrix reducing M to Hamiltonian-Hessenberg form must satisfy the set of nonlinear equations (4.17), and (4.17) is a necessary and sufficient condition for the existence of an orthogonal symplectic matrix performing the mentioned reduction. Note that the proof above is still valid for $2i-1$ replaced by i, for $i = 1\colon 2n-2$. A sufficient condition for the existence of a vector x satisfying (4.17) will be mentioned in § 4.1.3.

Similar results exist for symplectic matrices (Ammar and Mehrmann, 1991; Mehrmann, 1991).

Actually, the same set of equations has to be satisfied even if non-orthogonal symplectic transformations were used (Raines and Watkins, 1992). Thus, any method for reducing a matrix to Hamiltonian-Hessenberg form must, at least implicitly, solve the nonlinear equations (4.17). To be more specific we need several definitions.

A subspace $\mathcal{I} \in \mathbb{R}^{2n}$ is called *isotropic* if $x^T Jy = 0$ for all $x, y \in \mathcal{I}$. A *Lagrangian subspace* is an isotropic subspace of $\mathbb{R}^{2n}$ of maximal dimension, i.e., not included in an isotropic subspace of larger dimension. A suitable result is that the n-dimensional stable invariant subspace of M is a Lagrangian subspace. Lagrangian subspaces are important for algebraic Riccati equations because the subspace spanned by the columns of $[\, I \;\; X^T \,]^T$ (or $[\, -I \;\; X^T \,]^T$) is Lagrangian if and only if X is symmetric. Note that the subspace spanned by the first (or the last) n columns of a symplectic matrix is Lagrangian, because $S^T JS = J$. Let $S \in \mathcal{S}_{2n}$, $s_i := s_{:,i}$; the nth *Krylov subspace* generated by s_1 for M,

$$\mathcal{K} := \mathcal{R}(s_1, Ms_1, \ldots, M^{n-1}s_1), \tag{4.18}$$

will generically have rank n.

Theorem 4.3 *Let $M \in \mathbb{R}^{2n\times 2n}$ be Hamiltonian, $S \in \mathcal{S}_{2n}$, and assume that $\widetilde{M} := S^{-1}MS$ is in Hamiltonian-Hessenberg form, and that $\mathcal{K}$ in (4.18) has dimension n. Then $\mathcal{K}$ is Lagrangian.*

The proof will be through if showing that, for $j = 1{:}\,n$,

$$\mathcal{R}(s_1, Ms_1, \ldots, M^{j-1}s_1) = \mathcal{R}(s_1, s_2, \ldots, s_j),$$

because then $\mathcal{K} = \mathcal{R}(s_1, \ldots, s_n)$ is Lagrangian. The equality of these subspaces is established by induction on j, using the first $n-1$ columns of the equation $MS = S\widetilde{M}$, as in the usual Hessenberg case (§ 2.2.3). (Matrix $\widetilde{M}$ is upper Hessenberg in its first $n-1$ columns.)

Clearly, $\mathcal{K}$ is Lagrangian only if nonlinear equations like (4.17) are satisfied by $x = s_1$. Other important results will be given in § 4.1.3.

The considerations above justify the research effort being directed to finding other canonical forms and algorithms, and accepting the use of non-orthogonal transformations. For instance, Bunse-Gerstner (1986) and Della-Dora (1975) studied general QR-like methods, in which the QR decompositions are replaced by other decompositions. One of the proposed algorithms (Bunse-Gerstner and Mehrmann, 1986), called the *SR algorithm*, also needs some non-orthogonal symplectic transformations. Another, more efficient, algorithm (Bunse-Gerstner, Mehrmann and Watkins, 1989) exclusively uses non-orthogonal transformations and lays the basis for the implementations to be discussed in Section 4.3. A variant of the SR algorithm based on Gaussian elimination is used, combined with the symmetric updating method in (Byers and Mehrmann, 1986) and with Van Loan's algorithm (Van Loan, 1984) to economically obtain accurate approximations of the eigenvalues of M. In conjunction with Newton's algorithm in a defect correction procedure (Mehrmann and Tan, 1988), this approach is very efficient and accurate.

Several definitions should precede the presentation below. A matrix $M = (M_{ij})$, $M_{ij} \in \mathbb{R}^{n\times n}$, $i,j = 1{:}\,2$, is called *J-Hessenberg (matrix)* if M_{11}, M_{21} and M_{22} are upper triangular matrices, and M_{12} is in UHF. The matrix M is *unreduced* if M_{21} is nonsingular and M_{12} is in unreduced UHF, i.e. it has no zero elements on the subdiagonal. The matrix M is called *J-triangular* if M_{11}, M_{12}, M_{21} and M_{22} are upper triangular, and M_{21} has zero diagonal elements, and M is *J-tridiagonal* if M_{11}, M_{21} and M_{22} are diagonal, and M_{12} is tridiagonal. It follows that a J-Hessenberg Hamiltonian matrix Q is J-tridiagonal and Hamiltonian, that is Q_{21}, $Q_{11} = -Q_{22}$ are diagonal, and Q_{12} is symmetric tridiagonal.

Let $M \in \mathbb{R}^{2n\times 2n}$. A decomposition $M = SR$, where $S \in \mathcal{S}_{2n}$ and R is J-triangular, is called an *SR decomposition* of M. Denote

$$\begin{aligned} \widehat{P} &:= [e_1, e_3, \ldots, e_{2n-1}, e_2, e_4, \ldots, e_{2n}], \\ K(M,x,j) &:= [x, Mx, M^2x, \ldots, M^{j-1}x], \end{aligned}$$

where $K(M,x,j)$ is the *Krylov matrix* of dimension $2n \times j$. Note that if R is a J-triangular matrix, then $\widehat{P}R\widehat{P}^T$ is upper triangular, and if T is an (unreduced) J-Hessenberg matrix, then $\widehat{P}T\widehat{P}^T$ is an (unreduced) matrix in UHF. From this it follows that the product of two J-triangular matrices is J-triangular. There are the following results (Bunse-Gerstner and Mehrmann, 1986; Elsner, 1979).

Theorem 4.4 *Let $M \in \mathbb{R}^{2n\times 2n}$ be nonsingular. Then there is a matrix $S \in \mathcal{S}_{2n}$ and a J-triangular matrix R so that $M = SR$ if and only if all leading principal minors of even dimension of the matrix $\widehat{P}M^TJM\widehat{P}^T$ are nonzero.*

Theorem 4.5 *Let $M \in \mathbb{R}^{2n\times 2n}$, $S \in \mathcal{S}_{2n}$, $s_1 := Se_1$, $K := K(M, s_1, 2n)$.*

(i) Let K be nonsingular. If $K\widehat{P} = SR$ is an SR decomposition, then $T = S^{-1}MS$ is an unreduced J-Hessenberg matrix.

(ii) If $T = S^{-1}MS$ is a J-Hessenberg matrix, then $K\widehat{P}$ has an SR decomposition, $K\widehat{P} = SR$. If, in addition, T is unreduced, then R is nonsingular.

(iii) Let $\widetilde{S} \in \mathcal{S}_{2n}$ be so that $\widetilde{S}e_1 = \lambda s_1$, $\lambda \in \mathbb{R} \setminus \{0\}$, and let $T = S^{-1}MS$ and $\widetilde{T} = \widetilde{S}^{-1}M\widetilde{S}$ be J-Hessenberg matrices, where T is unreduced. Then there is a matrix D, defined as

$$D = \begin{bmatrix} C & N \\ 0 & C^{-1} \end{bmatrix} \in \mathcal{S}_{2n},$$

where $C = diag(c_1, \ldots, c_n)$, $N = diag(n_1, \ldots, n_n)$, such that $S = \widetilde{S}D$ and $T = D^{-1}\widetilde{T}D$.

Theorems 4.4 and 4.5 characterize those matrices which do not admit SR decompositions, or transformations into J-Hessenberg form, respectively. Using Theorem 4.4, it can be shown that the set of matrices with no SR decomposition has the Lebesgue measure zero. Indeed, it follows that there is a finite number of parameters $k \in \mathbb{R}$ for which $M - kI$ has no SR decomposition, since the leading principal minor of order $2i$ of $\widehat{P}(M-kI)^TJ(M-kI)\widehat{P}^T$, $i \in \{1, \ldots, n\}$, is a polynomial of degree $4i$

in k. Hence the set of matrices of order $2n$ having an SR decomposition is dense in $\mathbb{R}^{2n\times 2n}$. The SR decomposition of a matrix M is essentially *unique*, in the sense that if $M = SR$ and $M = \widetilde{S}\widetilde{R}$ are two such decompositions, then there is a matrix D as in Theorem 4.5, such that $S = \widetilde{S}D$ and $R = D^{-1}\widetilde{R}$ (Bunse-Gerstner and Mehrmann, 1986). Moreover, it can be shown (Bunse-Gerstner, Mehrmann and Watkins, 1989) that for any nonderogatory (§ 2.2.1) Hamiltonian matrix M, without purely imaginary eigenvalues, all vectors s_1, except for a set of zero measure, allow the transformation of M into J-Hessenberg form.

Generally, one cannot accomplish the SR decomposition of a matrix M only using orthogonal symplectic matrices, since a necessary and sufficient condition for this is that M has the form

$$M = \begin{bmatrix} A & B \\ -B & A \end{bmatrix} \widetilde{R},$$

where $A, B \in \mathbb{R}^{n\times n}$ and $\widetilde{R}$ is J-triangular.

The SR decomposition and the usage of symplectic transformations for reduction to J-Hessenberg form (or J-tridiagonal form, respectively) make the fundamentals for a symplectic QR-like method applied to a (Hamiltonian) matrix. Between the reduction of a matrix to J-Hessenberg form and the SR decomposition there is exactly the same relation as that between the reduction to the UHF and the QR algorithm. Specific algorithms will be discussed in Section 4.3.

Other results concerning the structure-preserving methods for solving algebraic matrix Riccati equations, including the discrete-time case and generalizations to matrix pencils, can be found, for instance, in (Bunse-Gerstner, Byers and Mehrmann, 1989a; Bunse-Gerstner and Mehrmann, 1989; Flaschka, Mehrmann and Zywietz, 1991; Lin, 1989; Mehrmann, 1991). More general structured eigenvalue problems are dealt with in (Bunse-Gerstner, Byers and Mehrmann, 1989b, 1992). Structure-preserving orthogonal algorithms for computing the eigenvalues of symplectic pencils are described in (Lin, 1987; Mehrmann, 1991; Patel, 1992).

4.1.3 Multishift method

Recently, Ammar, Benner and Mehrmann (1993) proposed an algorithm for solving continuous-time algebraic Riccati equations using only orthogonal symplectic transformations. This algorithm actually finds a Lagrangian subspace. It is based on the following result (Ammar and Mehrmann, 1991).

Theorem 4.6 *Let $M \in \mathbb{R}^{2n\times 2n}$ be a Hamiltonian matrix, and $\mathcal{I}$ be an isotropic M-invariant subspace of $\mathbb{R}^{2n}$. Then, any $x \in \mathcal{I}$ with $\|x\|_2 = 1$ satisfies (4.17).*

Indeed, $M^k x =: y \in \mathcal{I}$, $k = 1{:}\,2n$, because $\mathcal{I}$ is M-invariant, and $x^T J y = 0$, because $\mathcal{I}$ is isotropic.

A vector x satisfying the conditions in Theorem 4.6 does not determine an unreduced Hamiltonian-Hessenberg matrix, but it can be used to find a stable invariant subspace. The next theorem (Ammar and Mehrmann, 1991) makes it clear.

Theorem 4.7 *Let $M \in \mathbb{R}^{2n\times 2n}$ be a Hamiltonian matrix with eigenvalues $\lambda_1, \ldots, \lambda_{2n}$ (with multiplicities counted). Assume that there exists an n-dimensional real Lagrangian M-invariant subspace $\mathcal{I}$, corresponding to the eigenvalues $\lambda_1, \ldots, \lambda_n$. Then*

$$(M - \lambda_1 I_{2n}) \cdots (M - \lambda_n I_{2n}) e_1$$

is contained in the Lagrangian M-invariant subspace corresponding to the eigenvalues $\lambda_{n+1}, \ldots, \lambda_{2n}$.

Indeed, the matrix $(M - \lambda_1 I_{2n}) \cdots (M - \lambda_n I_{2n})$ has rank less than or equal to n, and is real, because the Lagrangian subspace is real (i.e. the set $\{\lambda_1, \ldots, \lambda_n\}$ is closed under complex conjugation). Therefore, there is an orthogonal matrix Q such that

$$Q^T (M - \lambda_1 I_{2n}) \cdots (M - \lambda_n I_{2n}) = \begin{bmatrix} R_{11} & R_{12} \\ 0 & 0 \end{bmatrix},$$

where R_{11} is an $n \times n$ upper triangular matrix. So,

$$(M - \lambda_1 I_{2n}) \cdots (M - \lambda_n I_{2n}) Q = Q \begin{bmatrix} \widetilde{R}_{11} & \widetilde{R}_{12} \\ 0 & 0 \end{bmatrix},$$

and, by partitioning $Q = \begin{bmatrix} Q_1 & Q_2 \end{bmatrix}$, with $Q_1, Q_2 \in \mathbb{R}^{2n\times n}$, it follows that the columns of Q_1 form a basis for the M-invariant subspace corresponding to the eigenvalues $\lambda_{n+1}, \ldots, \lambda_{2n}$, which is the Lagrangian subspace $J\mathcal{I}$. Thus, $Q_1 e_1 \in J\mathcal{I}$, and the property follows.

Note that there exist Lagrangian M-invariant subspaces when M is dichotomic, or has purely imaginary eigenvalues with even multiplicity. Such matrices appear in certain H^∞ optimal control problems (Clements and Glover, 1989).

The result above suggests an algorithm for computing a particular Lagrangian M-invariant subspace. This algorithm is based on another canonical form, the so-called Paige-Van Loan (PVL) form. A Hamiltonian matrix

$M = (M_{ij})$ is in *Paige-Van Loan (PVL) form* if M_{11} is in UHF, M_{21} is diagonal, and $M_{12} = M_{12}^T$. There is no structure-preserving QR algorithm for the PVL form, because PVL is not a true Hessenberg form, contrary to the Hamiltonian-Hessenberg form. Indeed, the similarity transformation $\bar{P}$ that reverses the last n rows and columns yields a matrix with non-zero elements on positions $(n+i, n-i+1)$, $i = 1{:}n$. Several illuminating connections between the PVL form, Lagrangian subspaces and Hamiltonian-Hessenberg form are illustrated in the following theorem, adapted from (Raines and Watkins, 1992).

Theorem 4.8 *Let $M \in \mathbb{R}^{2n\times 2n}$ be Hamiltonian, $S \in \mathcal{S}_{2n}$, and assume that $\widetilde{M} := S^{-1}MS$ is in PVL form, and that $\mathcal{K}$ in (4.18) is Lagrangian (of dimension n). Then*

(i) $\widetilde{M}$ is in Hamiltonian-Hessenberg form.

(ii) $\mathcal{R}(s_1, Ms_1, \ldots, M^j s_1) = \mathcal{R}(s_1, \ldots, s_{j+1})$, $j = 1{:}n-1$.

The proof is by induction on j. Given $j \le n-1$, suppose that for $k \le j-1$,

$$\widetilde{m}_{n+k,k} = 0 \quad \text{and} \quad \mathcal{R}(s_1, \ldots, M^k s_1) = \mathcal{R}(s_1, \ldots, s_{k+1}). \tag{4.19}$$

But

$$s_j^T JMs_j = s_j^T JS\widetilde{m}_j = s_j^T Js_{n+j}\widetilde{m}_{n+j,j} + \sum_{i=1}^{j+1} s_j^T Js_i\widetilde{m}_{ij} = \widetilde{m}_{n+j,j},$$

because $MS = S\widetilde{M}$ and $S^TJS = J$ (hence $s_j^T Js_{n+j} = 1$, $s_j^T Js_i = 0$, for $i = 1{:}j+1$). Now, from (4.19), written for $k = j-1$, it follows that $s_j \in \mathcal{R}(s_1, \ldots, M^{j-1}s_1) \subseteq \mathcal{K}$. So, $Ms_j \in \mathcal{R}(Ms_1, \ldots, M^j s_1) \subseteq \mathcal{K}$. Therefore, $\widetilde{m}_{n+j,j} = s_j^T JMs_j = 0$, because $\mathcal{K}$ is isotropic. Clearly, s_j can be expressed as

$$s_j = \alpha_0 s_1 + \cdots + \alpha_{j-1}M^{j-1}s_1,$$

where $\alpha_{j-1} \ne 0$, because

$$s_j \notin \mathcal{R}(s_1, \ldots, s_{j-1}) = \mathcal{R}(s_1, \ldots, M^{j-2}s_1).$$

Equating the jth columns of $MS = S\widetilde{M}$, and using $\widetilde{m}_{n+j,j} = 0$, we have

$$\begin{aligned} \alpha_0 Ms_1 + \cdots + \alpha_{j-1}M^j s_1 &= Ms_j = S\widetilde{m}_j \\ &= s_1\widetilde{m}_{1j} + \cdots + s_{j+1}\widetilde{m}_{j+1,j}, \end{aligned}$$

hence

$$s_{j+1}\widetilde{m}_{j+1,j} = \alpha_0 Ms_1 + \cdots + \alpha_{j-1}M^j s_1 - s_1\widetilde{m}_{1j} - \cdots - s_j\widetilde{m}_{jj}.$$

But $s_i \in \mathcal{R}(s_1, \ldots, M^{i-1}s_1)$, for $i = 1\colon j$, and so,

$$s_{j+1}\widetilde{m}_{j+1,j} = \sum_{i=0}^{j} \gamma_i M^i s_1,$$

where $\gamma_j = \alpha_{j-1} \neq 0$. Since s_1, Ms_1, ..., $M^j s_1$ are linearly independent, we deduce that $\widetilde{m}_{j+1,j} \neq 0$. Thus, $s_{j+1} \in \mathcal{R}(s_1, \ldots, M^j s_1)$, and so, $\mathcal{R}(s_1, \ldots, s_{j+1}) \subseteq \mathcal{R}(s_1, \ldots, M^j s_1)$. These two subspaces must be equal because s_1, ..., s_{j+1} are linearly independent.

The general case when $\mathcal{K}$ has dimension $k \leq n$ is proven in essentially the same way (Raines and Watkins, 1992). The generalizations of Theorems 4.3 and 4.8 are as follows.

Theorem 4.9 *Let $M \in \mathbb{R}^{2n\times 2n}$ be Hamiltonian, $S \in \mathcal{S}_{2n}$, and assume that $\mathcal{K}$ in (4.18) has dimension $k \leq n$, and $\widetilde{M} := S^{-1}MS$ is in PVL form with $\widetilde{m}_{n+j,j} = 0$, for $j = 1\colon k-1$. Then $\mathcal{K}$ is isotropic.*

Theorem 4.10 *Let $M \in \mathbb{R}^{2n\times 2n}$ be Hamiltonian, $S \in \mathcal{S}_{2n}$, and assume that $\widetilde{M} := S^{-1}MS$ is in PVL form, and that $\mathcal{K}$ in (4.18) is isotropic and invariant. Then*

(i) $\widetilde{m}_{n+j,j} = 0$, for $j = 1\colon k$.

(ii) $\mathcal{R}(s_1, Ms_1, \ldots, M^j s_1) = \mathcal{R}(s_1, \ldots, s_{j+1})$, $j = 1\colon k-1$.

Note that $\mathcal{K}$ needs be M-invariant when $k < n$. The results presented above will be used in Section 4.4.

4.2 Matrix Sign Function Algorithm

The matrix sign function method presented in § 4.1.1 will now be specialized to an algorithm for solving continuous-time algebraic matrix Riccati equations. Computational issues, as well as the scope of the matrix sign function-based Riccati solvers are also discussed.

4.2.1 Algorithmic templates

Let us show how the process (4.7) can be used to solve *continuous-time* Riccati equations. Note that if $X \in \mathbb{R}^{n\times n}$ is a solution of the Riccati equation (4.1), then

$$\begin{bmatrix} A & -F \\ -Q & -A^T \end{bmatrix} = \begin{bmatrix} I & 0 \\ X & I \end{bmatrix} \widetilde{G} \begin{bmatrix} I & 0 \\ -X & I \end{bmatrix}, \tag{4.20}$$

holds, where

$$\widetilde{G} := \begin{bmatrix} A - FX & -F \\ 0 & -(A - FX)^T \end{bmatrix}, \qquad F := BR^{-1}B^T. \tag{4.21}$$

(This follows immediately by performing the multiplications in the right hand side of (4.20) and taking the symmetry of matrices F and X into account. Note that $\widetilde{G}$ in (4.21) differs from $\widetilde{G}$ in (4.8), but not qualitatively.) If, in addition, X is a stabilizing solution, then the matrix $A - FX$ is stable, as from (4.3) one obtains that $-FX = BH$. Note that the matrix in the left hand side of (4.20) is just the Hamiltonian matrix G from (4.2) and that, in the right hand side,

$$\begin{bmatrix} I & 0 \\ -X & I \end{bmatrix} = \begin{bmatrix} I & 0 \\ X & I \end{bmatrix}^{-1}.$$

According to Theorem 1.16 (§ 1.1.2), the assumption that X is a stabilizing solution implies that $\text{sign}(G)$ is defined, so taking the structure of (4.20) into account, we obtain

$$\text{sign}(G) =: \begin{bmatrix} W_{11} & W_{12} \\ W_{21} & W_{22} \end{bmatrix} = \begin{bmatrix} I & 0 \\ X & I \end{bmatrix} \begin{bmatrix} -I & Z \\ 0 & I \end{bmatrix} \begin{bmatrix} I & 0 \\ -X & I \end{bmatrix}. \tag{4.22}$$

Since matrices G and $\text{sign}(G)$ commute, from (4.20)–(4.22) we have

$$\widetilde{G} \begin{bmatrix} -I & Z \\ 0 & I \end{bmatrix} = \begin{bmatrix} -I & Z \\ 0 & I \end{bmatrix} \widetilde{G},$$

and equating the $(1\colon n, n+1\colon 2n)$ submatrices of both sides of this relation, it follows that Z is the solution of the Lyapunov equation

$$(A - FX)Z + Z(A - FX)^T = 2F.$$

From (4.22), we have

$$\begin{bmatrix} W_{11} + W_{12}X & W_{12} \\ W_{21} + W_{22}X & W_{22} \end{bmatrix} = \begin{bmatrix} -I & Z \\ -X & XZ + I \end{bmatrix}, \tag{4.23}$$

hence, denoting by N and M the matrices formed by the first and last n columns of the matrix $\text{sign}(G) + I_{2n}$, respectively, we deduce that X is determined by the relation $MX = -N$. This relation is equivalent to a system of $2n^2$ linear algebraic equations with n^2 unknowns. The system is compatible (consistent), because

$$\begin{aligned} [\,-X \;\; I_n\,]M &= [\,-X \;\; I_n\,] \begin{bmatrix} W_{12} \\ W_{22} + I_n \end{bmatrix} \\ &= [\,-X \;\; I_n\,] \begin{bmatrix} Z \\ XZ + 2I_n \end{bmatrix} = 2I_n, \end{aligned}$$

(hence rank$(M) = n$), and it can be solved using QR decomposition without pivoting.

Matrix G in (4.2) is Hamiltonian. Since the multiplication of a matrix by a constant, inversion of a matrix and the addition of two matrices preserve the Hamiltonian property, it follows that all the iterates of the process (4.7) will be Hamiltonian matrices. As $(JZ_k)^{-1}J = Z_k^{-1}$, where J has been defined in (4.4) and, by definition, the matrix JZ_k is symmetric, it follows that in (4.7) we can replace the inversion of a general matrix by that of an indefinite symmetric matrix. The following algorithm is obtained.

Algorithm signm : $(X, k) = \texttt{signm}(\ A, B, Q, R, k_{max}\ )$

/* Solve continuous-time ARE (4.1) using matrix */
/* sign function method. */

/* 1. Construct the matrix G from (4.2). */

$$F = BR^{-1}B^T; \quad W = \begin{bmatrix} A & -F \\ -Q & -A^T \end{bmatrix}; \quad J = \begin{bmatrix} 0 & I_n \\ -I_n & 0 \end{bmatrix};$$

/* 2. Iterate. */

FOR $k = 1\colon k_{max}$,

/* Update W_k in W. */

$d = |\det(JW)|^{1/(2n)}$; $V = d(JW)^{-1}$; $Z = W/d$; $\kappa = \text{cond}(JW)$;
$W = Z - (Z - VJ)/2$; $s = ||Z - VJ||$;
IF $s/2 \le \varepsilon_M ||Z||$ OR $k \ge itma$ AND
$s \le 2n\varepsilon_M \kappa ||V||$, THEN GO TO 4;

END FOR

"No convergence"; /* 3. No convergence. */

/* 4. Solve in X a compatible least squares problem. */

$$(R, U) = \texttt{QR}(\begin{bmatrix} W_{12} \\ W_{22} + I_n \end{bmatrix}); \quad X \leftarrow -U^T \begin{bmatrix} W_{11} + I_n \\ W_{21} \end{bmatrix};$$

$X = \texttt{solve}(\ R, X\ )$;

Algorithm **signm** and the presentation above differ from those in (Byers, 1987); the same matrix G employed by Algorithm **Schv** (§ 3.3.1) can be directly used here.

4.2.2 Computational issues

Note that two convergence criteria are used in Algorithm **signm**. The first criterion terminates the calculations when the iterates W_k stop being significantly modified. The second criterion takes into account unavoidable rounding-errors produced during the calculation of the inverse of JW. The relative errors are $0(\varepsilon_M \text{cond}(JW))$, hence the iteration must be stopped when the computed correction is of this order. In practice, the first criterion sometimes acts too late (or never), and the second one, too soon. Consequently, a compromise had to be reached. Byers (1987) proposed to use the value 8 for *itma*; our experiments let us observe that a less value would help.

Generally, rounding-errors in the computations and stopping the iterative process after a finite number of iterations will affect the computed matrix $\widehat{X}$. It is better to refine $\widehat{X}$ and estimate the error (Mehrmann and Tan, 1988). An improvement can be made by symmetrization of the result, i.e. by computing $(\widehat{X} + \widehat{X}^T)/2$, which is just the Frobenius projection of $\widehat{X}$ onto the space of symmetric matrices. The defect correction procedure `RicCorr`, described in § 1.5.4, can then be employed. If E is the solution of the Riccati equation (1.132), then $X := \widehat{X} + E$ satisfies the given equation (4.1). In this manner, an approximate solution $\widehat{X}$ can be improved by solving (1.132) in E and replacing $\widehat{X}$ by $\widehat{X} + E$. Usually, the correction E is small. Any algorithm, none the less Algorithm `signm`, can be used for solving equation (1.132). However, the option for Newton's algorithm, initialized for $E = 0$ (hence with $H = 0$), is a good one. In practice, the defect correction phase can repeatedly take place, since rounding-errors will affect the calculation of E, obtaining a matrix $\widehat{E}$. If the first significant bits of $\widehat{E}$ are correct, then $\widehat{X} + \widehat{E}$ is a more accurate solution than $\widehat{X}$. In this way, if the initial Riccati equation is not too ill conditioned, then the achievable accuracy is only limited by the precision of machine arithmetic, the conditioning of equation and the precision used for the residual computation. Numerical stability can thus be obtained. Often, the limit accuracy is achieved after one or two corrections. Even if no improvement is obtained, the error estimate, $X - \widehat{X} = E \approx \widehat{E}$, is of correct order.

The subroutines `DSIFA` (or `DSICO`), `DSIDI`, `DQRDC` and `DQRSL` in the LINPACK package (Dongarra et al., 1979), or their counterparts in the LAPACK package (Anderson et al., 1992), may be used for implementing Algorithm `signm`. The first two subroutines factor and invert an indefinite symmetric matrix, respectively, and the last two routines compute and apply the QR decomposition (§ 1.4.3), respectively. Clearly, the multiplications by J are not actually performed, but the elements of W, V and Z are rearranged. The matrix W may conveniently be represented as the lower triangle of the symmetric matrix JW. The upper triangle can be used by `DSIFA` and `DSIDI` to compute $(JW)^{-1}$ and $\det(JW)$.

In implementing Algorithm `signm` the main points were to reduce the memory needed for arrays and increase efficiency as much as possible, by performing column-wise computations, or exploiting parallelism. If the LAPACK routines are used, then good efficiency for several categories of

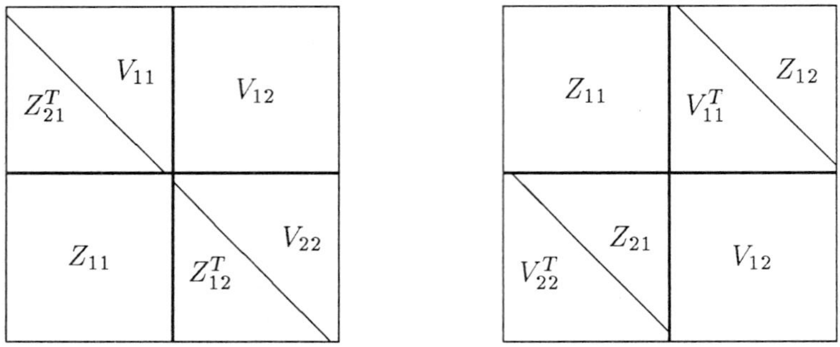

Fig. 4.1 Contents of the array A before and after the second restructuring.

advanced computer architectures is implicitly obtained.

Almost all relevant information is processed in an array A that has $2n$ rows and columns. At the beginning and the end of each iteration k, A contains the matrix W_k (W in Algorithm signm). The data stored in A are restructured so that the upper triangle contains the upper triangular part of the symmetric matrix JW, and the strict lower triangle together with a working array W contain information enabling to reconstruct the matrix W. The inverse of JW is obtained in the same space, using the combination DSIFA + DSIDI. (Note that DSIDI also gives the inertia (the numbers of positive, zero and negative eigenvalues) of JW. If $0 \in \lambda(JW)$, then $0 \in \lambda(W)$, and this allows to detect an error situation.) The relevant portions of the matrices V and Z are scaled by d and $1/d$, respectively, and then, $||V||_1$ and $||Z||_1$ are computed, using column operations and the previously obtained results. The content of array A is once more restructured, for the following processing convenience. The structure of A before and after the second restructuring is illustrated on left and on right of Fig. 4.1, respectively.

V_{11} and V_{22} in Fig. 4.1 denote the upper triangles, and Z_{21}^T and Z_{12}^T denote the strict lower triangles of the $n \times n$ symmetric matrices V_{11}, V_{22}, Z_{21} and Z_{12}, respectively. This is suggested by the position of the diagonal lines in the figure. In addition, the array W contains in order the diagonal elements of the matrices Z_{21} and $-Z_{12}$. In a loop, one computes the 1-norm of the first n columns of $Z - VJ$ and the first n columns of the updated matrix W, and in another loop the same calculations are performed for the last n columns. The symmetry and Hamiltonian structure are always taken into account (W, Z, VJ and $Z - VJ$ are Hamiltonian, and V is symmetric), including, for instance, the calculations of the norms of matrices V, Z and $Z - VJ$. The number $\kappa = \text{cond}(JW)$ is given by the product $||V||_1 \, ||Z||_1$, but it is not actually evaluated. Byers (1987) used its estimation, computed by the `DSICO` routine in LINPACK, but to use `DSIFA` would be better in this context. The rest of calculations closely follow Algorithm `signm`. We chose the value 4 for the parameter *itma*, since our experiments had shown that this value was suitable, enabling a quick detection of the convergence and at the same time granting sufficient accuracy (if combined with the defect correction procedure previously described). It is advisable to monitor the condition numbers of the iterates W_k and return the maximal condition number encountered.

There are several possibilities for implementing the calculation of the residual Y, defined by (1.130), of the Riccati equation solution, $\widehat{X}$. One of them uses the optimal gain matrix H in (4.3), which is often needed in practice. In this case, one computes the upper triangle of the matrix $A^T X$, to which the upper triangle of $X(A + BH) + Q$ is then added. The matrix $A + BH$ is obtained using matrix-matrix operations (see (1.105) in § 1.4.1). The computation of the residual Y and the improvement of the computed solution with the defect correction procedure is illustrated in Example 4.1.

Algorithm `signm` needs a space of $4n^2$ memory locations, hence with $0.5n^2$ locations less than that required by Algorithm `Ntnc` (§ 2.5.1).

Algorithm `signm` requires about $(4n^3 + 10n^2)k + 5n^3$ *flops*, where k the number of iterations executed. The first term represents the effort spent at stage 2, and the second term, the effort spent at stage 4. Note that an iteration costs about three times less than an iteration of Newton's algorithm. In our experiments (Sima, 1992b, 1993a), the number of iterations needed to obtain convergence of the algorithm to machine precision (in double precision), was generally very small (at most 11, and usually 6 iterations).

It is important that a similar algorithm be devised for the discrete-time case. The general, non-symplectic case is suggested in (Bai and Demmel, 1994; Bai, Demmel and Gu, 1994), but exploiting the symplectic structure is desirable.

4.2.3 Applicability and limitations

Algorithm `signm` can be used only for continuous-time matrix Riccati equations. To construct the matrix G, needed for initializing Algorithm `signm`, one assumes that matrix R is nonsingular.

If Algorithm `signm` terminates without achieving convergence, then X contains the result computed after k iterations of the algorithm. (In the implementation described at § 4.2.2, that result is returned in the first n rows and columns of array `A`.)

Since `signm` does not ensure the symmetry of the computed solution and, on each iteration of the algorithm, one computes the inverse of an indefinite symmetric matrix, it is advisable to employ symmetrization and iterative refinement of the solution. Algorithm `Ntnc` can be used for this purpose.

Let $\widehat{X}$ be the computed solution, and $\widehat{H}$ the gain matrix obtained from (4.3) with X replaced by $\widehat{X}$. Algorithm `Ntnc`, initialized by $A + B\widehat{H}$ (instead of A), Y (instead of Q), and $H = 0$, gives the solution E of (1.132) and the corresponding gain matrix, H_e. So, $\widehat{X} + E$ and $\widehat{H} + H_e$ are better approximations than $\widehat{X}$ and $\widehat{H}$ of the solution of Riccati equation (4.1) and of the gain matrix (4.3), respectively. Sometimes, in practice, the computed residual Y is not a positive semi-definite matrix; consequently, the positivity test described at § 2.5.2 should not be used.

The procedure above of symmetrizing and refining an approximate solution can also be applied to the methods using the (generalized) Schur vectors (§ 3.3.1, § 3.4.1).

We conclude this section by giving some examples of matrix sign algorithm use. The first example is in fact a template algorithm useful for solving a frequently encountered control problem. Other examples report numerical results obtained for specific problems.

Example 4.1 The following algorithm computes the solution of the algebraic matrix Riccati equation (4.1), and the corresponding optimal gain matrix H, using matrix sign function method. The system output is weighted with the matrix $\widetilde{Q}$ in the quadratic performance index, hence output regulation (Problem 1.3 in § 1.2.1) is intended. A proportional-plus-integral optimal control law (Problem 1.4) is used. It is assumed that the matrices A, B, C, $\widetilde{Q}$ and $\widehat{Q}$ in (1.66) in § 1.2.1, of dimensions $n \times n$, $n \times m$, $l \times n$, $l \times l$ and $l \times l$, respectively, as well as R, of dimension $m \times m$, are given. The computed approximate solutions X and H are then improved using symmetrization and *one iteration* of the defect correction procedure with Newton's algorithm (§ 1.5.4, § 2.5.1).

Algorithm sgn : $(X, H, k, \widetilde{k}) = \textbf{sgn}(\ A, B, C, \widetilde{Q}, \widehat{Q}, R\ )$

/* Solve continuous-time ARE (4.1) for Problem 1.4 */
/* using matrix sign function method and Newton correction. */

/* 1. Initialize parameters. */

$\varepsilon = \varepsilon_M^{1/2}; \quad k_{max} = 30;$

/* 2. Construct matrices for proportional-plus-integral control. */

$$A \leftarrow \begin{bmatrix} A & 0 \\ -C & 0 \end{bmatrix}; \quad B \leftarrow \begin{bmatrix} B \\ 0 \end{bmatrix}; \quad C \leftarrow \begin{bmatrix} C & 0 \\ 0 & I_l \end{bmatrix}; \quad Q = \begin{bmatrix} \widetilde{Q} & 0 \\ 0 & \widehat{Q} \end{bmatrix};$$

/* 3. Compute state weighting matrix. Save A. */

$Q \leftarrow C^TQC; \quad \widetilde{A} = A;$

/* 4. Solve Riccati equation. */
$(X, k) = \texttt{signm}(\ \widetilde{A}, B, Q, R, k_{max}\ );\quad X \leftarrow (X + X^T)/2;$

/* 5. Use one iteration of the defect correction procedure. */
$\widetilde{H} = -R^{-1}B^TX;\quad \widetilde{A} = A + B\widetilde{H};$
$Y = A^TX + XA - XBR^{-1}B^TX + Q;\quad H = 0;$
$(\widetilde{X}, H, \widetilde{k}) = \texttt{Ntnc}(\ \widetilde{A}, B, Y, R, H, k_{max}, \varepsilon\ );\quad X \leftarrow X + \widetilde{X};\quad H \leftarrow H + \widetilde{H};$

If matrix $\widehat{Q}$ is not given, then $l = 0$, and the construction at step 2 is not made. □

Example 4.2 Algorithm `sgn` in Example 4.1 was used for the turbogenerator model in (Hung and MacFarlane, 1982) and the weighting matrices in Example 3.9 (§ 3.3.3). The results X and H computed by `signm` were identical to those listed previously. Algorithm `signm` converged in 11 iterations. All the elements of the residual matrix Y were less than 10^{-6} in magnitude; the 1-norm of this matrix was 1.68×10^{-6}; dividing it by the norm of solution, we obtained 7.54×10^{-9}. Algorithm `Ntnc` performed two iterations; the computed correction had elements less than 10^{-7} (in magnitude), and the final value of the convergence criterion was 9.19×10^{-10}. The residual 1-norm was 5.78×10^{-12} and the relative residual 2.59×10^{-14}. For $\varepsilon = 5 \times 10^{-12}$, `Ntnc` converged in three iterations, with the final value of the criterion 2.02×10^{-16}. □

For other numerical examples, which we tested with Algorithm `signm`, the number of iterations was less than above (at most 7).

Example 4.3 Consider the continuous-time optimal problem defined by the matrices in Example 2.1 (§ 2.2.1). This example has already been studied in (Ammar, Benner and Mehrmann, 1993; Arnold and Laub, 1984; Bai, Demmel and Gu, 1994). It is known that the separation of stable and unstable eigenvalues of the corresponding matrix G, diminishes as ε approaches zero. (Note that ε has another meaning than in Algorithm `sgn`.) Table 4.1 gives the number of iterations of Algorithm `signm`, and the 1-norm of the residual Y of the computed solution $\widehat{X}$ for various values of ε. It is apparent that the accuracy of the result deteriorates as ε is reduced. The number of iterations does not increase as ε diminishes, contrary to the results reported in (Bai, Demmel and Gu, 1994), where the general process (4.6) was used. This behaviour is probably motivated by the use of scaling strategy and by exploiting the problem structure. We have also applied the defect correction procedure, initialized by the computed solution $\widehat{X}$. Fully accurate residuals have been obtained, but for $\varepsilon \leq 10^{-4}$ three iterations of the procedure (summing up to five Newton iterations) were needed. □

Table 4.1 Number of Algorithm **signm** iterations and the residual 1-norm for Example 4.3.

ε	k	$\|Y\|_1$
1	7	3.18×10^{-14}
10^{-1}	8	2.44×10^{-14}
10^{-2}	7	2.27×10^{-12}
10^{-3}	7	1.97×10^{-10}
10^{-4}	6	1.96×10^{-08}
10^{-5}	6	1.21×10^{-06}
10^{-6}	6	1.87×10^{-04}
10^{-7}	6	1.45×10^{-02}

Example 4.3 illustrates the difficulties encountered by the matrix sign function algorithm for problems with closed-loop poles near the imaginary axis.

4.3 Structure-Preserving QR-Type Algorithms

Among the structure-preserving QR-type algorithms proposed so far for solving general continuous-time Riccati equations, based on the methods discussed in § 4.1.2, we consider a variant of the SR algorithm using non-orthogonal symplectic transformations. This algorithm is illustrative for our purposes. Moreover, most of the transformation matrices involved can easily be replaced by orthogonal matrices; however, it is still impossible to restrict all transformations to orthogonal transformations. The SR variant dealt with here has greater potential efficiency than the "orthogonal" variant, and it has proven effective in solving many linear-quadratic optimization problems.

4.3.1 Algorithmic templates

We shall first describe an algorithm that computes a J-triangular form of a Hamiltonian matrix M, based on elementary symplectic transformations with Gaussian elimination matrices, as defined below. Let $k \in \{1, \ldots, n\}$, $v \in \mathbb{R}$, $w = (0, \ldots, 0, w_{k+1}, \ldots, w_n)^T \in \mathbb{R}^n$. Define

$$G_1(k, v) = \begin{bmatrix} I_n & V_1 \\ 0 & I_n \end{bmatrix}, \qquad V_1 = v(e_{k-1}e_k^T + e_k e_{k-1}^T); \quad (4.24)$$

$$G_2(k,v) = \begin{bmatrix} I_n & V_2 \\ 0 & I_n \end{bmatrix}, \qquad V_2 = ve_ke_k^T; \tag{4.25}$$

$$G_3(k,w) = \begin{bmatrix} L & 0 \\ 0 & L^{-T} \end{bmatrix}, \qquad L = I_n + we_k^T. \tag{4.26}$$

Note that a matrix of the form

$$\begin{bmatrix} L & V \\ 0 & L^{-T} \end{bmatrix}$$

is symplectic if $L^{-1}V$ is symmetric. Consequently, G_1, G_2 and G_3 are symplectic. Moreover, they are easily invertible, since

$$G_i^{-1}(k,v) = \begin{bmatrix} I_n & -V_i \\ 0 & I_n \end{bmatrix}, \qquad i = 1{:}\,2,$$

$$G_3^{-1}(k,w) = \begin{bmatrix} L^{-1} & 0 \\ 0 & L^T \end{bmatrix}, \qquad L^{-1} = I_n - we_k^T.$$

These matrices can be used to annihilate certain elements of a vector $x \in \mathbb{R}^{2n}$. So, if for given k and x, we have $x_{n+k-1} = 0$ only if $x_k = 0$, then choosing $v = 0$ if $x_k = 0$ and $v = -x_k/x_{n+k-1}$, otherwise, it follows that $y := G_1(k,v)x$ has the property that $y_k = 0$. Similarly, choosing $v = 0$ if $x_k = 0$ and $v = -x_k/x_{n+k}$, otherwise, then $(G_2(k,v)x)_k = 0$. For numerical stability, the latter definition of v would be used only if $|x_k| \le |x_{n+k}|$. Also, to annihilate $x_{n+k} \neq 0$, one chooses $v = -x_{n+k}/x_k$, so that $(G_2^T(k,v)x)_{n+k} = 0$. The matrix G_3 can be used for simultaneously annihilating $n-k$ elements of a vector $x \in \mathbb{R}^{2n}$. Assuming that $|x_k| \ge |x_{k+j}|$, $j = 1{:}\,n-k$, by choosing $w_i = 0$ if $x_k = 0$ and $w_i = -x_i/x_k$, otherwise, $i = k+1{:}\,n$, then it follows that $(G_3(k,w)x)_i = 0$, $i = k+1{:}\,n$. To annihilate elements from the second half of x, one should use $G_3^T(k,w)$, where $w_i = x_{n+i}/x_{n+k}$, for $i = k+1{:}\,n$.

As in the case of Gaussian elimination (§ 1.4.3), pivoting is needed for numerical stability. The permutation matrices serving this goal must also be symplectic. One makes use of

$$P_{k,l} = \begin{bmatrix} P & 0 \\ 0 & P \end{bmatrix},$$

where P interchanges the rows (and columns) k and l. Specifically, the permutation $P_{k,l}$ interchanges the elements x_k and x_l, where x_l is the element with maximal modulus among $x_k, \ldots, x_n$, for $k < n$, and among $x_k, \ldots, x_{2n-1}$, for $k \in \{n+1, \ldots, 2n-1\}$. We denote $P(k,x)$ such a permutation. Since the matrices G_1, G_2 and G_3 may be ill conditioned (if $|v|$

or $\|w\|$ are large), it is necessary to monitor their conditioning and provide remedies for unacceptable situations. We shall revert to this question.

We can now formulate the following template algorithm.

Algorithm Jtrid : $(M,S) = \texttt{Jtrid(}\ M,S\ \texttt{)}$

```
/* Reduce a Hamiltonian matrix M ∈ IR^{2n×2n} to J-tridiagonal */
/* form using elementary symplectic transformations; */
/* update matrix S ∈ IR^{2n×2n}. */

FOR j = 1:n − 1,
   IF  j < n − 1,  THEN
                                    /* 1. Annihilate m_{n+j+2:2n,j}. */
      T = G_3^T(j + 1, m_{n+j+2:2n,j}/m_{n+j+1,j})P(n + j + 1, Me_j);
      M ← TMT^{-1};   S ← ST^{-1};
   END IF
                                    /* 2. Annihilate m_{n+j+1,j}. */
   T = G_2^T(j + 1, −m_{n+j+1,j}/m_{j+1,j});   M ← TMT^{-1};   S ← ST^{-1};
   IF  j < n − 1,  THEN
                                    /* 3. Annihilate m_{j+2:n,j}. */
      T = G_3(j + 1, −m_{j+2:n,j}/m_{j+1,j})P(j + 1, Me_j);
      M ← TMT^{-1};   S ← ST^{-1};
   END IF
                                    /* 4. Annihilate m_{j+1,j}, if possible. */
   IF  m_{j+1,j} ≠ 0 AND m_{n+j,j} = 0,  THEN
                                    /* Use a random transformation. */
      T = G_2^T(j, −rand(1, 1))G_3(j, (1, rand(n − j, 1)^T)^T);
      M ← TMT^{-1};   S ← ST^{-1};   (M, S) = Jtrid( M, S );
   ELSE
      T = G_1(j + 1, −m_{j+1,j}/m_{n+j,j});   M ← TMT^{-1};   S ← ST^{-1};
   END IF
   IF  j < n − 1,  THEN
                                    /* 5. Annihilate m_{n+j+2:2n,n+j}. */
      T = G_3^T(j + 1, m_{n+j+2:2n,n+j}/m_{n+j+1,n+j})P(n + j + 1, Me_{n+j});
      M ← TMT^{-1};   S ← ST^{-1};
   END IF
                                    /* 6. Annihilate m_{n+j+1,n+j}. */
   T = G_2^T(j+1, −m_{n+j+1,n+j}/m_{j+1,n+j});   M ← TMT^{-1};   S ← ST^{-1};

   IF  j < n − 1,  THEN
                                    /* 7. Annihilate m_{j+2:n,n+j}. */
```

$T = G_3(j+1, -m_{j+2:n,n+j}/m_{j+1,n+j})P(j+1, Me_{n+j})$;
$M \leftarrow TMT^{-1}$; $S \leftarrow ST^{-1}$;
END IF
END FOR

Remark 4.1 The first $j+1$ zero components of w in the G_3-type transformations have been omitted, for convenience. When applied to a general matrix, Algorithm `Jtrid` accomplishes the reduction to J-Hessenberg form. When applied to a Hamiltonian matrix, it is assumed that advantage is taken of the special structure, to reduce the computational effort. Clearly, the elementary transformations of type G_2 and G_3, used at steps 1–3 and 5–7 can be replaced by the orthogonal Householder and Givens transformations in (4.10) and (4.11), respectively. Algorithm `rand(` n, m `)`, used at step 4 when the reduction cannot be continued, generates a random $n \times m$ real matrix with elements uniformly distributed in the open interval $(0, 1)$. The random technique is motivated by the fact that the set of (Hamiltonian) matrices which cannot be reduced to the J-Hessenberg (J-tridiagonal) form, has the Lebesgue measure zero (see Theorem 4.5 and the associated comments in § 4.1.2). Note that Algorithm `Jtrid` can be used recursively if the condition at step 4 is satisfied. Theoretically this does not yield an infinite process (Bunse-Gerstner, Mehrmann and Watkins, 1989). □

More comments about Algorithm `Jtrid` orderly follow. First, note that step 1 annihilates the elements $(n+j+2, j), \ldots, (2n, j)$ of matrix M in a stable way. Step 2 zeroes the element $(n+j+1, j)$, and step 3 zeroes the elements $(j+2, j), \ldots, (n, j)$. At step 4, it is decided that either M cannot directly be reduced to J-tridiagonal form, or $m_{j+1,j}$ should be annihilated. (In the first case, a random symplectic matrix is used and the entire reduction is restarted. Random symplectic transformations are also used in practice if the condition number of any of the transformation matrices involved exceeds a certain specified bound.) Similarly, steps 5, 6 and 7 annihilate the elements $(n+j+2, n+j), \ldots, (2n, n+j)$, the element $(n+j+1, n+j)$, and the elements $(j+2, n+j), \ldots, (n, n+j)$, respectively. In this way, matrix M is reduced to the desired form, the remaining zeros (not explicitly zeroed) appearing automatically. The various transformations T are not actually computed, but they are applied to M, ensuring similarity, and then accumulated in S. If, on input into Algorithm `Jtrid`, one sets $S = I_{2n}$, then on output, S contains the transformation matrix that reduces M to J-tridiagonal form. If S is chosen so that $Se_1 = \lambda e_1$, then this property is preserved by the algorithm.

The reduction of matrix M using Algorithm `Jtrid` requires about $4n^3$ *flops*, and the accumulation of transformations needs $6n^3$ *flops*.[1] One can

[1] For comparison purposes, the corresponding reduction stage to the Hessenberg form

gain in efficiency if, instead of accumulation, one directly updates the matrices corresponding to the solution of Riccati equation, exploiting the symmetry (Byers and Mehrmann, 1986). In this case, not the entire matrix $S = (S_{ij})$, $i, j = 1{:}2$, has to be stored, but only S_{11}^{-1}, $X = -S_{21}S_{11}^{-1}$ and $Y = S_{11}^{-1}S_{12}$, where X and Y are symmetric matrices, since S is symplectic. We shall prove that all the information needed about S can be reconstructed from the matrices S_{11}^{-1}, X and Y, which require only half of the memory space for S.

Let us first prove the symmetry of X and Y. From $S^T JS = J$, it follows

$$\begin{aligned} S_{11}^T S_{21} &= S_{21}^T S_{11}, \\ S_{22}^T S_{12} &= S_{12}^T S_{22}, \\ S_{11}^T S_{22} &= I_n + S_{21}^T S_{12}. \end{aligned} \tag{4.27}$$

If S_{11} is nonsingular, then the first relation is equivalent to $S_{21}S_{11}^{-1} = S_{11}^{-T}S_{21}^T = (S_{21}S_{11}^{-1})^T$, hence $X = X^T$. In addition, from the third relation (4.27), we obtain $S_{22} = S_{11}^{-T} + S_{11}^{-T}S_{21}^T S_{12}$; replacing it in the second relation, we have

$$S_{22}^T S_{12} = S_{11}^{-1}S_{12} + S_{12}^T S_{21} S_{11}^{-1} S_{12}, \tag{4.28}$$

from which it follows that $Y = Y^T$, since $S_{22}^T S_{12}$ and $S_{21}S_{11}^{-1}$ are symmetric.

Now, let us see how S_{11}^{-1}, X and Y can be updated after applying transformations with matrices G_1, G_2 and G_3 from (4.24)–(4.26), at the same time justifying for the necessity of updating Y. Let $\widetilde{S}$ be the updated matrix S after such a transformation. In case of updating with G_1, we have the relation

$$\widetilde{S} = S \begin{bmatrix} I & -V_1 \\ 0 & I \end{bmatrix} = \begin{bmatrix} S_{11} & -S_{11}V_1 + S_{12} \\ S_{21} & -S_{21}V_1 + S_{22} \end{bmatrix},$$

hence

$$\widetilde{S}_{11}^{-1} = S_{11}^{-1}, \qquad \widetilde{X} = X, \qquad \widetilde{Y} = S_{11}^{-1}(-S_{11}V_1 + S_{12}) = Y - V_1. \tag{4.29}$$

In case of updating with G_3, we obtain

$$\widetilde{S} = S \begin{bmatrix} L^{-1} & 0 \\ 0 & L^T \end{bmatrix} = \begin{bmatrix} S_{11}L^{-1} & S_{12}L^T \\ S_{21}L^{-1} & S_{22}L^T \end{bmatrix},$$

hence

$$\widetilde{S}_{11}^{-1} = LS_{11}^{-1}, \qquad \widetilde{X} = X, \qquad \widetilde{Y} = LYL^T. \tag{4.30}$$

---

in Algorithm **Schv** needs about $\frac{40}{3}n^3$ *flops*, plus $\frac{16}{3}n^3$ *flops*, for accumulation; the original SR algorithm (Bunse-Gerstner and Mehrmann, 1986) requires about $\frac{20}{3}n^3$ *flops* for the reduction.

A similar result corresponds to updating with G_3^T. (The relations are obtained from (4.30), by replacing L by L^T.) Finally, in case of updating with G_2^T, $\widetilde{S}$ has the form

$$\widetilde{S} = S \begin{bmatrix} I & 0 \\ -V_2 & I \end{bmatrix} = \begin{bmatrix} S_{11} - S_{12}V_2 & S_{12} \\ S_{21} - S_{22}V_2 & S_{22} \end{bmatrix},$$

with $V_2 = ve_k e_k^T$. Denote $\alpha := 1 - ve_k^T Y e_k$. Using a well-known formula for updating the inverse of a matrix after a rank one modification, specialized to our case, we obtain

$$\begin{aligned} \widetilde{S}_{11}^{-1} &= (S_{11} - S_{12}ve_k e_k^T)^{-1} \\ &= S_{11}^{-1} + vS_{11}^{-1}S_{12}e_k e_k^T S_{11}^{-1}/(1 - ve_k^T S_{11}^{-1}S_{12}e_k) \\ &= S_{11}^{-1} + \frac{v}{\alpha} Y e_k e_k^T S_{11}^{-1}, \end{aligned} \tag{4.31}$$

and

$$\begin{aligned} \widetilde{X} &= -(S_{21} - S_{22}V_2)\widetilde{S}_{11}^{-1} \\ &= X - \frac{v}{\alpha} S_{21}S_{11}^{-1}S_{12}e_k e_k^T S_{11}^{-1} + \frac{v}{\alpha} S_{22}e_k e_k^T S_{11}^{-1} \\ &= X + \frac{v}{\alpha} S_{11}^{-T} e_k e_k^T S_{11}^{-1}, \end{aligned} \tag{4.32}$$

$$\widetilde{Y} = \widetilde{S}_{11}^{-1}S_{12} = Y + \frac{v}{\alpha} Y e_k e_k^T Y. \tag{4.33}$$

Note that $\alpha \neq 0$ was assumed in (4.31); if $\alpha = 0$, then $S_{11} - S_{12}ve_k e_k^T$ is singular. To compute $\widetilde{X}$ in (4.32) we used the identities $ve_k^T Y e_k = 1 - \alpha$ and $S_{11}^{-T} = S_{22} - S_{21}S_{11}^{-1}S_{12}$; the last identity follows from the first and the last relation (4.27). The usage of (4.31)–(4.33) is not proper if $|v/\alpha|$ is large (for instance, if α is (almost) negligible compared with v, to machine precision), since a significant loss of accuracy can occur in this case. We shall discuss later how this difficulty can be circumvented.

Using the formulas (4.29)–(4.33), the cost of updating the matrices X, Y and S_{11}^{-1} in Algorithm `Jtrid`, adapted for this case, is $\frac{10}{3}n^3$ *flops*, hence almost half of the initial value. In both evaluations, we have not considered the initialization $S = I$, hence $S_{11}^{-1} = I$, $X = Y = 0$, that would significantly reduce the computational effort.

The use of the updating relations has also the advantage of preserving the symmetry of X at each stage of the algorithm. This statement is valid for the next algorithms too, so that the Riccati equation solution should be obtained directly, without evaluating $-S_{21}S_{11}^{-1}$.

We will assume in the sequel that Algorithm `Jtrid` updates the matrices S_{11}^{-1}, X, and Y. Moreover, for later use, a facility to deal with the submatrices $(d_1\colon d_2, d_1\colon d_2)$ in A, Q, and F, and annihilate a specified number s of subdiagonal elements, is added. So, a more detailed calling statement for Algorithm `Jtrid` is

$$(A, Q, F, \bar{S}, X, Y) = \texttt{Jtrid}(\ A, Q, F, \bar{S}, X, Y, d_1, d_2, s\ );$$

where $\bar{S}$ stands for S_{11}^{-1}. Some of these refinements are not too difficult to be incorporated. For instance, for operating on the submatrices $(d_1\colon d_2, d_1\colon d_2)$, the index of the FOR loop in Algorithm `Jtrid` should take the values $j = d_1\colon d_2 - 1$, the bound $n-1$ in the IF statements should be replaced by $d_2 - 1$, and the last $n - d_2$ elements of vector w in G_3-type transformations should be set to zero.

The algorithm for reducing a Hamiltonian matrix to J-tridiagonal form can be used to compute the RSF, the same way as the reduction to UHF has been previously used in the QR algorithm (§ 2.2.3). The most general procedure performing this computation by exclusively using symplectic transformations which preserve the Hamiltonian structure, is the following one.

Procedure HamSR

/* 1. [Reduction of M to J-tridiagonal form M_1.] */

$M_1 = S_0^{-1} M S_0, \qquad S = S_0, \qquad S_0 \in \mathcal{S}_{2n}, \qquad M_1$ J-tridiagonal.

/* 2. [SR iteration.] */

FOR $k = 1\colon\ldots,$

Find the SR decomposition, $p_k(M_k) = S_k R_k$, of a polynomial p_k in M_k.

$M_{k+1} = S_k^{-1} M_k S_k, \qquad S \leftarrow S S_k, \quad S_k \in \mathcal{S}_{2n}, \qquad M_{k+1}$ J-tridiagonal.

END FOR

Procedure `HamSR` is the basis for the SR algorithm. A specialization of this algorithm for solving continuous-time algebraic Riccati equations, called Algorithm `RicSR`, will be presented below. All computations can be performed using the transformations G_1, G_2, G_3 and P only.

The first stage of this procedure is accomplished using Algorithm `Jtrid`. We shall briefly describe the processing done at an iteration of stage 2. In the first place, note that all generated iterates M_k are in J-tridiagonal form. (If $p_k(M_k)$ is nonsingular, then this can easily be shown, since $p_k(M_k)$ and M_k commute and the matrices S_k and R_k are nonsingular, hence

$$M_{k+1} = S_k^{-1} M_k S_k = R_k p_k(M_k)^{-1} M_k p_k(M_k) R_k^{-1} = R_k M_k R_k^{-1};$$

if M_k is a J-Hessenberg matrix, then $R_k M_k R_k^{-1}$ is also J-Hessenberg, hence if M_k is Hamiltonian too, the result will be a J-tridiagonal matrix.) So, the form of an iterate will be

$$M_k = \left[\begin{array}{cccc|cccc} a_1 & & & & c_1 & b_1 & & \\ & \ddots & & & b_1 & \ddots & \ddots & \\ & & \ddots & & & \ddots & \ddots & b_{n-1} \\ & & & a_n & & & b_{n-1} & c_n \\ \hline q_1 & & & & -a_1 & & & \\ & \ddots & & & & \ddots & & \\ & & \ddots & & & & \ddots & \\ & & & q_n & & & & -a_n \end{array}\right].$$

The polynomials $p_k(x)$ are usually chosen as either $x - \lambda_k$, where λ_k is the last diagonal element of M_k ($\lambda_k = M_k(2n, 2n)$), or $(x - \lambda_k)(x - \bar{\lambda}_k)$, where $\lambda_k, \bar{\lambda}_k \in \lambda(M_k(2n-1\colon 2n, 2n-1\colon 2n))$. However, if $\lambda \in \lambda(M) \setminus \mathbb{R}$, where M is a (real) Hamiltonian matrix, then $-\lambda$, $\bar{\lambda}$, $-\bar{\lambda} \in \lambda(M)$; hence, one can use a polynomial of degree four,

$$p_k(x) = (x - \lambda_1)(x - \lambda_2)(x - \lambda_3)(x - \lambda_4),$$

where λ_1 to λ_4 are the eigenvalues of a submatrix of the form

$$\begin{bmatrix} a_{n-1} & 0 & c_{n-1} & b_{n-1} \\ 0 & a_n & b_{n-1} & c_n \\ q_{n-1} & 0 & -a_{n-1} & 0 \\ 0 & q_n & 0 & -a_n \end{bmatrix}.$$

The convergence of the SR algorithm can be proven as for the QR algorithm. Under certain conditions, $\{M_k\}$ tends to a J-triangular and J-tridiagonal, hence upper triangular matrix ($q_i \to 0$, $b_i \to 0$). In general, however, the sequence merely converges to a matrix with blocks of 4×4 submatrices, each one having the spectrum of the form $\{\lambda, -\lambda, \bar{\lambda}, -\bar{\lambda}\}$, where $\lambda \in \mathbb{C} \setminus \{\mathbb{R} \cup \mathbb{R}j\}$, i.e. not all elements b_i will necessarily tend to zero. If some eigenvalues are real, then 2×2 blocks will also appear.

Due to the uniqueness of the reduction to J-tridiagonal (or generally, J-Hessenberg) form, according to Theorem 4.5, the algorithm can be used without explicitly computing the decomposition $p_k(M_k) = S_k R_k$; indeed, it is sufficient to calculate the first column of $p_k(M_k)$, as in the implicit QR algorithm. In general, for large values of k, M_k and $p_k(M_k)$ (with usual p_k) have SR decompositions (since $\{M_k\}$ tends to a J-tridiagonal matrix). But for small values of k, it is possible that the matrix $p_k(M_k)$ has no SR decomposition, or be close to a matrix $\widetilde{M}_k$ for which $p_k(\widetilde{M}_k)$ has no SR decomposition. This fact will be revealed by obtaining a large condition number for a transformation G_i, $i \in \{1,2,3\}$. If a condition number exceeds a certain bound, c_b, the current iteration is abandoned and a random shift λ_k (in the explicit variant of the algorithm), or a random symplectic matrix (in the implicit variant) is used. One speaks about an *exceptional* shift or transformation.

Consider now the implicit variant of the algorithm. The following computations must be performed at each iteration k. The first column u of $p_k(M_k)$ is computed. Then, a symplectic matrix $\widetilde{S}$ that reduces u to a multiple of e_1 is determined, the matrix $\widetilde{S}M_k\widetilde{S}^{-1}$ is calculated and, finally, the symplectic reduction of $\widetilde{S}M_k\widetilde{S}^{-1}$ to J-tridiagonal form, $M_{k+1} = \widehat{S}\widetilde{S}M_k\widetilde{S}^{-1}\widehat{S}^{-1}$, is performed, with the matrices S_{11}^{-1}, X and Y updated accordingly; if S_{11}^{-1} is (nearly) singular, then an exceptional transformation can be used.

If p_k is the four degree polynomial above, then $u \in \mathcal{R}(e_1, e_2, e_3)$, the transformation $\widetilde{S}$ is simply computed, and $\widetilde{S}M_k\widetilde{S}^{-1}$ is a J-tridiagonal Hamiltonian matrix, except for six extra nonzero elements in the upper-left corner of each submatrix $M_{ij} \in \mathbb{R}^{n\times n}$, $i,j = 1{:}2$. To reduce this matrix to J-tridiagonal form, a specialization of Algorithm `Jtrid`, that exploits the particular structure, can be envisaged. In this manner, the computing effort per iteration (for reduction) is $0(n)$, and not $0(n^2)$, as for the QR algorithm. However, the updating of S_{11}^{-1}, X and Y requires an $0(n^2)$ effort.

At the beginning of each iteration, one checks if some elements b_j are negligible, i.e. smaller (in modulus) than $\varepsilon := \varepsilon_M\|M\|_1$. If $|b_j| < \varepsilon$, then the eigenvalue problem can be decomposed into two smaller subproblems which are solved consecutively.[2] The iterative process continues until a complete splitting into 2×2 or 4×4 subproblems succeeds, hence, until at least every second element b_j is negligible.

Algorithm `i4SR` below performs a quadruple SR step on the current J-tridiagonal matrix formed by the submatrices $(l{:}i, l{:}i)$ in A, Q, and F. (For the moment, we may assume that $l = 1$, $i = n$.) It uses two input parameters, defined as $\rho := |\lambda_i|^2$, $\sigma := \Re(\lambda_i)^2 - \Im(\lambda_i)^2$.

Algorithm i4SR : $(A, Q, F, \bar{S}, X, Y) =$ `i4SR`$(\ A, Q, F, \bar{S}, X, Y, l, i, \rho, \sigma\ )$

/* Perform a quadruple SR step on the J-tridiagonal matrix defined */
/* by $A, Q, F \in \mathbb{R}^{n\times n}$, using elementary symplectic transformations; */
/* update matrices $\bar{S}$, X and $Y \in \mathbb{R}^{n\times n}$. */

/* 1. Compute the three non-zero elements of the first column u of */

---

[2] Sometimes, a less stringent negligibility criterion is needed in practice (see Algorithm `SRiter` below), particularly for problems with multiple eigenvalues, and/or when Algorithm `sqred` is used for computing them in advance.

```
        /* (M̃ − λ_i I)(M̃ − λ̄_i I)(M̃ + λ_i I)(M̃ + λ̄_i I), where M̃ */
        /* is formed by the submatrices (l: i, l: i) of A, Q and F. */
α = a_{ll}^2 + f_{ll} q_{ll};   δ = a_{l+1,l+1}^2 + f_{l+1,l+1} q_{l+1,l+1};   u = 0;
u_l = α(α − 2σ) + f_{l+1,l}^2 q_{ll} q_{l+1,l+1} + ρ^2;
u_{l+1} = f_{l+1,l} q_{ll}(α + δ − 2σ);   u_{l+2} = f_{l+2,l+1} f_{l+1,l} q_{ll} q_{l+1,l+1};
w = u;   w_l = 0;
        /* 2. Apply a G_3 transformation annihilating u_{l+1} and u_{l+2}. */
(A, Q, F, S̄, X, Y) = update( A, Q, F, S̄, X, Y, G_3(l, −w/u_l)P(l, u), l, i );
(A, Q, F, S̄, X, Y) = Jtrid( A, Q, F, S̄, X, Y, l, i, 2 );
```

Algorithm **i4SR** calls Algorithm `update`, for updating the matrices A, Q, F, S_{11}^{-1} ($\bar{S}$), X and Y, using one of the transformations G_i (and P, for G_3). Obviously, A, Q, and F are updated by implicitly applying the similarity transformation T, defined by G_i (and possibly P), on the J-tridiagonal matrix M, TMT^{-1}. The formulas (4.29)–(4.33) are used for updating S_{11}^{-1}, X and Y.

A similar, but simpler algorithm performs a double SR step; the calling statement is:

$$(A, Q, F, \bar{S}, X, Y) = \texttt{i2SR}(\ A, Q, F, \bar{S}, X, Y, l, i, \rho\ );$$

The next stage is the separation of the stable M-invariant subspace. Let

$$N_j = \begin{bmatrix} a_j & c_j \\ q_j & -a_j \end{bmatrix} \tag{4.34}$$

be a resulted 2×2 submatrix. Its eigenvalues are $\lambda = (a_j^2 + c_j q_j)^{1/2}$ and $-\lambda$. If $q_j = 0$ and $a_j = -\lambda$, the spectrum has already taken the desired form and no processing is needed.

If $q_j = 0$ and $a_j = \lambda$, then according to some known results related to the ordering of eigenvalues (§ 2.2.4), the desired form is to be obtained using a transformation of type G_2, so chosen that the second element of $G_2^T \begin{bmatrix} c_j & -2a_j \end{bmatrix}^T$ is zero. It follows that $v = 2a_j/c_j$. Indeed, for this value of v, it is easy to verify that

$$\begin{bmatrix} 1 & 0 \\ v & 1 \end{bmatrix} \begin{bmatrix} a_j & c_j \\ 0 & -a_j \end{bmatrix} \begin{bmatrix} 1 & 0 \\ -v & 1 \end{bmatrix} = \begin{bmatrix} -a_j & c_j \\ 0 & a_j \end{bmatrix}.$$

Finally, in the general case ($q_j \neq 0$), the desired form can be obtained by applying a G_2-type transformation, so chosen that $G_2^T \begin{bmatrix} a_j - \lambda & q_j \end{bmatrix}^T$ has the second element zero. Therefore, $v = -q_j/(a_j - \lambda)$ and, by simple manipulations, we obtain

$$\begin{bmatrix} 1 & 0 \\ v & 1 \end{bmatrix} \begin{bmatrix} a_j & c_j \\ q_j & -a_j \end{bmatrix} \begin{bmatrix} 1 & 0 \\ -v & 1 \end{bmatrix} = \begin{bmatrix} -\lambda & c_j \\ 0 & \lambda \end{bmatrix}, \tag{4.35}$$

with $\lambda = (a_j^2 + c_j q_j)^{1/2}$.

Let us now consider a 4×4 resulted submatrix

$$N_j = \begin{bmatrix} a_j & 0 & c_j & b_j \\ 0 & a_{j+1} & b_j & c_{j+1} \\ q_j & 0 & -a_j & 0 \\ 0 & q_{j+1} & 0 & -a_{j+1} \end{bmatrix}, \qquad b_j \neq 0. \tag{4.36}$$

Let $\lambda, \mu \in \lambda(N_j)$, with $\Re(\lambda) > 0$, $\Re(\mu) > 0$ ($\mu = \bar{\lambda}$). (Remember that in our context, all eigenvalues must have nonzero real parts.) It is easy to verify that

$$K := (N_j - \lambda I_4)(N_j - \mu I_4) = -(\lambda + \mu) \begin{bmatrix} x & x & x & x \\ x & x & x & x \\ q_j & 0 & x & x \\ 0 & q_{j+1} & x & x \end{bmatrix}$$

and $\operatorname{rank}(K) \leq 2$. Assume first that $q_j q_{j+1} \neq 0$. If the elements of K in the positions (3,1), (2,1) and (4,2) are successively annihilated using transformations of type G_2, G_3 and G_2, respectively, then the result has the form

$$TK = \begin{bmatrix} x & x & x & x \\ 0 & x & x & x \\ 0 & 0 & 0 & 0 \\ 0 & 0 & 0 & 0 \end{bmatrix},$$

where T is the product of the three transformations. The zero in the position (3,2) is preserved due to the transformations made, and the special structure of K; the zeros of the submatrix $(3{:}4, 3{:}4)$ follow from the rank condition, since the (1,1) and (2,2) elements are nonzero, as q_j and q_{j+1} are. In this way, the last two rows of matrix T generate the left N_j-invariant subspace corresponding to the eigenvalues λ and μ. Therefore,

$$TN_jT^{-1} = \begin{bmatrix} \Delta & \Pi \\ 0 & -\Delta^T \end{bmatrix},$$

where $\lambda(\Delta) = \{-\lambda, -\mu\}$.

If $q_j q_{j+1} = 0$, we can without reducing the generality assume that $q_{j+1} = 0$. (Otherwise, the rows and columns of N_j are permuted accordingly.) If $a_{j+1} < 0$, then the $(j+1)$th column of the current transformation matrix S belongs to the stable invariant subspace, and hence, removing the second and fourth rows and columns of N_j, the remaining 2×2 problem is dealt with as shown above. If $a_{j+1} > 0$, then $\lambda = (a_j^2 + c_j q_j)^{1/2}$ and $\mu = a_{j+1}$ are the eigenvalues with positive real parts. Let

$$G_2^T = \begin{bmatrix} 1 & 0 \\ v & 1 \end{bmatrix}$$

be a symplectic matrix, such chosen that (4.35) holds. Then,

$$G_4 N_j G_4^{-1} = \begin{bmatrix} -\lambda & 0 & x & x \\ -b_j v & a_{j+1} & x & x \\ 0 & 0 & \lambda & b_j v \\ 0 & 0 & 0 & -a_{j+1} \end{bmatrix}, \qquad G_4 := \begin{bmatrix} 1 & 0 & 0 & 0 \\ 0 & 1 & 0 & 0 \\ v & 0 & 1 & 0 \\ 0 & 0 & 0 & 1 \end{bmatrix}.$$

By adopting a transformation L (in a G_3-type transformation), so that

$$L \begin{bmatrix} -\lambda & 0 \\ -b_j v & a_{j+1} \end{bmatrix} L^{-1} = \begin{bmatrix} -\lambda & x \\ 0 & a_{j+1} \end{bmatrix},$$

it follows that

$$G_3 G_4 N_j G_4^{-1} G_3^{-1} = \begin{bmatrix} -\lambda & x & x & x \\ 0 & a_{j+1} & x & x \\ 0 & 0 & \lambda & 0 \\ 0 & 0 & x & -a_{j+1} \end{bmatrix}.$$

The jth column of the current S matrix now belongs to the stable invariant subspace. The jth and $(n+j)$th rows and columns of M_k can then be eliminated and the desired result obtained by transforming the submatrix

$$\begin{bmatrix} a_{j+1} & x \\ 0 & -a_{j+1} \end{bmatrix},$$

as shown above.

Now, we have to specify how the eigenvalues λ and μ can be determined. We know that $\lambda(N_j) = \{\lambda, \mu, -\lambda, -\mu\}$, where $\mu = \bar{\lambda}$, hence $\lambda(N_j^2) = \{\lambda^2, \mu^2, \lambda^2, \mu^2\}$, $\mu^2 = \bar{\lambda}^2$. But it is easy to check that $\lambda(N_j^2) = \lambda(V) \cup \lambda(V)$, where

$$V := \begin{bmatrix} a_j^2 + q_j c_j & q_{j+1} b_j \\ q_j b_j & a_{j+1}^2 + q_{j+1} c_{j+1} \end{bmatrix}.$$

Hence λ^2 and μ^2 are the roots of the quadratic equation

$$\begin{aligned} x^2 \quad - \quad & (a_j^2 + q_j c_j + a_{j+1}^2 + q_{j+1} c_{j+1}) x + \\ & (a_j^2 + q_j c_j)(a_{j+1}^2 + q_{j+1} c_{j+1}) - q_j q_{j+1} b_j^2 = 0. \end{aligned}$$

The discussion above is summarized in Algorithm `spl4x4` below (where $i = l + 1$).

Algorithm spl4x4 : $(A, Q, F, \bar{S}, X, Y) =$ `spl4x4`$(A, Q, F, \bar{S}, X, Y, l, i)$

/* Move two stable eigenvalues of the current J-tridiagonal matrix */

```
/* defined by the submatrices (l: i, l: i) in A, Q, F ∈ IR^{n×n}, to the */
/* left-top part, using elementary symplectic transformations; */
/* update matrices S̄, X and Y ∈ IR^{n×n}. */

α = a_{ll}^2 + f_{ll} q_{ll};   δ = a_{ii}^2 + f_{ii} q_{ii};   w = f_{il}^2 q_{ll} q_{ii};
p = (αδ − w)^{1/2};   s = −(α + δ + 2p)^{1/2};
IF  q_{ll} q_{ii} ≠ 0,  THEN
        /* 1. Compute the needed part of K := (N_j − λI_4)(N_j − μI_4) */
        /* using approximate quantities related to unstable eigenvalues. */
   k_{11} = α + s a_{ll} + p;   k_{12} = f_{il} q_{ii};   k_{21} = f_{il} q_{ll};
   k_{22} = δ + s a_{ii} + p;   k_{31} = s q_{ll};   k_{32} = 0;   k_{41} = 0;   k_{42} = s q_{ii};
                          /* 2. Annihilate k_{31} using a transformation G_2^T. */
   (A, Q, F, S̄, X, Y) = update( A, Q, F, S̄, X, Y, G_2^T(l, −k_{31}/k_{11}), l, i );
   k_{32} = −k_{31} k_{12}/k_{11};
                          /* 3. Annihilate k_{21} using a transformation G_3. */
   v_{1:n} = 0;   v_{l:i} = k_{1:2,1};   w = v;   w_l = 0;   T = G_3(l, −w/v_l)P(l, v);
   (A, Q, F, S̄, X, Y) = update( A, Q, F, S̄, X, Y, T, l, i );
   IF  |k_{21}| > |k_{11}|,  THEN  k_{1,1:2} ↔ k_{2,1:2};   k_{32} ↔ k_{42};
   k_{22} ← k_{22} − w_i k_{12}/w_l;
                          /* 4. Annihilate k_{42} using a transformation G_2^T. */
   (A, Q, F, S̄, X, Y) = update( A, Q, F, S̄, X, Y, G_2^T(i, −k_{42}/k_{22}), l, i );
ELSE
   IF  |q_{ii}| > |q_{ll}|,  THEN
                                  /* 5. Interchange, to get the other case. */
     w = 0;   w_{l:i} = (1, 2)^T;
     (A, Q, F, S̄, X, Y) = update( A, Q, F, S̄, X, Y, P(l, w), l, i );
   END IF
   IF  a_{ii} > 0,  THEN
                          /* 6. Annihilate q_{ll} using a transformation G_2^T. */
     IF  α ≤ 0,  THEN  EXIT
     α ← α^{1/2};   a_2 = a_{ll} + α sign(a_{ll});
     IF  a_{ll} > 0,  THEN  a_2 = −f_{ll} q_{ll}/a_2;
     v_{1:n} = 0;   v_l = α + a_{ii};   v_i = −q_{ll} f_{il}/a_2;   w = v;   w_l = 0;
     (A, Q, F, S̄, X, Y) = update( A, Q, F, S̄, X, Y, G_2^T(l, −q_{ll}/a_2), l, i );
     T = G_3(l, −w/v_l)P(l, v);
     (A, Q, F, S̄, X, Y) = update( A, Q, F, S̄, X, Y, T, l, i );
     a_1 = −2a_{ii};
                          /* 7. Annihilate a_1 using a transformation G_2^T. */
     (A, Q, F, S̄, X, Y) = update( A, Q, F, S̄, X, Y, G_2^T(l, −a_1/f_{ii}), l, i );
   ELSE
```

```
                    /* 8. Move the remaining eigenvalue. */
    (A, Q, F, S̄, X, Y) = spl2x2( A, Q, F, S̄, X, Y, l, i );
  END IF
END IF
```

A similar, but simpler algorithm, called **spl2x2**, moves one stable eigenvalue of the current trailing J-tridiagonal matrix to the left-top part. Note that it can be called by **spl4x4**.

We can now state the algorithm performing the iterative part of the SR algorithm (stage 2 in Procedure **HamSR**). For efficiency reasons, it is assumed that the eigenvalues of the given Hamiltonian matrix are known. How these eigenvalues can be obtained will be described after the algorithm presentation.

Algorithm SRiter : $X =$ **SRiter**($A, Q, F, \bar{S}, X, Y, \Lambda$)

/* Reduce the J-tridiagonal matrix defined by $A, Q, F \in \mathbb{R}^{n\times n}$ */
/* to ordered RSF, using elementary symplectic transformations; */
/* update the matrices $\bar{S}$, X and $Y \in \mathbb{R}^{n\times n}$, returning the */
/* stabilizing ARE solution in X. */

$\underline{l} = 1$; $itn = 30n$; $i = n$; $\varepsilon_d \le \varepsilon_M^{1/2}$; $itr = 0$;

```
        /* 1. Start main loop: i decreases from n to 1 in steps of 1 or 2. */
WHILE i ≥ l̲ AND itr ≤ itn, DO
                    /* 2. Perform SR iterations on J-tridiagonal matrix */
                          /* until a 2 × 2 or 4 × 4 submatrix splits off. */
  itr ← itr + 1;
  l = l̲;                         /* Look for a small subdiagonal element. */
  FOR k = i: −1: l + 1,
     γ = |a_{k−1,k−1}| + |a_{kk}| + |f_{k−1,k−1}| + |f_{kk}|;
     IF |f_{k,k−1}| ≤ ε_d γ, THEN EXIT LOOP;
  l = k;  IF l > l̲, THEN f_{l,l−1} = 0;          /* f_{l,l−1} negligible. */
  IF l = i, THEN
          /* A 2 × 2 submatrix splits off; move the stable eigenvalue. */
     (A, Q, F, S̄, X, Y) = spl2x2( A, Q, F, S̄, X, Y, l, i );   i ← i − 1;
  ELSE IF l = i − 1, THEN

          /* A 4 × 4 submatrix splits off; move the stable eigenvalues. */
     (A, Q, F, S̄, X, Y) = spl4x4( A, Q, F, S̄, X, Y, l, i );   i ← i − 2;
  ELSE

          /* Active submatrices are (·)_{l:i,l:i}; l = l̲, or f_{l,l−1} negligible. */
                                        /* 3. Perform the SR iteration. */
     ρ = |λ_i|²;   σ = ℜ(λ_i)² − ℑ(λ_i)²;
```

```
        IF  ℑ(λ_i) = 0,  THEN
                                                /* Use double shift. */
          (A, Q, F, S̄, X, Y) = i2SR( A, Q, F, S̄, X, Y, l, i, ρ );
        ELSE
                                             /* Use quadruple shift. */
          (A, Q, F, S̄, X, Y) = i4SR( A, Q, F, S̄, X, Y, l, i, ρ, σ );
        END IF
    END IF
END WHILE
```

Other negligibility criteria were also tried, but did not work better. Note that the computation of RSF and the separation of stable invariant subspace are not done consecutively. Once a 2×2 or 4×4 submatrix N_j in (4.34) or (4.36) has split off, its stable eigenvalues are moved to the left-top corner of the Hamiltonian matrix.

The convergence rate of the SR algorithm can be improved by replacing the eigenvalues λ_i, $i = 1\!:\!4$, which define the polynomials p_k, by the corresponding "exact" eigenvalues of matrix M, computed in advance. For Hamiltonian matrices, the eigenvalues can be economically computed with both relative and absolute good accuracy, using the following procedure (Van Loan, 1984), which reduces the square of a Hamiltonian matrix M.

Procedure psqred

/* 1. [Construction of the square of Hamiltonian matrix M.] */
$W = M^2$.

/* 2. [Reduction of W to Hamiltonian-Hessenberg form.] */

$$S^T W S = \begin{bmatrix} \bar{Q} & R \\ 0 & -\bar{Q}^T \end{bmatrix}, \qquad S \in \mathcal{S}_{2n}, \quad S^T S = I_{2n}, \qquad \bar{Q} \text{ in UHF}.$$

The existence of an orthogonal symplectic matrix S is theoretically guaranteed. In the detailed algorithm below, called Algorithm `sqred`, orthogonal transformation matrices defined by (4.10) and (4.11) are used. Note that the matrix in the right hand side at step 2 of Procedure `psqred` is in a special Hamiltonian-Hessenberg form, where $w_{2n,n} = 0$. Indeed, matrix W is given by

$$W = \begin{bmatrix} A^2 + FQ & AF - FA^T \\ QA - A^T Q & QF + (A^T)^2 \end{bmatrix},$$

where the (2,1) submatrix is skew-symmetric, i.e. $(QA - A^T Q)^T = -(QA - A^T Q)$. (The (1,2) submatrix, $AF - FA^T$, is also skew-symmetric.) Consequently, its diagonal elements are zero. If the reduction to the Hamiltonian-Hessenberg form is applied to W, then the (2,1) submatrix will be zeroed. The matrix W is not actually constructed in Algorithm `sqred`, but all the necessary operations are implicitly executed on the $n \times n$ submatrices of the Hamiltonian matrix M. In an implementation it is sufficient to store the submatrix $M(1{:}n, 1{:}n)$ and the lower (or upper) triangles of $M(1{:}n, n+1{:}2n)$ and $M(n+1{:}2n, 1{:}n)$. Additional information about computations will be given at § 4.3.2.

Algorithm sqred : $(A, Q, F) = \texttt{sqred}(\ A, Q, F\ )$

/* Reduce the square of a Hamiltonian matrix $M \in \mathbb{R}^{2n \times 2n}$, defined */
/* by A, Q, $F \in \mathbb{R}^{n \times n}$, to Hamiltonian-Hessenberg form */
/* using orthogonal symplectic transformations. */

FOR $k = 1{:}n-1$,

IF $k < n-1$, THEN

/* 1. Annihilate the elements $k+2{:}n$ of $Q(Ae_k) - A^T(Qe_k)$. */

$w = q_{k+1:n,:}a_{:,k} - a_{:,k+1:n}q_{:,k};\quad U = \texttt{Hous}(\ w\ );$

/* Apply $H(k+1, U)$ to M (implicitly) by similarity. */

$a_{k+1:n,:} \leftarrow U a_{k+1:n,:};\quad a_{:,k+1:n} \leftarrow a_{:,k+1:n}U;$
$q_{k+1:n,:} \leftarrow U q_{k+1:n,:};\quad q_{:,k+1:n} \leftarrow q_{:,k+1:n}U;$
$f_{k+1:n,:} \leftarrow U f_{k+1:n,:};\quad f_{:,k+1:n} \leftarrow f_{:,k+1:n}U;$

END IF

/* 2. Annihilate $a_2 := (Q(Ae_k) - A^T(Qe_k))_{k+1}$, */
/* modifying $a_1 := (A(Ae_k) + F(Qe_k))_{k+1}$. */

$a_1 = a_{k+1,:}^T a_{:,k} + f_{k+1,:}^T q_{:,k};\quad a_2 = q_{k+1,:}^T a_{:,k} - a_{:,k+1}^T q_{:,k};$

IF $a_2 \neq 0$, THEN

$(c, s) = \texttt{rot}(\ a_1, a_2\ );\quad w = f_{k+1,:};\quad \alpha = c a_{k+1,k+1} + s f_{k+1,k+1};$

$f_{k+1,:} \leftarrow c f_{k+1,:} - s a_{:,k+1};\quad \begin{bmatrix} a_{k+1,:} \\ q_{k+1,:} \end{bmatrix} \leftarrow \begin{bmatrix} c & s \\ -s & c \end{bmatrix} \begin{bmatrix} a_{k+1,:} \\ q_{k+1,:} \end{bmatrix};$

$w_{k+1} = f_{k+1,k+1};\quad f_{k+1,k+1} \leftarrow -s a_{k+1,k+1} + c f_{k+1,k+1};$

$q_{k+1,k+1} \leftarrow c q_{k+1,k+1} - s\alpha;\quad a_{:,k+1} \leftarrow c a_{:,k+1} + s w;$

END IF

IF $k < n-1$, THEN

/* 3. Annihilate the elements $k+2{:}n$ of $A(Ae_k) + F(Qe_k)$. */

$w = a_{k+1:n,:}a_{:,k} + f_{k+1:n,:}q_{:,k};\quad U = \texttt{Hous}(\ w\ );$

/* Apply $H(k+1, U)$ to M (implicitly) by similarity. */

$a_{k+1:n,:} \leftarrow U a_{k+1:n,:};\quad a_{:,k+1:n} \leftarrow a_{:,k+1:n}U;$
$q_{k+1:n,:} \leftarrow U q_{k+1:n,:};\quad q_{:,k+1:n} \leftarrow q_{:,k+1:n}U;$
$f_{k+1:n,:} \leftarrow U f_{k+1:n,:};\quad f_{:,k+1:n} \leftarrow f_{:,k+1:n}U;$

END IF

END FOR

Algorithm `sqred` needs about $\frac{52}{3}n^3$ *flops*. Note that `sqred` does not return the upper Hessenberg matrix $\bar{Q}$, but it can be computed as $\bar{Q} = A^2 + FQ$, where A, F, and Q are the results produced by `sqred`. The eigenvalues of the matrix M can be obtained as follows

$$\lambda_i = (\lambda_i(\bar{Q}))^{1/2}, \qquad \lambda_{n+i} = -\lambda_i, \qquad i = 1\colon n,$$

where $\lambda_i(\bar{Q})$, $i = 1\colon n$, are computed by the QR algorithm, applied on the matrix $\bar{Q}$ in UHF.

Summarizing all the above, the following algorithm is obtained.

Algorithm RicSR : $\boxed{X = \texttt{RicSR}(\ A, B, Q, R\ )}$

/* Solve continuous-time ARE (4.1) using symplectic transformations */

/* 1. Reduce the square of Hamiltonian matrix $M = \begin{bmatrix} A & F \\ Q & -A^T \end{bmatrix}$. */

$F = BR^{-1}B^T$; $\widetilde{A} = A$; $\widetilde{Q} = Q$; $\widetilde{F} = F$;

$(\widetilde{A}, \widetilde{Q}, \widetilde{F}) =$ `sqred(` $\widetilde{A}, \widetilde{Q}, \widetilde{F}$ `);` $\bar{S} = \widetilde{A}^2 + \widetilde{F}\widetilde{Q}$;

/* 2. Compute the stable eigenvalues of M. */

$(\bar{S}, \Lambda) =$ `mSchur(` $\bar{S}$ `);` $\lambda_i = -(\lambda_i)^{1/2}$, $i = 1\colon n$;

/* 3. Initialize the matrices: $X = 0$, $Y = 0$, $S_{11}^{-1} = I_n$. */

$X = 0$; $Y = 0$; $\bar{S} = I_n$;

/* 4. Reduce M to J-tridiagonal form and update S_{11}^{-1}, X and Y. */

$(A, Q, F, \bar{S}, X, Y) =$ `Jtrid(` $A, Q, F, \bar{S}, X, Y, 1, n, n-1$ `);`

/* 5. Apply SR algorithm and update S_{11}^{-1}, X and Y. */

$X =$ `SRiter(` $A, Q, F, \bar{S}, X, Y, \Lambda$ `);`

Details about the implementation of Algorithm `RicSR` will be given in the next subsection.

4.3.2 Computational issues

Algorithm `RicSR` has very good convergence properties. It was experimentally noted that the average number of SR iterations per eigenvalue is usually 1. Bunse-Gerstner, Mehrmann and Watkins (1989) estimated that the algorithm required about $28n^3$ *flops*. The memory space needed by `RicSR` for the data is of $6n^2 + m^2 + nm + 5n$ locations.

If the norm of the residual (1.130) (§ 1.5.4) of the computed solution $\widehat{X}$ is too large, one can use Newton's method or another method in the defect correction procedure, to improve the solution. In this case it is possible to admit a large tolerance for the condition numbers of transformation matrices, for avoiding unnecessary exceptional transformations.

It is worth mentioning that Algorithm **RicSR** can be implemented so that most calculations (theoretically all of them) are performed using orthogonal symplectic transformations. The G_2-type and G_3-type transformations should be replaced by Givens symplectic and Householder symplectic transformations, respectively (Bunse-Gerstner and Mehrmann, 1986). The only difficulty is that no finite general algorithm is yet known that would reduce the initial Hamiltonian matrix to J-tridiagonal form, exclusively using such transformations. Orthogonal algorithms have been proposed for the special case of single-input or single-output systems (Byers, 1986; Mehrmann, 1988). The use of orthogonal transformations would provide for good numerical properties, but the required computing effort would increase much if compared to the variant presented above.

Quite recently structure-preserving methods have also been proposed for *descriptor systems*. In this case, one works with matrix pencils, and the QZ algorithm (§ 3.2.2) must be adapted. In addition, there are some variants for discrete-time Riccati equations, which extract the deflating subspaces of symplectic pencils. But, to our knowledge, the implementations of such techniques are either non-existent or insufficiently tested.

Some details about implementation will follow. After initializing the performance parameters, **RicSR** constructs the lower triangle of the matrix $F := BR^{-1}B^T$ and computes the so-called *deflation tolerance* ε_d, used in the negligibility tests, if no value is given on input. Then **RicSR** prepares the call of Algorithm **sqred**, taking into account that **sqred** changes the contents of the input arrays. In order to reduce the memory requirements, matrix A is saved in an array `W` and in the strict upper triangular parts of the arrays `F` and `Y`, that is $\mathtt{W}(j) = a_{jj}$, $j = 1\colon n$; $\mathtt{F}(i,j) = a_{ij}$, $i = 1\colon j-1$, $j = 1\colon n$; $\mathtt{Y}(j,i) = a_{ij}$, $i = j+1\colon n$, $j = 1\colon n-1$. It is worth mentioning that the implementation of **sqred** uses the lower triangles of the matrices F and Q in packed storage (column-wise).

The results produced by **sqred** are used to compute the eigenvalues of the Hamiltonian matrix. To simplify processing, and since there is memory space available (for later calculations), the packed lower triangles of the matrices F and Q, are restructured to the two-dimensional storage; F is entirely stored in the array `X`, and the lower triangle of Q is stored in the corresponding portion of `Y`. The matrix $V := A^2 + FQ$, in UHF, is computed and $\lambda(V)$ is then determined. Since $\lambda(M^2) = \lambda(V)$, it follows that $\lambda_i(M) = -(\lambda_i(V))^{1/2}$ and $\lambda_{n+i}(M) = -\lambda_i(M)$, $i = 1\colon n$. The components of indices $n+i$ are not stored. In previous implementations (Sima, 1992a, 1993b), the eigenvalues with negative real parts, $\lambda_i(M)$, $i = 1\colon n$ (representing the optimal system poles), were arranged in the increasing order of the moduli of real parts, and in case of equality, in the increasing order of the moduli of imaginary parts. Our latest experiments (Sima, 1994, 1995a) have proven that it is more efficient to arrange the eigenvalues in the decreasing order of their moduli, and finally this strategy has been adopted. (The number of iterations have halved for some examples, if compared with that for the previously used strategy.) An explanation could be that this ordering strategy agrees on the usual order in which the QR-like algorithms tend to find eigenvalues.

Next, `RicSR` reconstructs the matrix A from the information stored in `W` and in the strict upper triangles of `F` and `Y`, and initializes the array `SINV` to I, and the lower triangles of the arrays `X` and `Y` to 0. Algorithms `Jtrid` and `SRiter`, which perform the initial reduction to J-tridiagonal form and the SR algorithm, respectively, are then called; at the same time, the stable eigenvalues of M are moved by `SRiter` onto the first n positions. Finally, the information regarding performances is stored in `W`.

The implementation of Algorithm `sqred` closely follows the template given in § 4.3.1. Orthogonal transformations—Householder transformations U, and standard plane rotations G—are exclusively used. Only matrix A and the lower triangles (packed column-wise) of the matrices F and Q, are stored. There is a loop for $k = 1{:}\,n-1$. The elements $k+1{:}\,n$ of the vector $Q(Ae_k) - A^T(Qe_k)$ are first computed in the working array `W` and the Householder transformation U, that modifies the element $k+1$ and annihilates the elements $k+2{:}\,n$, is determined. The transformation U is applied on A, in the form $A \leftarrow UAU$, and then U is similarly applied on Q and F. Next, `sqred` computes $a_1 := (A(Ae_k) + F(Qe_k))_{k+1}$ and $a_2 := (Q(Ae_k) - A^T(Qe_k))_{k+1}$, determines the rotation G that modifies a_1 and zeroes a_2, and applies G on A ($A \leftarrow GAG^T$) and, similarly, on F and Q. (We assume here that G is embedded into the $\mathbb{R}^{n\times n}$ space.) To complete these operations, it is necessary that the $(k+1)$th row of F is saved before in the array `W` and the element $a_{k+1,k+1}$ is saved in a local variable. Finally, a Householder transformation U, that modifies the element $k+1$ and annihilates the elements $k+2{:}\,n$ of the vector $A(Ae_k) + F(Qe_k)$ (computed in `W`), is determined and applied as above.

The implementation of Algorithm `Jtrid` includes facilities for working only with the submatrices $(d_1{:}\,d_2, d_1{:}\,d_2)$ in A, F and Q, and for annihilating at most s sub- and superdiagonal elements on each row and column of these matrices. The variable s has the value $d_2 - 1$ at the initial reduction to the J-tridiagonal form, or when some exceptional transformations are used, but during the SR algorithm, s may have the values 1 or 2. (Initially, $d_1 = 1$, $d_2 = n$.) The calculation and application of the symplectic matrices G_1, G_2, P and G_3 take place by calling specialized routines. These routines estimate the condition number of the transformation matrix, test its size and accumulate the decimal logarithm of the product of condition numbers for all transformations performed up to the current call. (The logarithm is used because the evaluation of the product could produce overflows.) If any of the transformations involved in the iterative process is too ill conditioned, then `Jtrid` resorts to an exceptional transformation (actually representing the product of a random G_2-type transformation and a random G_3-type transformation). At most 20 exceptional transformations are allowed at a single call of `Jtrid`. It must also be emphasized here that the called routines can work with $(d_1\colon d_2, d_1\colon d_2)$ submatrices and exclusively use the lower triangles of the arrays `Q, F, X` and `Y`. However, the indices vary from 1 to n when the matrices X, Y and S_{11}^{-1} are updated.

Algorithm `SRiter` applies the iterative SR algorithm. At the beginning of each iteration, `SRiter` tests if there is a negligible subdiagonal element in F, and, if this is the case, the problem is split into two subproblems. (Note that for problems of large dimension and with multiple eigenvalues, it is recommended that the deflation tolerance, ε_d, used in the negligibility tests, be set to a value about the square root of the relative machine precision, ε_M.) Once a submatrix N_j in (4.34) or (4.36), of order two or four, respectively, has been detected, `SRiter` applies the transformations needed for moving the stable eigenvalues to the upper-left corner. (This is equivalent to finding one vector or two vectors, respectively, of a basis for the stable M-invariant subspace.) One more remark is that the algorithm for the 4×4 submatrices also deals with the case when $\lambda(N_j) = \{\lambda, \mu, -\lambda, -\mu\}$, $\lambda, \mu \in \mathbb{R}$. Clearly, if the convergence has not been met for any submatrix N_j at the current iteration in `SRiter`, then the iterative process should continue by using a double or quadruple shift, calling Algorithms `i2SR` and `i4SR`, respectively. The "exact" eigenvalues, computed by `RicSR` after the call of `sqred`, are used here. The corresponding routines apply an SR step; the nonzero elements of the first column of matrix $p_k(M_k)$ are computed, then the symplectic transformation reducing this column to e_1 is determined, and finally, the matrix resulted after applying this transformation, is reduced to J-tridiagonal form.

4.3.3 Applicability and limitations

Algorithm `RicSR` may only be used provided that the Hamiltonian matrix is dichotomic, i.e. it has no eigenvalues on the imaginary axis in $\mathbb{C}$. This is a common limitation of most Riccati solvers. Moreover, it is assumed that matrix R is positive definite. The latter condition is required merely to compute the matrix $F := BR^{-1}B^T$. The algorithm may be easily modified to replace the inverse of R by a pseudoinverse.

Using symplectic transformations certifies that the Hamiltonian structure is preserved during the entire computational process, hence the eigenvalues appear in pairs $(\lambda, -\lambda)$ and, in case of dichotomy, there are exactly

n stable eigenvalues. Due to rounding-errors, the above statements are not generally true if non-symplectic methods, like those presented in § 2.5.1, § 3.3.1, and § 3.4.1, are used.

The usage of orthogonal symplectic transformations in `sqred` guarantees good numerical properties. Large eigenvalues (in magnitude), computed from the results returned by `sqred`, present high accuracy.

For $n \leq 2$, `RicSR` determines the solution in zero iterations, by merely performing the separation of the stable part of spectrum of the Hamiltonian matrix. For $n > 2$, the use of the eigenvalues computed based on Algorithm `sqred` as (double or quadruple) implicit shifts, enabled that the convergence of the SR algorithm be met in a small number of iterations. In our preliminary experiments (Sima, 1992a, 1993b), the average number of iterations per eigenvalue (considering n eigenvalues) was 1. With the latest implementation, we often observed an average of 0.5 iterations per stable eigenvalue (Sima, 1994, 1995a, 1995b). The standard QR-based algorithm `Schv` (§ 3.3.1), required about 4–6 times more iterations than `RicSR`. However, the execution times for solving problems of large dimension have usually been longer with `RicSR` than with Algorithm `Schv`. This can be explained by the complexity of the code implementing `RicSR`, and the associated large overhead involved.

Algorithm `RicSR` computes an approximate symmetric solution $\widehat{X}$ of the Riccati equation (4.1). As Gaussian elementary transformations are used, the computed approximation may be quite inaccurate, if the transformation matrices used in the process have large condition numbers. Therefore, it is generally indicated to apply the defect correction procedure, described in § 1.5.4. In the examples discussed below, an iteration of this procedure, based on Newton's algorithm, was performed. Newton's process has always converged (with the usual tolerance $\varepsilon = 10^{-6}$) in two—or rarely—three iterations, even if the approximation $\widehat{X}$ had only the first significant digit correct. Generally, the norm of residual of $\widehat{X}$ was $0(\varepsilon_M^{1/2})$. When the defect correction procedure is used, the maximal allowable value of the condition numbers (c_b) may be large (for instance, $10^{-2}/\varepsilon_M$), to avoid unnecessary exceptional transformations. In the tests we performed, the number of exceptional transformations was very small (often 0).

Our numerical results have also shown that, without defect correction, the accuracy of Algorithm `RicSR` does not equal that of Algorithm `Schv` (or `signm`), especially for models of large dimension or ill-conditioned. On the other hand, the use of the defect correction procedure makes the accuracies of all these algorithms be comparable. Obviously, Algorithm `RicSR` should not be involved in problems known to be very ill conditioned, such as Example 6 in (Laub, 1979) for $n = 21$, $q_{11} = 10^4$.

We conclude this section by exemplifying the use of structure-preserving algorithms. The first example is in fact a template algorithm for solving a frequently encountered control problem. Other examples report numerical results obtained for specific problems.

Example 4.4 The following algorithm computes an approximate solution of the algebraic matrix Riccati equation (4.1), and the corresponding optimal gain matrix, H, using Gaussian symplectic transformations method, and then it improves the results by an iteration of the defect correction procedure with Newton's algorithm. The system output is weighted with the matrix $\widetilde{Q}$ in the quadratic performance index, hence output regulation (Problem 1.3 in § 1.2.1) is intended. A proportional-plus-integral optimal control law (Problem 1.4) is used. It is assumed that the matrices A, B, C, $\widetilde{Q}$ and $\widehat{Q}$ in (1.66) in § 1.2.1, of dimensions $n \times n$, $n \times m$, $l \times n$, $l \times l$ and $l \times l$, respectively, as well as R, of dimension $m \times m$, are given.

Algorithm SR : $\boxed{(X, H, \widetilde{k}) = \texttt{SR}(\ A, B, C, \widetilde{Q}, \widehat{Q}, R\ )}$

/* Solve continuous-time ARE (4.1) for Problem 1.4 */
/* using Gaussian symplectic transformations method */
/* and Newton correction. */

/* 1. Initialize parameters. */

$\varepsilon = \varepsilon_M^{1/2}; \quad k_{max} = 30;$

/* 2. Construct matrices for proportional-plus-integral control. */

$$A \leftarrow \begin{bmatrix} A & 0 \\ -C & 0 \end{bmatrix}; \quad B \leftarrow \begin{bmatrix} B \\ 0 \end{bmatrix}; \quad C \leftarrow \begin{bmatrix} C & 0 \\ 0 & I_l \end{bmatrix}; \quad Q = \begin{bmatrix} \widetilde{Q} & 0 \\ 0 & \widehat{Q} \end{bmatrix};$$

/* 3. Compute state weighting matrix. Save A and Q. */

$Q \leftarrow C^TQC; \quad \widetilde{A} = A; \quad \widetilde{Q} = Q;$

/* 4. Solve Riccati equation. */

$X = \texttt{RicSR}(\ \widetilde{A}, B, \widetilde{Q}, R\ );$

/* 5. Use one iteration of the defect correction procedure. */

$\widetilde{H} = -R^{-1}B^TX; \quad \widetilde{A} = A + B\widetilde{H};$
$Y = A^TX + XA - XBR^{-1}B^TX + Q; \quad H = 0;$
$(\widetilde{X}, H, \widetilde{k}) = \texttt{Ntnc}(\ \widetilde{A}, B, Y, R, H, k_{max}, \varepsilon\ ); \quad X \leftarrow X + \widetilde{X}; \quad H \leftarrow H + \widetilde{H};$

If matrix $\widehat{Q}$ is not given, then $l = 0$, and the construction at step 2 is not made. □

Example 4.5 Algorithm SR in Example 4.4 was used for the interconnected power system in (Hung and MacFarlane, 1982), with $n = 7$, $m = 2$, $l = 3$, described by the matrices

$$A = \begin{bmatrix} -.04165 & .00 & 4.92 & -4.92 & .00000 & .00 & .00 \\ -5.21000 & -12.50 & .00 & .00 & .00000 & .00 & .00 \\ .00000 & 3.33 & -3.33 & .00 & .00000 & .00 & .00 \\ .54500 & .00 & .00 & .00 & -.54500 & .00 & .00 \\ .00000 & .00 & .00 & 4.92 & -.04165 & .00 & 4.92 \\ .00000 & .00 & .00 & .00 & -5.21000 & -12.50 & .00 \\ .00000 & .00 & .00 & .00 & .00000 & 3.33 & -3.33 \end{bmatrix},$$

$$B = \begin{bmatrix} 0 & 12.5 & 0 & 0 & 0 & 0 & 0 \\ 0 & 0 & 0 & 0 & 0 & 12.5 & 0 \end{bmatrix}^T,$$

$$C = \begin{bmatrix} 1 & 0 & 0 & 0 & 0 & 0 & 0 \\ 0 & 0 & 0 & 1 & 0 & 0 & 0 \\ 0 & 0 & 0 & 0 & 1 & 0 & 0 \end{bmatrix},$$

and the quadratic performance index defined by the matrices $\widetilde{Q} = I_3$ and $R = 100I_2$ (without error integrators). The results obtained are:

$$X = \begin{bmatrix} .5821 & .0945 & .4653 & .0431 & -.1347 & -.0033 & -.0564 \\ .0945 & .0456 & .1717 & -.2203 & -.0033 & -.0048 & -.0181 \\ .4653 & .1717 & .6805 & -.6875 & -.0564 & -.0181 & -.0819 \\ .0431 & -.2203 & -.6875 & 4.3064 & -.0431 & .2203 & .6875 \\ -.1347 & -.0033 & -.0564 & -.0431 & .5821 & .0945 & .4653 \\ -.0033 & -.0048 & -.0181 & .2203 & .0945 & .0456 & .1717 \\ -.0564 & -.0181 & -.0819 & .6875 & .4653 & .1717 & .6805 \end{bmatrix},$$

$$H = \begin{bmatrix} -.0118 & -.0057 & -.0215 & .0275 & .0004 & .0006 & .0023 \\ .0004 & .0006 & .0023 & -.0275 & -.0118 & -.0057 & -.0215 \end{bmatrix}.$$

Above, H is the optimal gain matrix and X is the Riccati equation (4.1) solution, reached through 3 iterations of Algorithm **RicSR**. The 1-norm of the residual of computed solution was 6.78×10^{-7}, and the relative residual was 1.09×10^{-7}. Some performance statistics are given in Table 4.2. The eigenvalues of the optimal (closed-loop) system are:

$$\{-1.6815, -1.3866 \pm 2.2079j, -.5623 \pm 3.1319j, -13.1443, -13.1622\}.$$

Algorithm **Ntnc** with $\varepsilon = 10^{-6}$ then reduced the residual and relative residual to 9.65×10^{-16} and 1.55×10^{-16}, respectively, after having performed two iterations. □

Example 4.6 Using Algorithm **SR** for the chemical reactor model in Example 2.5 in § 2.5.3 ($n = 4$, $m = 2$, $l = 2$), and the quadratic performance index defined by the matrices $\widetilde{Q} = \widehat{Q} = I_2$, $R = 100I_2$, we obtained the results listed below:

$$X = \begin{bmatrix} 63.3241 & 10.2177 & 45.1674 & -23.8206 & -4.5368 & .9664 \\ 10.2177 & 3.4175 & 7.5223 & -2.8706 & -.4927 & -2.2582 \\ 45.1674 & 7.5223 & 32.2611 & -16.8068 & -3.1083 & .6648 \\ -23.8206 & -2.8706 & -16.8068 & 10.2998 & 3.7302 & 2.0161 \\ -4.5368 & -.4927 & -3.1083 & 3.7302 & 6.4133 & 8.5448 \\ .9664 & -2.2582 & .6648 & 2.0161 & 8.5448 & 22.8617 \end{bmatrix},$$

$$H = \begin{bmatrix} -.8228 & -.2469 & -.6028 & .2369 & .0209 & .0978 \\ 1.4210 & .2367 & 1.0149 & -.5287 & -.0978 & .0209 \end{bmatrix}.$$

Algorithm **RicSR** performed 3 iterations. The 1-norm of the residual of computed solution was 1.12×10^{-11}, and the relative residual was 7.56×10^{-14}. The eigenvalues of the optimal (closed-loop) system are:

$$\{-.0373, -.4346 \pm .3702j, -1.9971, -5.0867, -8.6686\}.$$

Two iterations of Algorithm **Ntnc** then reduced the residual and relative residual to 1.49×10^{-13} and 1.01×10^{-15}, respectively. □

Example 4.7 Using Algorithm **SR** in Example 4.4 for $A = \widetilde{A}^T$, $B = \widetilde{C}^T$, where $\widetilde{A}$ and $\widetilde{C}$ are the matrices of the chemical reactor model in Example 2.5, and the quadratic performance index defined by the matrices $Q = I_4$, $R = 10I_2$, after 2 iterations, we obtained:

$$X = \begin{bmatrix} 35.1533 & -2.7940 & 4.6750 & 1.1334 \\ -2.7940 & .6189 & 1.2246 & 1.5254 \\ 4.6750 & 1.2246 & 9.2419 & 8.8470 \\ 1.1334 & 1.5254 & 8.8470 & 8.9270 \end{bmatrix},$$

$$H = \begin{bmatrix} -3.8695 & .3095 & -.5070 & -.1053 \\ .2794 & -.0619 & -.1225 & -.1525 \end{bmatrix}.$$

The 1-norm of the residual of computed solution was 1.17×10^{-10}, and the relative residual was 2.67×10^{-12}. The eigenvalues of the optimal (closed-loop) system are:

$$\{-.1980, -2.0443, -5.0671, -8.6716\}.$$

Two iterations of Algorithm **Ntnc** then reduced the residual and relative residual to 1.42×10^{-14} and 3.25×10^{-16}, respectively. This example actually determines the Kalman filter corresponding to the chemical reactor model, Q and R being the covariance matrices of the state and output noises, respectively. The result X represents the covariance matrix of the state estimation error, and H^T is the filter matrix. □

Table 4.2 Performance statistics for Example 4.5, Example 4.6, and Example 4.7.

	Ex. 4.5	Ex. 4.6	Ex. 4.7
log(product of condition numbers)	8.05D+01	6.90D+01	2.15D+01
The largest condition number	3.62D+07	2.04D+04	1.36D+02
The largest updating factor	1.89D+05	9.88D+02	3.43D+01
Iterations/stable eigenvalues	4.29D-01	5.00D-01	5.00D-01

For all examples above, no exceptional transformations were needed on the initial reduction, or in the iterative process.

The results for other test problems were similar. For instance, for the turbogenerator model in Example 3.9 (§ 3.3.3), with $\widetilde{Q} = I_2$, $R = 10I_2$, Algorithm `RicSR` performed 5 iterations; the largest condition number was 1.17×10^8, and the 1-norm of the residual of computed solution was 2.55×10^{-8}; Newton's process required two iterations, and reduced the residual norm to 2.31×10^{-13}.

4.4 Multishift Algorithm

In this final section, we present an "orthogonal" algorithm, the so-called *multishift algorithm*, for solving continuous-time algebraic Riccati equations. Related computational and applicability issues are briefly described.

4.4.1 Algorithmic templates

As mentioned in § 4.1.3, the key idea of the multishift method proposed in (Ammar and Mehrmann, 1991), and basically implemented as described in (Ammar, Benner and Mehrmann, 1993), is to compute a basis for the stable invariant subspace using a reduction to the *Paige-Van Loan (PVL) form* (Paige and Van Loan, 1981), defined by

$$S^T MS := \begin{bmatrix} T & W \\ D & -T^T \end{bmatrix},$$

where T is in the upper Hessenberg form, W is a symmetric matrix, and D is a diagonal matrix. Any Hamiltonian matrix can be reduced to this form in $0(n^3)$ arithmetic operations using *orthogonal* symplectic similarity transformations. However, this form is not invariant under the Hamiltonian

QR-algorithm, hence it cannot be used for a structure-preserving QR-like algorithm. But PVL form can be employed for computing a basis for a Lagrangian subspace, and this, in turn, yields the stabilizing solution of an algebraic Riccati equation.

Algorithm `PVL` below performs the reduction of the Hamiltonian matrix M associated with a continuous-time Riccati equation to Paige-Van Loan form. It is similar to Algorithm `sqred` in § 4.3.1, because of the same transformations used. The difference lies in that these transformations are here defined directly in terms of the A, Q and F matrices, since the Hamiltonian matrix itself, not its square, must be transformed. Moreover, the transformations applied must also be accumulated, in order to obtain a basis for the stable invariant subspace. The fact that orthogonal symplectic transformations are only involved makes halve the computational effort and storage needed for determining such a basis. Indeed, as already mentioned in § 4.1.2, any orthogonal symplectic matrix $S = (S_{ij})$, $S_{ij} \in \mathbb{R}^{n\times n}$, $i, j = 1\!:2$, satisfies $S_{22} = S_{11}$ and $S_{12} = -S_{21}$; therefore, the first (or last) n rows or columns of S determine it altogether. The first n columns of a given orthogonal symplectic matrix S will be updated in the algorithms that follow. Obviously, when a Householder symplectic matrix $H(k, u)$ in (4.10) postmultiplies S, then

$$S_{11} \leftarrow S_{11}U, \qquad S_{21} \leftarrow S_{21}U.$$

Similarly, when a Jacobi symplectic matrix $J(k, \gamma, \sigma)$ in (4.11) postmultiplies S, then the updated submatrices are

$$\widetilde{S}_{11} = S_{11}\Gamma + S_{21}\Sigma, \qquad \widetilde{S}_{21} = S_{21}\Gamma - S_{11}\Sigma.$$

The latter relations suggest the updating formulas below, applied for the specified k,

$$t = s_{1:n,k}, \quad s_{1:n,k} \leftarrow \gamma s_{1:n,k} + \sigma s_{n+1:2n,k}, \quad s_{n+1:2n,k} \leftarrow \gamma s_{n+1:2n,k} - \sigma t.$$

For later use, a facility for working with the trailing principal submatrices in A, Q, and F is added.

First we give an algorithm that annihilates certain elements of a vector $x \in \mathbb{R}^{2n}$ using orthogonal symplectic transformations. The transformations are also applied to A, Q, F, and S.

Algorithm colred : $(A, Q, F, S) =$ `colred(` v, w, k, A, Q, F, S `)`

/* Annihilate $v_{k+1:n}$ (if $k < n$) and $w_{k:n}$, $v, w \in \mathbb{R}^n$, using */
/* orthogonal symplectic transformations, which are applied to the */
/* Hamiltonian matrix M, defined by A, Q, $F \in \mathbb{R}^{n\times n}$; */
/* update $S \in \mathbb{R}^{2n\times n}$. */

IF $k < n$, THEN

/* 1. Annihilate $w_{k+1:n}$ and update v. */

$(U, r) =$ `Hous`($w_{k:n}$); $\quad v_{k:n} \leftarrow U v_{k:n}$;

/* Apply $H(k, U)$ to M (implicitly) by similarity. */

$a_{k:n,:} \leftarrow U a_{k:n,:}$; $\quad a_{:,k:n} \leftarrow a_{:,k:n} U$;

$q_{k:n,:} \leftarrow U q_{k:n,:}$; $\quad q_{:,k:n} \leftarrow q_{:,k:n} U$;

$f_{k:n,:} \leftarrow U f_{k:n,:}$; $\quad f_{:,k:n} \leftarrow f_{:,k:n} U$;

$s_{:,k:n} \leftarrow s_{:,k:n} U$; $\quad w_k = -r$; /* Apply $H(k, U)$ to S. */

END IF

/* 2. Annihilate w_k, modifying v_k. */

IF $w_k \neq 0$, THEN

$(\gamma, \sigma) =$ `rot`(v_k, w_k); $\quad t = f_{k,:}$; $\quad \alpha = \gamma a_{kk} + \sigma f_{kk}$;

$f_{k,:} \leftarrow \gamma f_{k,:} - \sigma a_{:,k}$; $\quad \begin{bmatrix} a_{k,:} \\ q_{k,:} \end{bmatrix} \leftarrow \begin{bmatrix} \gamma & \sigma \\ -\sigma & \gamma \end{bmatrix} \begin{bmatrix} a_{k,:} \\ q_{k,:} \end{bmatrix}$;

$t_k = f_{kk}$; $\quad f_{kk} \leftarrow -\sigma a_{kk} + \gamma f_{kk}$;

$q_{kk} \leftarrow \gamma q_{kk} - \sigma\alpha$; $\quad a_{:,k} \leftarrow \gamma a_{:,k} + \sigma t$;

$t = s_{1:n,k}$; $\quad s_{1:n,k} \leftarrow \gamma s_{1:n,k} + \sigma s_{n+1:2n,k}$; $\quad s_{n+1:2n,k} \leftarrow \gamma s_{n+1:2n,k} - \sigma t$;

END IF

IF $k < n$, THEN

/* 3. Annihilate $v_{k+1:n}$. */

$(U, r) =$ `Hous`($v_{k:n}$);

/* Apply $H(k, U)$ to M (implicitly) by similarity. */

$a_{k:n,:} \leftarrow U a_{k:n,:}$; $\quad a_{:,k:n} \leftarrow a_{:,k:n} U$;

$q_{k:n,:} \leftarrow U q_{k:n,:}$; $\quad q_{:,k:n} \leftarrow q_{:,k:n} U$;

$f_{k:n,:} \leftarrow U f_{k:n,:}$; $\quad f_{:,k:n} \leftarrow f_{:,k:n} U$;

$s_{:,k:n} \leftarrow s_{:,k:n} U$; /* Apply $H(k, U)$ to S. */

END IF

Using Algorithm `colred`, the reduction to PVL form can be accomplished by few algorithmic statements.

Algorithm PVL : $(A, Q, F, S) =$ `PVL`(A, Q, F, S, l)

/* Reduce a Hamiltonian matrix $M \in \mathbb{R}^{2n \times 2n}$, defined by */
/* A, Q, $F \in \mathbb{R}^{n \times n}$, to PVL form using orthogonal symplectic */
/* transformations; (the submatrices $(l\!:n, l\!:n)$ only are reduced;) */
/* update $S \in \mathbb{R}^{2n \times n}$. */

FOR $k = l\!:n - 1$,

$v = a_{:,k}$; $\quad w = q_{:,k}$; $\quad (A, Q, F, S) =$ `colred`($v, w, k + 1, A, Q, F, S$);

END FOR

Note that, if the given matrix S has the property that $Se_1 = e_1$, then this property is preserved, because the first column of S is not modified.

The reduction to PVL form is used iteratively for computing a particular Lagrangian M-invariant subspace, according to the results presented in Theorems 4.7 and 4.8 in § 4.1.3. These theorems apply whenever a Lagrangian M-invariant subspace exists. It is necessary to know the unstable eigenvalues of the Hamiltonian matrix. Algorithm `sqred` in § 4.3.1 can be used to compute these eigenvalues. The algorithm for obtaining a basis for the Lagrangian subspace can be summarized as follows:

Algorithm Lagran : $(A, Q, F, S) =$ `Lagran`(A, Q, F, Λ)

/* Compute a basis $S \in \mathbb{R}^{2n\times n}$ for the stable Lagrangian subspace */
/* of a Hamiltonian matrix $M \in \mathbb{R}^{2n\times 2n}$, defined by */
/* A, Q, $F \in \mathbb{R}^{n\times n}$, using orthogonal symplectic transformations */
/* and the unstable eigenvalues contained in Λ. */

/* 1. Initialize S and parameters controlling iteration. */

$s_{1:n,1:n} = I_n; \quad s_{n+1:2n,1:n} = 0;$
$l = 1; \quad \ell = 0; \quad \mathit{first} = \mathrm{TRUE};$
$\varepsilon = n\varepsilon_M(\|A\|_F^2 + \|Q\|_F^2 + \|F\|_F^2)^{1/2}; \quad \varepsilon_d = \varepsilon/\varepsilon_M^{1/2};$

/* 2. Iterate. */

WHILE $\ell < n$, DO

IF $\mathit{first} = \mathrm{FALSE}$, THEN
FOR $k = l+1\colon n$,
IF $|q_{kk}| < \varepsilon$, THEN $q_{kk} = 0$; ELSE EXIT LOOP;
$\ell = k - 1;$
FOR $k = l+1\colon n-1$,
IF $k > \ell$, THEN EXIT LOOP;
IF $|a_{k+1,k}| \le \varepsilon_d(|a_{kk}| + |a_{k+1,k+1}|)$, THEN $a_{k+1,k} = 0; \quad l = k;$
END FOR
ELSE
$\mathit{first} = \mathrm{FALSE};$
END IF

/* 3. Construct the current multishift vector v. */

$G = \begin{bmatrix} a_{l:n,l:n} & f_{l:n,l:n} \\ q_{l:n,l:n} & -a_{l:n,l:n}^T \end{bmatrix};$
$v = (G - \lambda_1 I)\cdots(G - \lambda_n I)e_1; \quad v \leftarrow v/\|v\|;$
$w_{l:n} = v_{1:n-l+1}; \quad w_{n+l:2n} = v_{n-l+2:2(n-l+1)};$

/* 4. Reduce v to e_1 using orthogonal symplectic transformations, */
/* and update A, Q, F, and S. */

$(A, Q, F, S) =$ `colred(` $w_{1:n}, w_{n+1:2n}, l, A, Q, F, S$ `);`

/* 5. Reduce the current Hamiltonian matrix G to PVL form */
/* using orthogonal symplectic transformations, */
/* and update A, Q, F, and S. */

$(A, Q, F, S) =$ `PVL(` A, Q, F, S, l `);`

END WHILE

Only orthogonal symplectic transformations are used in the above algorithm, hence the Hamiltonian structure is preserved. In addition, the method allows eigenvalues to be on the imaginary axis (if they present even multiplicity). The latter property is not met within other algorithms.

The calculation of the "multishift" vector v can be made as in Algorithm **mSchur** in § 2.2.3. The difference is that the matrix premultiplying v is a Hamiltonian matrix represented by the submatrices $(l{:}\,n, l{:}\,n)$ of A, Q, and F. Clearly, matrix G is not explicitly formed. We omit any further details.

Algorithm **Lagran** is a *multishift algorithm* in the sense that the computed eigenvalues are simultaneously used to isolate the desired invariant subspace. In exact arithmetic, the Lagrangian subspace will be so computed. In practice, due to the rounding-errors, the (2,1) block of the transformed Hamiltonian matrix can be far from zero, hence only an approximation to the desired Lagrangian subspace is obtained. Note that usually the eigenvalues used as shifts are not exact, because the Hamiltonian matrix is squared up in **sqred**. As to the SR algorithms, one difficulty associated with Algorithm **Lagran** in practice is that of deciding on the moment of performing a *deflation*, which is critical for speed and accuracy. A practical approach is to use Algorithm **Lagran** iteratively, as a defect correction method (Mehrmann and Tan, 1988). The key fact is that the algorithm preserves the eigenvalues, so that it can also be applied for the subsequent iterations.

The approach used employs the so-called *implicit defect correction iterations*, to be distinguished from the "explicit" defect correction iterations, which can be performed with any other method after a fairly accurate solution has been reached. An ingenious device for avoiding serious loss of accuracy associated with the computation of residual and the solution of the linear system which gives the matrix X, is described in (Ammar, Benner and Mehrmann, 1993). Specifically, if $S = (S_{ij}) \in \mathcal{S}_{2n}$, $S^T S = I_{2n}$, $S_{ij} \in \mathbb{R}^{n \times n}$, is such that the first n columns of S form an approximate basis for the stable invariant subspace of M, then $\widetilde{X} = -S_{21} S_{11}^{-1}$ is an approximate solution of the algebraic Riccati equation (4.1). Then the residual involved in the defect correction procedure is given by

$$\widetilde{Q} = Q - A^T S_{21} S_{11}^{-1} - S_{11}^{-T} S_{21}^T A - S_{11}^{-T} S_{21}^T F S_{21} S_{11}^{-1}.$$

Multiplying this equation by S_{11}^T from the left and by S_{11} from the right, we obtain $\widehat{Q} := S_{11}^T \widetilde{Q} S_{11}$, which is just the (2,1) block of the transformed Hamiltonian matrix, $\widehat{M} = S^T M S$. Therefore, $\widehat{Q}$ is directly available from the computations, and no linear system needs be solved to obtain $\widetilde{X}$. The defect correction procedure is applied through the residual $\widehat{Q}$, and the solution of the linear system to obtain X comes in only after the correction procedure has converged.

The approach just described requires a basis for an approximate Lagrangian subspace be updated during the iterative process. Taking into account that Algorithm **Lagran** initializes S at step 1, we have to accumulate the symplectic transformations used. Let S be the current basis, and $\widetilde{S}$ be its correction returned by **Lagran** at the current iteration. Then, the updated basis can be computed using the formulas

$$T = S_{11}\widetilde{S}_{11} - S_{21}\widetilde{S}_{21}, \qquad S_{21} \leftarrow S_{21}\widetilde{S}_{11} + S_{11}\widetilde{S}_{21}, \qquad S_{11} = T,$$

which provide the first n columns of the matrix product $S\widetilde{S}$.

The complete algorithm for solving Riccati equations is given below.

Algorithm Ricmsh : $X = \texttt{Ricmsh}(\ A, B, Q, R\ )$

/* Solve continuous-time ARE (4.1) using orthogonal */
/* symplectic transformations (multishift method). */

/* 1. Reduce the square of Hamiltonian matrix $M = \begin{bmatrix} A & F \\ Q & -A^T \end{bmatrix}$. */

$F = BR^{-1}B^T;\quad \varepsilon = n\varepsilon_M(\|A\|_F^2 + \|Q\|_F^2 + \|F\|_F^2)^{1/2};$
$\widetilde{A} = A;\quad \widetilde{Q} = Q;\quad \widetilde{F} = F;$
$(\widetilde{A}, \widetilde{Q}, \widetilde{F}) = \texttt{sqred}(\ \widetilde{A}, \widetilde{Q}, \widetilde{F}\ );\quad \widetilde{S} = \widetilde{A}^2 + \widetilde{F}\widetilde{Q};$

/* 2. Compute the unstable eigenvalues of M. */

$(\widetilde{S}, \Lambda) = \texttt{mSchur}(\ \widetilde{S}\ );\quad \lambda_i = (\lambda_i)^{1/2},\ \ i = 1{:}n;\quad \Lambda = \lambda_{1:n};$

/* 3. Initialize S. */

$s_{1:n,1:n} = I_n;\quad s_{n+1:2n,1:n} = 0;$

/* 4. Apply implicit defect correction iterations. */

WHILE $\|Q\| \geq \varepsilon$, DO

$(A, Q, F, \widetilde{S}) = \texttt{Lagran}(\ A, Q, F, \Lambda\ );$
$T = s_{1:n,:}\widetilde{s}_{1:n,:} - s_{n+1:2n,:}\widetilde{s}_{n+1:2n,:};$
$s_{n+1:2n,:} = s_{n+1:2n,:}\widetilde{s}_{1:n,:} + s_{1:n,:}\widetilde{s}_{n+1:2n,:};\quad s_{1:n,:} = T;$

END WHILE

/* 5. Solve matrix equation $S_{11}^T X = -S_{21}^T$ using the LU factorization. */
$X =$ `solve(` $s_{1:n,1:n}^T, -s_{n+1:2n,1:n}^T$ `);`

Alternatively, matrix X at step 5 can be computed using QR factorization.

4.4.2 Computational issues

Algorithm `Ricmsh` has good convergence properties. There are optimal problems for which a single iteration in Algorithm `Ricmsh` is enough for granting sufficient accuracy. However, for problems with large dimension and with multiple eigenvalues, the basis for the Lagrangian subspace computed at the first iteration may be a poor approximation of the required basis. The available implementation[3] employs implicit defect correction iterations for improving accuracy.

The implementation uses a less stringent deflation criterion than the usual one in the QR algorithm (§ 2.2.3), replacing the relative machine precision by its square root. This helps efficiency, and sometimes identify the need for a deflation. Another implementation aspect is the use of Givens plane rotations instead of Householder transformations, for annihilating matrix elements. The reason for this is that the elements of the multishift vector v in Algorithm `Lagran` got very small in magnitude after one iteration step, and Givens rotations scored much better. The normalization of v is applied after each premultiplication by matrix $(G - \lambda_i I)$. The matrices Q and F are stored in a symmetric packed form.

4.4.3 Applicability and limitations

The multishift algorithm seems to be the most general available algorithm for solving continuous-time algebraic matrix Riccati equations. It can solve problems with closed-loop eigenvalues on the imaginary axis, provided that they appear with even multiplicity.

Algorithm `Ricmsh` is usually very efficient. See, for example, the comparative results in (Sima, 1994, 1995b). It is often faster than Algorithm `Schv`. However, for large problems, having multiple eigenvalues, its efficiency does not equal that of `Schv`. One reason is the large number of implicit defect correction iterations needed for convergence. Moreover, with increasing n, the number of shifts becomes too large, and the computational effort needed for reducing the Hamiltonian matrix extends accordingly. Note that a limited number of shifts is used in the LAPACK routines.

---

[3] This implementation was due to Peter Benner from Technical University Chemnitz-Zwickau.

It must be emphasized that Algorithm **Ricmsh** exploits the Hamiltonian structure, while Algorithm **Schv** does not.

Some numerical results obtained using Algorithm **Ricmsh** will now be given. (Clearly, we can employ an algorithm similar to Algorithm **SR** in Example 4.4, calling **Ricmsh** instead of **RicSR**.)

Example 4.8 Let us consider the continuous-time tubular ammonia reactor model in (Patnaik, Viswanadham and Sarma, 1980), described by the matrices

$$A = \begin{bmatrix} -4.019 & 5.120 & .000 & .00 & -2.082 & .0 & .0 & .0 & .87 \\ -.346 & .986 & .000 & .00 & -2.340 & .0 & .0 & .0 & .97 \\ -7.909 & 15.407 & -4.069 & .00 & -6.450 & .0 & .0 & .0 & 2.68 \\ -21.816 & 35.606 & -.339 & -3.87 & -17.800 & .0 & .0 & .0 & 7.39 \\ -60.196 & 98.188 & -7.907 & .34 & -53.008 & .0 & .0 & .0 & 20.40 \\ .000 & .000 & .000 & .00 & 94.000 & -147.2 & .0 & 53.2 & .00 \\ .000 & .000 & .000 & .00 & .000 & 94.0 & -147.2 & .0 & .00 \\ .000 & .000 & .000 & .00 & .000 & 12.8 & .0 & -31.6 & .00 \\ .000 & .000 & .000 & .00 & 12.800 & .0 & .0 & 18.8 & -31.60 \end{bmatrix},$$

$$B = \begin{bmatrix} .010 & .003 & .009 & .024 & .068 & 0 & 0 & 0 & 0 \\ -.011 & -.021 & -.059 & -.162 & -.445 & 0 & 0 & 0 & 0 \\ -.151 & .000 & .000 & .000 & .000 & 0 & 0 & 0 & 0 \end{bmatrix}^T,$$

$$C = \begin{bmatrix} 1 & 0 & 0 & 0 & 0 & 0 & 0 \\ 0 & 0 & 0 & 0 & 1 & 0 & 0 \end{bmatrix},$$

and the quadratic performance index defined by the matrices $\widetilde{Q} = 50I_9$ and $R = I_3$ (without error integrators). Note that the discrete-time system corresponding to the continuous-time model above has already been used in Example 3.10 in § 3.3.3. The results obtained are:

$$X = \left[\begin{array}{rrrrrrrrr}
50.7310 & .2307 & 7.7307 & -2.0760 & -3.7324 & -.1989 & -.0101 & -1.6281 & -1.3428 \\
.2307 & 109.2282 & 11.3522 & 6.0920 & -5.3411 & .4207 & .0168 & 3.6994 & 3.3125 \\
7.7307 & 11.3522 & 11.5653 & 1.8959 & -2.9575 & -.0149 & -.0013 & -.0836 & .0120 \\
-2.0760 & 6.0920 & 1.8959 & 6.2428 & -2.2521 & .0137 & .0000 & .1599 & .2637 \\
-3.7324 & -5.3411 & -2.9575 & -2.2521 & 2.2741 & .1370 & .0259 & .3349 & .2690 \\
-.1989 & .4207 & -.0149 & .0137 & .1370 & .2206 & .0549 & .1810 & .0565 \\
-.0101 & .0168 & -.0013 & .0000 & .0259 & .0549 & .1698 & .0166 & .0030 \\
-1.6281 & 3.6994 & -.0836 & .1599 & .3349 & .1810 & .0166 & 1.4083 & .5281 \\
-1.3428 & 3.3125 & .0120 & .2637 & .2690 & .0565 & .0030 & .5281 & 1.0908
\end{array}\right],$$

$$H = \left[\begin{array}{rrrrrrrrr}
-.2740 & -.2152 & -.0598 & -.0113 & -.0206 & -.0088 & -.0017 & -.0207 & -.0212 \\
-.9782 & 1.5762 & -.0032 & .2261 & .3194 & .0690 & .0117 & .2298 & .2179 \\
7.6604 & .0348 & 1.1673 & -.3135 & -.5636 & -.0300 & -.0015 & -.2458 & -.2028
\end{array}\right].$$

Above, H is the optimal gain matrix and X is the solution of the Riccati equation (4.1), reached through Algorithm **Ricmsh** after one iteration of implicit defect correction. The reciprocal of the condition number of the linear system solved for X was about 2.61×10^{-3}. The 1-norm of the residual of computed solution was 2.57×10^{-9}, and the relative residual was 1.84×10^{-11}. The eigenvalues of the optimal (closed-loop) system are:

$$\begin{aligned}\lambda &= \{-147.1219, -153.1738, -56.1788, -37.5229, \\ &\qquad -15.4661, -.9319, -5.4134, -3.5673 \pm .2579\}.\end{aligned}$$

Three iterations of Procedure **RicCorr**, summing up to six iterations of Algorithm **Ntnc** (for $\varepsilon = 10^{-6}$), then reduced the residual and relative residual to 2.20×10^{-13} and 1.58×10^{-15}, respectively. □

Table 4.3 Number of implicit defect correction iterations for Algorithm `Ricmsh` and the residual 1-norm for Example 4.9.

ε	k	$\|\|Y\|\|_1$
1	1	5.50×10^{-15}
10^{-1}	1	1.03×10^{-15}
10^{-2}	1	1.47×10^{-15}
10^{-3}	1	1.73×10^{-15}
10^{-4}	1	2.52×10^{-15}
10^{-5}	1	3.26×10^{-15}
10^{-6}	3	1.89×10^{-15}
10^{-7}	9	3.22×10^{-15}

Example 4.9 Consider the continuous-time optimal problem defined by the matrices in Example 2.1 (§ 2.2.1). This example has already been studied in (Ammar, Benner and Mehrmann, 1993; Arnold and Laub, 1984; Bai, Demmel and Gu, 1994), and the results obtained with Algorithm `signm` are summarized in Table 4.1, § 4.2.3. It is known that the associated Hamiltonian matrix tends to one having a pair of complex conjugate eigenvalues on the imaginary axis, as ε approaches zero. Table 4.3 gives the number of implicit defect correction iterations performed by Algorithm `Ricmsh`, and the 1-norm of the residual Y of the computed solution $\widehat{X}$, for various values of ε. It is apparent that the accuracy of the result does not deteriorate as ε is reduced, contrary to the behaviour observed for Algorithm `signm`. Note that the number of implicit defect correction iterations, k, is greater than 1 only for $\varepsilon \leq 10^{-6}$. □

Example 4.9 illustrates the ability of Algorithm `Ricmsh` to cope with optimal problems having closed-loop poles near the imaginary axis.

References

Ammar, G., Benner, P., and Mehrmann, V. (1993). A multishift algorithm for the numerical solution of algebraic Riccati equations, *Electronic Trans. Numer. Anal.*, *1*: 33–48.

Ammar, G., and Mehrmann, V. (1991). On Hamiltonian and symplectic Hessenberg forms, *Lin. Alg. Appl.*, *149*: 55–72.

Anderson, E., Bai, Z., Bischof, C., Demmel, J., Dongarra, J., Du Croz, J., Greenbaum, A., Hammarling, S., McKenney, A., Ostrouchov, S., and Sorensen, D. (1992). *LAPACK Users' Guide*, SIAM, Philadelphia.

Arnold, W. F., III, and Laub, A. J. (1984). Generalized eigenproblem algorithms and software for algebraic Riccati equations, *Proc. IEEE*, *72*(*12*): 1746–1754.

Bai, Z., and Demmel, J. W. (1994). Design of a parallel nonsymmetric eigenroutine toolbox, Part II, Technical report, Department of Mathematics, University of Kentucky, Lexington, KY 40506, 12 Feb.

Bai, Z., Demmel, J. W., and Gu, M. (1994). Inverse free parallel spectral divide and conquer algorithms for nonsymmetric eigenproblems, Research Report 94-01, Department of Mathematics, University of Kentucky, Lexington, KY 40506, 9 Feb. Also Report UCB//CSD-94-793, University of California at Berkeley, and Report LBL-34969, Lawrence Berkeley Laboratory.

Balzer, L. A. (1980). Accelerated convergence of the matrix sign function method of solving Lyapunov, Riccati and other matrix equations, *Int. J. Control*, *32*(*6*): 1057–1078.

Barraud, A. Y. (1979). Investigations autour de la fonction signe d'une matrice. Application à l'équation de Riccati, *R.A.I.R.O. Automatique/Systems Analysis and Control*, *13*(*4*): 335–368.

Barraud, A. Y. (1980). Produit étoile et fonction signe de matrice. Application à l'équation de Riccati dans le cas discret, *R.A.I.R.O. Automatique/Systems Analysis and Control*, *14*: 55–85.

Beavers, A. N., and Denman, E. D. (1974a). A computational method for eigenvalues and eigenvectors of a matrix with real eigenvalues, *Numer. Math.*, *21*: 389–396.

Beavers, A. N., and Denman, E. D. (1974b). A new solution method for matrix quadratic equations, *Mathematical Biosciences*, *20*: 135–143.

Beavers, A. N., and Denman, E. D. (1974c). Asymptotic solutions to the matrix Riccati equation, *Mathematical Biosciences*, *20*: 339–344.

Bunse-Gerstner, A. (1986). Matrix factorizations for symplectic QR-like methods, *Lin. Alg. Appl.*, *83*: 49–77.

Bunse-Gerstner, A., Byers, R., and Mehrmann, V. (1989a). "Numerical methods for algebraic Riccati equations", Lecture Notes of the Workshop on "The Riccati Equation in Control, Systems, and Signals", Como, Italy (S. Bittanti, ed.), Pitagora Editrice, Bologna, pp. 107–115.

Bunse-Gerstner, A., Byers, R., and Mehrmann, V. (1989b). A quaternion QR algorithm, *Numer. Math.*, *55*: 83–95.

Bunse-Gerstner, A., Byers, R., and Mehrmann, V. (1992). A chart of numerical methods for structured eigenvalue problems, *SIAM J. Matrix Anal. Appl.*, *13(2)*: 419–453.

Bunse-Gerstner, A., and Mehrmann, V. (1986). A symplectic QR like algorithm for the solution of the real algebraic Riccati equation, *IEEE Trans. Automat. Control*, *AC-31(12)*: 1104–1113.

Bunse-Gerstner, A., and Mehrmann, V. (1989). The HHDR algorithm and its application to optimal control problems, *R.A.I.R.O. Automatique, Productique, Informatique Industrielle*, *APII 23*: 305–329.

Bunse-Gerstner, A., Mehrmann, V., and Watkins, D. (1989). An SR algorithm for Hamiltonian matrices, based on Gaussian elimination, *Methods of Operations Research*, *58*: 339–358.

Byers, R. (1986). A Hamiltonian QR-algorithm, *SIAM J. Sci. Stat. Comput.*, *7(1)*: 212–229.

Byers, R. (1987). Solving the algebraic Riccati equation with the matrix sign function, *Lin. Alg. Appl.*, *85(1)*: 267–279.

Byers, R., and Mehrmann, V. (1986). "Symmetric updating of the solution of the algebraic Riccati equation", Proceedings of X Symposium on Operations Research 1985 (Beckmann, Gaede, Ritter, and Schneeweiss, eds.), volume 54 of *Methods of Operations Research*, Anton Hain, Königstein, pp. 117–125.

Clements, D. J., and Glover, K. (1989). Spectral factorization via Hermitian pencils, *Lin. Alg. Appl.*, *122/123/124*: 797–846.

Della Dora, J. (1975). Numerical linear algorithms and group theory, *Lin. Alg. Appl.*, *10*: 267–283.

Denman, E. D., and Beavers, A. N. (1976). The matrix sign function and computations in systems, *Appl. Math. Comp.*, *2*: 63–94.

Dongarra, J. J., Bunch, J. R., Moler, C. B., and Stewart, G. W. (1979). *LINPACK Users' Guide*, SIAM, Philadelphia.

Elsner, L. (1979). On some algebraic problems in connection with general eigenvalue algorithms, *Lin. Alg. Appl.*, *26*: 123–138.

Flaschka, U., Mehrmann, V., and Zywietz, D. (1991). An analysis of structure preserving numerical methods for symplectic eigenvalue problems, *R.A.I.R.O. Automatique, Productique, Informatique Industrielle, APII 25*: 165–190.

Gardiner, J. D., and Laub, A. J. (1986). A generalization of the matrix sign function solution for algebraic Riccati equations, *Int. J. Control*, *44*: 823–832.

Golub, G. H., and Van Loan, C. F. (1983). *Matrix Computations*, The Johns Hopkins University Press, Baltimore, Maryland.

Howland, J. L. (1983). The sign matrix and the separation of matrix eigenvalues, *Lin. Alg. Appl.*, *49*: 221–232.

Hung, Y. S., and MacFarlane, A. G. J. (1982). *Multivariable Feedback: A Quasi-Classical Approach*, volume 40 of *Lect. Notes in Control and Information Sciences*, Springer-Verlag, Berlin.

Laub, A. J. (1979). A Schur method for solving algebraic Riccati equations, *IEEE Trans. Automat. Control*, *AC-24(6)*: 913–921.

Laub, A. J., and Meyer, K. (1974). Canonical forms for symplectic and Hamiltonian matrices, *Celest. Mechanics*, *9*: 213–238.

Lin, W.-W. (1987). A new method for computing the closed-loop eigenvalues of a discrete-time algebraic Riccati equation, *Lin. Alg. Appl.*, *96*: 157–180.

Lin, W.-W. (1989). An SDR algorithm for the solution of the generalized algebraic Riccati equation, *IEEE Trans. Automat. Control*, *AC-34(8)*: 875–879.

Mattheys, R. L. (1978). Stability analysis via the extended matrix sign function, *Proc. IEE*, *125(3)*: 241–243.

Mehrmann, V. (1988). A symplectic orthogonal method for single input or single output discrete time optimal quadratic control problems, *SIAM J. Matrix Anal. Appl.*, *9(2)*: 221–247.

Mehrmann, V. (1991). *The Autonomous Linear Quadratic Control Problem. Theory and Numerical Solution*, volume 163 of *Lect. Notes in Control and Information Sciences* (M. Thoma and A. Wyner, eds.), Springer-Verlag, Berlin.

Mehrmann, V., and Tan, E. (1988). Defect correction methods for the solution of algebraic Riccati equations, *IEEE Trans. Automat. Control*, *AC-33*(*7*): 695–698.

Paige, C., and Van Loan, C. F. (1981). A Schur decomposition for Hamiltonian matrices, *Lin. Alg. Appl.*, *41*: 11–32.

Patel, R. V. (1992). "On computing the eigenvalues of a symplectic pencil", Proceedings of the 31st Conference on Decision and Control, Tucson, AZ, December 1992, pp. 1921–1926.

Patnaik, L. M., Viswanadham, N., and Sarma, I. G. (1980). Computer control algorithms for a tubular ammonia reactor, *IEEE Trans. Automat. Control*, *AC-25*(*4*): 642–651.

Raines, A. C., III, and Watkins, D. S. (1992). A class of Hamiltonian-symplectic methods for solving the algebraic Riccati equation, Technical report, Washington State University, Pullman, WA.

Roberts, J. (1980). Linear model reduction and solution of the algebraic Riccati equation by the use of the sign function, *Int. J. Control*, *32*: 667–687.

Sima, V. (1992a). Algorithm GRICSR: Solving continuous-time algebraic Riccati equations using Gaussian symplectic transformations, *Studies in Informatics and Control*, *1*(*3*): 237–254.

Sima, V. (1992b). Algorithm SIGNM: Solving continuous-time algebraic Riccati equations using matrix sign function method, *Studies in Informatics and Control*, *1*(*4*): 335–355.

Sima, V. (1993a). "Iterative improvement of Riccati equations solutions", Preprints of the 9th International Conference on Control Systems and Computer Science (CSCS 9), May 25–28, 1993, Bucharest, volume 1, Politehnica University of Bucharest, Faculty of Control and Computers, Bucharest, pp. 6–14.

Sima, V. (1993b). "Computational experience in solving algebraic Riccati equations using Gaussian symplectic transformations", Preprints of The 12-th IFAC World Congress, 19th–23rd July 1993, Sydney, volume 6, The Institution of Engineers, Australia, pp. 433–436.

Sima, V. (1994). Evaluation of structure preserving algorithms for solving linear-quadratic optimization problems, Final Report, Fixed Contribution Contract No. CIPA3511CT920870 between the European Economic Community and the Universität Bielefeld, Federal Republic of Germany, Commission of the European Communities, Directorate-General XIII, Brussels, 24 June, 1994.

Sima, V. (1995a). "Numerical solution of algebraic Riccati equations using Gaussian symplectic transformations", Preprints of the 10th International Conference on Control Systems and Computer Science CSCS 10, May 24–26, 1995, volume 1, "Politehnica" University of Bucharest, Faculty of Control and Computers, Bucharest, pp. 49–53.

Sima, V. (1995b). "Computational experience with structure preserving algorithms for solving algebraic Riccati equations", Proceedings of the third European Control Conference ECC 95, September 5–8, 1995, Roma, Italy (A. Isidori, S. Bittanti, E. Mosca, A. De Luca, M. D. Di Benedetto and G. Oriolo, eds.), volume 4, Part one, pp. 3212–3217.

Van Loan, C. F. (1984). A symplectic method for approximating all the eigenvalues of a Hamiltonian matrix, *Lin. Alg. Appl.*, *61*: 233–251.

Appendix A
Comparison of Riccati Solvers

This appendix presents some results of a numerical investigation of various algorithms for solving continuous-time algebraic matrix Riccati equations,[1] discussed in previous chapters: Newton algorithm, Schur and generalized Schur algorithms, matrix sign function algorithm, the SR algorithm based on Gaussian symplectic transformations, and the multishift algorithm using orthogonal transformations.[2] Defect correction is also used to improve the computed solutions accuracy.

Our investigation has mainly been limited to comparing accuracy and efficiency on a conventional serial computer—an IBM PC 486 DX2S computer. For getting relevant results, we tried to use similar programming conventions, and the same low-level routines for implementing all algorithms. Except for the multishift algorithm, the EISPACK (Garbow et al.,

[1] Part of the research summarized in this appendix was supported by Commission of the European Communities, Directorate-General XIII, PECO'92 Action, in the framework of "Community's Action in Science and Technology with Central and Eastern European Countries", under the **Fixed Contribution Contract** No. CIPA3511CT920870 between the European Economic Community and Bielefeld University, The Federal Republic of Germany.

[2] The code for the multishift algorithm used in our tests has been developed by Peter Benner. This code incorporates the SQRED modules implemented by Ralph Byers and Eric Barth from University of Kansas. Their permission of using all these codes is much acknowledged.

1977) and LINPACK (Dongarra et al., 1979) routines (and their modifications) have been used. Implementations based on the newly developed, public-domain linear algebra package LAPACK (Anderson et al., 1992), for all the algorithms considered, have not been made yet. (The multishift algorithm calls LAPACK subroutines.) Adaptation of the available codes for incorporating LAPACK components is under way, and an expert driver for solving Riccati equations has been designed (Sima, 1994). By using such LAPACK routines as building blocks, advantage can be taken of modern computer architectures.

The main purpose of our investigation was to compare the efficiency and accuracy of the algorithms described, for some typical classes of problems. Some "real-world" problems have been included, but they are of small dimension. Other automatic-generable problems have been considered, to exercise the algorithms behaviour for various values of n. Results for few ill-conditioned problems are given, because no algorithm can guarantee accurate results for such problems.

Each example in the following tables has a name composed of a root and a suffix indicating its dimension. The examples having the "EX4L", "EX5L" and "EX6L" roots are examples 4 to 6 in (Laub, 1979). (All examples "EX6L" are solved for $q_{11} = 10^4$, $R = 1$.) "EX2M" is the second example in (Mehrmann and Tan, 1988). "REAC", "ENERG" and "TGEN" are models of a chemical reactor, an interconnected power system, and a turbogenerator, respectively, in (Hung and MacFarlane, 1982), with performance index matrices $Q = C^T C$, and $R = 100I_2$, $R = I_2$, and $R = 100I_2$, respectively. "HELIC" is the helicopter model in (Yue, Postlethwaite and Padfield, 1989), and "AMMON" is the continuous-time tubular ammonia reactor in (Patnaik, Viswanadham and Sarma, 1980). "RAND" examples are some randomly generated models with random matrices A, B, and Q (using random variables from a uniform (0, 1) distribution), R being set to the identity matrix. For $n \leq 20$, matrix Q is initially generated as upper triangular; $Q^T Q$ is computed and, then, the diagonal is scaled by $10n$. For $n > 20$, Q is generated as a diagonal matrix, with random diagonal elements.

Note that "REAC", "ENERG", "TGEN", "HELIC", and "AMMON" have also been used in previous sections of this book. No results obtained by Algorithms `Ntnc` and `gSchv`, and no results in "HELIC" and "AMMON" examples, have been included in (Sima, 1994, 1995).

Since the exact solution to most of the examples is unknown, results computed by a "hybrid" technique have been used as reference, to assess the algorithms accuracy. Specifically, the results denoted by X in the tables below were computed by the stability-enhancing scaling procedure in

(Kenney, Laub and Wette, 1989), see Procedure `SclSch` in § 3.3.1,[3] followed by the defect correction procedure in (Mehrmann and Tan, 1988), see Procedure `RicCorr` in § 1.5.4, using Newton's process as Riccati solver, implemented as described in (Sima, 1990). The results denoted by $\widehat{X}$ were computed using any of the algorithms discussed, with or without defect correction. All computations were done in double precision on an IBM PC 486 DX2S computer (66 MHz, with mathematical co-processor).

Table A.1 presents the absolute errors in $\widehat{X}$ (in the absolute norm) for various algorithms, without defect correction. The relative errors had the same magnitude or a smaller one. In case of the very ill-conditioned example EX6L21, Algorithm `Stac`, called by `Ntnc`, did not succeed to yield a stabilizing gain matrix. Example HELIC could not be solved by Algorithm `RicSR`. The results for Algorithm `RicSR` were obtained without any explicit condition bound: the algorithm itself decides to perform an exceptional transformation when needed, if a transformation matrix is too ill conditioned. (No exceptional transformations were needed in the iterative process.) The default deflation tolerance, $\varepsilon_d = \sqrt{\varepsilon_M} \approx 10^{-8}$, was used, except for the examples EX6L10 and TGEN10, solved by $\varepsilon_d = 10^{-10}$. (A smaller deflation tolerance might marginally deliver more accurate results for certain examples, but often it asks for additional SR iterations and results in an increase of condition numbers and absolute errors.) A similar "relaxed" deflation criterion was used for Algorithm `Ricmsh`. Table A.1 shows that the accuracy of Algorithm `RicSR` does not equal that of Algorithm `Schv`, especially for models of large dimension or ill conditioned;[4] this is caused by the use of non-orthogonal transformations and possible large condition numbers of the transformations involved. The accuracy of other algorithms is comparable. It should be mentioned, however, that in general it is mandatory to use an implicit defect correction procedure in Algorithm `Ricmsh`. Otherwise, the results could be very inaccurate. (For instance, the absolute errors for the examples EX4L29 and EX4L39 would be 6.09×10^{-2} and 10.9, respectively.)

Table A.2 presents the absolute errors in $\widehat{X}$ (in the absolute norm) with defect correction. The value $\varepsilon = 10^{-8}$ was employed in the convergence criterion for Newton's process. With few exceptions, all algorithms were to take two defect correction iterations (the second one being often incomplete) and two Newton iterations. Table A.2 shows that, except for example EX6L21, use of the defect correction procedure makes the Algorithm `RicSR` accuracy be comparable to that of other algorithms.

Table A.3 and Table A.4 present the number of iterations and execu-

[3] The value trace($\widehat{X}$), where $\widehat{X}$ is the solution delivered by a previous computation with Algorithm `Schv`, was used in our calculations as a scaling factor for each example.

[4] The results for example EX6L21 are not significant.

Table A.1 Absolute errors without defect correction.

Example	$\|X - \widehat{X}\|$					
	Ntnc	signm	gSchv	Schv	RicSR	Ricmsh
EX4L9	1.33D-14	3.55D-15	4.62D-14	1.53D-14	1.10D-12	7.11D-15
EX4L19	2.31D-14	1.07D-14	3.87D-14	3.38D-14	9.11D-11	8.35D-14
EX4L29	3.02D-14	1.60D-14	2.21D-14	6.04D-14	1.12D-08	9.77D-14
EX4L39	3.55D-14	1.78D-14	6.57D-14	4.88D-14	1.61D-07	2.45D-12
EX5L8	7.77D-16	4.44D-16	5.83D-16	3.33D-16	8.65D-13	3.89D-16
EX5L16	7.77D-16	3.33D-16	8.33D-16	5.00D-16	3.72D-11	3.26D-16
EX5L32	1.39D-15	6.11D-16	7.77D-16	4.44D-16	9.20D-07	2.45D-15
EX5L40	8.88D-16	9.99D-16	5.55D-16	4.44D-16	5.40D-05	3.91D-15
EX2M15	1.67D-15	5.55D-16	1.33D-15	2.22D-15	2.28D-07	1.55D-15
EX6L5	6.73D-11	9.91D-11	1.37D-05	5.50D-06	1.45D-03	7.09D-11
EX6L10	7.57D-08	1.26D-03	7.09D-02	5.92D-03	1.65D+02	3.24D-06
EX6L21	–	1.58D+01	8.69D+01	2.26D+02	3.99D+08	9.44D-01
REAC4	2.34D-13	7.60D-13	9.45D-13	7.11D-13	4.49D-12	4.05D-13
ENERG7	3.33D-15	1.44D-15	2.66D-15	7.99D-15	9.94D-09	1.78D-15
HELIC8	8.64D-12	9.09D-13	4.90D-11	3.76D-11	–	1.15D-11
AMMON9	8.24D-13	1.71D-13	2.87D-11	1.17D-11	2.86D-02	7.22D-10
TGEN10	2.46D-14	1.06D-13	8.37D-14	1.80D-13	1.00D-09	2.66D-13
RAND10	6.04D-14	1.95D-14	1.02D-12	9.09D-13	1.99D-07	1.07D-12
RAND20	3.13D-13	2.53D-12	1.10D-10	2.15D-10	1.53D-07	1.86D-10
RAND30	2.01D-14	1.97D-14	2.04D-14	2.16D-14	3.54D-08	2.56D-14
RAND40	1.61D-14	1.59D-14	2.40D-14	1.82D-14	3.63D-03	3.29D-14

Table A.2 Absolute errors with defect correction.

Example	$\|X - \widehat{X}\|$					
	Ntnc	signm	gSchv	Schv	RicSR	Ricmsh
EX4L9	4.44D-16	1.78D-15	1.78D-15	1.78D-15	1.78D-15	3.33D-15
EX4L19	1.78D-15	1.78D-15	1.78D-15	1.78D-15	1.78D-15	1.51D-14
EX4L29	1.78D-15	1.78D-15	1.78D-15	1.78D-15	1.78D-15	2.22D-14
EX4L39	1.78D-15	1.78D-14	2.66D-15	3.55D-15	2.66D-15	3.46D-14
EX5L8	1.11D-16	5.55D-17	1.11D-16	1.11D-16	5.55D-17	2.22D-16
EX5L16	5.55D-17	3.33D-16	5.55D-17	5.55D-17	1.67D-16	3.89D-16
EX5L32	1.11D-16	6.11D-16	1.11D-16	5.55D-17	3.33D-16	8.08D-16
EX5L40	1.11D-16	9.99D-16	1.11D-16	5.55D-17	2.78D-16	1.05D-15
EX2M15	3.33D-16	5.55D-16	2.22D-16	3.33D-16	2.22D-16	1.55D-15
EX6L5	6.73D-11	2.73D-12	1.82D-12	1.82D-12	1.82D-12	1.14D-10
EX6L10	7.57D-08	2.63D-08	3.13D-08	2.46D-08	2.81D-08	1.49D-04
EX6L21	–	9.48D-02	2.28D-01	2.05D+00	5.21D+08	1.77D+00
REAC4	2.34D-13	4.83D-13	4.90D-13	5.68D-13	3.55D-14	1.78D-13
ENERG7	3.33D-15	1.44D-15	2.66D-15	7.99D-15	2.66D-15	1.78D-15
HELIC8	3.41D-13	1.14D-13	2.27D-13	1.14D-13	–	1.14D-13
AMMON9	4.26D-14	2.23D-14	7.11D-15	4.26D-14	2.13D-14	2.30D-14
TGEN10	2.46D-14	1.23D-14	1.54D-14	3.82D-14	1.38D-14	5.92D-13
RAND10	6.04D-14	1.95D-14	7.11D-15	7.11D-15	5.33D-15	6.93D-14
RAND20	3.13D-13	2.84D-13	1.14D-13	5.26D-13	1.42D-13	2.46D-12
RAND30	2.01D-14	1.97D-14	2.04D-14	2.16D-14	2.00D-14	2.56D-14
RAND40	1.61D-14	1.59D-14	2.40D-14	1.82D-14	1.65D-14	1.71D-14

tion times in seconds, respectively, for all algorithms investigated, with no defect correction. The number of iterations have not been counted for Algorithm `gSchv`. For Algorithm `Ricmsh` we listed the number of implicit defect correction iterations. It is useful to compare the number of iterations for `Ntnc` and `signm`, as well as for `Schv` and `RicSR`. It is apparent that `Ntnc` usually required more iterations than `signm`; this is also reflected in the execution times. Note that the execution times for `Ntnc` are in fact comparable to those for `gSchv`. On the other hand, Algorithm `RicSR` generally performs a much smaller number of iterations than `Schv` does, deflating one, or most frequently, two stable eigenvalues per iteration. This illustrates the potential effectiveness of Algorithm `RicSR`. A similar behaviour was observed with even smaller values of ε_d, like 10^{-12}. However, the execution times for examples EX4L, EX5L and RAND (with large values of n) are longer than those for Algorithm `Schv`. One reason for the inefficiency of Algorithm `RicSR` is that the same level 1 BLAS-based routine is called for both the initial reduction of M and the subsequent reductions in the QR-like iterations (but dealing with two or three off-diagonals only). Moreover, the present implementation cannot benefit from the possible splitting of the Hamiltonian matrix into submatrices of order larger than two or four, since the eigenvalues currently used as shifts in the QR-like process may not belong to the current trailing submatrix. Future implementation should take these remarks into account. Algorithm `Ricmsh` has execution times better than those for Algorithm `RicSR`, is more reliable and delivers more accurate results. The most efficient algorithm is Algorithm `signm`. It was also very accurate, at least for the examples used in our tests. For problems with associated Hamiltonian matrices having eigenvalues close to imaginary axis in $\mathbb{C}$, Algorithm `Ricmsh` is the only reasonable choice.

The results show that the structure exploiting/preserving algorithms—especially the matrix sign function algorithm and the multishift algorithm—do score well in practice. Algorithm `RicSR` is less accurate and reliable; it can be used for solving some on-line control problems where accuracy is not of much concern, or certain large problems where the Hamiltonian matrix is in the J-tridiagonal form. The matrix sign function algorithm is very efficient, accurate and reliable. It is strongly recommended for solving problems with large dimensions, which other algorithms could not solve.

Other issues worth considering in a comparison are generality, memory requirements, and availability of these algorithms. Algorithm `Schv` is the most general, for it can easily be adapted to solve nonsymmetric Riccati equations. But it needs almost twice the memory space which Algorithm `signm` does. Algorithm `signm` is probably the most widely available, because it is essentially based on a symmetric indefinite solver only. However, with the release of LAPACK, Algorithm `Schv` can also be considered

Table A.3 Number of iterations.

Example	Iterations				
	Ntnc	signm	Schv	RicSR	Ricmsh
EX4L9	8	7	26	4	1
EX4L19	8	7	52	9	1
EX4L29	8	7	76	15	2
EX4L39	8	7	101	20	6
EX5L8	6	6	17	5	1
EX5L16	7	6	46	13	2
EX5L32	9	6	85	30	5
EX5L40	10	6	112	45	7
EX2M15	8	7	54	11	1
EX6L5	11	6	40	2	1
EX6L10	8	7	50	4	1
EX6L21	–	9	71	10	1
REAC4	6	7	10	2	1
ENERG7	4	8	28	3	1
HELIC8	20	7	28	–	1
AMMON9	6	7	25	5	1
TGEN10	17	10	33	5	2
RAND10	12	7	34	7	1
RAND20	15	8	60	17	2
RAND30	10	9	115	15	4
RAND40	12	9	143	27	6

Table A.4 Execution times.

Example	Time (seconds)					
	Ntnc	signm	gSchv	Schv	RicSR	Ricmsh
EX4L9	.66	.22	.66	.38	.33	.22
EX4L19	4.45	1.32	5.49	2.63	3.35	1.54
EX4L29	14.23	3.96	12.47	8.13	14.89	6.71
EX4L39	31.80	8.68	38.78	19.34	37.52	33.67
EX5L8	.28	.11	.32	.22	.28	.16
EX5L16	2.25	.71	2.31	1.37	2.26	1.37
EX5L32	19.89	4.51	16.26	9.39	21.15	17.41
EX5L40	40.38	8.13	28.29	17.96	51.08	41.96
EX2M15	2.96	.72	2.52	1.48	1.93	1.04
EX6L5	.16	.06	.33	.22	.06	.11
EX6L10	.88	.27	1.26	.77	.54	.33
EX6L21	–	1.65	6.15	4.34	4.55	2.36
REAC4	.06	.06	.11	.06	.00	.00
ENERG7	.16	.16	.33	.22	.16	.11
HELIC8	.99	.17	.60	.33	–	.16
AMMON9	.44	.17	.55	.33	.33	.22
TGEN10	1.21	.33	.77	.49	.39	.39
RAND10	1.37	.27	.93	.55	.49	.28
RAND20	9.89	1.75	5.55	3.30	4.12	3.95
RAND30	23.95	5.55	20.05	13.51	15.27	14.12
RAND40	63.11	12.19	51.46	33.50	55.53	46.25

as largely available.

References

Anderson, E., Bai, Z., Bischof, C., Demmel, J., Dongarra, J., Du Croz, J., Greenbaum, A., Hammarling, S., McKenney, A., Ostrouchov, S., and Sorensen, D. (1992). *LAPACK Users' Guide*, SIAM, Philadelphia.

Dongarra, J. J., Bunch, J. R., Moler, C. B., and Stewart, G. W. (1979). *LINPACK Users' Guide*, SIAM, Philadelphia.

Garbow, B. S., Boyle, J. M., Dongarra, J. J., and Moler, C. B. (1977). *Matrix Eigensystem Routines—EISPACK Guide Extension*, volume 51 of *Lect. Notes in Computer Science*, Springer-Verlag, Berlin.

Hung, Y. S., and MacFarlane, A. G. J. (1982). *Multivariable Feedback: A Quasi-Classical Approach*, volume 40 of *Lect. Notes in Control and Information Sciences*, Springer-Verlag, Berlin.

Kenney, C., Laub, A. J., and Wette, M. (1989). A stability-enhancing scaling procedure for Schur-Riccati solvers, *Systems Control Lett.*, *12*: 241–250.

Laub, A. J. (1979). A Schur method for solving algebraic Riccati equations, *IEEE Trans. Automat. Control*, *AC-24(6)*: 913–921.

Mehrmann, V., and Tan, E. (1988). Defect correction methods for the solution of algebraic Riccati equations, *IEEE Trans. Automat. Control*, *AC-33(7)*: 695–698.

Patnaik, L. M., Viswanadham, N., and Sarma, I. G. (1980). Computer control algorithms for a tubular ammonia reactor, *IEEE Trans. Automat. Control*, *AC-25(4)*: 642–651.

Sima, V. (1990). Comparison of some algorithms to solve the algebraic Riccati equations, *Studies and Researches in Computers and Informatics*, *1(2)*: 77–96.

Sima, V. (1994). Evaluation of structure preserving algorithms for solving linear-quadratic optimization problems, Final Report, Fixed Contribution Contract No. CIPA3511CT920870 between the European Economic Community and the Universität Bielefeld, Federal Republic of Germany, Commission of the European Communities, Directorate-General XIII, Brussels, 24 June, 1994.

Sima, V. (1995). "Computational experience with structure preserving algorithms for solving algebraic Riccati equations", Proceedings of the third European Control Conference ECC 95, September 5–8, 1995, Roma, Italy (A. Isidori, S. Bittanti, E. Mosca, A. De Luca, M. D. Di Benedetto and G. Oriolo, eds.), volume 4, Part one, pp. 3212–3217.

Yue, A., Postlethwaite, I., and Padfield, G. (1989). H^∞ design and the improvement of helicopter handling qualities, *Vertica*, *13*(*2*): 119–132.

Appendix B

Notation and Abbreviations

This appendix tabulates main and consistently used notation and abbreviations.

B.1 Notation

$a := b$	a is defined as being equal to b.
$a =: b$	b is defined as being equal to a.
ε_M	Relative machine precision, $\varepsilon_M := b^{1-t}$, where b is the base for floating-point representation, and t is the number of mantissa digits in base b.
$\mathrm{fl}(\alpha)$	Value of floating-point representation of α.
ε_u	The smallest positive representable number such that $\mathrm{fl}(1+\varepsilon_u) > 1$.
$\mathbb{R}$ ($\mathbb{C}$)	The field of real (complex) numbers.
$\mathbb{R}^{m\times n}$ ($\mathbb{C}^{m\times n}$)	The set of real (complex) matrices with m rows and n columns.
$X^{1/2}$	Square root of X, a scalar or a square matrix.
j (sometimes j)	The purely imaginary unit, $\mathrm{j} := (-1)^{1/2}$.
$\Re(\lambda)$	The real part of $\lambda \in \mathbb{C}$.

$\Im(\lambda)$	The imaginary part of $\lambda \in \mathbb{C}$.
$\bar{\lambda}$	The complex conjugate of $\lambda \in \mathbb{C}$, $\bar{\lambda} = \Re(\lambda) - \Im(\lambda)\mathrm{j}$.
$\|\lambda\|$	Modulus of λ, $\|\lambda\| = (\Re(\lambda)^2 + \Im(\lambda)^2)^{1/2}$.
$\{\alpha : p\}$	The set of numbers α with property "p".
$\mathbb{C}^-$	The stability domain: $\mathbb{C}^- := \{z \in \mathbb{C} : \Re(z) < 0\}$ in case of continuous-time dynamical systems; $\mathbb{C}^- := \{z \in \mathbb{C} : \|z\| < 1\}$ in case of discrete-time dynamical systems.
$\mathrm{sign}(x)$	Function which is 1 if $x > 0$, and -1 if $x < 0$.
$\arg\min\{F(x) : x \in \mathcal{X}\}$	Value of $x \in \mathcal{X}$ which minimizes function F.
$m \bmod n$	The remainder of the integer division of m to n.
$i = n{:}\,l{:}\,m$	Index i takes the values n, $n + l$, ..., m ($m \bmod l = n$). If l is omitted, $l = 1$ is assumed. If $l > 0$ and $m < n$, or if $l < 0$ and $m > n$, then the set of values is empty.
$i, j = 1{:}\,n$	Equivalent to $i = 1{:}\,n$ and $j = 1{:}\,n$.
$0(\varepsilon)$	Quantity of order ε, i.e. $\lim_{\varepsilon \to 0} \|0(\varepsilon)/\varepsilon\| < \infty$.
A^T	The transpose of matrix $A \in \mathbb{R}^{m \times n}$.
A^H	The conjugate transpose of matrix $A \in \mathbb{C}^{m \times n}$.
A^{-1}	The inverse of matrix A.
$A^\dagger$	The Moore-Penrose generalized inverse of A.
A^{-T}	Inverse of the transpose of matrix A.
A^{-H}	Inverse of the conjugate transpose of matrix A.
$x = (x_1, \ldots, x_n)^T$	Vector x with components x_i, $i = 1{:}\,n$.
$A = (a_{ij})$	Matrix A with entries a_{ij}.
$A = (a_1, \ldots, a_n)$	Matrix A with columns $a_1, \ldots, a_n$.
$A(i{:}\,j, k{:}\,l)$ (or $a_{i:j,k:l}$)	Submatrix formed by rows $i{:}\,j$ and columns $k{:}\,l$ of matrix A.
$\mathrm{diag}(a_1, \ldots, a_n)$	Diagonal matrix having $a_1, \ldots, a_n$ as diagonal elements (scalars or matrices).
I	Identity matrix of order resulting from context.
I_n	Identity matrix of order n, $I_n = \mathrm{diag}(\underbrace{1, \ldots, 1}_{n})$.
e_i	The ith column of identity matrix.
$x \perp y$	The vectors x and y in $\mathbb{C}^n$ are orthogonal, i.e. $$x^H y := \sum_{i=1}^{n} \bar{x}_i y_i = 0.$$
$\mathcal{R}(a_1, \ldots, a_n)$	The subspace generated (spanned) by the vectors $a_1, \ldots, a_n \in \mathbb{R}^m$.

$\mathcal{R}(A)$	The range space of matrix $A \in \mathbb{C}^{m\times n}$, spanned by columns of A: $\mathcal{R}(A) := \{\, y : Ax = y, x \in \mathbb{C}^n \,\}$.
$\mathcal{N}(A)$	Null space of matrix $A \in \mathbb{C}^{m\times n}$: $\mathcal{N}(A) := \{\, x : Ax = 0, x \in \mathbb{C}^n \,\}$.
$\mathcal{X}^\perp$	Orthogonal complement of the subspace $\mathcal{X}$: any $x \in \mathbb{C}^n$ can be written as $x = y + z$, $y \in \mathcal{X}$, $z \in \mathcal{X}^\perp$, and $y^H z = 0$ for any $y \in \mathcal{X}$ and $z \in \mathcal{X}^\perp$.
$\dim(\mathcal{X})$	Dimension of subspace $\mathcal{X}$.
$\Vert x\Vert_p$	The p-norm of vector $x \in \mathbb{C}^n$. Examples: $\Vert x\Vert_1 = \sum_{i=1}^{n} \vert x_i\vert$; $\Vert x\Vert_2 = \left(\sum_{i=1}^{n} \vert x_i\vert^2\right)^{1/2}$; $\Vert x\Vert_\infty = \max_{i=1:n}\{\, \vert x_i\vert \,\}$.
$\Vert A\Vert_p$	The p-norm of matrix $A \in \mathbb{C}^{m\times n}$ subordinate to vector p-norm: $\Vert A\Vert_p = \max_{\Vert x\Vert_p=1}\{\, \Vert Ax\Vert_p \,\}$. Usually, $p = 1, 2, \infty$: $\Vert A\Vert_1 = \max_{j=1:n}\left\{ \sum_{i=1}^{m} \vert a_{ij}\vert \right\}$; $\Vert A\Vert_2 = \sigma_{\max}(A)$; $\Vert A\Vert_\infty = \max_{i=1:m}\left\{ \sum_{j=1}^{n} \vert a_{ij}\vert \right\}$.
$\Vert A\Vert_F$	The Frobenius norm of matrix $A \in \mathbb{C}^{m\times n}$: $\Vert A\Vert_F = \left(\sum_{j=1}^{n}\sum_{i=1}^{m} \vert a_{ij}\vert^2 \right)^{1/2}$.
$\Vert A\Vert_a$	The absolute norm of matrix $A \in \mathbb{C}^{m\times n}$: $\Vert A\Vert_a = \max_{i,j}\{\, \vert a_{ij}\vert \,\}$.
$\det(A)$	Determinant of the square matrix A.
$\mathrm{tr}(A)$	Trace of the square matrix $A \in \mathbb{C}^{n\times n}$: $\mathrm{tr}(A) = a_{11} + \cdots + a_{nn}$.
$\mathrm{rank}(A)$	Rank of matrix A: $\mathrm{rank}(A) = \dim(\mathcal{R}(A))$.
$\mathrm{cond}(A)$	Condition number of matrix A: $\mathrm{cond}(A) = \Vert A\Vert\,\Vert A^{-1}\Vert$.
$A > 0$ $(A \geq 0)$	Matrix A, $A = A^H \in \mathbb{C}^{n\times n}$, is positive definite (semi-definite), i.e.: $x^H Ax > 0$ $(x^H Ax \geq 0)$, $\forall\, x \in \mathbb{C}^n$, $x \neq 0$.
$\lambda(A) = \{\, \lambda_1, \ldots, \lambda_n \,\}$	Collection of eigenvalues of matrix $A \in \mathbb{R}^{n\times n}$: $\lambda(A) = \{\, \lambda \in \mathbb{C} : \det(A - \lambda I_n) = 0 \,\}$.
$\lambda_i \in \lambda(A)$	The ith eigenvalue of matrix A.
$\lambda(A, B)$	Collection of generalized eigenvalues of matrix pencil (A, B):

	$\lambda(A, B) = \{ \lambda \in \mathbb{C} : \det(A - \lambda B) = 0 \}$.
$\sigma(A)$	Collection of singular values of matrix A: $\sigma(A) = \{ \sigma_i : \sigma_i = (\lambda_i(A^H A))^{1/2} \}$.
$\sigma_i(A)$	The ith singular value of matrix A.
$\sigma_{\max}(A)$	The maximal singular value of matrix A.
$u \leftarrow v$	The value of variable v is put into the location corresponding to variable u.
$u \leftrightarrow v$	The values of variables u and v are interchanged.
EXIT LOOP	The current (innermost) loop is exited.

B.2 Abbreviations

ARE	Algebraic Riccati equation.
flops	Standard floating-point operation: $a_{ij} \leftarrow a_{ij} + t \times a_{ik}$.
GRSF	Generalized upper real Schur form. A matrix pencil (A, B) is in GRSF if A is in RSF, and B is upper triangular (with non-negative diagonal elements).
JCF	Jordan canonical form. A matrix is in JCF if it is block-diagonal, with Jordan blocks on the diagonal.
LQG	Linear-quadratic-Gaussian.
PVL	Paige-Van Loan (form).
QTF	Quasi-triangular form. A matrix is in QTF if it is block-upper triangular, and each diagonal block has order at most two.
RSF	Upper real Schur form. A real matrix is in RSF if it is in QTF and all its diagonal blocks of order two correspond to complex conjugate eigenvalues.
SVD	Singular value decomposition.
UHF (HF)	Upper Hessenberg form (Hessenberg form). A matrix A is in UHF if $a_{ij} = 0$, for $i > j + 1$.

Index of Algorithms Definitions

Index